Springer Water

The book series Springer Water comprises a broad portfolio of multi- and interdisciplinary scientific books, aiming at researchers, students, and everyone interested in water-related science. The series includes peer-reviewed monographs, edited volumes, textbooks, and conference proceedings. Its volumes combine all kinds of water-related research areas, such as: the movement, distribution and quality of freshwater; water resources; the quality and pollution of water and its influence on health; the water industry including drinking water, wastewater, and desalination services and technologies; water history; as well as water management and the governmental, political, developmental, and ethical aspects of water.

Victor Alcaraz Gonzalez ·
René Alejandro Flores Estrella ·
Andreas Haarstrick · Victor Gonzalez Alvarez
Editors

Wastewater Exploitation

From Microbiological Activity to Energy

Springer

Editors
Victor Alcaraz Gonzalez
Departamento de Ingeniería Química
Universidad de Guadalajara
Guadalajara, Mexico

Andreas Haarstrick
Leichtweiß Institut
TU Braunschweig
Braunschweig, Germany

René Alejandro Flores Estrella
Ciencias de la Sustentabilidad
El Colegio de la Frontera Sur
Tapachula, Mexico

Victor Gonzalez Alvarez
Departamento de Ingeniería Química
Universidad de Guadalajara
Guadalajara, Mexico

ISSN 2364-6934 ISSN 2364-8198 (electronic)
Springer Water
ISBN 978-3-031-57737-6 ISBN 978-3-031-57735-2 (eBook)
https://doi.org/10.1007/978-3-031-57735-2

This Springer imprint is published by the registered company Springer Nature Switzerland AG
The registered company address is: Gewerbestrasse 11, 6330 Cham, Switzerland

Foreword

The years from 2015 to 2023 were the warmest on record, according to six major global temperature datasets compiled by the World Meteorological Organization (WMO), which show that greenhouse gas concentrations continued to rise, and heat accumulated. Moreover, the increase in global average temperature in 2022 was about 1.15 [1.02–1.27] °C higher than preindustrial levels (1850–1900).[1] These greenhouse gas emissions are closely related to human activities and demographic projections by the United Nations[2] suggest that there will be 9.7 billion people in the world by 2050, with 2022 predicted to have seen the 8 billionth person.

The UNEP 2011 report "Decoupling natural resource use and environmental impacts from economic growth" reminds us that while the twentieth century has seen spectacular growth in life expectancy and human development, so too has resource extraction: extraction of construction materials has increased 34-fold since the beginning of the twentieth century, extraction of minerals and ores 27-fold, fossil fuel consumption 12-fold and biomass consumption 3.6-fold.

The IPCC recommends leaving 60% of oil and gas reserves and 90% of coal reserves in the ground by 2050 [1]. There is a generally recognized planetary limit that identifies the need to conserve at least 75% of the Earth's natural ecosystems, although only 62% remain [2]. These forecasts are accompanied by a widespread awareness of planetary limits. The current pressure on resources and associated waste flows offers no prospect of sustainability in either space or time. In its report "What a Waste 2.0: A global Snapshot of solid waste to 2050", published in 2018, the World Bank predicts that global waste production will increase by 70% between now and 2030 because of rapid urbanization and population growth, i.e. 3.4 billion tons over the next three decades, compared with 2.01 billion tons in 2016. The problems of waste production and management offer humankind the opportunity to redefine economic growth by moving from a linear economy to a circular economy and trying to gain better control of its flows. How can the circular economy and efficient waste

---

[1] https://public.wmo.int/en/media/press-release/past-eight-years-confirmed-be-eight-warmest-record.

[2] https://www.un.org/fr/global-issues/population.

management and waste valorization could help to solve the problem? This is the topic of this book addressing research challenges in bioprocess engineering and bioenergy.

The different chapters detail the intricate realm of bioprocess engineering and bioenergy, addressing critical challenges at the forefront of energy production and merging scientific domains to propel sustainable solutions while minimizing environmental impacts. The ensuing synopsis encapsulates the diverse challenges and strides outlined across the chapters, navigating this intricate landscape:

1. ***Bioreactor Design and Optimization***: The quest to bridge the gap between laboratory experimentation and industrial-scale production demands innovative strategies for dynamic process control. Seamlessly adapting to varying conditions while maintaining consistent yields poses an ongoing challenge.
2. ***Biohydrogen Production and Optimization***: The dynamic interplay of microbial consortia dynamics stands at the forefront of biohydrogen production challenges. Unlocking the full potential necessitates innovative strategies for managing these intricate interactions, alongside identifying sustainable carbon sources.
3. ***Bioprocess Coupling for Biohydrogen Production***: The integration of diverse hydrogen production platforms beckons researchers to navigate uncharted waters, fusing microbial dynamics and process engineering. As microbial communities interplay, understanding their harmonious synergy emerges as a research imperative.
4. ***Microbial Fuel Cells and Bioelectrochemical Systems***: The journey toward efficient microbial–electrode interactions unfolds, demanding ingenious approaches to enhance electron transfer efficiency and ensure system stability. Novel electrode materials emerge as a potential game-changer, with ongoing exploration for optimal performance.
5. ***Harvesting Biofuels with METs***: Microbial electrochemical technologies stand poised to revolutionize biofuel harvesting, yet unraveling the intricate interplay between microorganisms and electrodes remains an ongoing challenge. Designing optimal electrode configurations and elucidating microorganism-electrode interactions are focal points.
6. ***Biodiesel Production and Catalyst Development***: Catalyzing advancements in biodiesel production calls for an intricate dance between catalyst efficiency and feedstock diversification. Researchers strive to unearth innovative pathways for boosting biodiesel yields while ensuring superior quality.
7. ***Bioethanol Production and Process Optimization***: Grasping the nuances of inhibition mechanisms and harnessing real-time process optimization strategies are paramount to attaining peak bioethanol production efficiency. These challenges underscore the intricate balance between biological systems and process engineering.
8. ***Kinetic Modeling for ABE Fermentation***: Peering into the complex metabolic pathways of ABE fermentation necessitates the development of predictive kinetic models. These models, striving for accuracy, form a cornerstone for refining fermentation strategies and advancing technology.

9. ***Algal Biofuel Production***: The potential of algal biofuels hinges on tackling multifaceted challenges, from meticulous strain selection and genetic engineering to optimizing cultivation systems. Unlocking higher yields of algal biomass and lipid content remains a paramount goal.
10. ***Microalgae-Based Diesel Production***: Propelling microalgae-based diesel into the limelight requires surmounting hurdles related to economic feasibility and efficient lipid extraction methods. Researchers delve into uncharted territories, seeking robust approaches for sustainable diesel production.
11. ***Bioconversion of Industrial CO_2into Synthetic Fuels***: The transformation of industrial CO_2 into viable synthetic fuels demands a delicate balance between energy efficiency and renewable integration. Achieving this synergy remains a cornerstone challenge, with microbial electrochemical technologies as a promising avenue.
12. ***Evolution of the Biorefinery Concept***: As biorefineries evolve, striking a harmonious balance between technological advancement and sustainability becomes paramount. Comprehensive assessments of environmental, social, and technical dimensions guide this evolution, aiming to seamlessly integrate Circular Economy principles.

In unity, these chapters illuminate the vibrant landscape of bioprocess engineering and bioenergy, underscoring the significance of interdisciplinary collaboration, innovation, and sustainable practices. Researchers ardently pursue these challenges, driving the paradigm shift toward a greener energy frontier and envisioning a world where bioenergy plays a pivotal role in shaping a sustainable future.

Jean-Philippe Steyer
Research Director at INRAE
Laboratoire de Biotechnologie de l'Environnement (LBE), INRAE
University of Montpellier
Narbonne, France

References

1. Welsby D, Price J, Pye S et al (2021) Unextractable fossil fuels in a 1.5 °C world. Nature 597:230–234
2. Steffen W, Broadgate W, Deutsch L, Gaffney O, Ludwig C (2015) The trajectory of the Anthropocene: the great acceleration. Anthropocene Rev 2(1):81–98

Preface

Without a doubt, water and wastewater management are among the most significant challenges of the twenty-first century, and there is also that the challenges posed by climate change will become even greater. Besides, current fossil energy sources are finite, and they are not going to last long. Thus, the role of wastewater as a source of energy is becoming even more critical.

Unfortunately, most efforts, especially in developing countries but also in the so-called developed countries, have been less than optimal or not optimal at all. There are still too many people who must live without clean water and decent sanitation. According to UN-Water,[3] today, 2.2 billion people lack access to safely managed drinking water and wastewater and 4.2 billion people lack safely managed sanitation services. The question of why this is so, and why in many cases in developing countries can be answered in large part mainly by the fact that in these countries, political systems (prevail that are composed) consist of extractive political institutions and extractive economic action. However, the scarcely practiced conservation of resources, which is to be considered in this context, can also be traced back to industrialized countries.

Unfortunately, an acute lack of capacity is constraining water resources development and management in all its facets across most developing countries, particularly in sub-Saharan Africa and South and South-Eastern Asia but also Latin America and Middle East North Africa. Human resource shortages are reported in all key areas, including agriculture and irrigated farming, water-related risk management, water and sanitation services, wastewater treatment, recycling and reuse technologies, and

[3] www.unwater.org/water-facts/.

desalination. The currently prevailing economic action does not prioritize sustainability and resource conservation in the agenda. There is a need to put one's own house in order.

Guadalajara, Mexico — Victor Alcaraz Gonzalez
Tapachula, Mexico — René Alejandro Flores Estrella
Guadalajara, Mexico — Victor Gonzalez Alvarez
Braunschweig, Germany — Andreas Haarstrick

About This Book

With all the current efforts to use non-fossil sources as a starting point for future energy solutions, consideration is also being given to using microbial activities as a direct or indirect source of energy production. This ranges from the use of algae as biomass or as H_2 producers, anaerobic microorganisms to produce methane, hydrogen, and even electricity directly.

This book deals with both theoretical and technical possibilities of using anaerobic microorganisms in combination with wastewater as a substrate source to produce biofuels and bioenergy in the form of biomass, CH_4 and H_2 as well as the corresponding power densities and electricity quantities in economically justifiable processes. Unique process facilities are widely addressed; however, special interest is also placed in biorefinery and circular economy-related concepts. The theoretical background, as well as application examples, is presented.

Contents

Part VIII [illegible]

About the Editors

Victor Alcaraz Gonzalez is Chemical Engineer by Universidad de Guadalajara (UdeG, Mexico) in 1991, and received his Ph.D. degree in Engineering Sciences from Université de Perpignan in France in 2001. Since 2002, he is a member of the (Mexican) National Researchers System that is founded by the Mexican National Council of Humanities, Science and Technology (CONAHCyT). Currently, he is Full Professor Researcher at the Department of Chemical Engineering, at UdeG, and a founder member of the Bioprocess Engineering Research Group (CA-496) at the same university. The current research topics and interests of Dr. Victor Alcaraz Gonzalez deal with modeling and control of bioprocesses for wastewater treatment, focusing on obtaining bioenergy and added-value products from wastewater.

René Alejandro Flores Estrella was born in Mérida, Yucatán, México. He received the B.Sc. degree in Chemical Engineering from the Universidad Autónoma de Yucatán (UADY, Mexico) in 2002, the M.Sc. degree from Instituto Tecnológico de Celaya (TecNM Celaya, Mexico) in 2007, and Ph.D. degree from Instituto Potosino de Investigación Científica y Tecnológica A.C. (IPICYT, Mexico) in 2013. He is a member of the National Researchers System that is founded by the Mexican National Council of Humanities, Science and Technology (CONAHCyT). Currently, he is Full Researcher at the Department of Sustainability, at El Colegio de la Frontera Sur (ECOSUR, Mexico).

His scientific interests include mathematical modeling, dynamics, and control of bioprocesses.

Andreas Haarstrick studied Chemistry at the Technische Universität Braunschweig, Germany (1983–1989) and received his doctorate in 1992 in bioengineering. Since 2006, he is Professor for Bioprocess Engineering at the TU Braunschweig. His teaching and research cover modeling biological and chemical processes in heterogeneous systems, development of models predicting pollutant reduction in and emission behavior of landfills, growth kinetics at low substrate concentrations under changing environmental conditions, advanced oxidation processes (AOP), and groundwater management. Since 2012, he is the managing director of the DAAD exceed-Swindon Project dealing with sustainable water management in developing countries.

Victor Gonzalez Alvarez holds a B.S. degree from the University of Guadalajara, Mexico and a Ph.D. degree from the University of Minnesota, USA, both in chemical engineering. His research interests focused on the design and implementation of nonlinear robust monitoring and control techniques in chemical and biochemical processes. He has authored and co-authored more than 200 research papers in peer-review journals and conference proceedings, edited 3 books, and held 3 patents. He has supervised 30 Ph.D. and 72 M.S. students. Prof. Gonzalez Alvarez was the Dean of the University of Guadalajara (UDG, Mexico) Institute of Technology (2004–2010). Later, he was in charge of the (UdeG) Graduate School and Research System and was the liaison of UdeG with the Regional and National Councils of Humanities, Science and Technology (COECyTJal and CONAHCyT, respectively) and his duties focused on the quality assurance of all graduate programs and the establishment of technology transfer policies at UdeG.

Part I
Introduction

Chapter 1
General View on Synergies and Trade-Offs Using Wastewater and Anaerobic Processes for Current in the Form of Biomass, CH_4 and H_2 as Well as Energy Production Systems

Victor Alcaraz-Gonzalez, René Alejandro Flores-Estrella, Victor Gonzalez-Alvarez, and Andreas Haarstrick

Abstract More than ever before in human history, climate change, which is largely caused by humans, will pose serious existential challenges for the first time. In particular, the future energy supply of mankind will play a decisive role. Fossil fuels can no longer be a future option unless the risk of further excessive use is ignored and accepted. There are sufficient alternatives available to replace fossil fuels. Renewable energies, including biomass, can replace almost 100 % of fossil fuels in an intelligent mix of wind, solar, tidal, and geothermal energy. In addition, efficient and optimised interaction is possible when combined with artificial intelligence. This chapter focuses on the use of biological processes and their contribution to future energy management and highlights synergies and trade-offs. The aim is to make it clear that with the help of these technologies, mankind can certainly keep the complex Earth system from becoming completely out of balance.

Keywords Renewable energy · Biological processes · Synergies · Trade-offs

V. Alcaraz-Gonzalez
Universidad de Guadalajara—CUCEI, Blvd. Marcelino Garcia Barragán 1451, Col Olímpica, 44430 Guadalajara, Jalisco, México
e-mail: victor.agonzalez@academicos.udg.mx

R. A. Flores-Estrella · V. Gonzalez-Alvarez
Departamento Ciencias de La Sustentabilidad, El Colegio de La Frontera Sur, Carretera Antiguo Aeropuerto Km. 2.5, C.P. 30700 Tapachula, Chiapas, México
e-mail: rene.flores@ecosur.mx

A. Haarstrick (✉)
Leichtweiss-Institut Für Wasserbau, Technische Universität Braunschweig, Beethovenstr. 51 a, 38106, Braunschweig, Germany
e-mail: a.haarstrick@tu-braunschweig.de

V. Alcaraz Gonzalez et al. (eds.), *Wastewater Exploitation*, Springer Water,
https://doi.org/10.1007/978-3-031-57735-2_1

1.1 Introduction

According to conservative estimates, the world's energy demand will have roughly tripled by the end of this century [1]. Society's dependence on fossil fuels for this energy poses one of the greatest challenges to global environmental sustainability and economic stability. Urgently needed transformation processes in technology, industry, and society for a complete shift away from fossil fuels are currently making only slow progress. The climatic changes taking place worldwide leave little time to buffer extreme impacts and avoid predicted tipping points. The transition to new forms of energy is imperative and inescapable.

A look at photosynthesis-driving microorganisms gives an idea of the potential to use solar energy efficiently with the help of sophisticated technologies. The amount of solar energy that strikes the Earth every hour (~4.3×10^{20} J) is approximately equal to the total amount of energy that is consumed on the planet every year. Therefore, capturing even a small fraction of the available solar energy could make a significant contribution to global energy needs. The type of technologies using microorganisms, what can be established at which location, in what scope, and in what capacity depends, among other things, on the socio-economic and political framework conditions as well as the investment potential available. Another factor that should not be underestimated is the quality and degree of expansion of existing infrastructure for storing and transmitting energy in situations where fossil fuels are no longer used— If we stay with photosynthesis for a moment, it remains to be emphasised that the understanding of photosynthesis represents an enormous treasure trove of knowledge on which to build energy conversion cleanly and sustainably. For example, the 3D crystal structure of photosystem II from the cyanobacterium *Thermosynechococcus elongatus* provides a detailed view of the water-splitting apparatus that could be used as a basis to produce hydrogen from water [2, 3]. It is known that anaerobic photosynthetic bacteria can use the excess reducing power produced in the light to evolve hydrogen. In addition, photosynthetic microorganisms can be used to power solar-driven microbial fuel cells. Furthermore, the metabolic capabilities of photosynthetic microorganisms could also be redirected to generate high-energy chemicals (such as alcohols, alkanes, and fats) that currently serve as feedstocks for the chemical industry.

1.2 Capability of Microorganisms

The adaptability of microorganisms is enormous. Billions of microorganisms populate every niche of the Earth, many of which have untapped potential to help solve the global energy challenge. To grow in unusual environments, microorganisms have evolved unique metabolic strategies to extract energy from the available nutrients and generate various potentially useful byproducts. In many cases, these microorganisms work cooperatively in communities where their concerted activities perform functions that would not be possible in the absence of their partners. Current and further research at the theoretical/model-based level as well as empirical studies are essential to acquire a systems level understanding of energy capture and its transformation to direct the reaction products into pathways that produce alternative fuels or sequester greenhouse gases. Anhuai et al. [4] reported a pathway that demonstrates a role of light in non-phototrophic microbial activity. Visible light-excited photoelectrons from metal oxide, metal sulfide, and iron oxide stimulated the growth of chemoautotrophic and heterotrophic bacteria. The measured bacterial growth was dependent on light wavelength and intensity, and the growth pattern matched the light absorption spectra of the minerals. Similar observations were obtained in a natural soil sample containing both bacteria and semiconducting minerals. Results from this study provide evidence for a newly identified, but possibly long-existing pathway, in which the metabolisms and growth of non-phototrophic bacteria can be stimulated by solar light through photocatalysis of semiconducting minerals.

Another example is the microbiological formation of molecular hydrogen, which is considered nowadays as a promising tool to produce "green" hydrogen. The release of electrons in metabolic oxidation processes can also be used to produce electricity in microbiological fuel cells or for reducing protons, with the consequent hydrogen production in microbial electrolysis cells. The ability to produce H_2 has been observed in a relatively large number of microbial species, including very different taxonomic and physiological types. The representative organisms can be divided into four categories: *(i)* heterotrophic anaerobic species whose growth is inhibited by molecular oxygen, *(ii)* heterotrophic facultative anaerobic species that typically contain cytochromes and evolve H_2 from formate, *(iii)* heterotrophic strict anaerobe strains that contain a cytochrome; this species normally use sulfate as the terminal oxidant for energy-yielding, cytochrome-linked an-aerobic oxidations (for example, of lactate), but certain strains can liberate H_2 from pyruvate and formate when sulfate is absent, and *(iv)* photosynthetic microorganisms, both bacteria and algae, which can produce H_2 by a light-dependent mechanism.

Current research in the field of microbial electrolysis cells (MEC) focuses on issues of modelling metabolic processes of hydrogen production, use in technical systems and their modelling. The optimisation of growth and environmental conditions for the microorganisms plays an essential role here. The main interest here is in the maximum yield of H_2 and the associated production of free electrons for conversion into electricity. The reactions catalysed at the anode electrode by electroactive microorganisms take place within an active biofilm. There, the removal of organic compounds from nutrient-rich wastewater takes place during the oxidation of organic matter under anaerobic conditions. Protons, electrons and CO_2 are produced during this process. The electrons migrate from the anode surface to the cathode by applying a small additional electrical potential from the outside. Protons (hydrogen ions) migrate through a membrane towards the cathode electrode, where the electrical current (electrons) is used to reduce the protons to biohydrogen. However, despite the recent interest in using MEC for wastewater treatment, their development has not yet moved beyond laboratory and pilot-scale implementations, and their adoption by the industry remains still limited.

The use of biomass for microbiological processes to generate energy sources such as CH_4 poses a dilemma. The biomass potential of energy crops is inextricably linked to food demand and how it will be met. Although agricultural productivity has improved, there is uncertainty about the extent of these improvements. The issue of land use and the maintenance of soil quality and ecosystem services as well as biodiversity around it produces several concerns. Moreover, these issues are compounded by the problems posed by climate change and the rising world population.

The question of which system will ultimately be used with the least implications for sustainability, biodiversity and resource conservation must be considered and answered on a decentralised basis in each case and whenever possible, all possible trade-offs must be systematically analysed in advance. These must not remain empty phrases, because the urgency and seriousness of the situation forbid this; ultimately, massive existential decisions are at stake.

1.3 Synergies and Trade-Offs—the Way to Go

Research and experimentation have the potential to narrow down some areas of uncertainty and create evidence. This can also answer the question of which system is the best and causes the least damage. It must be clear that there will be no man-made system without uncertainties. There will always be uncertainties that cannot be resolved, and trade-offs that will always be contested. However, the claim to reduce uncertainties to an acceptable level remains justified. The use of biomass and suitable microorganisms to tackle the future energy issue certainly has synergistic effects that can be of real benefit to humans and nature.

Looking at the process line of the future use and linkage of renewable energies (Fig. 1.1), synergy effects can be created with electricity generation through microbial activity based on renewable biosubstrates that do not originate from energy crops from monocultures and additional land consumption in conjunction with the use of primary energies.

It has been known for decades that a wide variety of energy carriers (H_2, CH_4, biodiesel, ethanol, etc.) can be obtained from microbial activity. Special attention is currently being paid to the production and use of H_2 as well as the direct use of microbial activity and conversion into electrical energy in microbial fuel cells. Wastewater from sewage treatment plants often comes into focus as a possible substrate source. Although the investigations in this area are still largely in the laboratory and in some cases in the pilot phase, there is an awareness of the potential and is continuing to work on reaching the pilot phase or industrial scale to a greater extent.

As an example, the annual volume of wastewater in Germany is about 10 km^3. The annual energy demand of municipal sewage treatment plants is 4.4 GWh. Considering the energy content of the substrates in the wastewater of approx. 3.3 kWh/m^3, the

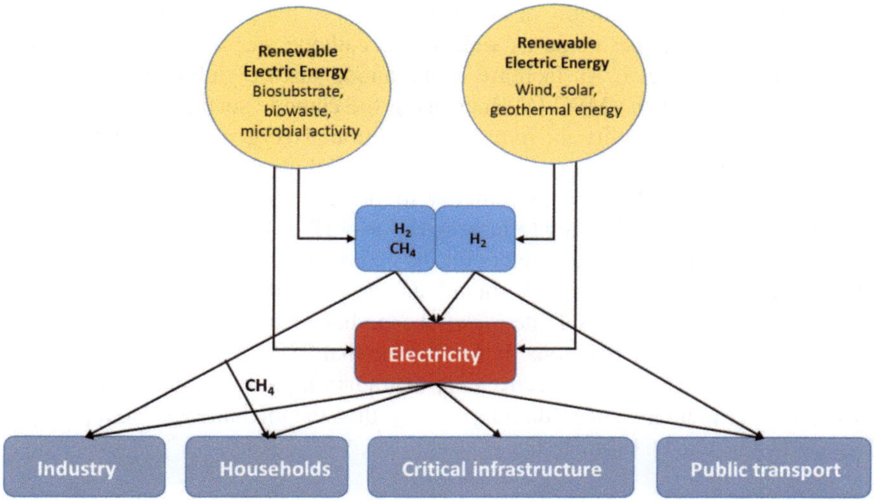

Fig. 1.1 Principle outline of the combination of microbial and primary energy sources with complete exclusion of fossil energy sources. (own drawing)

annual energy content is 33 GWh. This illustrates the potential that can microbially be used for conversion into electrical energy [5]. It may be that the combustion of biofuels or the use of other alternative energy sources for direct use by the end user do not constitute the energy paradigm in the future. More efficient can be to generate electricity from these sources, standardize technologies and procedures, and then distribute it evenly for all kinds of human activities. This book explores the full path of the first part on the biological side, i.e., the generation of biofuels (and even electricity directly) from organic substrates, mainly wastewater by microbial activity. This approach is of considerable social and economic importance and once again underlines the synergy potential as expressed in Fig. 1.1.

As trade-off, it is often discussed to what extent the combination of the energy resources and supply systems, as shown in this figure, can guarantee a reliable conversion into electrical energy without, for example, the so-called blackouts in the power supply. One solution lies in the diversity of combinations and in striving for decentralised energy supply solutions. Large power grids and power generation systems are more susceptible to disruptions than decentralised units which enable faster mutual compensation in emergencies. Another disadvantage is a functioning transport infrastructure (often in developing countries) for the collection and transport of biowaste (sewage sludge, kitchen waste, other plant waste, etc.) to the nearest facility. Of course, the CO2 balance is still negative until the transport is also based on renewable energy sources. Current research, especially in the field of MFCs, is focused on increasing the efficiency and electrical power density of such cells. Furthermore, intending to pilot and industrial scales, keeping the optimised process parameters constant is still a challenge to keep the biological processes stable over time.

If we look again specifically at the optimization of the power yield of an MFC, the dimensioning of the anode as well as the cathode surface has turned out to be a sensitive parameter. An increase of the anode surface yielded an increase of the power density of the MFC [6]. Increasing the cathode surface area resulted in increased long-term stability and improved reduction of oxygen as a rate-determining step. Furthermore, connecting the cathodes in parallel resulted in a reduction in the internal resistance of the MFC system, due to the relationship between voltage and resistance ($U = R \times I$), leading to increased power ($P = U \times I = R \times I^2$). Further, two other not insignificant parameters for the activity of the microorganisms are the COD content and the C:N ratio. It should also be noted that in addition to the CO_2 from the exoelectrogenic COD degradation, methane can also be produced, quasi unintentionally, by side reactions. If we assume that 25% of the COD entering the MFC is removed via the exoelectrogenic degradation pathway and the remaining 75% via the fermentative degradation pathway, then the methane gas yield per kg of degraded COD can be estimated.

1.4 Concluding Remark

One can guess from the discussion that the use of microorganisms for energy conversion into electrical energy and production of sustainable energy sources such as H_2, CH_4 or biofuels from sustainable biomass will continue to play a major role. They will have their place in the network of sustainable systems through which the energy supply of mankind should and must be guaranteed in the future. As mentioned at the beginning of this chapter, we will certainly not be able to eliminate all uncertainties and problems with the help of renewable energies and environmentally friendly technologies. However, with the help of these technologies, we can certainly keep the complex Earth system from becoming completely unbalanced.

References

1. Kober T, Schiffer HW, Densing M, Panos E (2019). Global energy perspectives to 2060 – WEC's World Energy Scenarios 2019. Energy Strategy Rev 31:100523
2. Ferreira KN, Iverson TM, Maghlaoui K, Barber J, Iwata S (2004) Architecture of the photosynthetic oxygen-evolving center. Science 303:1831–1838
3. Loll B, Kern J, Saenger W, Zouni A, Biesiadka J (2009) Towards complete cofactor arrangement in the 3.0 Å resolution structure of photosystem II. Nature 438:1040–1044
4. Anhuai L, Yan L, Song J, Xin W, Xiao-Lei W, Cuiping Z, Yan L, Hongrui D, Ruixia H, Ming L, Changqiu W, Yueqin T, Hailiang D (2012) Nat Commun 3:768
5. Statistik Abwasserentsorgung (2017) BMUV (German Federal Ministry for the Environment, Nature Conservation, Nuclear Safety and Consumer Protection)
6. Hiegemann H, Lübken M, Wichern M, Schulte P, Schmelz KG, Gredigk-Hoffmann S (2015) Optimierung der mikrobiellen Brennstoffzellentechnik für den Kläranlagen-betrieb. Gewässerschutz–Wasser–Abwasser 236:20/1–20/15

Part II
Anaerobic Digestion

Chapter 2
Decreasing the Retention Time as a Way for Stabilizing Anaerobic Digestion Processes

Jérôme Harmand

Abstract Operating Diagrams (OD) are a particularly elegant way of graphically representing the asymptotic behavior(s) of a dynamic system. Provided that the model of the system of interest is well known, they enable the asymptotic behavior of the system to be determined as a function of the process input parameters. As we shall see, they can also be used to discover new system properties, or at the very least, devise new control strategies that would not otherwise have been discovered. In this way, we have discovered that it is sometimes possible to stabilize a multi-stage bioprocess by increasing the dilution ratio, rather than lowering it, as it is classically the case.

Keywords Anaerobic digestion · Nitrification · Chemostat · Multi-stability · Stabilization

2.1 Introduction

As most biological processes, anaerobic digestion (AD) processes can be formalized as a multi-stage process: a succession of different functional microbial species act in cascade in which the product of a stage is the substrate for the next one. In case of organic/hydraulic overloads, or still if changes in environmental conditions arise, such processes may destabilize in the sense some metabolites accumulate within a given stage yielding possible severe changes, as for instance in the pH. To deal with such unexpected situations, control systems may be used. For observers and control design purposes, a one, two, three or even more multistage formalization, as the well-known ADM1 is used: such formalizations take the form of deterministic dynamic models consisting of a set of differential equations describing the dynamics

J. Harmand (✉)
Laboratory of Environmental Biotechnology (LBE), National Research Institute for Agriculture, Food and Environment (INRAE), University of Montpellier, Avenue des étangs, 11100 Narbonne, France
e-mail: jerome.harmand@inrae.fr

V. Alcaraz Gonzalez et al. (eds.), *Wastewater Exploitation*, Springer Water,
https://doi.org/10.1007/978-3-031-57735-2_2

of the AD main state variables [1–6]. To study such models, better characterize and understand their asymptotic properties and dynamic, an important tool is to build their Operating Diagram (OD), see for example Pavlou's work, who was the first to propose the use of this method to graphically represent long-term behaviour of bioreactor dynamics [7]. As pointed out by Sari and Benyahia [8], the OD has the operating parameters as its coordinates and the different regions defined in it correspond to qualitatively different asymptotic behaviors. The operating diagram is a bifurcation diagram that shows how the system behaves when we vary the control parameters. This diagram also shows how extensive the parameter region is, where some asymptotic behaviors occur. Surprisingly, it is not yet systematically used in practice, although it is a powerful graphical method to synthesize the main asymptotic properties of a system and a very interesting tool to interpret mathematical properties related to biological processes. Recently, while mathematically investigating a well-known two-stage AD model—called AM2 (cf. [3])—we discovered a very intriguing and remarkable property: under certain conditions, a multistage biological system can be stabilized by decreasing the hydraulic retention time (HRT) instead of increasing it, or by increasing the dilution rate instead of decreasing it. More than this very surprising and counterintuitive result itself, it is the interest of OD as such that we want to highlight here below as a tool to discover and interpret new results about multistage biological systems.

This chapter is organized as follows. First, the basic properties of the chemostat with a Haldane-like kinetic are investigated. Then, the OD of the AM2 model is presented. In the Results and Discussion section, we explain how the OD may be used to better understand the main dynamics of a biological system. Finally, conclusions are drawn.

2.2 Materials and Methods

2.2.1 The OD of the Chemostat Model

The well-known chemostat model is described by the following system of ordinary differential equations:

$$\begin{cases} \dot{S} = (S_{in} - S)D - \mu(S)X \\ \dot{X} = (\mu(S) - D)X \end{cases} \tag{2.1}$$

where, S and X are the concentrations in substrate and biomass respectively, D the dilution rate, S_{in} the input substrate concentration and $\mu(S)$ the specific growth rate.

First, since our goal is to study the mathematical properties of the system, the yield coefficient has been set to 1, which is always possible by changing the variables appropriately [9]. Second, it should be mentioned that such a model has been shown to be suitable for describing the AD process with appropriate parameters and under certain

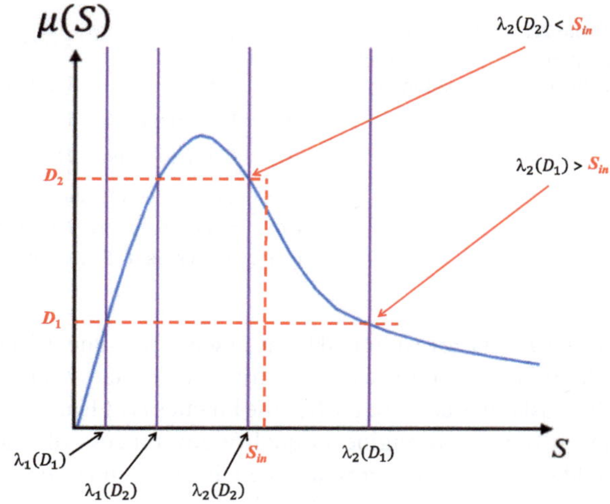

Fig. 2.1 Typical shape of a Haldane function

conditions [1]. Thirdly, a continuously operated stirred tank reactor (homogeneously mixed reactor with constant volume; chemostat) represents the best option for biogas production, as it provides the best biogas production compared to semi-continuous or batch modes of operation with fixed control inputs (constant parameters) [10–12].

For the substrate dependence of the growth rate μ under substrate limiting or growth limiting conditions, the following Haldane relation is used in the model.

$$\mu(S) = \mu_0 \frac{S}{S + K_S + S^2/K_I} \tag{2.2}$$

where μ_0 is the maximum growth rate without inhibition, while K_S and K_I are the limitation and inhibition coefficients, respectively.

An example of Haldane function is plotted in Fig. 2.1.

The properties of the chemostat model considering a Haldane kinetic have been extensively studied and are well-known, including some extensions notably with respect to the kinetic function considered, when the dilution term (*D*) is not the same in both equations, if the biomass is retained or still if mortality is considered [9]. If the kinetic function is of the Haldane type (meaning that a much larger class of functions than those satisfying Eq. (2.2) is to be considered) the following boundary conditions are to be considered: (i) $\mu(S) = 0$, (ii) μ increases until a maximum is reached, and then (iii) decreases as it proceeds. These considerations are summarized again in Table 2.1.

This table reads as follows:

- The first line gives conditions of existence of steady states.
- When they exist, the nature of the stability of steady states are indicated either as unstable or locally exponentially stable (LES).

Table 2.1 Asymptotic properties of the chemostat with Haldane-type kinetic

S_{in}	$S_{in} < \lambda_1(D)$	$\lambda_1(D) < S_{in} < \lambda_2(D)$	$\lambda_2(D) < S_{in}$
E_0	LES	Unstable	LES
E_1	Does not exist	LES	LES
E_2	Does not exist	Does not exist	Unstable

The notations $\lambda_1(D)$ and $\lambda_2(D)$ are the values of S such that $\mu(S) = D$, cf. Figure 2.1 (by convention they equal infinity if D does not cross $\mu(S)$), from [9]

Remark These results call for an essential comment for process engineers. Unlike the Monod function, for which positive steady state is always stable when $D < \mu(S_{in})$, the washout is attracting when the kinetics is of Haldane-type. This calls into question the way process engineers qualify any process of interest. Conventionally, in the field of dynamic systems, a system is often said to be stable. However, in the field of process engineering, a system that is not functioning—a process that goes towards the washout—is readily described as "unstable". On the one hand, it is important to remember that it is the equilibrium point around which the system operates that is stable or not. On the other hand, with Haldane kinetics, we end up with a washout steady state which is stable. Thus, from the point of view of the dialectics adopted in process engineering, it seems important to precise that a process is unstable when the steady state corresponding to the washout is stable.

It should be noted, of course, that the operating diagram (OD) depends strongly on the model parameters. Using the information compiled in Table 2.1, the OD can be constructed using geometric methods. However, a simple "algorithm" can also be used, where the dilution rate D is varied from 0 to a certain maximum and the corresponding properties of the associated steady states are read off in Table 2.1. For $D = D_1$, $\lambda_1(D_1)$ and $\lambda_2(D_1)$ are defined. Thus, for each value of S_{in}, we can browse the line $D = D_1$ and read which steady state exist and are stable according to the relative position of S_{in} with respect to $_1(D_1)$ and $_2(D_1)$. Depending on the nature of the steady state, different colors are used. According to the Haldane kinetics and the aforementioned specifications, the operating diagram was created (Fig. 2.2):

Each colored area then defines a region. The obtained regions denoted $J_{i,i=1..3}$ are summarized in Table 2.2:

Based on the existence conditions of the steady states and the nature of their stability, OD can be created using either geometric or analytical approaches [8, 13]. However, when the system becomes more complex and not all this information is available, numerical approaches can be followed, as in [14, 15].

In addition to the theoretical background, it is important to consider the way in which the operating diagram OD may be handled by the user:

- Once the control/operating parameters $\{D, S_{in}\}$ have been set, they define a functioning mode characterized by a point situated in one of the areas $J_{i,i=1..3}$. Referring to Table 2.2, the user knows immediately the possible steady states and their stability properties.

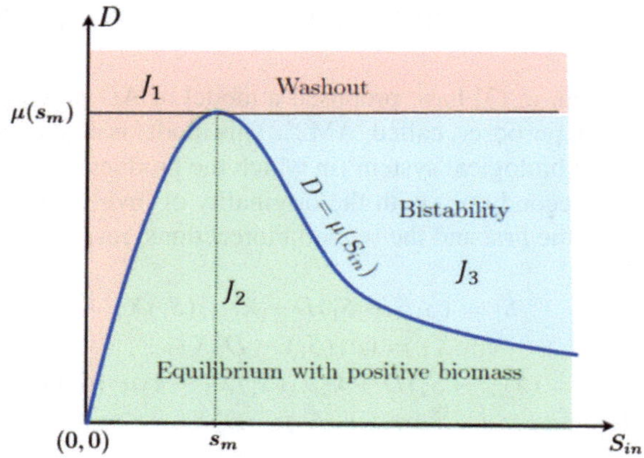

Fig. 2.2 OD of the chemostat model with a Haldane kinetic (modified from [9])

Table 2.2 Steady state existence and stability for a chemostat with Haldane-like kinetic

Region	E_0	E_1	E_2	Color
J_1	S			Red
J_2	U	S		Green
J_3	S	S	U	Blue

U and S stand for Unstable and LES, respectively

- If one of the operating parameters varies with time—sufficiently slowly to be assume we remain close to a steady state—the user has simply to browse the OD following a line and is able to discover bifurcations of steady states. For instance, consider Fig. 2.2 and assume that (i) S_{in} remains constant, (ii) $S_{in} > s_m$ (s_m being the maximum of μ) and that (iii) D increases slowly. Furthermore, assume D to be small enough such that the initial functioning point in the J_2 area where the seady state E_1 is stable, a steady state with positive biomass. Increasing D will yield the system to attain J_3, an area where the system exhibits bistability. Increasing even more D will definitely destabilize the process in bringing the system to a region where E_0, corresponding to the washout of the biomass, is stable.
- As it is well-known (cf. Table 2.2), a chemostat with a Haldane-type kinetic exhibits bistability as long as $\lambda_2(D) < S_{in}$. Assume we initialize the system with a couple of operating parameters $\{D, S_{in}\}$ such that these conditions holds (the system evolves in the J_3 area where there is precisely bistability). Furthermore, assume we initialize the system with initial conditions $\{S_0, X_0\}$ in the attraction basin of E_0. We immediately see from the OD plotted in Fig. 2.2 that the only way to stabilize the system and avoid washout is to decrease D enough so that we return the system to the J_2 area. Notice that an alternative is to decrease S_{in}.

2.2.2 The Operating Diagram of the AM2 Model (from [8])

In 2001, Bernard et al. in [3] have proposed a model of AD processes specifically developed for control purposes, called 'AM2'. This model is a general mass balance model of a two-stage biological system (in which the product of the first reaction is the substrate of the second one) with the originality of involving both Monod and Haldane kinetics for the first and the second bioreactions, respectively:

$$\begin{cases} \dot{S}_1 = (S_{1in} - S_1)D - k_1\mu_1(S_1)X_1 \\ \dot{X}_1 = (\mu_1(S_1) - D)X_1 \\ \dot{S}_2 = (S_{2in} - S_2)D + k_2\mu_1(S_1)X_1 - k_3\mu_2(S_2)X_2 \\ \dot{X}_2 = (\mu_2(S_2) - D)X_2 \end{cases} \tag{2.3}$$

where the notations are those commonly used, $\mu_1(S_1) = \mu_{max1}\frac{S_1}{S_1+K_{S1}}$ and $\mu_2(S_2) = \mu_0\frac{S_2}{S_2+K_{S2}+S_2^2/K_I}$ which attains its maximum for $S_2 = S_2^M$.

Notice that the original model involves αD instead of D on the dynamics of X_i but, for the sake of simplicity we will simply consider here that $\alpha = 1$. For an extensive study of such a model with Monod and Haldane-like kinetics, mortality and different retention terms on X_i, the reader may refer to [8, 16, 17].

For such systems, notice that there may have more than two inputs. In such a case, it may be problematic to represent the different regions of the input space. If there are three inputs, here $\{D, S_{1in}, S_{2in}\}$, we can use 3D-plots, cf. for instance Sari and Benyahia [8]. However, such figures can be difficult to interpret. Instead, we can decide to fix some of the inputs to a constant value, for instance S_{2in} and represented the OD in the $\{S_{1in}, D\}$ plane.

When studying the system described by the dynamic system (3), it can be shown that the system can have at most six steady states. Taking into account such a system describes two biological reactions in series, from a biological viewpoint, they can be easily interpreted. They are denoted and defined as E_1^0 (washout of both biomasses), E_1^i (washout of X_1: as $\mu_2(S_2)$ is not monotonous, there are two points such that $\mu_2(S_2) = D$), E_2^0 (washout of X_1) and E_2^i (positive steady staes: again, since $\mu_2(S_2)$ is not monotonous, there are two points such that $\mu_2(S_2) = D$). If μ_1 and μ_2 are known, depending on the inputs and on the parameters of the model, they can be computed and the nature of their stability established. The components of the steady states are denoted as $\left\{S_1^*, X_1^*, S_2^{i*}, X_2^{i*}\right\}$ (by convention, we ordered $S_2^{1*} < S_2^M < S_2^{2*}$).

In order to get straight to the heart of the matter, we leave aside the more technical considerations and simply report the most striking results below. To do so, let us first introduce the following notations: $D_1 = \mu_{max1}$, $D_2 = \mu_2\left(S_2^M\right)$ and $H_2(D) = S_2^{2*}(D) + \frac{k_2}{k_1}S_1^*(D)$ which is defined for $0 \leq D \leq min(D_1, D_2)$.

Using these notations, Sari and Benyahia have shown that the 'shapes of the OD of system (3)' fall within three possible cases defined by:

- Case (A): $D_1 > D_2$ and $\frac{dH_2}{dD} < 0$;
- Case (B): $D_1 > D_2$ and H_2 non monotonous;
- Case (C): $D_1 < D_2$.

Sometimes, a very interesting shape for the bistability region arises; more particularly, we may observe the appearance of an area where positive equilibria are stable between the regions where the system exhibits bistability and the region where the washout is the only stable steady state. For the model (3), It never happens in case (A), it may happen but it is not systematic in case (B) and it is always the case in case (C). This very interesting phenomena has been shown to arise in a number of papers of the literature. For instance, in Fig. 2.3, we reported the OD of the system described by model (3) but with a Contois kinetic in the first reaction instead of a Monod function and with mortality terms in both X_1 and X_2 dynamics. More precisely, the general model under interest was:

$$\begin{aligned} \dot{S}_1 &= (S_{1in} - S_1)D - \mu_1(S_1, X_1)X_1 \\ \dot{X}_1 &= (\mu_1(S_1, X_1) - D_1)X_1 \\ \dot{S}_2 &= (S_{2in} - S_2)D + \mu_1(S_1)X_1 - \mu_2(S_2)X_2 \\ \dot{X}_2 &= (\mu_2(S_2) - D_2)X_2 \end{aligned} \tag{2.4}$$

with $\mu_1(S_1, X_1) = \mu_{max1}\frac{S_1}{S_1+K_{X1}X_1}$ and a choice of the parameters of the model such that these phenomena indeed arise.

In this example, Fig. 2.3 represents the operating diagram of considered model in the plan $\{S_{1in}, D\}$ for a fixed value of S_{2in}. Without entering into technical details for our purpose, we simply introduce the notations used: A_1 (in green) is the stability region of the washout E_1^0 (all biomasses are null), A_5 (in blue) is the stability of a steady state with non zero X_1 but without X_2 (washout), A_7 (in dark blue) is the

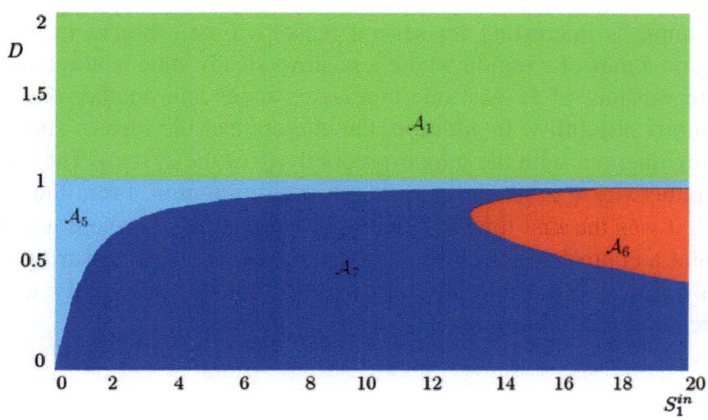

Fig. 2.3 OD of the chemostat model with a Haldane kinetic (from [14])

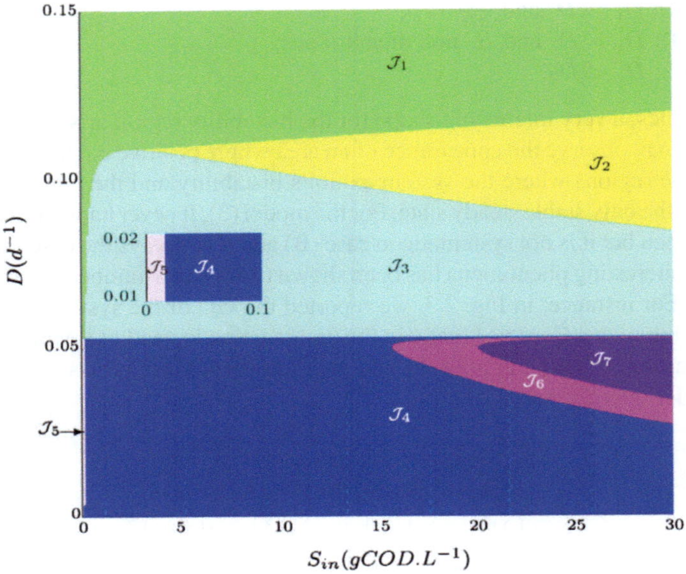

Fig. 2.4 OD of the model proposed in [3] (from [15])

stability regions of positive steady-states and A_6 is the bistabilty region. As well as highlighted here above, for some values of S_{1in}, a part of the regions A_7 separate the bistability area A_6 from A_1.

Remark Another example of such situation for multi-stage AD models can be found in the literature. The example reported by [15] proposed the OD plotted in Fig. 2.4.

This example is interesting for several reasons. First, observe in Fig. 2.4 that we have again a part of a region where a positive steady state is attractive between areas where washout of at least one biomasses arises and another one where the system exhibits bistability. In addition, the authors had the idea of superimposing the operating diagram with the biogas productivity of the system. They discovered that the productivity was maximum in the bistability region. From the best of our knowledge, it was the first time authors highlight a situation where it is necessary to implement a control system in order to optimize AD while ensuring the global stability of the positive steady state in closed loop. Or said differently: avoiding the washout steady state to be attractive.

2.3 Results and Discussion

2.3.1 Interpreting ODs

In this part, we show how the OD can be used to synthesize a way of stabilizing a multi-stage biological system when conditions are such that a part of a region where the positive steady state is stable appears between the stability region of the washout and the bistability region. We conjecture that a necessary condition for such phenomena to appear is the presence, at one another step than the first one, of a biomass exhibiting a Haldane-type kinetic.

Let us come back to the example studied by Hanaki et al., in [14]. Consider D to be a control and let us fix S_{2in} and let $S_{1in} = 18g/L$. The corresponding OD is plotted in Fig. 2.5.

Let us browse the control slowly from a small ($D < d_1$) to a high value ($D \gg d_2$). In such a situation, we go through the following regions in considering successive equilibria when increasing D: $A_7 \rightarrow A_6 \rightarrow A_5 \rightarrow A_1$, cf. Figure 2.5. The OD thus allows us to see the appearance/disappearance of steady states as a function of operating parameters, here the control D. As long as D is small enough, the quantity of substrate entering the second step of the reaction is important since the high retention time leaves the time needed to X_1 to convert S_1 into S_2: the system is in the region A_7 where the positive steady state is the only one which exists, and it is thus stable. As D increases, the size of the attraction basin of this equilibrium decreases until it reaches a critical value (corresponding to the point p_1 in Fig. 2.5), which corresponds to the frontier between regions A_7 and A_6. Increasing a little bit more D the system enters then in the region A_6. When $D = d_2$, the quantity of available

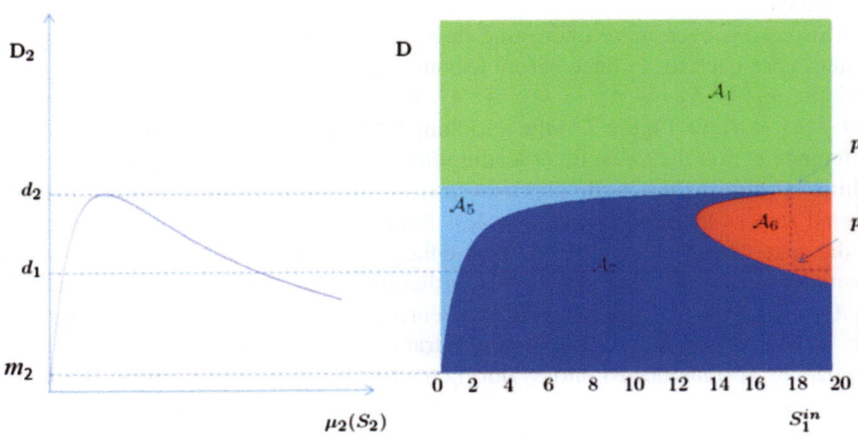

Fig. 2.5 OD of a two-stage biological system for studying the case $S_{1in} = 18g/L$. The notation D_2 stands for $\alpha D + m_2$ where m_2 is the mortality of X_2 (modified from [14])

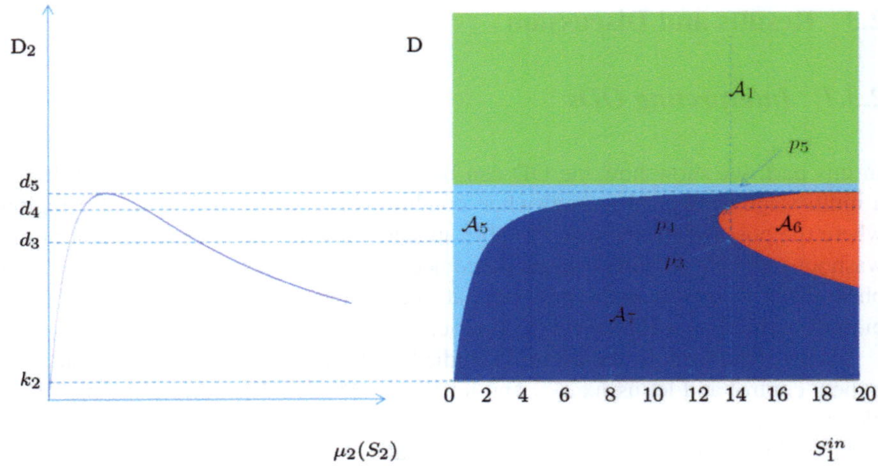

Fig. 2.6 OD of a two-stage biological system for studying the case $S_{1in} = 14g/L$ (modified from [14])

resource necessary for X_2 to grow may become limiting for some initial conditions, leading the system to enter a bistability zone. With the values of the parameters chosen, further increasing D definitely leads to the washout of X_2: the system enters into A_5 in crossing the point p_2 in Fig. 2.5. Finally, if D attains the critical value corresponding to the maximum growth rate of X_1, it also goes to extinction: the system enters in the region A_1.

Now, instead of $S_{1in} = 18g/L$ let us consider $S_{1in} = 14g/L$. The OD is plotted again in Fig. 2.6 together with the plots necessary to interpret the encountered phenomena.

This case is even more interesting than before. Indeed, when D inscreases, the system goes back to A_7 once before leaving it definitely in browsing the following regions: $A_7 \to A_6 \to A_7 \to A_5 \to A_1$. While D is small enough (i.e. such that $\alpha D + m_2 < d_3$, cf. Figure 2.6) the reasoning remains the same than before: the only difference is that the value of D leading the system to enter into A_6 through p_3 is a little bit higher than in the previous case ($d_1 < d_3$). It is due to the fact that the second step of the reaction receives less input from the first step when compared to the case where $S_{1in} = 18g/L$, thus enlarging the attraction basin of the stable positive equilibrium. Then, when D is further increased, the system enters back into A_7 through point p_4 instead of directly entering A_5 as it was the case before. In fact, it depends on how the input concentration of the second step which includes the part of S_1 transformed into S_2 during the first reaction is affected by D. On the one hand, if we are in a 'flat' zone of the growth rate $\mu_2(S_2)$ assuming we consider concentrations at the right of the maximum of $\mu_2(S_2)$, a small variation of D (and thus of D_2) will change very much l_2 (the largest solution of the equation $\mu_2(S_2) = D_2$) while what enters into the second reaction may actually remains almost constant. On the other hand, if D is such that D_2 crosses $\mu_2(S_2)$ in a sharper zone of the Haldane-type function (typically around the inflexion point, still considering S_2 evolves at concentrations such as we are on the right of the maximum of $\mu_2(S_2)$), a small change in D (and thus on D_2) will much more affect l_2 than before. In any of these situations, the relative positions of the largest solution of the equation $\mu_2(S_2) = D_2$ and of what enters into the second reaction stage (sum of what has been converted by X_1 and what enters through S_{2in}) determine whether the system will evolve in the region A_7 or A_6 and we can observe, as D increases (at a value such $\alpha D + m_2 = d_4$) a return of the system from A_7 to A_6 through p_3 and then back into A_7 through p_4.

2.3.2 *Stabilizing an Unstable Multi-stage Biological System Using ODs*

Now, consider again Fig. 2.3. Assume the set $\{S_{1in}, D\}$ is such that the system evolves in A_6 in a zone for which the value of S_{1in} is such that if we increase D, we first enter into A_7 before entering in A_5. For clarifying this, let us simply assume that $S_{1in} = 14g/L$. Furthermore, assume that the initial conditions are such that, in this bistability region, this is the steady state with $X_2 = 0$ which is attractive. The question is 'what are the options to stabilize the system?'. Obviously, from the analysis we performed here above, and at the opposite to the case of the mono-stage chemostat model, we have now two options and no longer only one: the classical option is to decrease D sufficiently such that the system leaves A_6 in entering in the region A_7. However, with respect to the analysis there is another option which consists in…increasing D! Indeed, if we increase D such as we leave A_6 and we do not enter A_1, then we are in the region A_7 where the positive steady state is the only stable steady state: the system is stabilized.

The graphical interpretation of the possibility of stabilizing a biological process in increasing the dilution rate is given in Fig. 2.7. With respect to the single chemostat, the main point to keep in mind to understand this phenomena is that, in a multistage biological system, the 'input concentration' $\tilde{\mathrm{S}}_{\mathrm{in}}$ that enters into the next reaction stage is not a constant, as in the chemostat, but it varies with D as well as λ_2 with respect to which it has to be compared to study the stability of the steady states (Fig. 2.8).

Remember from Table 2.2 that as long as $\lambda_2(D) < S_{in}$—here $\lambda_2(D) < \tilde{S}_{in}(D)$—the steady state corresponding to the washout of X—here the steady state corresponding to the washout of X_2 is stable. Depending on the rates at which λ_2 and $\tilde{S}_{in}$ vary with D one with respect to the other, one may observe a bifurcation corresponding to the transition from regions of bistability to a region where the positive steady state is stable while D increase. It is precisely what is illustrated in Fig. 2.7; when it is increased, then both λ_2 and $\tilde{S}_{in}$ decrease, but not at the same rate. And it may happen that $\lambda_2(D)$ becomes greater than $\tilde{S}_{in}$ which stabilizes the system rendring the washout unstable.

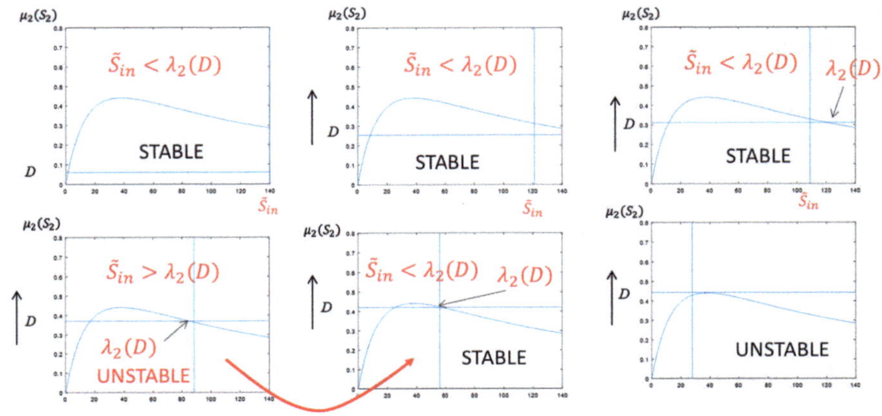

Fig. 2.7 Nature of the stability of the positive equilibrium: graphical interpretation of the strategy consisting in increasing the dilution rate to stabilize a two-stage biological system where $\tilde{S}_{in}(D)$ is the total concentration entering the second reaction step (the sum of what is converted within the first reaction and what comes into the reaction through S_{2in}) and l_2 is the second solution of the equation $\mu_2(S_2) = D$. For the sake of simplicity, $\alpha = 1$ and we did not consider mortality. Notice that both $\tilde{S}_{in}$ and l_2 are functions of D

In [18], we have already proposed numerical simulations with the aforementioned strategy for the AM2 model. The simulations are described below.

2.3.3 Limits of the Approach

- First, are the conditions for the existence of the highlighted phenomena restrictive? In other terms, are they encountered often or are they rare? In the AM2 model these phenomena appears when we are in case (C) in [8]: it means that the maximum activity of the second biological stage is greater than the maximum activity of the first stage. In this sense, the approach could eventually be of interest for systems where the first step is limiting, for instance, when the substrate to be degraded is slowly biodegradable. Yet, it is rather in the other direction, when the first stage is faster than the second, that we tend to use modeling for control synthesis purposes. Indeed, if this is the case, it leaves open the possibility for metabolites to accumulate in the second stage of the reaction, hence the need to worry about system control. If the second stage is faster than the first one, there may not be much point in worrying about control. However, it should be stressed here that the condition only applies to maximum rates parameters, not to the kinetics as a whole. This criticism must therefore be put into perspective.
- Second, one may question the actual applicability of the approach with respect to uncertainties in bioreactor modeling. Indeed, it is the rule rather than the exception that biological kinetics are unknown and sometimes difficult to model. The size

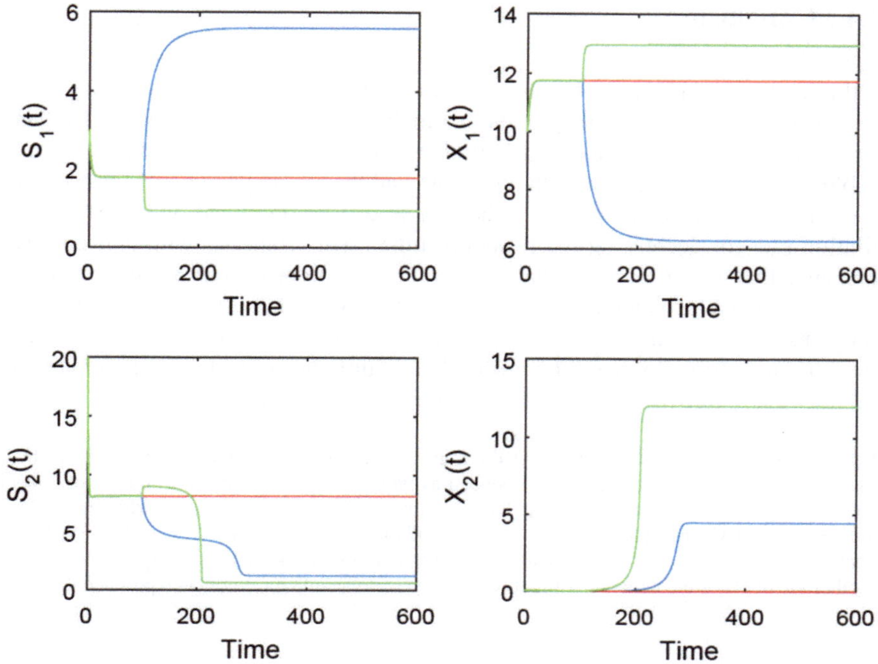

Fig. 2.8 Dynamical simulation of the two-step system. The trajectories of the system with a constant dilution rate ($D = 0.7$) are plotted in red: the washout of X_2 is globally attracting. In blue, with the same initial conditions, the dilution rate is changed to 0.8 at $t = 100$ and then switched back to $D = 0.7$ at $t = 400$: the system is now globally attracted by a positive steady state. In green, again for the same initial conditions, the trajectories are plotted when the classical strategy is used: the dilution rate is changed to 0.6 at $t = 100$ and then switched back to $D = 0.7$ at $t = 400$ (from [18])

of the region that can occur between the bistability region and the region where the washout is stable can be extensive, depending on the parameter values of the model.

- Thirdly, the question arises whether the proposed approach leads to a better productivity in biogas. Unfortunately, the simulations performed showed that there was no improvement in biogas production. Moreover, if we increase the D, we dilute S_2. But at the same time, we also dilute the biomass. Increasing D, while allowing theoretically to stabilize the system, can be risky, and the control is probably not robust with respect to this point.

2.4 Conclusion

In this chapter, using the operational diagrams, we have presented a surprising property of multistage biological systems in which one of the stages is of the Haldane type. More precisely, it is a case where part of the region where the equilibrium is stable is between the regions with bistability and the region where only the washout is stable. Because of this property, the system can be stabilized by increasing the dilution rate (or decreasing the residence time). This contrasts with conventional strategies that aim to decrease the dilution rate (in order to increase the residence time), which leads to an unintuitive control strategy. Without the use of operational diagrams, the above findings would never have been obtained. However, this methodology is susceptible to modeling errors. In addition, unfortunately, the simulations did not yield improvements in system performance, e.g., gas productivity.

Acknowledgements I thank the Euro-Mediterranean research network TREASURE within which a large part of these results has been produced as well as my coauthors of a number of papers from those part of the presented results have been extracted, and more particularly Tewfik Sari, Alain Rapaport (who accepted to read the chapter and made a number of feedbacks: thanks!), Denis Dochain, Radhouane Fekih Salem, Nahla Abdellatif and Zoubida Mghazli. Finally, thank you also to other members of the network without whom the results would have not been produced.

References

1. García-Diéguez C, Bernard O, Roca E (2013) Reducing the anaerobic digestion model no. 1 for its application to an industrial wastewater treatment plant treatingwinery effluent wastewater. Biores Technol 132:244–253. https://doi.org/10.1016/j.biortech.2012.12.166
2. Ghouali A, Tewfik T, Harmand J (2015) Maximizing biogas production from the anaerobic digestion. J Process Control 36:79–88. https://doi.org/10.1016/j.jprocont.2015.09.007
3. Bernard O, Hadj-Sadock Z, Dochain D, Genovesi A, Steyer J-P (2001) Dynamical model development and parameter identification for an anaerobic wastewater treatment process. Biotech Bioeng 75:424–438. https://doi.org/10.1002/bit.10036
4. Alcaraz-Gonzalez V, Harmand J, Rapaport A, Steyer JP, Gonzalez-Alvarez V, Pelayo-Ortiz C (2002) Software sensors for highly uncertain WWTPs: a new approach based on interval observers. Water Res 36:2515–2524. https://doi.org/10.1016/s0043-1354(01)00466-3
5. Weedermann M, Seo G, Wolkowics GSK (2013) Mathematical model of anaerobic digestion in a chemostat: Effects of syntrophy and inhibition. J Biol Dyn 7:59–85. https://doi.org/10.1080/17513758.2012.755573
6. Batstone D, Keller J, Angelidaki I, Kalyhuzhnyi S, Pavlosthathis S, Rozzi A, Sanders W, Siegrist H, Vavilin V (2002) Anaerobic digestion model no.1 (ADM1), IWA Publishing, London
7. Pavlou S (1999) Computing operating diagrams of bioreactors. J Biotechnol 71:7–16. https://doi.org/10.1016/s0168-1656(99)00011-5
8. Sari T, Benyahia B (2021) The operating diagram for a two-step anaerobic digestion model. Nonlinear Dyn 105:2711–2737. https://doi.org/10.1007/s11071-021-06722-7
9. Harmand J, Lobry C, Rapaport A, Sari T (2017) The chemostat: mathematical theory of microorganism cultures. Wiley ISTE Editions
10. de Gooijer DW, Bakker AM, Beeftink HH, Tramper J (1996) Bioreactors in series: an overview of design procedures and practical applications, Enzyme Microbial Tech 18(3):202–219https://doi.org/10.1016/0141-0229(95)00090-9

11. Dali-Youcef M, Rapaport A, Sari T (2022) Performance study of two serial interconnected Chemostats with mortality. Bulletin Math Biol 84(110). https://doi.org/10.1007/s11538-022-01068-6
12. Dali-Youcef M, Sari T (2023) The productivity of two serial chemostats. Int J Biomath 16(06):2250113. https://doi.org/10.1142/S1793524522501133
13. Nouaoura S, Fekih-Salem R, Abdellatif N, Sari T (2022) Operating diagrams for a three-tiered microbial food web in the chemostat. J Math Biol 85(44). https://doi.org/10.1007/s00285-022-01812-5
14. Hanaki M, Harmand J, Mghazli Z, Rapaport A, Sari T, Ugalde P (2021) Mathematical study of a two-stage anaerobic model when the hydrolysis is the limiting step. Processes 9(11):2050–2067. https://doi.org/10.3390/pr9112050
15. Khedim Z, Benyahia B, Cherki B, Sari T, Harmand J (2018) Effect of control parameters on biogas production during the anaerobic digestion of protein-rich substrates. Appl Math Model 61:351–376. https://doi.org/10.1016/j.apm.2018.04.020
16. Benyahia B, Sari T, Cherki B, Harmand J (2012) Bifurcation and stability analysis of a two step model for monitoring anaerobic digestion processes. J Process Control 22(6):1008–1019. https://doi.org/10.1016/j.jprocont.2012.04.012
17. Sari T (2022) Best operating conditions for biogas production in some simple anaerobic digestion models. Processes 10(258). https://doi.org/10.3390/pr10020258.
18. Harmand J, Dochain D, Rapaport A (2020) Increasing the dilution rate can globally stabilize two-step biological systems. J Process Control 95:67–74. https://doi.org/10.1016/j.jprocont.2020.08.009

Part III
Dark Fermentation

Chapter 3
Microbial Population Dynamics in Continuous Hydrogen Production Systems by Dark Fermentation of Tequila Vinasse

Oscar Aguilar-Juárez, Luis Arellano-García, and Elizabeth León-Becerril

Abstract Bioenergy production from agro-industrial waste has been a topic of interest for many years and now, with technological developments, and the identification of microbial species involved in the biodegradation of substrates, it is possible to establish population dynamics in waste biodegradation processes that promote a circular economy. This work presents a comparative study of continuous hydrogen production systems by dark fermentation of tequila vinasse emphasizing the analysis on the operational conditions and the microbial population dynamics. In particular, the comparative study aims to understand the synergy between the biohydrogen production in continuous systems, and the microbial population dynamics in systems with or without immobilized biomass. Interactions between H_2-producing bacteria (HPB) (e.g., *Clostridium, Enterobacter*), lactic acid bacteria (LAB) (e.g., *Lactobacillus, Sporolactobacillus*, *Streptococcus*), and acetic acid bacteria (AAB) (e.g., *Acetobacter*), are described considering the impact on hydrogen production. The study includes experiments in CSTR reactors, where the effect of increasing organic load; the use of carriers; type of inoculum; hydraulic residence time; feed concentration and type of vinasse is evaluated in terms of the biohydrogen production rate and composition of the biogas generated. The results show clear tendencies of hydrogen production and presence of specific microbial populations in the continuous dark fermentation of tequila vinasse. Also, it was observed that productivity of hydrogen is governed by the organic loading rate in combination with key hydrogen-producing bacterial genera while vinasse from different tequila manufacture processes did not affect the dominant bacterial genera in fermentation.

O. Aguilar-Juárez (✉) · L. Arellano-García · E. León-Becerril
Centro de Investigación y Asistencia en Tecnología y Diseño del Estado de Jalisco A.C, Av. Normalistas 800, Colinas de La Normal, 44270 Guadalajara, Jalisco, México
e-mail: oaguilar@ciatej.mx

L. Arellano-García
e-mail: larellano@ciatej.mx

E. León-Becerril
e-mail: eleon@ciatej.mx

V. Alcaraz Gonzalez et al. (eds.), *Wastewater Exploitation*, Springer Water,
https://doi.org/10.1007/978-3-031-57735-2_3

Keywords Bioenergy · Agro-industrial waste · Continuous reactors · *Clostridium* · *Sporolactobacillus* · *Prevotella*

Abbreviations

CSTR	Continuous stirred tank reactor
DF	Dark fermentation
HAc	Acetate
HBu	Butyrate
HFor	Formate
HLac	Lactate
HPB	Hydrogen producing bacteria
HPR	Hydrogen production rate (L H_2/L-d)
HRT	Hydraulic retention time (h)
HVa	Valerate
LAB	Lactic acid bacteria
NCP	Non-cooking process with water diffusers to obtain agave sugars
OLR	Organic loading rate (g COD/L-d)
SCA	Short carboxylic acids
TOC	Total organic carbon
TrP	Traditional process to obtain agave sugars by cooking

3.1 Introduction

Nowadays, research is addressing priority issues to provide answers to counter the progressive climate and environmental crisis. Process intensification promoting circular economy and concepts for action at the local level is of high importance. This also includes sustainable measures to promote the resilience of economies. For example, optimizing material and energy flows, minimizing waste, and recycling, can be part of the solution that increases this resilience.

In Jalisco, Mexico, one of the most important industries, is addressed to traditional spirits production such as tequila. Tequila is manufactured from the distillation of fermented sugars extracted from the mature stems of *Agave tequilana* Weber var. Azul. During the distillation process, organic waste is generated, known as vinasse. Tequila vinasse is currently generated at a rate of 430 million liters (about 113,593,960 gal) per month considering a mean waste generation of 10–12 L of vinasse per liter of tequila [1]. Vinasse is a high-strength residue, presenting a dark color, with an acid pH, a high organic matter content of 30–70 g of chemical oxygen demand (COD) per liter, high content of solids, and rich in phenolic compounds and contains residual fibers, agave polymeric carbohydrates, and fermentable sugars [2].

To date, alternative uses of vinasse include irrigation of agave and sugarcane fields [3] and nowadays it is being used also as component of agave bagasse compost. However, the high organic load and mineral content counteracts the fertilizing character of vinasse promoting accumulation of metal cations and reducing diversity and concentration of mycorrhizal fungi [4]. Despite these characteristics, vinasse is used for irrigation or discharged into surface waters provoking malodorous emissions and oxygen depletion [5].

The increasing global demand for tequila has accelerated both production and waste generation rates, with faster alternative processes such as cooking agave in autoclaves or leaching agave carbohydrates in continuous diffusers. Two of the main processes for sugar extraction in the tequila manufacture: cooking agave and use of diffusers are named here as traditional process (TrP) and non-cooking process (NCP), respectively. Each process eventually generates vinasse with different compositions, where for example it is known that among different processes, TrP produces vinasse with the highest values in COD, biochemical oxygen demand (BOD), and total organic carbon (TOC) [6].

Physicochemical characteristics of the tequila vinasse

Table 3.1 shows the main characteristics of the vinasse used as a reference for this work [7, 8, 12]. It is important to note that tequila vinasse tends to present heterogeneous characteristics, even in the same factory. Indeed, for TrP and NCP vinasses there are differences in the values of the physicochemical properties, however, similar main characteristics emerge (low pH, high organic load, high sugar concentration and high concentration of total phenols) [2].

3.1.1 Bioenergy

Hydrogen is an efficient and clean energy carrier gas, generated in the dark fermentation of organic waste used as source of fermentable carbohydrates, as tequila vinasse. It is possible to optimize hydrogen production by pretreating the inoculum or operation variables, depending on the characteristics of the organic waste. The wastewater from this dark fermentation is rich in fatty acids, which can be used in a further step to produce methane (biogas) as an additional energy source.

Biogas production from anaerobic treatment of vinasse can support environmental protection, for example, by displacing methane and hydrogen for heating in the distillation process or commercialized to other industries. Direct methane recovery from anaerobic digestion of tequila vinasse would allow biogas recovery in a single process. However, it was recently demonstrated that hydrolysis and dark fermentation of tequila vinasse, as the first of a two-step process, is advantageous due to the additional generation of hydrogen and obtaining an effluent rich in volatile acids, which favor methane formation in the anaerobic digestion as second step [8–11].

The research works review in this chapter aims at comparing the bacterial population dynamics and overall performance in the dark fermentation of tequila vinasse analyzing the main trends related to:

Table 3.1 Physicochemical characteristics of the TrP and NCP tequila vinasses used for the systems

Parameter	TrP vinasse			NCP vinasse
	System 1		System 2	System 3
	Vinasse A	Vinasse B		
Total COD (g/L)	42.2 ± 0.6	51.5 ± 1.5	57.6	35.5 ± 5.1
Soluble COD (g/L)	38.0 ± 0.4	48.9 ± 1.0		
TOC (g/L)				12.5 ± 2.8
pH	3.9 ± 0.2	3.7 ± 0.1	3.6	3.5 ± 0.3
TS (g/L)	29.4 ± 0.51	42.8 ± 2.9		23.5 ± 4.2
TDS (g/L)	24.3 ± 0.8	25.5 ± 0.9		
TSS (g/L)	5.1 ± 0.5	16.3 ± 0.2		
TVS (g/L)	26.7 ± 0.7	38.2 ± 1.5		21.6 ± 4.5
VSS (g/L)	4.9 ± 0.5	13.7 ± 0.9		
Total nitrogen (mg/L)	69.3 ± 1.2	210.0 ± 10.0		110.0 ± 30.0
Total phosphorous (mg/L)	1298.3 ± 98.3	730.0 ± 90.0		1170.0 ± 43.0
Total carbohydrates (g/L)	14.4 ± 0.2	18.5 ± 1.7	17.6 ± 0.6	8.25 ± 2.01
Total reducing sugars (g/L)	10.0 ± 0.1	10.6 ± 0.8		
Total phenols (mg GAE/L)	547.7 ± 28.1	1430 ± 120		140.0 ± 40.0
Potassium (mg/L)	539.4	309.6		
Sulfur (mg/L)	61.4	68.7		
Magnesium (mg/L)	258.2	171.9		
Zinc (mg/L)	3.8	1.2		
Sodium (mg/L)	23.2	25.2		
Iron (mg/L)	22.7	39.1		
Calcium (mg/L)	566.0	345.2		
Manganese (mg/L)	<1.0	3.4		
Nickel (mg/L)	<1.0	<1.0		
Copper (mg/L)	<1.0	16.9		
Molybdenum (mg/L)	<1.0	<1.0		
Lactate (g/L)	2.5	1.9	2.6	
Acetate (g/L)	2.2	1.5	1.9	

(continued)

Table 3.1 (continued)

Parameter	TrP vinasse			NCP vinasse
	System 1		System 2	System 3
	Vinasse A	Vinasse B		
Butyrate (g/L)	Negligible	Negligible	Negligible	
Propionate (g/L)	0.3	0.3		
Formate (g/L)			4.0	
Valerate (g/L)			1.1	

TS total solids, *TSS* total suspended solids, *TVS* total volatile solids, *VSS* volatile suspended solids

a. Changes in operation conditions such as hydraulic retention time (HRT), and influent concentration, both impacting on the organic loading rate (OLR) to reactors;
b. Biomass state (suspended or immobilized);
c. Type of vinasse (TrP, NCP).

3.2 Materials and Methods

3.2.1 Dark Fermentation Reactors

3.2.1.1 System 1

A 3.0-L continuous stirred tank reactor (CSTR) with a working volume of 1.0 L (Applikon Biotechnology, The Netherlands) was operated continuously for 61 days (about 2 months) in seven periods (I-VII) at decreasing hydraulic retention times (HRT) of 24–20, 18, 16, 16, 12 and 4 h, keeping the substrate concentration constant at 42.5 g COD/L (Table 3.2). The CSTR was operated in mesophilic conditions (T 35 ± 1 °C), at pH 5.5 ± 0.05, only during a brief time of period VI at pH 3.8. Further details can be found elsewhere [8].

3.2.1.2 System 2

A 3.0-L CSTR with 1.5 L of working volume (Applikon Biotechnology, The Netherlands) was operated continuously for 100 days in eight periods (I-VIII) at organic loading rates (OLR) between 40 and 168 g COD/L-d achieved by changing the concentration of substrate (Table 3.2), while maintaining HRT constant at 6 h. Additionally, different batches of vinasse were used to assess the impact of the variability of the substrate and the biomass immobilization effect was evaluated by adding 300 plastic Kaldness K1 carriers in the last three operation stages. Temperature and pH

Table 3.2 Operation conditions, dark fermentation performance and main bacterial genera in systems

Period	I	II	III	IV	V	VI a	VI b	VII	VIII
System 1									
HRT (h)	24	20	16	12	8	4	4	4	–
Feed (g COD/L)	24	42.5	42.5	42.5	42.5	42.5	42.5	42.5	–
OLR (g COD/L-d)	42.6	51.1	63.9	85.2	127.8	309.0	67.8	309.0	–
HPR (L H_2/L-d)	1.4 ± 0.4	3.0 ± 0.1	1.6 ± 1.2	4.2 ± 0.7	6.0 ± 1.1	12.4 ± 1.9	4.7 ± 2.3	11.9 ± 1.7	–
H_2 (%)	78 ± 5	73 ± 0	79 ± 4	79 ± 3	72 ± 0	77 ± 4	74 ± 2	79 ± 4	–
Bacterial genera (%)				CL (66), ETB (15), ACN (9)	CL (50), SPL (39), PPB (2)	CL (59), SPC (10), BT (10)	SPC (56), PPB (27), SPL (6)	CL (56), BT (20), SPL (4)	–
System 2									
HRT (h)	6	6	6	6	6	6		6	6
Feed (g COD/L)	10	20	37–40	20	30	26.2		20	30
OLR (g COD/L-d)	40	80	148–160	80	120	105		80	120
HPR (L H_2/L-d)	0.5 ± 0.2	1.1 ± 0.0	0.1 ± 0.0	0.3 ± 0.0	0.3 ± 0.1	0.1 ± 0.1		0.4 ± 0.0	0.7 ± 0.0
H_2 (%)	51 ± 2	50 ± 2	17 ± 20	40 ± 6	44 ± 4	2 ± 2		43 ± 1	44 ± 3
Bacterial genera (%)	CL (81), SPL (11)	CL (82), SPL (11)	SPL (65), CL (19), ETG (3)	SPL (68), CL (25), ETG (1)	SPL (73), CL (17), ACB (3)	LAC (46), SPL (29), ACB (20)		SPL (51), CL (26), ACB (14)	SPL (69), CL (23), CP (3)
System 3									
HRT (h)	4.2	4.2	4.2	4.2	4.2	4.2		–	–
Feed (g COD/L)	10	15	20	25	30	29–32		–	–
OLR (g COD/L-d)	54.5	81.8	109.1	136.4	163.6	157.0–174.6		–	–
HPR (L H_2/L-d)		2.5	2.1	1.8	1.3	0.2		–	–

(continued)

Table 3.2 (continued)

Period	I	II	III	IV	V	VI a	VI b	VII	VIII
H_2 (%)		85	85	85	85	85		–	–
Bacterial genera (%)	SPL (45), CL (24), ACB (15)	SPL (54), CL (18), BFB (5), PVT (4)	SPL (34), CL (22), PVT (21), LAC (6)	SPL (33), PVT (29), CL (15) LAC (4)	SPL (34), CL (18), PVT (19), LAC (8)	SPL (33), LAC (23), CL (18) BFB (7)		–	–

Bacterial genera percentage refers to relative abundance. *ACB* Acetobacter, *ACN* Acinetobacter, *BFB* Bifidobacterium, *BT* Blautia, *CL* Clostridium, *CP* Caproiciproducens, *ETB* Enterobacter, *ETG* Ethanoligenens, *LAC* Lactobacillus, *PPB* Propionibacterium, *PVT* Prevotella, *SPC* Streptococcus, *SPL* Sporolactobacillus. In system 2, 300 plastic carriers Kaldness K1 were present in stages VI, VII, and VIII. In system 3, 150 plastic Kaldness K1 carriers were present in all stages were controlled at 35 ± 1 °C and 5.5 ± 0.05, respectively. Further details about this system can be found elsewhere [12].

3.2.1.3 System 3

A 1.0-L CSTR with 0.74 L of working volume was operated for 90 days in six stages (I-VI), with 150 Kaldness K1 carriers in suspension and operating with an HRT of 4.2 h. The first phase served to grow the inoculum in diluted vinasse, meanwhile in the following phases the OLR was increased by decreasing vinasse dilution until feeding undiluted vinasse in the last phase (Table 3.2). Temperature and pH were controlled at 37 ± 1 °C and 5.6 ± 0.2, respectively. Further details about this system can be found elsewhere [7].

3.2.2 Inoculum

In system 1 the CSTR was inoculated with 10% (*v/v*) of a specific inoculum (ATCC, PTA-124566) previously acclimated for 24 h [8]. This inoculum has been described to contain genera related to hydrogen-producing bacteria (HPB) such as *Clostridium* and *Enterobacter*, lactic acid bacteria (LAB) e.g., *Lactobacillus*, *Sporolactobacillus*, *Streptococcus* and acetic acid bacteria (AAB) such as *Acetobacter*.

In systems 2 and 3, granular sludge from an anaerobic digester fed with wastewater from a tequila factory was used as inoculum. Sludge was pretreated at 120 °C for 24 h to select the sporulating hydrogen-producing bacteria in the inoculum.

3.2.3 DNA Extraction and Sequencing

For system 1, microbial communities were assessed by Illumina MiSeq sequencing (Illumina, USA) as described by García-Depraect et al. [13]. Briefly, the isolation of Genomic DNA was performed with the MoBio Power Soil DNA Extraction Kit (Mo Bio Laboratories) following the supplier´s instructions. A NanoDrop-2000 spectrophotometer was used to determine the amount and quality of the extracted DNA. DNA samples were processed by RTL Genomics (RTL, USA) for 16 rRNA gene sequencing using the primers 28F (GAGTTTGATCNTGGCTCAG) and 388R (TGCTGCCTCCCGTAGGAGT).

In systems 2 and 3 DNA extraction was performed using the Fast DNA Spin Kit for Soil (MP Biomedicals, USA). DNA concentration was quantified with a Qubit 4 fluorimeter (Thermo Fisher Scientific, USA). DNA samples were sequenced using Illumina MiSeq technology at RTL Genomics (Texas Tech University, USA) by polymerase chain reaction (PCR) for amplification of 449 bp in the V3-V4 region of the 16S rRNA genes, using primer pair 357’F (CCTACGGNGGCWCCAG)/806’R

(GGACTACAVGGGGGTWTCTAATT) for generation of the bacterial sequence library [12].

3.2.4 Dark Fermentation Performance Evaluation

The following formulas were utilized to measure performance of dark fermentation in the different systems reviewed here:

- Organic Loading Rate (OLR) = S_{in}/HRT [g COD/L-d]
- Hydrogen production rate (HPR) = $V_{H2}/(V_r$*time) [NL H_2/ L-d]
- Volume of hydrogen (V_{H2}) = Total volume of gas*H_2%

Where S_{in} is the COD in the feed, HRT (h) is the hydraulic retention time, V_{H2} is the volume of hydrogen produced and Vr is the working volume in the fermenter. The volume of produced hydrogen was corrected to show the corresponding value in normal conditions (1 atm, 0 °C). Details on the chromatographic method utilized to measure the hydrogen content and other analytical methods can be found elsewhere [8, 12].

A hydrogen production rate (HPR) value was considered stable when complying with the following criteria:

- System 1: <10% variance in HPR for at least 3 consecutive HRT
- System 2: <15% variance in HPR for at least 12 consecutive HRT
- System 3: not defined

3.2.5 Statistical Analysis

Pearson's correlation analysis was performed with data of OLR (g COD/L-d), HPR (NL H_2/L-d), feed concentration (g COD/L), the composition of the gas (H_2%) and the relative abundance of the detected bacterial genera. A two-tailed test of significance was performed to indicate a significant correlation at the 0.05 level. The correlation analysis was carried out in the Origin 9.7 version, which was also used for creating the heatmap which color levels correspond to the Pearson's correlation coefficient.

3.3 Results and Discussion

3.3.1 Biohydrogen Production

Dark fermentation of tequila vinasse may promote treatment of residues in the tequila manufacture reducing the organic content of this liquid waste and producing in addition hydrogen as energy carrier. The works reviewed in this chapter are a compendium of results obtained under a common basis of operation conditions of pH (5.3–5.8), temperature (34–38 °C), and operating volume (0.74–1.5 L) in CSTR reactors. Vinasse generated in different tequila manufacture processes TrP and NCP (Table 3.1) with two types of inoculums (a specific inoculum or granular sludge from an anaerobic digester) with or without carriers (and under varied operating conditions HRT (from 4 to 24 h), feed concentration (from 10 to 42.5 g COD/L), OLR (from 40 to 309 g COD/L-d).

Thus, of all the conditions studied, the highest rate of biohydrogen production was obtained in stage VI-a of system 1 with a hydraulic retention time of 4 h and an organic load of 309 g COD/L-d, without carriers, and with TrP undiluted vinasse with 42.5 g COD/L in the feed. A specific inoculum was used, obtaining a biohydrogen production rate of 12.4 L H_2/L-d and a concentration close to 80% of hydrogen in the biogas, achieved with a bacterial population dominated by the genus *Clostridium* and accompanied by the genera *Streptococcus* and *Blautia*.

Despite the results being conclusive, it is important to make some observations. From system 1, we can identify that the extended hydraulic residence time is inversely proportional to the biohydrogen production rate. Hydraulic residence times of less than 12 h have better biohydrogen production rates.

In this way, systems 2 and 3 have hydraulic residence times of less than 12 h that remained constant in all stages with values of 6 and 4.2 h, respectively. System 3 had better hydrogen production rates on average than system 2 which used the same type of inoculum (granular sludge from an anaerobic digester).

All the systems evaluated were performed with increasing organic load rates, where systems 1 and 2 in the initial stages increased and then decreased their biohydrogen production rates, coincidentally the three systems presented a local maximum of the biohydrogen production rate at an organic loading rate between 80 and 85 g COD/L-d. In fact, in system 3, as the organic load rate increased, the biohydrogen production rate decreased proportionally.

It is important to note that systems 2 and 3 gradually increased the feed concentration (diluted vinasse to undiluted vinasse) while system 1 practically operated with undiluted vinasse except for the first stage.

Systems 2 and 3 had their maximum biohydrogen production rate of 1.1 L H_2/L-d and 2.5 L H_2/L-d respectively, at similar operating conditions of feed concentration of 20 and 15 g COD/L and organic load rate of 80 and 82 g COD/L-d, respectively. Likewise, with a dominant relative abundance of microorganisms of the genera *Clostridium* and *Sporolactobacillus*, although dominance of *Clostridium* (82%) in system 2 and *Sporolactobacillus* (54%) in system 3.

3.3.2 Short Carboxylic Acid Profiles

The TrP vinasse presents initial organic acids related to the fermentation of the organic matter from the process. Taking System 2 as a reference, the main initial organic acids of the vinasse were formate (4 g/L), lactate (2.6 g/L), and acetate (1.9 g/L); butyrate was not detected in any of the TrP vinasses (Table 3.1).

During dark fermentation, the main short carboxylic acids (SCA) were lactate (HLac), butyrate (HBu), acetate (HAc), and propionate (HPr), for System 1, accounting for 23.4–54.5, 26.8–44.1, 6.7–14.8, 2.9–25.7% of the total soluble metabolites, respectively (Table 3.3). Comparable results were observed for System 2, where principal SCA were HLac, HBu, HFor, and Hac with 21.6–58.6, 9.0–31.9, 12.5–37.2 and 8.6–27.8%, respectively (Table 3.3).

As the main SCA reflected in concentrations were HLac and HBu and minor changes in concentrations were HAc, HFor and HPr, suggesting that propiogenesis and homoacetogenesis were maintained at low levels regardless of the systems.

During the perturbations in System 1, the SCA showed a decrease in HLac, HAc and HBu, except during pH shock which led to increased levels of HPr. However, for

Table 3.3 Main short carboxylic acids produced in systems 1 and 2

System 1 period	I	II	III	IV	V	VIa	VIb	VII
HLac (g/L)	6.4 ± 0.5	6.3 ± 0.0	6.2 ± 0.4	5.9 ± 1.9	7.0 ± 0.0	4.7 ± 1.3	0.7 ± 0.1	1.8 ± 1.5
HAc (g/L)	2.7 ± 0.7	1.6 ± 0.0	2.7 ± 1.1	1.7 ± 0.9	1.9 ± 0.0	0.9 ± 0.3	0.02 ± 0.01	0.5 ± 0.3
HBu (g/L)	2.7 ± 1.3	4.8 ± 0.0	3.6 ± 0.8	3.8 ± 1.2	3.5 ± 0.0	3.1 ± 0.8	1.2 ± 0.1	2.8 ± 0.3
HPr (g/L)	0.0	0.6 ± 0.0	1.3 ± 0.4	0.3 ± 0.1	0.5 ± 0.0	1.3 ± 1.0	0.9 ± 0.1	1.7 ± 0.2
System 2 period	I	II	III	IV	V	VI	VII	VIII
HLac (g/L)	0.4 ± 0.0	1.5 ± 0.1	4.5 ± 0.7	2.5 ± 0.2	4.1 ± 1.1	2.3 ± 1.1	4.1 ± 0.1	3.1 ± 0.1
HAc (g/L)	0.5 ± 0.1	0.8 ± 0.1	1.5 ± 0.3	0.9 ± 0.1	1.1 ± 0.3	0.08 ± 0.4	0.6 ± 0.1	0.8 ± 0.0
HBu (g/L)	0.6 ± 0.2	1.3 ± 0.1	1.1 ± 1.1	0.9 ± 0.1	1.0 ± 0.1	0.5 ± 0.0	1.1 ± 0.1	1.4 ± 0.1
HFor (g/L)	0.3 ± 0.1	0.5 ± 0.1	2.5 ± 0.9	2.0 ± 0.2	2.5 ± 0.9	2.1 ± 1.6	1.2 ± 0.1	1.5 ± 0.1
HVa (g/L)**	ND	ND	1.1	0.7	0.9	0.6	0.3	0.6

**Influent, ND: not detected

the set of perturbations in OLR, type of vinasse and over-acidification, the stability of the process was recovered, highlighting the robustness of the system. The increment in OLR from 127.8 (period V) to 309 g COD/L-d (period VI-a) led to the highest HPR of 12.4 NL H_2/L-d, attributed to the dominant populations of HPB (biohydrogen-producing bacteria) as *Clostridium* in the reactor (detailed in Sect. 3.3). Due to the decrease in HRT (from 12 to 4 h), some species could be washed out from the reactor or remain at low relative abundance during the operation time. This change in the microbial populations is reflected in the SCA concentration, HLac diminishes by 33% with respect to period V, and higher HBu (up to 3.0 g/L) concentrations were found in the highest performance periods VI-a and VII. It is important to note the syntrophic interactions between LAB fermenting carbohydrates into lactate (conversion carbohydrate conversion up to 30%) and lactate utilizing HPB able to transform lactate into hydrogen and butyrate. Hydrogen production was found to be inversely correlated to lactate concentration suggesting a lactate-driven dark fermentation [14], Eq. (3.1):

$$CH_3CH(OH)COOH + 0.5CH_3COOH \rightarrow 0.75CH_3CH_2CH_2COOH + CO_2 + 0.5H_2 + 0.5H_2O \quad (3.1)$$

This metabolic pathway was also observed by System 2. During DF of vinasse, HLac could be produced by *Clostridium* species (Eqs. 3.2 and 3.3) from sugar-derived pyruvate as the node in the metabolic network [15] or by LAB from carbohydrate conversion, this later observed in the periods at high OLR (periods III, V, and VII) attained HLac concentrations up to 4 g/L.

$$C_6H_{12}O_6 + 2H_2O \rightarrow 2CH_3COOH + 2CO_2 + 4H_2 \quad (3.2)$$

$$C_6H_{12}O_6 \rightarrow CH_3CH_2CH_2COOH + 2CO_2 + 2H_2 \quad (3.3)$$

Interestingly, HBu was detected in effluent samples, produced either by the transformation of carbohydrates to pyruvate and acetyl-CoA (Eq. 3.3), and HLac and Hac to hydrogen reaction (Eq. 3.1). Indeed, the inverse correlation to HLac concentration with the high hydrogen production efficiency was found, suggesting the occurrence of lactate-driven DF. In period II, a low concentration of HLac (1.5 g/L) with an elevated level in HBu (1 g/L) generated the highest HPR in the whole experiment, 1.08 NL H_2/L-d. Also, for period IV, HBu concentration of 1.4 g/L reported the highest HPR (0.67 NL H_2/L-d) using immobilized biomass.

It is important to note the dominance of HBu-type metabolism through HLac-type fermentation as indicated by the mismatch between HPR and carbohydrate conversion, for the two Systems analyzed here.

Regarding HAc, in period VIII when carriers were added to support the biomass, a higher HPR and a more stable production of HAc were observed. Also, a consistent consumption of HFor was observed, the production of hydrogen and carbon dioxide from HFor by *Escherichia*, *Enterobacter*, and *Klebsiella* was reported [15]. Even when this study none of these bacterial genera were identified, comparable results have been reported for anaerobic fermentations of sweet potato fermentation slurry [16], starch [17], glucose [18], and tequila vinasse [15]. HVa was also detected in the influent, as *Clostridium* was identified in the microbial community (details in Sect. 3.3), it is suggested that metabolism and complete consumption of HVa was performed by this bacterial genus.

3.3.3 Bacterial Population Dynamics

In system 1, predominance of the genus *Clostridium* was observed in all operation periods, except when vinasse was transiently replaced by lactose decreasing acutely the productivity of hydrogen, when the main genera were *Streptococcus* followed by *Propionibacterium* (Table 3.2). In general, in system 1 notable H_2 productivities were observed through the operation periods, particularly in the last two periods with HPR close to 12 L H_2/L-d. This remarkable performance was related by the authors to the use of a specific consolidate inoculum, *i.e.,* ATCC- PTA-124566. Also, a 4 h HRT in the last operation periods might have exerted a selective pressure by washout over bacterial genera benefiting the hydrogen-producing *Clostridium* genus, which has a doubling time as low as 1 h [19].

In system 2, *Clostridium* was the dominating genus when OLR of 80 g COD/L-d or below were applied attaining the highest HPR of the experiment i.e., 1.1 L H_2/L-d. Feeding OLR above 148 g COD/L-d to the fermenter reduced 90% the H_2 production when *Sporolactobacillus* became the dominant bacterial genus (period III). Taking back OLR to 80 g COD/L-d increased HPR modestly to 27% of the maximum value, while genus *Sporolactobacillus* remained dominant. Finally, the addition of plastic carriers, in operation period VI initially decreased the HPR to the minimum value registered (0.1 L H_2/L-d) with *Lactobacillus* as the dominant genus. However, plastic carriers (periods VII, VIII) promoted robustness against high concentrations of feed, cataloged as inhibiting H_2 production in dark fermentation when above 20 g COD/ L [20].

In system 3, the highest mean HPR (2.5 L H_2/L-d) was obtained in operation period II, when the bacterial population was composed by *Sporolactobacillus* and *Clostridium.* Immobilization of biomass from the beginning of the experiment in system 3 seemed to promote HPR surpassing the values observed in system 2 (stage 2), at a shorter HRT (6.0 vs. 4.2 h). Increasing feed concentration, resulting in OLR above 100 g COD/L-d led to a HPR decrease in system 3, however not as severely as observed in system 2, at similar OLR with no plastic carriers (operation periods III, IV, V in both systems). The *Prevotella* genus was detected among the three more abundant bacterial genera in system 3 associated with OLR between 109 and 164 g COD/L-d. Suspended *Prevotella* became increasingly more abundant in periods with the lowest gas production in system 3, and it even became the most abundant bacterial genus in the biofilm immobilized on the Kaldness K1 carriers from period III (data not shown).

According to recent reports on dark fermentation of tequila vinasse, HPB species of the *Clostridium* genus commonly coexist with *Sporolactobacillus* and *Lactobacillus* in systems seeded with thermally pretreated anaerobic sludge [21, 22]. Moreover, the genera accompany *Clostridium* at OLR higher than 50 g COD/L-d irrespective of the carbohydrate source [12].

Previously, it was considered that LAB such as *Lactobacillus* competed with *Clostridium* for sugars in dark fermentation [23]. Also, it was demonstrated that the coculture of *Clostridium* with LAB *Sporolactobacillus* improved H_2 production by 50% compared to *Clostridium* single cultures [24]. Recent reports suggest that LAB coexists with *Clostridium* in converting sugars and cross-feeding lactate into hydrogen [25]. Here, the predominance of *Sporolactobacillus* in systems 2 and 3 strengthens this synergy between LAB and *Clostridium* to produce hydrogen through the lactate-acetate pathway, as discussed previously in Sect. 3.2.

3.3.3.1 Pearson's Correlation Analysis

The Pearson's correlation analysis showed that HPR was positively correlated (0.99) with OLR in system 1 and negatively correlated (−0.90) with feed concentration in system 3, while no significant correlation with OLR or feed concentration was found in system 2 (Fig. 3.1). This suggests that in dark fermentation systems dominated by the bacterial genus *Clostridium*, an increase in hydrogen production is expected after increasing the loading of the fermenter as typically observed in the literature [8, 12]. Contrarily, in fermentations dominated by lactic acid bacteria such as *Sporolactobacillus*, a decrease in hydrogen productivity is to be projected after increasing the loading.

In the correlation analysis, HPR was found to be positively correlated with genera *Blautia* in system 1 (0.89) and *Clostridium* in system 2 (0.73), while in system 3 the HPR was negatively correlated with the *Lactobacillus* genus (−0.96), which presence also negatively correlated with H_2 percentage in system 2. The previous notion that *Lactobacillus* affects HPR by sugar competition [23] is changing to now regard *Lactobacillus* as secondary fermenters compensating transient HPB depletion [25].

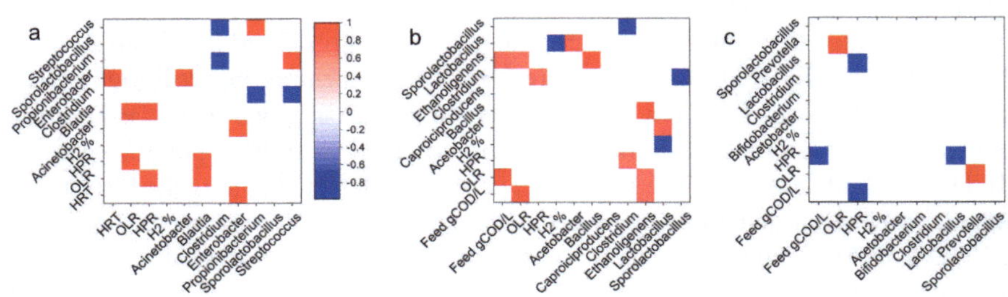

Fig. 3.1 Pearson's correlation analysis heatmap of dark fermentation outcomes in **a** system 1, **b** system 2 and **c** system 3

A positive correlation between HPR and *Blautia* in system 1 is notable since species such as *Blautia coccoides* are known to consume H_2 for growth (homoacetogen) [8, 26]. However, it has been found that HPB, homoacetogens and LAB can coexist while maintaining a suitable HPR, particularly at high OLR [25].

Feed concentration and OLR were found to be positively correlated with genera *Ethanoligenens* in system 2 (0.73) and *Prevotella* in system 3 (0.83), respectively. Other authors found *Ethanoligenens harbinense* and *Prevotella* as the most abundant HPB and non-HPB in continuous dark fermentation of tequila vinasse, in that order [21, 22]. *Prevotella* has been described also as contributing to H_2 production by helping the formation of the biofilm [27]. Abundance of other extremely specific genera was positively correlated only with an abundance of other key genera. For example, strong correlation was found in system 1 between *Acinetobacter* and *Enterobacter* (0.99), *Streptococcus* and *Propionibacterium* (0.98) and in system 2 between *Ethanoligenens* and *Bacillus* (0.93), *Lactobacillus* and *Acetobacter* (0.81).

The key HPB-*Clostridium* was negatively correlated with *Propionibacterium* (−0.99) and *Streptococcus* (−0.95) in system 1, and with *Sporolactobacillus* (−0.72) in system 2.

In terms of bacterial population, no differences were observed in systems fed either with TrP vinasse (system 2) or NCP vinasse (system 3) due to a domination of genus *Sporolactobacillus* was observed at similar feed concentration and OLR. This suggests that distribution of bacterial population in dark fermentation is more a result of the combination of HRT and feed concentration to attain a certain OLR than a consequence of vinasse composition (Tables 3.1 and 3.2).

3.4 Conclusions

Based on the results related to operation strategies in the three systems reviewed here, we conclude that suitable conditions useful to promote production of hydrogen with tequila vinasse as substrate include:

(a) Short HRT, as low as 4 h in CSTR with suspended biomass in combination with a consolidated H_2-producing inoculum such as ATCC-PTA-124566.
(b) Immobilization of biomass to provide robustness against high concentration in the feed avoiding inhibition by substrate.

Suitable hydrogen producing conditions were confirmed in the systems reviewed here by:

- The carboxylic acids profile pointing towards the importance of alternative H_2-producing routes such as the lactate-acetate pathway.
- The presence of key bacterial genera such as *Clostridium* and *Sporolactobacillus* could be positively associated following the lactate-acetate pathway as well.

Acknowledgements This work was supported by Fondo de Sustentabilidad Energética SENER-CONACYT (Mexico), grant number 247006 Gaseous Biofuels Cluster.

References

1. CRT (2020) Consejo Regulador del Tequila (Tequila Regulatory Council). Datos y estadísticas. Retrieved from https://www.crt.org.mx/EstadisticasCRTweb/. https://www.crt.org.mx/EstadisticasCRTweb/. Accessed on 9 Apr 2020
2. López-López A, Davila-Vazquez G, León-Becerril E et al (2010) Tequila vinasses: generation and full-scale treatment processes. Rev Environ Sci Bio/Technol 9:109–116. https://doi.org/10.1007/s11157-010-9204-9
3. Iñiguez G, Acosta N, Martinez L et al (2005) UTILIZACIÓN DE SUPBRODUCTOS DE LA INDUSTRIA TEQUILERA. PARTE 7. COMPOSTAJE DE BAGAZO DE AGAVE Y VINAZAS TEQUILERAS. J Revista internacional de contaminación ambiental 21:37–50
4. Sanchez-Lizarraga AL, Arenas-Montaño V, Marino-Marmolejo EN et al (2018) Vinasse irrigation: effects on soil fertility and arbuscular mycorrhizal fungi population. J Soils Sediments 18:3256–3270. https://doi.org/10.1007/s11368-018-1996-1
5. Christofoletti CA, Escher JP, Correia JE et al (2013) Sugarcane vinasse: Environmental implications of its use. Waste Manag 33:2752–2761. https://doi.org/10.1016/j.wasman.2013.09.005
6. Sánchez-Ureña SG, Bolaños-Rosales RE, Aguilar-Juárez O et al (2022) Tequila production process influences on vinasses characteristics. a comparative study between traditional process and non-cooked agave process. Waste Biomass Valorizat 13:3183–3195. https://doi.org/10.1007/s12649-022-01731-y
7. Muñoz-Estrada J (2021) Producción de hidrógeno y metano y su relación con poblaciones microbianas en un sistema anaerobio a partir de vinazas tequileras crudas. CIATEJ
8. García-Depraect O, Diaz-Cruces VF, Rene ER, León-Becerril E (2020) Changes in performance and bacterial communities in a continuous biohydrogen-producing reactor subjected to substrate- and pH-induced perturbations. Bioresour Technol 295:122182. https://doi.org/10.1016/j.biortech.2019.122182
9. García-Depraect O, Diaz-Cruces VF, León-Becerril E (2020) Upgrading of anaerobic digestion of tequila vinasse by using an innovative two-stage system with dominant lactate-type fermentation in acidogenesis. Fuel 280:118606. https://doi.org/10.1016/j.fuel.2020.118606
10. García-Depraect O, Muñoz R, van Lier JB, et al (2020) Three-stage process for tequila vinasse valorization through sequential lactate, biohydrogen and methane production. Bioresour Technol 307:123160. https://doi.org/10.1016/j.biortech.2020.123160
11. Buitrón G, Kumar G, Martinez-Arce A, Moreno G (2014) Hydrogen and methane production via a two-stage processes (H2-SBR + CH4-UASB) using tequila vinasses. Int J Hydrogen Energy 39:19249–19255. https://doi.org/10.1016/j.ijhydene.2014.04.139
12. Arellano-García L, Velázquez-Fernández JB, Macías-Muro M, Marino-Marmolejo EN (2021) Continuous hydrogen production and microbial community profile in the dark fermentation of tequila vinasse: Response to increasing loading rates and immobilization of biomass. Biochem Eng J 172:108049. https://doi.org/10.1016/j.bej.2021.108049
13. García-Depraect O, Valdez-Vázquez I, Rene ER, et al (2019) Lactate- and acetate-based biohydrogen production through dark co-fermentation of tequila vinasse and nixtamalization wastewater: Metabolic and microbial community dynamics. Bioresour Technol 282:236–244. https://doi.org/10.1016/j.biortech.2019.02.100
14. García-Depraect O, Muñoz R, Rodríguez E, et al (2021) Microbial ecology of a lactate-driven dark fermentation process producing hydrogen under carbohydrate-limiting conditions. Int J Hydrogen Energy 46:11284–11296. https://doi.org/10.1016/j.ijhydene.2020.08.209

15. García-Depraect O, León-Becerril E (2018) Fermentative biohydrogen production from tequila vinasse via the lactate-acetate pathway: operational performance, kinetic analysis and microbial ecology. Fuel 234:151–160. https://doi.org/10.1016/j.fuel.2018.06.126
16. Matsumoto M, Nishimura Y (2007) Hydrogen production by fermentation using acetic acid and lactic acid. J Biosci Bioeng 103:236–241. https://doi.org/10.1263/jbb.103.236
17. Baghchehsaraee B, Nakhla G, Karamanev D, Margaritis A (2009) Effect of extrinsic lactic acid on fermentative hydrogen production. Int J Hydrogen Energy 34:2573–2579. https://doi.org/10.1016/j.ijhydene.2009.01.010
18. Kim T-H, Lee Y, Chang K-H, Hwang S-J (2012) Effects of initial lactic acid concentration, HRTs, and OLRs on bio-hydrogen production from lactate-type fermentation. Bioresour Technol 103:136–141. https://doi.org/10.1016/j.biortech.2011.09.093
19. Sandoval-Espinola WJ, Chinn M, Bruno-Barcena JM (2015) Inoculum optimization of Clostridium beijerinckii for reproducible growth. FEMS Microbiol Lett 362: fnv164. https://doi.org/10.1093/femsle/fnv164
20. Elbeshbishy E, Dhar BR, Nakhla G, Lee H-S (2017) A critical review on inhibition of dark biohydrogen fermentation. Renew Sustain Energy Rev 79:656–668. https://doi.org/10.1016/j.rser.2017.05.075
21. Toledo-Cervantes A, Méndez-Acosta HO, Arreola-Vargas J et al (2022) New insights into microbial interactions and putative competitive mechanisms during the hydrogen production from tequila vinasses. Appl Microbiol Biotechnol 106:6861–6876. https://doi.org/10.1007/s00253-022-12143-2
22. Gabriel-Barajas JE, Arreola-Vargas J, Toledo-Cervantes A et al (2022) Prokaryotic population dynamics and interactions in an AnSBBR using tequila vinasses as substrate in co-digestion with acid hydrolysates of Agave tequilana var. Azul bagasse for hydrogen production. J Appl Microbiol 132:413–428. https://doi.org/10.1111/jam.15196
23. Park J-H, Kim D-H, Kim S-H et al (2018) Effect of substrate concentration on the competition between Clostridium and Lactobacillus during biohydrogen production. Int J Hydrogen Energy 43:11460–11469. https://doi.org/10.1016/j.ijhydene.2017.08.150
24. Park J-H, Kim D-H, Baik J-H et al (2021) Improvement in H2 production from Clostridium butyricum by co-culture with Sporolactobacillus vineae. Fuel 285:119051. https://doi.org/10.1016/j.fuel.2020.119051
25. Montoya-Rosales J de J, Ontiveros-Valencia A, Esquivel-Hernández DA, et al (2023) Metatranscriptomic Analysis Reveals the Coexpression of Hydrogen-Producing and Homoacetogenesis Genes in Dark Fermentative Reactors Operated at High Substrate Loads. Environ Sci Technol 57:11552–11560. https://doi.org/10.1021/acs.est.3c02066
26. Fuentes L, Palomo-Briones R, de Jesús M-R et al (2021) Knowing the enemy: homoacetogens in hydrogen production reactors. Appl Microbiol Biotechnol 105:8989–9002. https://doi.org/10.1007/s00253-021-11656-6
27. Cabrol L, Marone A, Tapia-Venegas E et al (2017) Microbial ecology of fermentative hydrogen producing bioprocesses: useful insights for driving the ecosystem function. FEMS Microbiol Rev 41:158–181. https://doi.org/10.1093/femsre/fuw043

Chapter 4
Practical Applications of Dark Fermentation for Hydrogen Production

Virginia Montiel-Corona and Germán Buitrón

Abstract Among the biological processes that produce hydrogen, dark fermentation is the most developed technology with the potential to be used on an industrial scale. However, there are still several challenges in implementing the industrial application of this process. Using residual carbon sources is crucial in ensuring economic viability and reducing carbon dioxide emissions during hydrogen production by dark fermentation. For this reason, in the last two decades, efforts have been intensified to improve the yields and volumetric hydrogen production rates from residual biomass. This chapter begins with a critical analysis of the advantages and disadvantages of using organic wastes (agricultural residues, food processing industries, organic fraction of municipal solid waste, and effluents from livestock farms), indicating the suitability of each one. Section 4.2 presents the types of reactors used for hydrogen production and the advantages and challenges of each design. The reactor configuration substantially impacts the retention of hydrogen-producing microorganisms and the mass transfer. Mass transfer is vital for the transport of the substrate toward the microorganisms and for the transfer of the hydrogen produced toward the headspace of the reactors for its recovery. The control of homoacetogenesis, one of the consuming pathways of the produced hydrogen, also depends on the reactor configuration and operating conditions. Section 4.3 presents the cases that have been scaled up and the challenges involved in going from laboratory to pilot scale. Finally, Sect. 4.4 shows the combination of dark fermentation with another process. It is discussed that dark fermentation can be coupled to anaerobic digestion for methane production from residual volatile fatty acids to increase the energy conversion efficiency. The mixture of hydrogen and methane (biohythane) is suitable for stabilizing combined heat and power units, improving combustion efficiency.

V. Montiel-Corona · G. Buitrón (✉)
Laboratory for Research on Advanced Processes for Water Treatment, Unidad Académica Juriquilla, Instituto de Ingeniería, Universidad Nacional Autónoma de México, Blvd. Juriquilla 3001, Querétaro 76230, México
e-mail: GBuitronM@iingen.unam.mx

V. Montiel-Corona
e-mail: VMontielC@iingen.unam.mx

V. Alcaraz Gonzalez et al. (eds.), *Wastewater Exploitation*, Springer Water,
https://doi.org/10.1007/978-3-031-57735-2_4

Keywords Agro-industrial wastes · Bio-hydrogen · Biohythane · Dark fermentation · Reactor configuration · Scale-up

4.1 Introduction

Dark fermentation (DF) is a process that produces hydrogen and various by-products from biomass. It is a climate-friendly alternative process to the use of fossil sources. Substances obtained through DF are mainly H_2, acetic acid, lactic acid, butyric acid, propionic acid, ethanol, and butanol. In the search for alternative energy sources to fossil fuels to mitigate global warming, research on H_2 production through DF has intensified over the last two decades [1]. DF has the potential to be used on an industrial scale. Low cost, low energy demand, and relatively elevated H_2 production rates (HPR) show the potential of DF. However, DF's H_2 production yield and energy conversion efficiency (ECE) are still around 30% [2, 3]. This challenge needs to be overcome to reach a profitable commercial application. Among the factors affecting yields and ECE are H_2-consuming metabolic pathways such as homoacetogenesis, methanogenesis, and sulfate reduction. Operating conditions have been established to control the last two; however, homoacetogenesis continues to be a challenge to overcome. The configuration of the reactors and the stirring system strongly influence this metabolic pathway, so it is crucial to consider when scaling the process. The use of organic wastes is decisive in ensuring the economic viability of DF. Among the potential organic materials, lignocellulose has also been targeted as a raw material for H_2 production. However, given the complex structure of lignocellulose, pretreatment is required to release carbohydrates, which increases production costs. The application of pretreatments also often implies the generation of inhibitory substances [4], so it is vital to analyze the technical and economic feasibility of using this type of biomass compared to other wastes of easy degradation, such as agro-industrial wastewater or organic fraction of municipal solid waste (OFMSW). The step from the laboratory scale to the pilot scale is crucial to consider operational parameters that, on a larger scale, can hinder the operation of the process compared to the laboratory scale, including feeding strategies, the establishment of anaerobic conditions, pH and temperature control, agitation, among other aspects. To improve the energy conversion efficiency, DF should be coupled with other processes where the volatile fatty acids (VFA) can be used to produce more energy or other value-added products under a biorefinery scheme in the framework of the circular economy to give economic and energy viability to the process. This chapter analyzes the aspects mentioned above for the scaling and application of this technology.

4.1.1 Hydrogen Production from Residual Wastes

Using organic wastes as the substrate for fermentative H_2 production reduces the production cost and achieves dual benefits: eco-friendly energy production (by CO_2 reduction emissions) and improved waste management [3]. Therefore, in the last two decades, the use of residual biomass as feedstock instead of simple sugars has increased considerably [1]. Among the residues that have been used for H_2 production can be found food wastes, waste-activated sludge, lignocellulose-based biomass, algae biomass [3], industrial waste and effluents like olive oil mill wastewater, palm oil mill effluent, and wastes from the dairy and brewery industries [5]. Carbohydrates are the preferred substrate for H_2 production by DF. Numerous studies have reported that H_2 production potential from carbohydrate-rich wastes is higher than lipids, proteins, or glycerol [6]. There are several aspects to consider when selecting suitable substrates for H_2 production, of which the availability of the residue, the carbohydrate content, and the biodegradability are the most important. Although many of the residues used are rich in carbohydrates, they are not readily biodegradable, so it is necessary to apply pretreatments to release simple carbohydrates or short oligomers. The type of pretreatment that must be used depends on the composition and chemical structure of the wastes; for example, lignocellulosic materials require complicated-severe pretreatments compared to other wastes such as food wastes or the organic fraction of municipal solid wastes (OFMSW) that only require mechanical pretreatment and, in other cases, effluents like cheese whey do not require at all. Although the biodegradability of lignocellulosic wastes is low, the carbohydrate content and availability could be high compared to other readily biodegradable wastes.

4.1.2 Lignocellulosic Wastes

The lignocellulosic biomass residues production is around 220 billion tons per year worldwide. It can be used for H_2 production [5] and is considered the fourth energy source after coal, petroleum, and natural gas [3]. It has been calculated that of the 1.2 billion tons of agricultural residues available worldwide, 16% of the total current H_2 demand could be obtained. If the amount of collectible agricultural waste reaches 1.74 billion tons, the percentage could be 42% [2]. Obtaining H_2 from biomass can be carried out by thermochemical and biological methods [7], but biological methods are preferred, given the environmental benefits [3]. The potential of lignocellulosic waste for energy supply is high; however, the structure and chemical composition complexity makes its direct application in biological systems difficult. Lignocellulosic biomass is composed of three main polymers: cellulose (30–70%), hemicellulose (15–30%), and lignin (10–25%) [7]. Although the monomers contained in cellulose and hemicellulose are highly biodegradable, the complex network associated with lignin and the high crystallinity of cellulose makes it difficult for the microorganisms to access these sugars [3], imposing the necessity for applying pretreatments to disrupt the complex structure and remove lignin [4]. Pretreatment of lignocellulose biomass accounts for about 20−30% of the overall process cost and up to 50% of the total capital investment of equipment [4, 7].

Physical (such as mechanical, thermal, ultrasound, microwave, electrokinetic), chemical (like acid, alkali, oxidative, organic-aqueous solvent), physicochemical (thermochemical and steam explosion, AFEX), and biological treatments (bacterial or fungal) are pretreatment in practice [8]. The simultaneous release of inhibitory by-products with sugars has been detected in all of them. Among the inhibitory substances released are furfural, 5-hydroxymethylfurfural, vanillin, syringaldehyde, phenolic compounds, formic acid, acetic acid, levulinic acid, ketones, and undissociated short-chain carboxylic acids [4, 8]. The adverse effects of these toxic substances on hydrogen-producing microorganisms limit the use of this abundant biomass for H_2 production at the commercial scale [8, 9]. Different strategies have been proposed to detoxify hydrolysates and avoid inhibiting toxic compounds in H_2 production. They include heating and evaporation, steam stripping, electrochemical, liquid–liquid, and liquid–solid extraction, treating with some chemicals, adsorption (activated carbon or resins), and enzymatic or microbial removal. Although current physicochemical methods for detoxification can be effective and fast in removing inhibitory components from hydrolysates, the drawback is the implementation of this additional step to the process, which increases operating costs, waste generation, and, in some cases, the loss of up to 30% of the sugars present in the hydrolysates [4].

Applying biological pretreatments to release sugars and lignin degradation generates low or null concentrations of inhibitors [10]. However, the drawbacks of these pretreatments are long pretreatment times (15–40 days), strict microbial growth conditions, and sugar consumption [7].

4.1.3 The Organic Fraction of Municipal Solid Wastes (OFMSW) and Food Wastes

The OFMSW and food wastes is another abundant source of biomass for H_2 production. The OFMSW represents, on average, 50% of municipal solid waste in the world [3], and due to population increase and urbanization, there will be an increase of 2.01 billion tons/year of municipal solid waste to 3.4 billion tons/year by 2050 [11]. Food wastes are a significant fraction of municipal solid wastes (15–60%) [6, 12]. One-third of all worldwide food is wasted annually, representing 1.3 billion tons of food waste/year [13]. Due to its high biodegradability, food waste has been regarded as ideal for hydrogen-producing microorganisms [14]. Food wastes have high volatile solids (85–95%), moisture (75–85%), and biodegradable carbohydrate content, which allows them to obtain high HPR [6]. The carbohydrate content of food wastes ranges from 30 to 70%, while sewage sludge and livestock waste contain less than 10% [6].

Despite the great potential of the OFMSW for gaseous biofuel production (H_2 or methane), almost all of them are disposed of in sanitary landfills, and only a small fraction is used in AD plants for biogas production [12]. As previously mentioned, the OFMSW comprises easily degradable carbohydrates, lipids, and proteins from processed food waste. Still, the OFMSW may contain garden and park wastes, paper, and raw vegetables, which contribute with 10–12% cellulose, 5–17% hemicellulose, and 5–9% lignin [15] which, like unprocessed meat and fat waste, may require some pretreatment to improve biodegradability. Mechanical, thermal, chemical, and biological pretreatments have been applied to OFMSW to enhance H_2 production [6]. Of the pretreatments, the application of mechanical pretreatment (grinding) is strictly necessary to give OFMSW homogeneity and facilitate the process. The application of thermal and chemical pretreatments has improved the production of H_2, but, as stated by Yun et al. [6], it seemed that the determining role of thermal and chemical pretreatment was not to increase the solubility of wastes but to favor the predominance of hydrogen-producing microorganisms [6]. For example, Abubackar et al. [16] used autoclaved and non-autoclaved fruit and vegetable wastes for H_2 production. The H_2 yield was 27.19 and 20.81 NmL H_2/gVS, respectively. The final VFA concentration for autoclaved waste was lower than non-autoclaved waste (17.4 and 23.4 mg/L), which means that the inoculum could solubilize-hydrolyze the fruit and vegetable wastes regardless of the pretreatment and the increase in H_2 production could be attributed to the microbial community composition as demonstrated by Kim et al. [17] and Jang et al. [18] who observed a predominance of lactic acid bacteria in untreated food wastes and H_2-producing *Clostridium spp* in alkali, acid, and thermal pretreated food wastes. Although applying pretreatments to OFMSW favors the predominance of H_2-producing bacteria, other reports have achieved stable and long-term production using OFMSW without applying additional pretreatments to the grinding [19, 20]. Therefore, unlike lignocellulosic waste, pretreatments may not be strictly necessary for the OFMSW, representing savings in the cost of the process.

4.1.4 Industrial Waste Effluents

Industrial effluents, rich in carbohydrates from the food or beverage industry, are other good options for H_2 production by DF. For example, dairy industries produce 2.5 L of waste effluents per liter of processed milk plus 10 L of cheese whey per kilogram of cheese made [21]. In the world, about 183 million tons of cheese whey are produced annually, and only 50% is processed into powder cheese whey and condensed cheese whey by big factories; the rest is discharged without treatment [22], representing a severe environmental problem due to its high COD and BOD concentration (50–120 and 27–60 g/L, respectively) and significant volume generated. The high organic load of these effluents makes it challenging to treat with a conventional activated sludge process. The high energy demand required for aeration is economically unsustainable [21], so DF is a better option. The cheese whey discharged is an abundant source of carbohydrates (lactose 44–46%) and other highly biodegradable components (proteins, lipids, vitamins, and minerals) for H_2 production [22]. Although the high biodegradability of cheese whey is an advantage over lignocellulosic waste, the low alkalinity and high COD concentration hinder the stability of DF or AD; also, a high concentration of volatile acids can be formed and inhibit the process [23]. One solution to this problem has been the co-fermentation of cheese whey with other residues that provide alkalinity as animal slurry [24] or directly use the fermented cheese whey [25].

The presence and dominance of lactic acid bacteria in the reactors fed with cheese whey inhibit H_2 production [26]. However, lactic acid produced by these bacteria can be a precursor for H_2 production [27]. García-Depraect et al. [28] observed the transformation of carbohydrates into lactate by lactic acid bacteria in the first stage; then, hydrogen-producing bacteria metabolized the produced lactate with acetate to generate H_2.

Effluents with the potential for H_2 production by DF are generated from the alcoholic beverage industry. The wine production process generates between 0.2 and 4 L of effluents [29]. Winery effluents were suitable substrates for biohydrogen production by DF due to the high carbohydrate content (around 33 g/L) and COD (252 g/L) [30]. The brewer industry generates 3–10 L of wastewater per liter of beer produced with COD soluble up to 50 g/L [31]. The distilled alcoholic beverages industry generates between 10 and 12 L of vinasses that are the final product of distillation and have been used for H_2 production [32]. These effluents contain significant concentrations of biodegradable carbohydrates, but they also contain inhibitory phenolic compounds needing detoxification; however, the loss of sugars due to detoxification can be up to 50%, as observed by García-Becerra et al. [33]. Combining vinasse with other residues to dilute the concentration of inhibitory compounds seems a better strategy [32].

An industrial effluent with high availability in biodiesel-producing countries is glycerol. For every 10 L of biodiesel produced, 1 L of glycerol is generated. Pure glycerol is highly biodegradable, and higher H_2 yields can be obtained in co-fermentation with other wastes [34]. However, the use of real biodiesel wastewater presents some drawbacks. The glycerol concentration in wastewater is low (4 g/L) and is accompanied by inhibitory substances such as methanol (128 g/L) and salts (12 g/L) [35], making the previous purification-separation of glycerol from the wastewater necessary. Other industrial effluents used for H_2 production have been olive oil mill wastewater, palm oil mill effluent, cassava, sugar beet, and sugarcane molasses. The variety and availability of agro-industrial effluents for H_2 production are high, and the process can be installed in the same factories that generate the effluents.

4.1.5 Wastes of Animal Origin and Sewage Sludge

Sewage sludge and animal wastes such as slurry or solid manure and wastewater from livestock or poultry are produced in large quantities and are highly polluting.

Their low carbohydrate and high nitrogen content, alkalinity, and pH content make them unsuitable for H_2 production [32]. Additionally, the high load of methanogenic and pathogenic microorganisms requires constant proliferation control. However, its use as a co-substrate for other carbohydrate-rich residues becomes essential due to the contribution of alkalinity, carbon, and nutrients to balance the process [24].

4.2 Considerations About Reactor Configuration

For H_2 production, reactors such as the continuous stirred-tank reactor (CSTR), up-flow anaerobic sludge blanket reactor (UASB), anaerobic fluidized bed reactor (AFBR), membrane bioreactor (MBR), trickling bed reactor (TBR), fixed bed reactor (FBR) and expanded granular sludge bed reactor (EGSBR) have been used [5, 36]. The characteristics of the substrate (concentration of solids, COD, inhibitors, and alkalinity), the retention of hydrogen-producing microorganisms, and mass transfer are important factors to consider for proper reactor selection.

The CSTR is an easy design to build and operate [2]. Its mechanical and continuous stirring system allows effective control of pH and temperature, providing a homogeneous distribution of the substrate towards the microorganisms [5], and, depending on the stirring speed, the removal of H_2 can be increased from the liquid media [37]. CSTR can process substrates with higher suspended solids concentration than UASB reactors. One disadvantage of CSTR is the low concentration of H_2-producing microorganisms, which is limited by the hydraulic retention time (HRT). When selecting HRT, special care must be taken to avoid biomass washout [38]. Despite this limitation, in this type of reactor, hydrogen productio rates (HPR) up to 40 L H_2/L-d have been reached using simple sugars such as galactose [39], 17 L H_2/L-d from sugarcane syrup [40], and 12 L H_2/L-d using tofu-processing waste [38].

UASB reactors are cylindrical vessels with a gas–liquid-solid separator at its top, which prevents biomass loss and allows large amounts of it to accumulate at the bottom, often in the form of granules [40]. In principle, these reactors were designed to retain slow-growing methanogenic archaea in methane production systems. Using these reactors in the H_2 production has made it possible to offset the solids retention time from the HRT; therefore, shorter HRT management has positively influenced the HPR. The high concentration of hydrogen-producing microorganisms allows for handling high organic loads, efficient COD removal, and greater tolerance to inhibitory substances from the substrate [41]. Mixing in the UASB reactor is produced by the release of H_2 bubbles produced by the biomass granules. Therefore, mechanical stirring is not necessary [5]. However, the poor level of agitation generated in the system is a limiting factor for good gas–liquid mass transfer and for maintaining a low H_2 partial pressure in the liquid [2]. Another disadvantage of these systems is the time required to start the system. The production of sufficient biomass and the formation of granules is a slow process that can require several months [5, 41]. Once the system reaches stability in biomass production, programmed purging is necessary to avoid excessive biomass accumulation. Another of the requirements for the proper functioning of these reactors is the low concentration of suspended solids that can be present in the feed (1- 3%). Otherwise, clogging problems may occur [42]. In UASB reactors, HPR of 52–56 L H_2/L-d has been reached using galactose at HRT of only 2 h [43]. When the complexity of the substrate increases, UASB must operate at higher HRT to degrade the substrate, decreasing the HPR. Mahmod et al. [42] used HRT of 6 h with a pretreated palm oil mill effluent obtaining an HPR of 11.7 L H_2/L-d.

The EGSB reactor is an analogous version of the UASB that operates at higher upflow velocities than the UASB to keep the granules of biomass distributed throughout the reactor, improving with them the transfer-contact of the substrate with the microorganisms and avoiding dead spaces or bypass flow that can decrease the reactor performance [44]. This type of reactor also works at short TRH and high organic loads, but one of the drawbacks is the energy cost of keeping the granules in suspension. With an HRT of two hours and OLR of 183 g sucrose/L-d, an HPR of 52 L H_2/L-d using sugarcane juice as substrate [44].

A membrane bioreactor (MBR) is another reactor configuration that seeks to maintain high concentrations of producing-hydrogen microorganisms inside the reactor. The reactors have a membrane that can be submerged inside or placed outside the reactor, preventing microorganisms' washout [5]. The main drawbacks of these reactors are the fouling, high cost, and maintenance of the membrane system, which prevents its scaling at an industrial level [45]. An alternative has been the implementation of the so-called dynamic membranes, in which the biofilm is formed on a cheap support material such as polyester mesh or filter cloth with a big pore size (10–100 microns), decreasing clogging problems [46]. HPR as high as the UASB has been reached in this type of laboratory-scale reactor. Park et al. [46] obtained an HPR of 51 L H_2/L-d at HRT of 3 h using glucose as a substrate. Another reactor that involves biofilm formation is the TBR, where microorganisms are attached to a supporting material, and the substrate trickles into the biofilm. The biofilm is not submerged in liquid, which promotes easy H_2 transfer from the fluid to the gas phase and low H_2 partial pressures [47]. Oh et al. [48] reached HPR de 25.3 L H_2/L-d using glucose, and Arreola-Vargas et al. [47] 20.16 L H_2/L-d; with complex substrates such as enzymatic oat straw hydrolysate, the HPR was 0.624 L H_2/L-d [47] and 2.25 L H_2/L-d with enzymatic hydrolysates of agave bagasse [36]. Despite maintaining a low H_2 partial pressure, the performance of these reactors is lower compared to UASB, MBR, or EGSB due to the low transfer of the substrate to the microorganisms [36].

FBR are systems with high performance for H_2 production. An immobilized-entrapped cell operated in a CSTR has allowed obtaining the highest HPR reached with model substrates. Pugazhendhi et al. [49] reported an HPR of 78 L H_2/L-d at HRT of 1.5 h using glucose as substrate with cells immobilized in alginate beads. Sivagurunathan et al. [43] obtained 55 L H_2/L-d using beverage wastewater. The immobilized-cell CSTR system allows high cell density and continuous and homogeneous mixing, good substrate transfer to microorganisms, and efficient pH control. The disadvantage of these systems is the high cost of the materials and process to immobilize the cells [46].

FBR is the most efficient system when all the different reactor configurations are compared using the same model substrate in lab-scale systems. However, when waste substrates are used, the nature of the substrate, the load of native microorganisms on the substrate, and the operating conditions will determine H_2 production. Homoacetogenesis is related to the metabolic pathways that are detrimental to H_2 production and which, among other things, is related to the reactor configuration. One condition that favors homoacetogenesis is the increase in the H_2 partial pressure in the liquid. Among the reactors used for H_2 production, the TBR is the only one to avoid H_2 accumulation in the liquid. Their configuration provides a high gas–liquid interphase that favors H_2 release. Montoya-Rosales et al. [36] compared the homoacetogenic activity in a TBR and CSTR, and the results confirmed that the TBR configuration was effective in keeping homoacetogenesis activity at lower levels than the CSTR. However, in the TBR, the HPR was lower (2.25 L H_2/L-d) than in CSTR (13 L H_2/L-d), which was related to the inadequate substrate distribution in the biofilm's inner layers. A high HPR can be achieved in the CSTR by a good distribution of the substrate and the implementation of an adequate mixing system that reduces the consumption of the H_2 produced, as was demonstrated by Montiel-Corona & Razo-Flores [37], who managed to control the H_2 consumption for the synthesis of homoacetogenic-acetic acid by increasing the stirring speed in a CSTR, increasing the stirring speed from 150 to 300 rpm allowed to double the HPR. In the UASB reactors, a consumption of 80% of the H_2 produced by homoacetogenesis was recorded by Buitrón et al. [30]. They used the increase in the recirculation rate of the headspace gas to improve the H_2 release from the liquid to the gas phase, which helped triple the HPR. In the dynamic membrane bioreactors, shear velocity is a critical condition for the homoacetogenesis control, and the homoacetogenic-acetic acid decreased from 0.29 to 0.21 mol acetate/mol glucose $_{consumed}$ by the increase of shear velocity from 2.41 to 6.75 m/h [50]. The production of homoacetogenic-acetic acid has been reported for all reactor designs, and although homoacetogenesis is one of the unresolved challenges of the DF, reactor design and operating conditions are decisive for controlling this unwanted metabolic pathway.

4.3 Scaling-Up of the Process

Several efforts have been made to scale up from laboratory to bench scale. However, there are few pilot-scale reactors, and very few pilot-scale reactors use real waste. Table 4.1 presents works that have used reactors with volumes greater than 200 L for H_2 production. According to Gosh [51], a pilot scale can be considered for reactors with 50 to 10,000 L capacity and industrial scale volumes greater than 10,000 L. To our knowledge, the reactors with the largest volume reported for DF is 100 m^3 [52]. For research or demonstrative purposes, the pilot scale is a helpful tool to refine and resolve various aspects before transitioning to an industrial scale. Among the challenges when scaling up the reactors is the feeding system process, considering equipment to store and pretreat the substrate and prepare the inoculum. The stirring system is related to mass transfer. Also, strategies to maintain the stability of H_2 production derived from the changes in the microbial community composition and an efficient temperature control system are needed.

Some pilot scale works used model carbon sources; although these works can be a reference point, they do not face real problems such as variation in substrate composition with time. For example, the variation in the C/N ratio of food residues influences H_2 production. Kim et al. [53] observed a 50% decrease in H_2 production yield when the C/N ratio passed from 20 to 30. A supplement rich in nitrogen was used to reestablish the C/N ratio. Another alternative to improve the nutrient balance is the co-digestion of two residues [54]. However, when scaling the process, the disposition and proximity of such residues must be considered to ensure the economic and practical viability of the process.

Other aspects to consider when using residues are the cost involved for storage and equipment necessary for conditioning wastes with high solid content. For example, food waste and OFMSW require grinding, sieving, and pressing to homogenize, increase solubility, facilitate mixing, and avoid clogging feed or recirculation pumps [53–56]. The storage of the OFMSW (−4 °C) has been suggested to prevent its degradation or colonization with unwanted microorganisms [55]. Kim et al. [53] used alkali pretreatment at pH 12.5 for one day with 6 N KOH solution, and Lee and Chung [56] applied a heat treatment (100 °C, 30 min). Among the less desired microorganisms frequently present in waste are lactic acid bacteria that have been considered detrimental to H_2 production. However, it has recently been observed that the H_2-producing bacteria can subsequently use the lactic acid produced by these bacteria to generate H_2 [28], so the stage of pretreatment-preservation of the substrate could be replaced by one in which the lactic acid pathway ferments the substrate such as food waste and cheese whey and later that pre-fermented substrate is fed to the H_2 producing reactor, this pre-fermentation is probably more economical than pretreating by thermal shock, keeping frozen or applying an alkaline treatment to the provided substrate.

On the other hand, once the systems are operating, they are not exempt from alternate pathways that consume H_2, as observed by Kim et al. [53], who solved the problem by applying an alkaline shock to the entire reactor content (150 L). Another method is the biokinetic control that, through TRH and OLR variations, it is also possible to control the proliferation of methanogenic archaea and favor the prevalence of H_2-producing bacteria. Increasing agitation speed decreases H_2 consumption by homoacetogenesis [37]. Also, recirculating digestate from a methanogenic reactor can provide sufficient alkalinity and keep the pH and H_2 production stable for long periods of operation, as reported by Cavinato et al. [19] and Micolucci et al. [57]. Table 4.1 shows that most of the reported studies operated continuously and for up to 300 days, indicating that it is possible to maintain the stability of the process, which is needed for the commercial application of this technology.

Within the configurations of reactors used at the pilot level (Table 4.1), the CSTR predominates, probably because of its uncomplicated design and, on the other hand, because the CSTR can provide the best stirring. Increasing the mixing speed of a 400 L CSTR from 10 to 30 rpm improved the HPR and yield from 5.2 to 28.9 L H_2/L-d

Table 4.1 Hydrogen production at pilot scale

Feedstock	Inoculum/ pretreatment	Reactor type and volume	Regime and operation time	Temperature/HTR	HPR	H_2 yield	Ref.
Glucose	Sludge from WWTP/100 °C for 30 min	Cubic fermenter with a baffled structure to improve mixing of 3,000 L	Continuous/53 d	35 °C/24 h	2.24 L H_2/L-d	0.59 mol H_2/mol glucose	[59]
Sucrose	Beach sludge/100 °C for 45 min	CSTR-Agitated granular sludge bed reactor of 400 L	Continuous/28 d	35 °C/6 h	28.9 L H_2/L-d	3.84 mol H_2/mol sucrose	[58]
Cane molasses	*Enterobacter cloacae* IIT-BT	CSTR of 10,000 L	Batch/1 d	37 °C	13.4 L H_2/L-d	362 L/kg $COD_{removed}$	[54]
Cheese whey	Sludge 105 °C for 2 h	Combined fluidized and trickling bed reactor/600 L	Continuous	60 °C/15 h	6.6 L H_2/L-d	1.93 mol H_2/mol hexose	[62]
Sugarcane distillery effluent	*C. freundii* 01, *E. aerogenes* E10 and *R. palustris* P2	CSTR of 100,000 L	Batch/40 h	Ambient	0.53 kg/100 m^3/h or 1.4 L H_2/L-d	2.76 mol H_2/mol glucose	[52]
Molasses	Sludge of municipal WWTP	CSTR with conical bottom/2000 L	Continuous/ 200 d	35 °C/4 h	5.57 L H_2/L-d	585 L/kg $COD_{removed}$	[63]
Citric acid wastewater	Facultative anaerobic inoculum	UASB of 50,000 L	Continuous/90 d	35–38 °C /12 h	0.72 L H_2/L-d	0.84 mol H_2/mol hexose	[60]
Food wastes	Indigenous bacterial communities from food wastes	CSTR of 200 L	Continuous/ 300 d	55 °C/79 h	0.65 L H_2/L-d	70–74 L H_2/kg TVS_{fed}	[57]

(continued)

Table 4.1 (continued)

Feedstock	Inoculum/ pretreatment	Reactor type and volume	Regime and operation time	Temperature/HTR	HPR	H_2 yield	Ref.
OFMSW	Anaerobic sludge/ 70 °C for 30 min	Dry up-flow intermittently stirred tank reactor of 500 L	Continuous / 230 d	35 °C/72 H	2.2 L H_2/L-d	2 mol H_2/mol CHO	[55]
Food waste	Sludge from an anaerobic digester	CSTR Anaerobic sequencing batch reactor of 230 L	Continuous/ 220 d	35 °C/36 h	NR	0.69 mol H_2/mol hexose	[53]
Food waste	Sludge from an anaerobic digester/ 80 °C for 20 min	CSTR of 500 L_180 rpm	Continuous/90 d	33 °C/21 h	3.88 L H_2/m^3-d	1.82 mol H_2/mol glucose	[56]

and from 2.34 to 3.84 mol H_2/mol sucrose, respectively [58]. HPR and yield were very low (2.24 L H_2/L–h and 0.59 mol H_2/mol glucose), although model substrates were used [58, 59]. That could be related to the low level of mixing in the reactor that uses a baffled structure.

Although this mixing method may not require high energy, it seems inefficient and could not be applied to effluents with a high solid concentration. Yang et al. [60] ensure complete mixing in a UASB by equipping the reactor with a distributed inflow and a mixer.

Regarding the inoculum, at an industrial level, using pure cultures is unfeasible due to the complications of preparing the necessary amount [54] and maintaining the culture's prevalence. Generally, pretreated sludge from wastewater treatment plants was used (Table 4.1). In some cases, starting the reactor without an external inoculum was possible. Micolucci et al. [57] reported using native bacterial communities from food waste. Facultative bacteria in the inocula could help maintain the anaerobic conditions necessary for H_2-producing strict anaerobic bacteria [61].

Another aspect to consider as the volume of the reactors increases is the energy demand to maintain the temperature in mesophilic or thermophilic conditions. The implementation of solar water heaters may be an option [59].

4.4 Coupling Dark Fermentation with Other Process

Although considerable progress has been made to scale up DF, the low energy conversion efficiencies are one of the most critical constraints for the commercial implementation of the process. The incomplete waste degradation and accumulation of VFA limit the H_2 yield to less than 30% of the theoretical value of 12 mol H_2/mol glucose [1, 2]. The only way for the commercial application of DF is under a biorefinery scheme where the by-products, in this case, the effluents rich in VFA, can be used to obtain more energy (through AD, photofermentation, or bioelectrochemical systems) as the substrate for biodegradable polymers production or being purified to obtain chemical building blocks [1]. Of the available options, the coupling of DF with AD is the most viable technology to be implemented on a large scale due to the degree of development, easy implementation, low-cost, high-energy recovery in the form of biohythane ($H_2 + CH_4$), and the growing interest in using hythane as fuel for the transport sector. Hythane is a mixture of H_2 and CH_4, where the H_2 content ranges from 10 to 25% by volume, and is used in transport vehicles with internal combustion engines [64]. The mixture of H_2 and CH_4 improves the fuel's flammability even at low temperatures, makes the engine easy to ignite, reduces combustion time, and reduces CO_2, CO, and NOx emissions [65]. The use of hythane dates from 1995 to the Montreal Hythane Bus Project. Afterward, Sweden, China, Italy, and India developed similar projects [66]. Companies like Toyota, Volvo, and Fiat have developed cars that run on hythane as fuel or hybrid systems (flex fuel) that allow gasoline or hythane. The United States of America and India are commercializing hythane as a transportation fuel [65, 66]. Currently, the H_2 in the hythane comes from fossil fuels, which does not represent a real environmental advantage. The production of hythane in the two-stage process, DF and AD, becomes relevant for its ecological benefits, in addition to the fact that this process could already be implemented in the industrial-scale plants of AD that currently operate in European countries, China, and the USA. AD systems for biogas production from organic waste are now in one stage [12]. All AD phases are carried out in a single reactor where hydrolytic-acidogenic bacteria and methanogenic archaea are held together in a delicate balance. Both groups differ in physiology, nutritional requirements, growth kinetics, and sensitivity to operating conditions [67]. For example, the pH in one-stage AD must be kept between 7 and 8, which is not adequate for the optimal growth of hydrolytic-acidogenic bacteria [67]. In the two-stage process, hydrolysis/acidogenesis (a phase that occurs at high rates) and methanogenesis (a phase that occurs at low rates) can be separated, with the possibility of optimizing each one [68]. The main advantages of the two-stage process are better stabilization of the process under two different optimum pH, better control of the methanogenic process in the face of organic overloads, digestion time reduction, and an increase in overall energy yield [37, 40, 69], and decreased consumption of alkaline substances to control pH [70].

During DF, a gradual decrease in pH takes place due to VFA production; when the pH is less than 4.5, H_2 production is inhibited [57], so it is necessary to maintain the pH between 5–6, which is an optimal range [19, 57]. An additional advantage of a two-stage AD is that the digestate from the methanogenic reactor can be used to control the pH in the acidogenic reactor [6]. Cavinato et al. [19] and Micolucci et al. [57] maintained the pH at 5.0–6.0 by recycling the effluent from the methanogenic reactor to the acidogenic reactor without adding any external buffer. The volume of recycled methane digestate must be carefully controlled to avoid an increase in the concentration of harmful ammonium in the system [57]. As mentioned above, reducing the partial pressure of H_2 during DF is vital to improve H_2 production. Recycling the methane and CO_2 stream from the methanogenic reactor to the acidogenic reactor is a good option [68].

Regarding energy conversion efficiency, two-stage AD substantially improves energy recovery and COD removal compared to one-stage AD. Schievano et al. [69] subjected different organic residues to an AD process in two stages and, in all cases, reported an increase from 8 to 43% in energy recovery compared to a single stage. De Gioannis et al. [67] and Lee and Chung [56] recovered 20 and 25% more energy from the food wastes that they submitted to the two-stage process, while Arreola-Vargas et al. [71] recovered 3.3 times more energy. Montiel-Corona and Razo-Flores [3] reported the production of 105 and 225 L CH_4/kg ST using agave bagasse, which represents 56% of the energy contained in this residue. Antonopoulou et al. [72] reported productions of 41 and 310 mL CH_4/g COD, equivalent to 71% of the energy content in food waste, and a COD removal efficiency of 94%.

The AD in two stages has been scaled to a pilot level. In some cases, the plants have operated for long periods, demonstrating the stability of the system and the feasibility of taking the process to a commercial scale. For example, Micolucci et al. [57] used a CSTR of 200 L for the DF stage and a CSTR of 600 L for the second stage. They implemented an automatic, economical, and fast system to control the volume of the digestate to be recycled from the AD to the DF reactor in such a way that they managed to maintain the pH and ammonium concentration at non-inhibitory levels for both stages, the operating time was 300 days using OFMSW as substrate at HTR of 3.3 and 12.5 d, respectively. The specific H_2 production was 70–74 L H_2/kg TVS_{fed}, and the specific methane production was 480–550 L CH_4/kg TVS_{fed}. Lee and Chung [56] operate a complete pilot plant for hythane production in two stages. The plant was equipped with a CSTR of 500 L for DF and a UASB of 2300 L for methanogenesis, storage, and biogas purification by adsorption and fuel cell generator systems. The plant was operated for 90 days using food waste as a substrate. The biogas production in the first stage was 2000–2500 L/d with 40–60% H_2, and in the second, the biogas production reached 12,000–14,000 L/d with 80% methane. Their economic analysis showed that the cost of producing H_2 or methane separately and in a two-stage process was similar. Still, the two-stage energy production had significantly greater potential for recovering energy than hydrogen or methane-only fermentation. Another two-stage system at the pilot scale called HyMeTek (Innovative Hydrogenation & Methanation Technology) has been constructed by Feng Chia University (Taiwan). The plant includes an H_2 production CSTR of 400 L and a UASB digestor of 2500 L. Food industrial wastewater was used as substrate. The HPR was 2.97 L H_2/L-d with an H_2 yield of 1.5 mol H_2/mol hexose at HRT of 9 h, the CH_4 production rate was 0.86 L CH_4/L-d and yield of 27.56 mL CH_4/g COD at HRT of 67 h [61]. Cavinato et al. [19] used CSTR and UASB reactors of 200 and 760 L, respectively, for the treatment of OFMSW without external inoculum, only the one coming from the residues and the pH was controlled by recycling digestate. The H_2 yield was 66.7 L H_2/kgTVS added, and the specific biogas production in the second stage was 720 L CH_4/kgTVS added.

Implementing this technology for biohythane production is independent of the degree of technological maturity. Pilot plants are already in operation, and scaling up to a commercial level depends on transitioning from using natural gas to hythane in the transport sector. As the demands for reducing greenhouse gas emissions increase, adopting more environmentally friendly technologies will increase.

4.5 Conclusions

Industrial effluents, rich in carbohydrates from the food or beverage industry, and the organic fraction of municipal solid waste are the most suitable options for H_2 production by DF. Lignocellulosic biomass still needs to reduce the operative cost related to pretreatment. The reactor configuration substantially impacts the retention of hydrogen-producing microorganisms and the mass transfer, which is crucial for the transport of the substrate toward the microorganisms and for the transfer of the hydrogen produced toward the headspace of the reactors for its recovery. The control of homoacetogenesis, one of the consuming pathways of the produced hydrogen, also depends on the reactor configuration and operating conditions. The process's scaling up still has challenges. Dark fermentation cannot be used alone but coupled with anaerobic digestion for methane production from residual volatile fatty acids. The mixture of hydrogen and methane (biohythane) can be suitable for stabilizing combined heat and power units, improving combustion efficiency. Pilot plants are already in operation, and scaling up to a commercial level will strongly depend on enforcing environmental policies related to renewable energy sources.

Acknowledgements Financial support provided by the DGAPA-UNAM (PAPIIT IT102522) and Instituto de Ingeniería through GII project 3406 is acknowledged. Virginia Montiel-Corona thanks CONAHCyT for the postdoctoral scholarship.

References

1. Lopez-Hidalgo AM, Smoliński A, Sanchez A (2022) A meta-analysis of research trends on hydrogen production via DF. Int J Hydrog Energy 47(27):13300–13339. https://doi.org/10.1016/j.ijhydene.2022.02.106
2. Dahiya S, Chatterjee S, Sarkar O, Mohan SV (2021) Renewable hydrogen production by dark-fermentation: current status, challenges and perspectives. Bioresour Technol 321:124354. https://doi.org/10.1016/j.biortech.2020.124354
3. Wang J, Yin Y (2018) Fermentative hydrogen production using various biomass-based materials as feedstock. Renew Sustain Energy Rev 92:284–306. https://doi.org/10.1016/j.rser.2018.04.033
4. Basak B, Jeon BH, Kim TH, Lee JC, Chatterjee PK, Lim H (2020) Dark fermentative hydrogen production from pretreated lignocellulosic biomass: effects of inhibitory by-products and recent trends in mitigation strategies. Renew Sustain Energy Rev 133:110338. https://doi.org/10.1016/j.rser.2020.110338
5. Łukajtis R, Hołowacz I, Kucharska K, Glinka M, Rybarczyk P, Przyjazny A, Kamiński M (2018) Hydrogen production from biomass using DF. Renew Sustain Energy Rev 91:665–694. https://doi.org/10.1016/j.rser.2018.04.043
6. Yun YM, Lee MK, Im SW, Marone A, Trably E, Shin SR, Kim MG, Cho SK, Kim DH (2018) Biohydrogen production from food waste: current status, limitations, and future perspectives. Bioresour Technol 248:79–87. https://doi.org/10.1016/j.biortech.2017.06.107
7. Saravanan A, Kumar PS, Jeevanantham S, Karishma S, Vo DVN (2022) Recent advances and sustainable development of biofuels production from lignocellulosic biomass. Bioresour Technol 344:126203. https://doi.org/10.1016/j.biortech.2021.126203
8. Beig B, Riaz M, Naqvi SR, Hassan M, Zheng Z, Karimi K, Pugazhendhi A, Atabani AE, Chi NTL (2021) Current challenges and innovative developments in pretreatment of lignocellulosic residues for biofuel production: a review. Fuel 287:119670. https://doi.org/10.1016/j.fuel.2020.119670
9. Dong H, Bao J (2010) Biofuel via biodetoxification. Nat Chem Biol 6:316–318. https://doi.org/10.1038/nchembio.355
10. Valdez-Vazquez I, Alatriste-Mondragón F, Arreola-Vargas J, Buitrón G, Carrillo-Reyes J, León-Becerril E, Mendez-Acosta HO, Ortíz I, Weber B (2020) A comparison of biological, enzymatic, chemical and hydrothermal pretreatments for producing biomethane from Agave bagasse. Ind Crops Prod 145:112160. https://doi.org/10.1016/j.indcrop.2020.112160
11. Wang S, Yang H, Shi Z, Zaini IN, Wen Y, Jiang J, Jönsson PG, Yang W (2022) Renewable hydrogen production from the organic fraction of municipal solid waste through a novel carbon-negative process concept. Energy 252:124056. https://doi.org/10.1016/j.energy.2022.124056

12. Clarke WP (2018) The uptake of AD for the organic fraction of municipal solid waste–push versus pull factors. Bioresour technol 249:1040–1043. https://doi.org/10.1016/j.biortech.2017.10.086
13. Martins I, Surra E, Ventura M, Lapa N (2022) BioH_2 from DF of OFMSW: effect of the hydraulic retention time and organic loading rate. Appl Sci 12(9):4240. https://doi.org/10.3390/app12094240
14. Ramos C, Buitrón G, Moreno-Andrade I, Chamy R (2012) Effect of the initial total solids concentration and initial pH on the biohydrogen production from cafeteria food waste. Int J Hydrogen Energy 37:13288–13295. https://doi.org/10.1016/j.ijhydene.2012.06.051
15. Panigrahi S, Dubey BK (2019) A critical review on operating parameters and strategies to improve the biogas yield from AD of organic fraction of municipal solid waste. Renew Energy 143:779–797. https://doi.org/10.1016/j.renene.2019.05.040
16. Abubackar HN, Keskin T, Yazgin O, Gunay B, Arslan K, Azbar N (2019) Biohydrogen production from autoclaved fruit and vegetable wastes by dry fermentation under thermophilic condition. Int J Hydrog Energy 44(34):18776–18784. https://doi.org/10.1016/j.ijhydene.2018.12.068
17. Kim DH, Jang S, Yun YM, Lee MK, Moon C, Kang WS, Kwak SS, Kim MS (2014) Effect of acid-pretreatment on hydrogen fermentation of food waste: microbial community analysis by next generation sequencing. Int J Hydrog Energy 39(29):16302–16309. https://doi.org/10.1016/j.ijhydene.2014.08.004
18. Jang S, Kim DH, Yun YM, Lee MK, Moon C, Kang WS, Kwak SS, Kim MS (2015) Hydrogen fermentation of food waste by alkali-shock pretreatment: microbial community analysis and limitation of continuous operation. Bioresour technol 186:215–222. https://doi.org/10.1016/j.biortech.2015.03.031
19. Cavinato C, Giuliano A, Bolzonella D, Pavan P, Cecchi F (2012) Bio-hythane production from food waste by DF coupled with AD process: a long-term pilot scale experience. Int J Hydrog Energy 37(15):11549–11555. https://doi.org/10.1016/j.ijhydene.2012.03.065
20. Lee DY, Xu KQ, Kobayashi T, Li YY, Inamori Y (2014) Effect of organic loading rate on continuous hydrogen production from food waste in submerged anaerobic membrane bioreactor. Int J Hydrog Energy 39(30):16863–16871. https://doi.org/10.1016/j.ijhydene.2014.08.022
21. Asunis F, De Gioannis G, Dessì P, Isipato M, Lens PN, Muntoni A, Polettini A, Raffaella P, Rossi A, Spiga D (2020) The dairy biorefinery: integrating treatment processes for cheese whey valorisation. J Environ Manage 276:111240. https://doi.org/10.1016/j.jenvman.2020.111240
22. Osorio-González CS, Gómez-Falcon N, Brar SK, Ramírez AA (2022) Cheese whey as a potential feedstock for producing renewable biofuels: a review. Energies 15(18):6828. https://doi.org/10.3390/en15186828
23. Lovato G, Albanez R, Stracieri L, Ruggero LS, Ratusznei SM, Rodrigues JAD (2018) Hydrogen production by co-digesting cheese whey and glycerin in an AnSBBR: temperature effect. Biochem Eng J 138:81–90. https://doi.org/10.1016/j.bej.2018.07.007
24. Marone A, Varrone C, Fiocchetti F, Giussani B, Izzo G, Mentuccia L, Rosa S, Signorini A (2015) Optimization of substrate composition for biohydrogen production from buffalo slurry co-fermented with cheese whey and crude glycerol, using microbial mixed culture. Int J Hydrog Energy 40(1):209–218. https://doi.org/10.1016/j.ijhydene.2014.11.008
25. Ordoñez-Frías EJ, Muñoz-Páez KM, Buitrón G (2023). Biohydrogen production from fermented acidic cheese whey using lactate: reactor performance and microbial ecology analysis. Int J Hydrog Energy (in press, July). https://doi.org/10.1016/j.ijhydene.2023.06.307
26. Castelló E, Braga L, Fuentes L, Etchebehere C (2018) Possible causes for the instability in the H_2 production from cheese whey in a CSTR. Int J Hydrog Energy 43(5):2654–2665. https://doi.org/10.1016/j.ijhydene.2017.12.104
27. Muñoz-Páez KM, Vargas A, Buitrón G (2023) Feedback control-based strategy applied for biohydrogen production from acid cheese whey. Waste Biomass Valoriz 14:447–460. https://doi.org/10.1007/s12649-022-01865-z
28. García-Depraect O, Castro-Muñoz R, Muñoz R, Rene ER, León-Becerril E, Valdez-Vazquez I, Kumar G, Reyes-Alvarado LC, Martínez-Mendoza LJ, Carrillo-Reyes J, Buitrón G (2021)

A review on the factors influencing biohydrogen production from lactate: the key to unlocking enhanced dark fermentative processes. Biores Technol 324:124595. https://doi.org/10.1016/j.biortech.2020.124595

29. Welz PJ, Holtman G, Haldenwang R, le Roes-Hill M (2016) Characterisation of winery wastewater from continuous flow settling basins and waste stabilization ponds over the course of 1 year: implications for biological wastewater treatment and land application. Water Sci Technol 74:2036–2050
30. Buitrón G, Muñoz-Páez KM, Quijano G, Carrillo-Reyes J, Albarrán-Contreras BA (2020) Biohydrogen production from winery effluents: control of the homoacetogenesis through the headspace gas recirculation. J Chem Technol Biotechnol 95(3):544–552. https://doi.org/10.1002/jctb.6263
31. Sganzerla WG, Sillero L, Forster-Carneiro T, Solera R, Perez M (2022) Determination of anaerobic Co-fermentation of brewery wastewater and brewer's spent grains for bio-hydrogen production. BioEnergy Res 16:1073–1083. https://doi.org/10.1007/s12155-022-10486-2
32. Tena M, Luque B, Perez M, Solera R (2020) Enhanced hydrogen production from sewage sludge by cofermentation with wine vinasse. Int J Hydrog Energy 45(32):15977–15984. https://doi.org/10.1016/j.ijhydene.2020.04.075
33. García-Becerra M, Macías-Muro M, Arellano-García L, Aguilar-Juárez O (2019) Bio-hydrogen production from tequila vinasses: effect of detoxification with activated charcoal on DF performance. Int J Hydrog Energy 44(60):31860–31872. https://doi.org/10.1016/j.ijhydene.2019.10.059
34. Silva FM, Oliveira LB, Mahler CF, Bassin JP (2017) Hydrogen production through anaerobic co-digestion of food waste and crude glycerol at mesophilic conditions. Int J Hydrog Energy 42(36):22720–22729. https://doi.org/10.1016/j.ijhydene.2017.07.159
35. García AB, Cammarota MC (2019) Biohydrogen production from pretreated sludge and synthetic and real biodiesel wastewater by DF. Int J Energy Res 43(4):1586–1596. https://doi.org/10.1002/er.4376
36. Montoya-Rosales J, Olmos-Hernández DK, Palomo-Briones R, Montiel-Corona V, Mari AG, Razo-Flores E (2019) Improvement of continuous hydrogen production using individual and binary enzymatic hydrolysates of agave bagasse in suspended-culture and biofilm reactors. Bioresour technol 283:251–260. https://doi.org/10.1016/j.biortech.2019.03.072
37. Montiel-Corona V, Razo-Flores E (2018) Continuous hydrogen and methane production from *Agave tequilana* bagasse hydrolysate by sequential process to maximize energy recovery efficiency. Bioresour Technol 249:334–341. https://doi.org/10.1016/j.biortech.2017.10.032
38. Kumar G, Sen B, Sivagurunathan P, Lin CY (2016) High rate hydrogen fermentation of cello-lignin fraction in de-oiled jatropha waste using hybrid immobilized cell system. Fuel 182:131–140. https://doi.org/10.1016/j.fuel.2016.05.088
39. Kim MS, Lee DY (2010) Fermentative hydrogen production from tofu-processing waste and anaerobic digester sludge using microbial consortium. Bioresour Technol 101:S48–S52. https://doi.org/10.1016/j.biortech.2009.03.040
40. Nualsri C, Kongjan P, Reungsang A (2016) Direct integration of CSTR-UASB reactors for two-stage hydrogen and methane production from sugarcane syrup. Int J Hydrog Energy 41(40):17884–17895. https://doi.org/10.1016/j.ijhydene.2016.07.135
41. Kongjan P, Inchan S, Chanthong S, Jariyaboon R, Reungsang A, Sompong O (2019) Hydrogen production from xylose by moderate thermophilic mixed cultures using granules and biofilm up-flow anaerobic reactors. Int J Hydrog Energy 44(6):3317–3324. https://doi.org/10.1016/j.ijhydene.2018.09.066
42. Mahmod SS, Azahar AM, Tan JP, Jahim JM, Abdul PM, Mastar MS, Anuar N, Yunus MFM, Asis AJ, Wu SY (2019) Operation performance of up-flow anaerobic sludge blanket (UASB) bioreactor for biohydrogen production by self-granulated sludge using pretreated palm oil mill effluent (POME) as carbon source. Renew Energy 134:1262–1272. https://doi.org/10.1016/j.renene.2018.09.062
43. Sivagurunathan P, Anburajan P, Kumar G, Kim SH (2016) Effect of hydraulic retention time (HRT) on biohydrogen production from galactose in an up-flow anaerobic sludge blanket

reactor. Int J Hydrogen Energy 41(46):21670–21677. https://doi.org/10.1016/j.ijhydene.2016.06.047
44. de Menezes CA, Silva EL (2019) Hydrogen production from sugarcane juice in expanded granular sludge bed reactors under mesophilic conditions: the role of homoacetogenesis and lactic acid production. Ind Crops Prod 138:111586. https://doi.org/10.1016/j.indcrop.2019.111586
45. Buitrón G, Muñoz-Páez KM, Hernández-Mendoza CE (2019) Biohydrogen production using a granular sludge membrane bioreactor. Fuel 241:954–961. https://doi.org/10.1016/j.fuel.2018.12.104
46. Park JH, Chandrasekhar K, Jeon BH, Jang M, Liu Y, Kim SH (2021) State-of-the-art technologies for continuous high-rate biohydrogen production. Bioresour technol 320:124304. https://doi.org/10.1016/j.biortech.2020.124304
47. Arreola-Vargas J, Alatriste-Mondragón F, Celis LB, Razo-Flores E, López-López A, Méndez-Acosta HO (2015) Continuous hydrogen production in a trickling bed reactor by using triticale silage as inoculum: effect of simple and complex substrates. J Chem Technol Biotechnol 90(6):1062–1069. https://doi.org/10.1002/jctb.4410
48. Oh YK, Kim SH, Kim MS, Park S (2004) Thermophilic biohydrogen production from glucose with trickling biofilter. Biotechnol Bioeng 88(6):690–698. https://doi.org/10.1002/bit.20269
49. Pugazhendhi A, Anburajan P, Park JH, Kumar G, Sivagurunathan P, Kim SH (2017) Process performance of biohydrogen production using glucose at various HRTs and assessment of microbial dynamics variation via q-PCR. Int J Hydrogen Energy 42(45):27550–27557. https://doi.org/10.1016/j.ijhydene.2017.06.184
50. Sim YB, Jung JH, Park JH, Bakonyi P, Kim SH (2020) Effect of shear velocity on dark fermentation for biohydrogen production using dynamic membrane. Bioresour technol 308:123265. https://doi.org/10.1016/j.biortech.2020.123265
51. Ghosh S (2022) Assessment and update of status of pilot scale fermentative biohydrogen production with focus on candidate bioprocesses and decisive key parameters. Int J Hydrogen Energy 47(39):17161–17183. https://doi.org/10.1016/j.ijhydene.2022.03.183
52. Vatsala TM, Raj SM, Manimaran A (2008) A pilot-scale study of biohydrogen production from distillery effluent using defined bacterial co-culture. Int J Hydrogen Energy 33(20):5404–5415. https://doi.org/10.1016/j.ijhydene.2008.07.015
53. Kim DH, Kim SH, Kim KY, Shin HS (2010) Experience of a pilot-scale hydrogen-producing anaerobic sequencing batch reactor (ASBR) treating food waste. Int J Hydrogen Energy 35(4):1590–1594. https://doi.org/10.1016/j.ijhydene.2009.12.041
54. Balachandar G, Varanasi JL, Singh V, Singh H, Das D (2020) Biological hydrogen production via dark fermentation: a holistic approach from lab-scale to pilot-scale. Int J Hydrogen Energy 45(8):5202–5215. https://doi.org/10.1016/j.ijhydene.2019.09.006
55. Elsamadony M, Tawfik A (2015) Potential of biohydrogen production from organic fraction of municipal solid waste (OFMSW) using pilot-scale dry anaerobic reactor. Bioresour technol 196:9–16. https://doi.org/10.1016/j.biortech.2015.07.048
56. Lee YW, Chung J (2010) Bioproduction of hydrogen from food waste by pilot-scale combined hydrogen/methane fermentation. Int J Hydrogen Energy 35(21):11746–11755. https://doi.org/10.1016/j.ijhydene.2010.08.093
57. Micolucci F, Gottardo M, Bolzonella D, Pavan P, Majone M, Valentino F (2020) Pilot-scale multi-purposes approach for volatile fatty acid production, hydrogen and methane from an automatic controlled anaerobic process. J Clean Prod 277:124297. https://doi.org/10.1016/j.jclepro.2020.124297
58. Lin PJ, Chang JS, Yang LH, Lin CY, Wu SY, Lee KS (2011) Enhancing the performance of pilot-scale fermentative hydrogen production by proper combinations of HRT and substrate concentration. Int J Hydrogen Energy 36(21):14289–14294. https://doi.org/10.1016/j.ijhydene.2011.04.147
59. Lu C, Wang Y, Lee DJ, Zhang Q, Zhang H, Tahir N, Jing Y, Liu H, Zhang K (2019) Biohydrogen production in pilot-scale fermenter: effects of hydraulic retention time and substrate concentration. J Clean Prod 229:751–760. https://doi.org/10.1016/j.jclepro.2019.04.233

60. Yang H, Shao P, Lu T, Shen J, Wang D, Xu Z, Yuan X (2006) Continuous bio-hydrogen production from citric acid wastewater via facultative anaerobic bacteria. Int J Hydrogen Energy 31(10):1306–1313. https://doi.org/10.1016/j.ijhydene.2005.11.018
61. Tapia-Venegas E, Ramírez-Morales JE, Silva-Illanes F et al (2015) Biohydrogen production by dark fermentation: scaling-up and technologies integration for a sustainable system. Rev Environ Sci Biotechnol 14:761–785. https://doi.org/10.1007/s11157-015-9383-5
62. Ren N, Li J, Li B, Wang Y, Liu S (2006) Biohydrogen production from molasses by anaerobic fermentation with a pilot-scale bioreactor system. Int J Hydrogen Energy 31(15):2147–2157. https://doi.org/10.1016/j.ijhydene.2006.02.011
63. Martinek S, Kastner V, Schnitzhofer W (2013) Efficient biohydrogen production from whey using a pilot scale carrierbased bioreactor system. Redbiogas. http://www.redbiogas.cl/wp-content/uploads/2013/07/IWA-11343.pdf
64. Sasidhar KB, Kumar PS, Xiao L (2022) A critical review on the two-stage biohythane production and its viability as a renewable fuel. Fuel 317:123449. https://doi.org/10.1016/j.fuel.2022.123449
65. Hans M, Kumar S (2019) Biohythane production in two-stage anaerobic digestion system. Int J Hydrogen Energy 44(32):17363–17380. https://doi.org/10.1016/j.ijhydene.2018.10.022
66. Bolzonella D, Battista F, Cavinato C, Gottardo M, Micolucci F, Lyberatos G, Pavan P (2018) Recent developments in biohythane production from household food wastes: a review. Bioresour technol 257:311–319. https://doi.org/10.1016/j.biortech.2018.02.092
67. De Gioannis G, Muntoni A, Polettini A, Pomi R, Spiga D (2017) Energy recovery from one-and two-stage anaerobic digestion of food waste. Waste Manage 68:595–602. https://doi.org/10.1016/j.wasman.2017.06.013
68. Algapani DE, Qiao W, di Pumpo F, Bianchi D, Wandera SM, Adani F, Dong R (2018) Long-term bio-H_2 and bio-CH_4 production from food waste in a continuous two-stage system: energy efficiency and conversion pathways. Bioresour technol 248:204–213. https://doi.org/10.1016/j.biortech.2017.05.164
69. Buitrón G, Kumar G, Martinez-Arce A, Moreno G (2014) Hydrogen and methane production via a two-stage processes (H_2-SBR + CH_4-UASB) using tequila vinasses. Int J Hydrogen Energy 39(33):19249–19255. https://doi.org/10.1016/j.ijhydene.2014.04.139
70. Jung KW, Moon C, Cho SK, Kim SH, Shin HS, Kim DH (2013) Conversion of organic solid waste to hydrogen and methane by two-stage fermentation system with reuse of methane fermenter effluent as diluting water in hydrogen fermentation. Bioresour Technol 139:120–127. https://doi.org/10.1016/j.biortech.2013.04.041
71. Arreola-Vargas J, Flores-Larios A, González-Álvarez V, Corona-González RI, Méndez-Acosta HO (2016) Single and two-stage anaerobic digestion for hydrogen and methane production from acid and enzymatic hydrolysates of Agave *tequilana* bagasse. Int J Hydrogen Energy 41(2):897–904. https://doi.org/10.1016/j.ijhydene.2015.11.016
72. Antonopoulou G, Stamatelatou K, Venetsaneas N, Kornaros M, Lyberatos G (2008) Biohydrogen and methane production from cheese whey in a two-stage anaerobic process. Ind Eng Chem Res 47:5227–5233. https://doi.org/10.1021/ie071622x

Chapter 5
Biohydrogen Production: A Focus on Dark Fermentation Technology

Jose Antonio Magdalena, Lucie Perat, Lucia Braga-Nan, and Eric Trably

Abstract Hydrogen is considered a promising alternative to fossil fuels due to its clean combustion and high-energy content compared to other resources (i.e., methane or ethanol). However, hydrogen is produced from fossil-energy based processes leading to considerable greenhouse gas (GHG) emissions. One of the main solutions to hasten the energy transition is to develop efficient green processes for hydrogen production. Biological-based technologies have specifically demonstrated to have the least impact on the environment due to the low energy requirements (mild temperatures and pressures) and scarce GHG emissions. Among biological processes, Dark Fermentation (DF) appears to be the most effective and feasible technology for efficient hydrogen production at an industrial scale. DF has the advantage of valorizing a wide range of carbohydrate-rich organic waste without light energy through microbial consortia. Nonetheless, this technology deserves in-depth research since it still presents limited yields and productivity that hamper its industrial development. This book chapter aims to give an overview of DF biohydrogen production, presenting the state of the art. In this sense, DF's underlying mechanisms of biohydrogen production are detailed. The influence of operational parameters was assessed focusing on optimizing biohydrogen production and favoring the growth of biohydrogen-producing bacteria. Finally, the different roles of the microbial communities participating in the DF process are unraveled.

Keywords Biohydrogen · Bioeconomy · Dark fermentation · Microbial communities

---

Jose Antonio Magdalena, Lucie Perat, Lucia Braga-Nan and Eric Trably authors contributed equally to the work.

J. A. Magdalena (✉) · L. Perat · L. Braga-Nan · E. Trably (✉)
INRAE, Université de Montpellier, LBE, 102 Avenue des Étangs, 11100 Narbonne, France
e-mail: jose.antonio-magdalena.cadelo@inrae.fr

E. Trably
e-mail: eric.trably@inrae.fr

J. A. Magdalena
Vicerrectorado de Investigación y Transferencia de la Universidad Complutense de Madrid, 28040 Madrid, Spain

V. Alcaraz Gonzalez et al. (eds.), *Wastewater Exploitation*, Springer Water,
https://doi.org/10.1007/978-3-031-57735-2_5

5.1 Dark Fermentation for Biohydrogen Production

Dark Fermentation (DF) is the most promising method for biohydrogen production from organic waste. During this process, organic residues that contain high amounts of organic matter, are converted into biohydrogen in the absence of light through a microbial consortium composed of obligate and facultative anaerobic fermentative bacteria. A wide amount of substrates can be subjected to DF, particularly carbohydrate-rich materials such as lignocellulosic biomass, food waste, or sugar-containing crop residues that are ideal for hydrogen conversion via this biological process.

The first degradation step involves the hydrolysis of complex organic matter (i.e., cellulose) into simple polymers or soluble molecules (Fig. 5.1). Second, acidogenic bacteria can transform carbohydrates, proteins, and long-chain fatty acids into Volatile Fatty Acids (VFAs), alcohols, and other soluble metabolic end-products (i.e., lactate) and biogas composed of carbon dioxide and biohydrogen. The released hydrogen accumulates in the reactor and can easily be consumed by microorganisms belonging to homoacetogens, methanogens, and sulphate-reducing bacteria groups. For this reason, different strategies applied to the microbiome or the operational conditions should be considered to avoid H_2 consumption.

Fermentation pathways for hydrogen production

Soluble carbohydrates undergo fermentation via glycolysis, where simple sugars such as glucose are converted to pyruvate by forming reduced Nicotinamide Adenine Dinucleotide (NADH), and Adenosine Triphosphate (ATP). Pyruvate is a pivotal intermediate molecule from which many different compounds are obtained. According to Fig. 5.2, there are three main metabolic pathways for hydrogen production: (i) The Pyruvate Ferredoxin OxidoReductase (PFOR) pathway, which is common in strict anaerobes; (ii) the Pyruvate Formate Lyase (PLF) pathway that can be found in facultative bacteria; and (iii) the NADH pathway [1]. Along the PFOR pathway and in the absence of oxygen, pyruvate is degraded, generating a reduced ferredoxin (Fd_{red}) and acetyl-CoA. Hydrogen generation occurs when the Fd_{red} transfers the electrons to protons (H^+) using a ferredoxin-dependent hydrogenase (Fd-FeFe). In the PFL pathway, pyruvate and coenzyme A are converted to formate and acetyl-CoA using a glycyl radical in a reaction catalyzed by the formate hydrogenlyase (FHL) complex. The aforementioned glycyl radical entails anaerobic or microaerobic conditions to avoid deactivating the enzymatic complex [2]. Subsequently, formate is fragmented into biohydrogen and carbon dioxide by a sequential formate dehydrogenase coupled with a hydrogenase (either FeFe or NiFe) [3]. Finally, the NADH-ferredoxin oxidoreductase oxidizes NADH and reduces ferredoxin, which results in hydrogen generation.

Hydrogenases

There are two basic types of hydrogenases, based on the metals in the active site and referred to as [FeFe] hydrogenases and [NiFe] hydrogenases. These enzymes

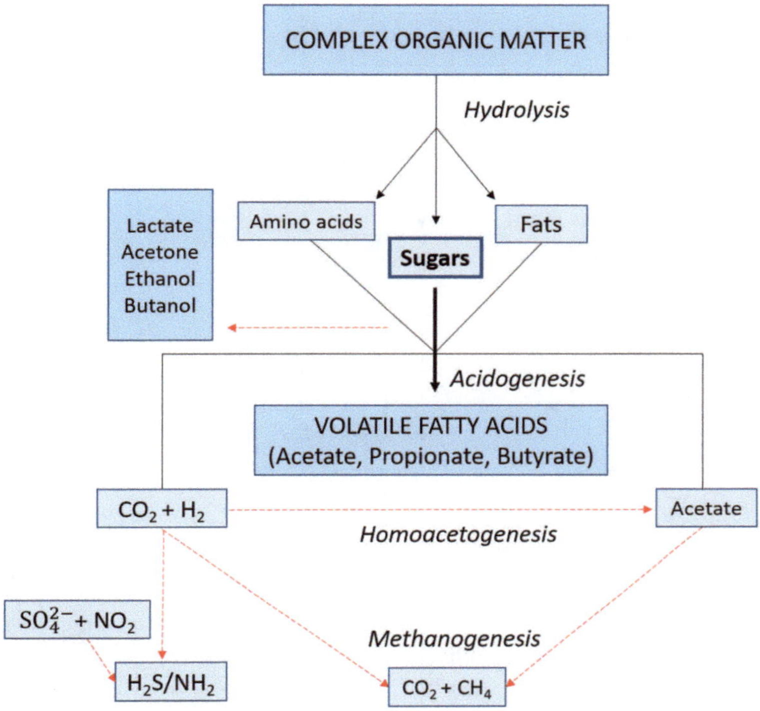

Fig. 5.1 Biochemical pathways occurring in DF. Red dash lines represent the pathways decreasing the hydrogen yield in the DF process

catalyze the hydrogen oxidation, (Eq. 5.1) under anaerobic conditions. Oxygen irreversibly damages some of them. A recent study pointed out the need for an in-depth investigation of the enzyme´s tolerance towards oxygen and their protective mechanisms [4].

$$2H^+ + 2e^- \rightarrow H_2 \quad (5.1)$$

Despite sharing an analogous activity, [FeFe]- and [NiFe]-hydrogenases also exhibit fundamental differences. In fact, [FeFe]-hydrogenases couple hydrogen evolution with the recycling of electron carriers that accumulate during the anaerobic metabolism, whereas [NiFe]-hydrogenases catalyse hydrogen oxidation as part of the energy metabolism [5]. Furthermore, [FeFe]-hydrogenases are found in prokaryotes and eukaryotes but not in archaea [5].

Fermentation products and hydrogen yields

The thermodynamics of the hydrogenase reaction hampers hydrogen production. This reaction involves the transfer of electrons from an electron carrier (e.g., Fd_{red} or NADH) to protons. Since hydrogen generation is thermodynamically favourable

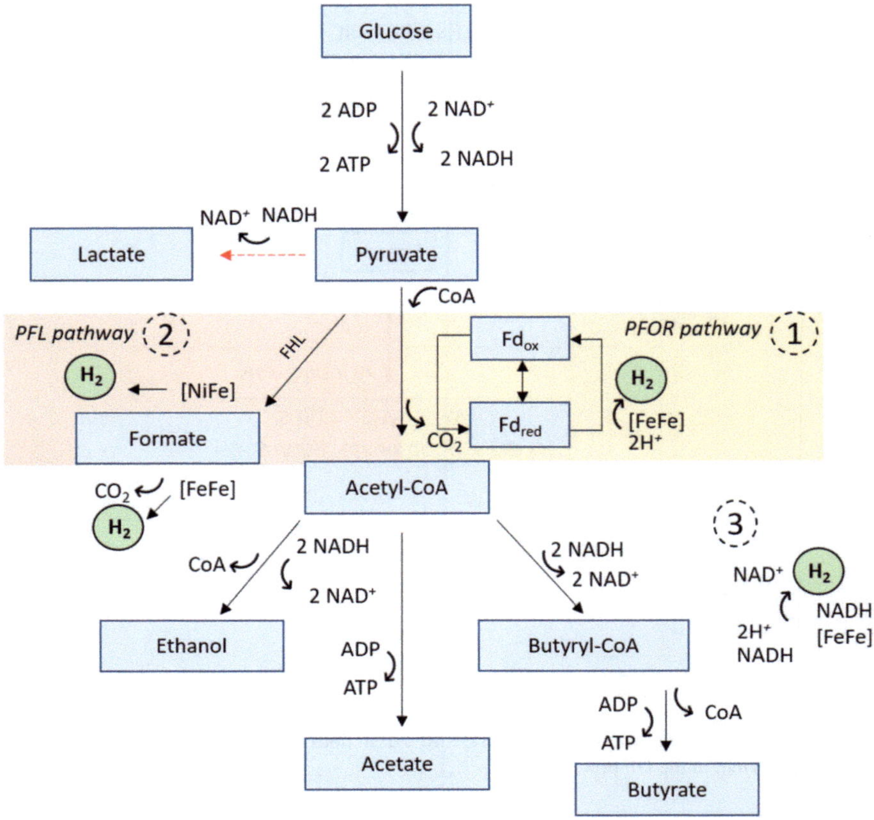

Fig. 5.2 Fermentation pathways for hydrogen production from glucose

only under low-partial pressure conditions, the hydrogen partial pressure plays a crucial role in hydrogen production and the type of co-metabolites formed. Hydrogen production in DF can occur when hydrogen partial pressures are below 30,397 Pa and 60.7 Pa, for Fd_{red} and NADH, respectively [6]. These authors pointed out that at partial hydrogen pressure below 60.7 Pa up to 4 mol of H_2 can be achieved, producing acetate as the main co-product considering glucose as substrate (Eq. 5.2).

$$C_6H_{12}O_6 + 2H_2O \rightarrow 4H_2 + 2CH_3COOH + 2\,CO_2 \tag{5.2}$$

On the other hand, hydrogen partial pressures above 60.7 Pa yield 2 mol H_2/mol glucose and butyrate accumulation via acetyl-CoA fermentation (Eq. 5.3).

$$C_6H_{12}O_6 \rightarrow 2\,H_2 + CH_3CH_2CH_2COOH + 2\,CO_2 \tag{5.3}$$

Additionally, acetate and butyrate could be further transformed into hydrogen by syntrophic oxidations (Eqs. 5.4 and 5.5); nonetheless, this reaction is thermodynamically unfavourable and can only occur when hydrogen partial pressure is reduced (by either hydrogen extraction or consumption via methanogens).

$$CH_3COOH + H_2O \rightarrow 4\,H_2 + 2\,CO_2 \tag{5.4}$$

$$CH_3CH_2CH_2COOH + 2\,H_2O \rightarrow 2\,H_2 + 2\,CH_3COOH \tag{5.5}$$

Overall, 12 $mol_{H2}/mol_{glucose}$ is the maximum theoretical yield obtained by considering reactions in Eqs. 5.2 and 5.4. In practice, however, much lower yields are obtained than reported in this chapter. In fact, the presence of other fermentation products, such as ethanol or lactate (Fig. 5.2), indicates the existence of other side reactions occurring during DF that outcompete the PFOR pathway. Lactate and ethanol pathways lead to a decrease in the overall hydrogen yields obtained. In addition, the hydrogen produced in the above metabolic pathways depends on the various operating parameters involved in the fermentation.

5.2 Operational Parameters

Biohydrogen production relies on biotic factors (i.e., inoculum and substrate nature) as well as on abiotic factors such as pH, temperature, hydrogen/carbon dioxide partial pressures, HRT (Hydraulic Retention Time), OLR (Organic Loading Rate), and reactor configuration. These main parameters, presented in Fig. 5.3, widely have been studied and optimized to improve biohydrogen yield in DF.

Substrate: macromolecular composition and C/N ratio

Substrate nature has a remarkable impact on hydrogen yield and production rate. Residual streams are increasingly being studied as feedstock for DF due to the low economic costs [7]. Nonetheless, organic matter complexity and the high variability in macromolecular composition of the residues (i.e., different carbohydrate, lipid, and protein contents) entail technical challenges. The carbohydrate content exhibited by the substrate, especially soluble and easily degradable sugars [8], has a crucial effect on the process [9]. In fact, it was observed that the hydrogen production potential was about 15 times lower for lipid-rich (fat meat and chicken skin) and 20 times lower for protein-rich substrates (egg and lean meat) compared to carbohydrate-rich substrates [10]. Lipid-rich substrates are unfavorable for hydrogen production due to the inhibition of Hydrogen-Producing Bacteria (HPB) by long-chain fatty acids (LCFA). Moreover, β-oxidation of LCFA into hydrogen and acetate is only possible at very low hydrogen partial pressures, representing a thermodynamic barrier [11, 12]. Furthermore, protein degradation through Stickland reactions is not associated with hydrogen production [11]. Therefore, co-digesting carbohydrate-rich substrates with

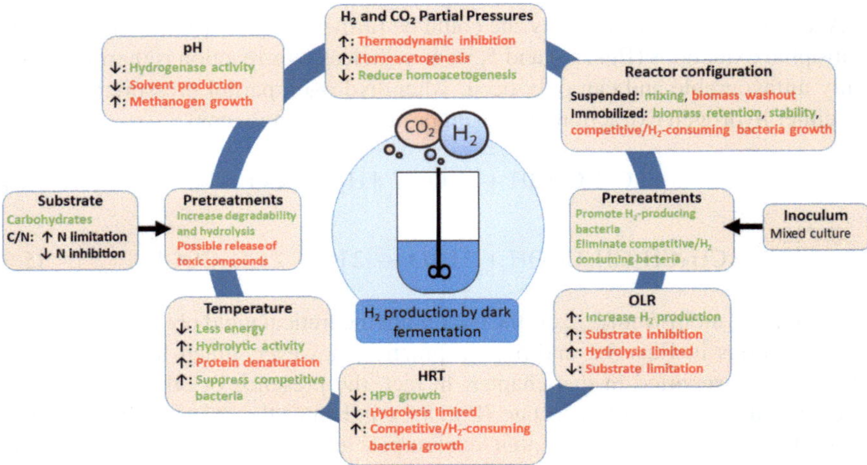

Fig. 5.3 Main operational parameters for hydrogen production by DF. Black arrows represent the increase/decrease of the operational parameter evaluated. Green and red colors indicate, respectively, positive and negative effects on hydrogen production. HPB: Hydrogen Producing Bacteria

an appropriate ratio of a protein-rich feedstock might improve hydrogen production by ensuring nutritional balance, with proteins acting as potential sources of nitrogen, phosphorus, and sulfur [13, 14]. Balancing the C/N ratio can favor hydrogen production and substrate degradation in co-fermentation [15]. In this sense, it was observed in batch reactors a maximum hydrogen yield of 134.4 mL_{H2}/g_{COD} for a glucose/peptone ratio of 3/2 ($[mg_{COD}/L]/[mg_{COD}/L]$) compared to a yield of 44.8 mL_{H2}/g_{COD} with glucose as sole substrate. Higher peptone content led to a decrease in hydrogen yield and cell growth [16]. Then, the C/N ratio must be well balanced to ensure enzyme synthesis and growth for N-deficient substrates without leading to nitrogen inhibition or excessive cellular maintenance [17]. For this reason, C/N values are not consensual and depend on substrate macromolecular composition, inoculum, reactor configuration, and operating conditions (ranging from 21 for food waste in batch reactors [15] to 137 on the up-flow fixed-bed anaerobic reactor for sucrose-based synthetic wastewater [17] or even 200 for wheat in batch experiments [18]).

Pretreatments applied to the substrate

Pretreatments are mainly applied on complex structures such as lignocellulosic substrates to increase their biodegradability and facilitate the hydrolytic step in DF. Pretreatments are mainly classified as mechanical, thermal, chemical, and biological [19, 20]. Thermal pretreatment is one of the most used to induce cell wall lysis, improve solubilization and subsequently increase organic matter availability. For instance, hydrogen yield increased from 5 NL_{H2}/kg_{VSfed} to 23 NL_{H2}/kg_{VSfed} when food waste was heated at 134 °C for 20 min [21]. The main drawback of thermal treatments is the considerable amount of energy needed [8].

In contrast, adding chemicals such as acid or base products does not require energy input and allows to break down the lignin-carbohydrates complexes [8]. For instance, hydrogen yield increased from 4.39 mL_{H2}/g to 72.21 mL_{H2}/g of initial dry grass by applying an optimal concentration of 4% HCl (v/v). In contrast, for the same substrate, the addition of 0.5% NaOH (v/v) improved the hydrogen yield from 4.39 mL_{H2}/g to 19.25 mL_{H2}/g of initial dry grass [22]. However, chemical products have environmental impacts that should be considered in terms of upscaling the process [23]. Furthermore, the harsh conditions imposed with thermal and chemical pretreatments can lead to the formation of toxic and inhibitory compounds (i.e., furan derivatives, phenolic compounds, lignin) that can limit hydrogen production [23, 24]. Comparative studies of different substrate pretreatments disagree on which pretreatment best favors hydrogen production, mainly due to the large variability of the conditions studied [19].

pH

pH is one of the most studied operating parameters for hydrogen production since it influences the hydrogenase activity, metabolic pathways and microbial populations [25]. Specifically, initial pH (for a batch reactor with no pH control) and the operating pH (when pH is controlled at a specific value throughout DF) should be clearly distinguished. For example, the study conducted by Khanal and co-authors assessed sucrose and starch in batch reactors with different initial pH values ranging from 4.5 to 6.5 [26]. Results showed that increasing the initial pH led to a decrease in specific hydrogen production potential for both substrates, with a maximal hydrogen production at initial pH of 4.5 (200 mL_{H2}/g_{CODin} and 110 mL_{H2}/g_{CODin} for sucrose and starch, respectively). The authors observed that higher pH drop was obtained for higher initial pH (from 5 to 6.5), associated with a shorter duration of production (40 vs 80 h for initial pH of 4.5). They suggested that for pH higher than 4.5, the hydrogen production was faster and was associated with VFAs production that rapidly reached inhibitory levels for HPB. Hydrogen production was slower for pH 4.5, likely due to the unsuitable starting environment resulting in a lag phase before adaptation [26]. Thus, pH control during DF is crucial to prevent pH drop due to VFAs accumulation. The pH in the plant must be above 4.5 to account for the optimal pH for hydrogenase activity [27] and to avoid a shift in metabolism to the production of solvents (e.g., butanol, acetone, and ethanol) [28]. In practice, a pH between 5.0 and 6.0 is considered as optimal for hydrogen production from food waste, while neutral pH favours hydrogen production from crop residues and animal manure [7]. From a practical point of view, an acidic pH value might also contribute to favour HPB against H_2-consuming methane producers. As a matter of fact, [29] observed a total inhibition of methanogens at pH 5.5 and operation at 70 °C. Overall, pH can be considered a control parameter, whose optimal value depends on the substrate, inoculum and reactor configuration.

Temperature

Similarly to pH, temperature values can affect hydrogen yields and the microbiome structure by triggering specifically metabolic pathways [7]. DF can be conducted at different temperature ranges: psychrophilic (0–25 °C), mesophilic (26–40 °C), thermophilic (41–65 °C), extreme thermophilic (66–80 °C), and hyperthermophilic (>80 °C) conditions, but most of the studies assessed DF under mesophilic conditions [7, 30]. A recent study identified the optimal temperature (35.2 °C) by response surface methodology with a pure culture of *Clostridium acetobutylicum* YM1 in batch conditions using de-oiled rice bran as substrate, resulting in a yield of 135.2 mL_{H2}/g_{sugar} [31]. In addition, hydrogen production from complex substrates (e.g., rice bran or crop residues) can be improved with thermophilic temperatures due to the enhancement of hydrolytic enzyme activity [32]. Illustratively, another study observed an improvement in hydrogen production rate from 57.65 to 97.08 mL H_2/h when increasing the temperature from 37 to 55 °C in batch fermenters using rice bran as substrate [33]. In contrast, too high temperatures with a non-specialized microbiome might adversely affect microbial activity due to the denaturation of proteins and enzymes [34]. Thus, temperature changes might contribute to metabolic and microbial modifications that can affect the metabolites produced during DF. For instance, Shin and co-authors observed that continuous hydrogen production using food waste as substrate (HRT of 5 days and pH 5.5) was higher for thermophilic (2.9 $mL_{H2}/(g_{VSS}$ h) at 55 °C) than mesophilic conditions (0.7 $mL_{H2}/(g_{VSS}$ h) at 35 °C) [35]. The production of methane and propionate (H_2 consuming reactions) was observed under mesophilic conditions, while these reactions were inhibited at 55 °C, favoring H_2 production. These results can be explained by the analysis of microbial composition. Several species not associated with hydrogen production were identified in the mesophilic DGGE PCR profile, while the microbial community under thermophilic conditions consisted mainly of HPB [35]. Overall, mesophilic conditions are preferred for DF because they require less energy and operational costs compared to extreme, hyper, and thermophilic conditions [36, 37]. Furthermore, reduced microbial diversity is exhibited at thermophilic conditions, which might harm substrate degradation when dealing with solid waste [38].

Hydrogen and carbon dioxide partial pressure

According to Henry's law, high hydrogen and carbon dioxide contents in the gas phase (i.e., high partial pressures) reflect high liquid concentration in the media, which can lead to end-product inhibition [39]. In particular, hydrogenase reduction by NADH is thermodynamically favorable only for hydrogen partial pressures lower than 60.7 Pa [40]. Reducing the hydrogen partial pressure by facilitating mass transfer from the liquid to the gas phase favours the hydrogen production [41]. Moreover, by decreasing hydrogen and carbon dioxide concentrations, H_2-consuming homoacetogenic activities could be reduced [42]. Several methods have been studied to decrease hydrogen and carbon dioxide partial pressures, such as gas release, vacuum, or sparging. For instance, continuous gas release improved hydrogen production from 36 to 108 mL H_2/Lh in continuous reactors and enhanced glucose degradation efficiency from 90 to 95% [43]. Consistently, [44] observed in an Up-flow Anaerobic Sludge Blanket reactor (UASB) an increase of hydrogen yield from 2.80 to 4.55 $mol_{H2}/kg_{CODremoved}$ at OLR 35 $kg_{COD}/(m^3d)$ and HRT 24 h when applying vacuum (reducing total pressure to 75,000 Pa). Following this trend, hydrogen yield increased from 0.85 to 1.43

$mol_{H2}/mol_{glucose}$ with N_2 sparging in the liquid phase (flow rate of 110 mL/min) [45]. However, gas sparging requires high purification costs due to hydrogen dilution in the outlet, hampering its application at an industrial scale [46]. As for the efficiency of extraction methods in reducing homoacetogenic activity, this pathway was still observed at a total pressure of only 81,000 Pa with a contribution of up to 30% of the acetate produced [47]. It was demostrated with isotope experiments that hydrogen is immediately consumed by homoacetogens, even when the hydrogen partial pressure is reduced by gas sparging [48]. Access to dissolved concentrations of hydrogen and carbon dioxide is thus of crucial importance to better understand these mechanisms. [49].

HRT (Hydraulic Retention Time)

HRT (time) is defined according to the following equation (Eq. 5.6), where V is the volume of the reactor and Q is the feeding flow rate.

$$HRT(h) = \frac{V(L)}{Q(L/h)} \tag{5.6}$$

HRT is linked to the microbial growth rate since slow-growing microorganisms do not thrive with a short HRT, being washed out from the continuous reactor. It has been widely shown that short HRTs (6–12 h) favour HPB while reducing microbial diversity and resulting in improved hydrogen productivity and yields [50–52]. Moreover, short HRT allows for the washout of hydrogen production competitors such as methanogens due to their low growth rates (0.2 h^{-1} [7]). In fact, it was shown a decrease in methane production from 110 to 27 mL CH_4/g VS when the HRT was reduced from 6 to 2 days, using household solid waste as substrate [29]. Additionally, this parameter impacts economic operation since it is related to the size of the fermenter and the quantity of organic matter handled during DF [38]. Optimal HRT values depend on the type of substrates employed since the degradation of complex organic matter is limited by the hydrolysis rate, and the HRT drives the contact time between organic matter and bacteria. Liu and co-authors showed that HRT values that are too short (less than two days) limit the hydrolysis of domestic waste and lead to a decrease in hydrogen yield [29]. Therefore, HRT must be adjusted to reach a compromise of both the biodegradability of the substrate employed and to promote a suitable environment for HPB.

Substrate concentration and Organic loading rate (OLR)

The organic loading rate (OLR, g/(L d)) defines the amount of organic matter fed per day per volume of fermenter according to the following equation (Eq. 5.7), where S is the substrate concentration in the inlet, and HRT is the hydraulic retention time:

$$OLR(g/Ld) = \frac{S(g/L)}{HRT(d)} \tag{5.7}$$

Increasing substrate concentration generally improves hydrogen yield and production rate and minimizes energy requirements [53]. [54] observed that increasing the initial starch concentration from 5 g_{COD}/L to 20 g_{COD}/L in batch reactors improved the hydrogen yield from 0.5 to 2.2 mol_{H2}/mol_{hexose}. Elreedy and co-authors worked in an anaerobic sequencing batch reactor (ASBR) with mixed culture at a psychrophilic temperature range from 15 to 25 °C using petrochemical wastewater containing mono-ethylene glycol (MEG) as substrate [55]. Similarly to [54], these authors found that the H_2 yield increased when the OLR increased from 134.32 ± 10.79 to 189.09 ± 22.35 mL_{H2}/g_{MEG} at 1 g_{COD}/(L d) and 4 g_{COD}/(L d), respectively [55]. However, a substrate concentration higher than the optimal value led to the inhibition of hydrogen production due to metabolite accumulation [56]. Changes in microbial composition were also observed in the case of substrate overloading. In this sense, it was pointed out that HPB were favored at an optimal OLR of 60 g_{COD}/(L d), resulting in a hydrogen yield of 1.09 mol_{H2}/mol_{hexose} added [57]. Nonetheless, a further increase to 120 g_{COD}/(L d) did not improve the hydrogen yield. Following this trend, another research carried out continuous anaerobic DF of two-phase olive mill residues in a mesophilic CSTR operated at increasing OLRs (up to 12.9 g_{COD}/(L d)) [58]. These authors observed a decrease in acetic acid (from 8 to 4 g/L) and hydrogen production (from 0.26 to 0.21 L_{H2}/(L d)) when the OLR increased from 9.6 to 12.9 g_{COD}/(L d). In this case, it was concluded that the highest OLR imposed exceeded the hydrolytic capacity of the system, worsening the process performance. However, there is no agreement regarding an optimal OLR value. For the same type of substrate (cheese whey) and identical conditions (CSTR, HRT 6 h, 37 °C, and pH 5.9), [56] observed an optimal OLR of 138.6 $g_{lactose}$/(L d) and a hydrogen yield of 2.8 $mol_{H2}/mol_{lactose}$. In contrast, [59] obtained an optimal OLR of 29 $g_{lactose}$/(L d) and a hydrogen yield of 2.17 mol_{H2}/mol_{hexose}. Therefore, as for HRT, an optimal OLR depends on the substrate nature, the microbiome in the inoculum, and the reactor configuration.

Bioreactor configuration

Reactors applied in DF can be classified according to the state of biomass: (i) suspended reactors (continuous stirred tank reactor (CSTR) and anaerobic sequencing batch reactor (ASBR)) or (ii) immobilized reactors (anaerobic fixed-bed reactor (AFBR) and up-flow anaerobic sludge blanket reactor (UASB)) [38]. CSTR is the most common suspended configuration assessed for hydrogen production, characterized by an HRT equal to the solid retention time (SRT) due to the complete mixing of biomass and substrate [60]. The dependence of SRT with HRT leads to biomass washout when too short HRTs are applied. Several strategies were developed to prevent cell washout and increase biomass retention, such as ASBR or immobilized reactors where SRT is decoupled from HRT. Kim and co-authors showed that optimal hydrogen production from food waste in ASBR (35 °C, pH 5.3) was achieved for 126 h SRT, and 33 h HRT (80.9 mL_{H2}/g_{VS} added) [61]. In CSTR with comparable conditions (household solid waste, pH 4.8–5.2, 37 °C, HRT and SRT of 48 h), it was shown a lower yield of 43 mL H_2/g VS added [62]. The higher hydrogen yield obtained in the immobilized reactor was explained by higher biomass concentrations, which confers more resistance to substrate inhibition at high OLR [63]. Following this trend, other authors compared CSTR and a continuously mixed immobilized sludge reactor (CMISR, with active carbon as a carrier, similar to AFBR) on hydrolyzed waste pastry at 35 °C and acidic pH (pH > 4). The maximum production rate of the CMISR (320 $mL_{H2}/(L\ h)$) remained stable at higher OLRs (32 and 40 $g_{COD}/(L\ d)$) compared to the CSTR (maximum at 277 $mL_{H2}/(L\ h)$ and OLR of 24 $g_{COD}/(L\ d)$ [64]. Furthermore, biomass can also be immobilized in biofilm or granular forms without external carriers. Specifically, UASB reactors based on granular biomass retention with a biomass blanket zone above a sludge bed were reported to exhibit better hydrogen production stability and substrate conversion stability at high organic loading rates concerning CSTR [65, 66]. Gavala et al. [66] observed that UASB (liquid upward velocity of 450 cm/h) at 35 °C and HRT of 2 h was more suitable for hydrogen production (19 $mmol_{H2}/(L\ h)$) than CSTR (8 $mmol_{H2}/(L\ h)$) operated under similar conditions. Nonetheless, the same study showed that the hydrogen yield was higher in the CSTR than the UASB for all HRT applied (respectively, an average yield of 1.7 $mol_{H2}/mol_{glucose}$ and 0.7 $mol_{H2}/mol_{glucose}$). For this reason, reactor configuration should carefully be chosen regarding the specific application. In general, immobilized reactors are now preferred over suspended configurations, although technical obstacles remain to be overcome. Further, coming studies should focus on hydrogen consumption, competitive bacteria growth in continuous operation, and the long startup periods needed for biofilm development and stabilization (e.g., in UASB) [67]. Therefore, the choice of reactor configuration must consider the differences between technical efficiency linked to hydrogen yield or economic efficiency related to production rate, as well as SRT and HRT that impact rector size and costs [66].

Inoculum source and pretreatments

The inoculum source in pure or mixed cultures can partially explain the differences observed in DF investigations. Several pretreatments have been applied on mixed cultures to favour hydrogen producers, such as heat-shock, acid, and alkaline treatments. Heat shock is the most common pretreatment applied because of its simplicity of execution and efficiency. [20] reviewed that the common temperature range and duration range for inoculum heat treatment varied between 65–100 °C and 15–120 min. [68] compared different heat shock strategies (T1: 90 °C for 10 min, T2: 105 °C for 120 min, and T3: 121 °C for 20 min) and showed that the T1 strategy best favoured the emergence of hydrogen producers (maximum yield of 4.75 $mmol_{H2}/g_{COD}$ at pH 6, compared with 2.78 and 4.47 $mmol_{H2}/g_{COD}$ for T2 and T3, respectively).

Regarding acid and alkali pretreatment methods, the aim is to apply extreme pH conditions on the inoculum. [69] observed that acid and alkali pretreatment on anaerobic seed sludge had a better positive influence on hydrogen production from food waste (similar yields of 150 mL_{H2}/g_{VS}) compared to heat-treatment (130 mL_{H2}/g_{VS}). Therefore, inoculum pretreatment methods seem to have different impacts on the inoculum source and microbial diversity. One of the biggest knowledge gaps regarding this topic is their suitability in a long-term operation. In this sense, some authors have pointed out the short-term effect that these pretreatments might have

[19, 70]. For this reason, combining them with appropriate operational conditions (short HRT) is recomended, which in turn might contribute to avoiding the growth of undesired microorganisms.

5.3 Microbial Communities in Dark Fermentation

Organic matter degradation in DF is driven by the interconnections between microbial species in absence of both oxygen and light. The microbial community of DF is composed of strictly anaerobic and facultative microorganisms. Bacteria and archaea capable of producing hydrogen are normally present in natural anaerobic environments such as wastewater treatment plants, AD facilities, and cow's rumen, among others [71]. Most microorganisms identified as HPB belong to the bacteria domain and have been deployed in DF either in pure or mixed cultures [72].

On the one hand, pure cultures often render more predictable hydrogen yields due to the presence of a sole bacterial strain in the reactor [73]. On the other hand, these systems show less robustness and higher sensitivity to disturbances due to potential contaminations [74]. Another important drawback of pure cultures is the high costs attributed to the culture requirements and maintenance. Substrate sterilization, together with the need for specific compounds for culture preparation, should be taken into account [75]. For this reason, waste cannot be used as substrates in pure cultures unless sterilized before DF. In contrast, no specific media is needed for mixed cultures, reducing the costs and effort for its maintenance. Mixed cultures can handle better the input of indigenous bacteria from the substrate, which makes it possible to directly use waste without sterilization [76]. However, using mixed cultures can make process control and hydrogen yield prediction more difficult.

Microorganisms can be divided into three different types when using mixed cultures in DF: (i) HPB, (ii) non-hydrogen producing bacteria, which can collaborate or compete with the HPB, and (iii) hydrogen-consuming microorganisms. Table 5.1 summarizes the main reactions that could take place during DF of the organic matter by mixed cultures.

Hydrogen producing bacteria (HPB)

As mentioned, mesophilic temperatures are often chosen for DF since they have been proven to be less costly to scale up the process [36]. Besides, most HPB bacteria described in the literature are mesophilic (optimal growth temperature between 25 °C and 45 °C) [80]. However, the temperatures tested to unravel bacterial roles have not been limited to the mesophilic range. Some psychrophilic microorganisms (optimal growth temperature between 0 and 25 °C) producing H_2 have also been identified [55]. Psychrophilic temperatures have usually been dismissed due to the lower yields associated with lower microbial growth rates. In contrast, high temperatures have been assessed to improve hydrogen yields. For instance, [81] cultivated *Thermotoga maritima* and *Thermotoga neapolitana* in pure batch cultures, using glucose as substrate and an incubation temperature of 80 °C [81]. These authors

Table 5.1 Principal reactions and microorganisms that take place in DF process when working with mixed cultures [37, 77–79]

Eqs	Reaction	Hydrogen producing reactions	Main microorganisms
8	Hydrogen production	$C_6H_{12}O_6 + H_2O \rightarrow CH_3CH_2OH + CH_3COOH + 2\,CO_2 + 2\,H_2$	*Clostridium* sp. (HPB)
9	Hydrogen production	$4\,C_6H_{12}O_6 + 2\,H_2O \rightarrow 3\,CH_3CH_2CH_2COOH + 2\,CH_3COOH + 8\,CO_2 + 10\,H_2$	*Enterobacteriaceae* (HPB)
		Reactions competing with hydrogen production	
10	Heterolactic fermentation	$C_6H_{12}O_6 \rightarrow C_3H_6O_3 + 1.5\,CH_3COOH$	Heterolactic bacteria
11	Heterolactic fermentation	$C_6H_{12}O_6 \rightarrow C_3H_6O_3 + CH_3CH_2OH + CO_2$	Heterolactic bacteria
12	Homolactic fermentation	$C_6H_{12}O_6 \rightarrow 2\,C_3H_6O_3$	Homolactic bacteria
13	Propionate production	$C_6H_{12}O_6 + 2\,H_2 \rightarrow 2\,CH_3CH_2COOH + 2\,H_2O$	Propionic bacteria
14	Propionate and acetate production	$1.5\,C_6H_{12}O_6 \rightarrow 2\,CH_3CH_2COOH + CH_3COOH + CO_2 + H_2O$	Propionic bacteria
15	Ethanol fermentation	$C_6H_{12}O_6 \rightarrow 2\,CH_3CH_2OH + 2\,CO_2$	*Enterobacteriaceae* (HPB)
16	Butanol production	$C_6H_{12}O_6 \rightarrow CH_3CH_2CH_2CH_2OH + 2\,CO_2 + H_2O$	*Clostridium* sp. (HPB)
		Reactions consuming hydrogen	
17	Sulfate reduction	$SO_4^{2-} + 4\,H_2 \rightarrow H_2S + 4\,H_2O$	Sulfate-reducing bacteria (SRB)
18	Hydrogenotrophic methanogenesis	$4\,H_2 + CO_2 \rightarrow CH_4 + 2\,H_2O$	Hydrogenotrophic methanogenic archaea
19	Homoacetogenesis	$4\,H_2 + 2\,CO_2 \rightarrow CH_3COOH + 2\,H_2O$	Homoacetogenic bacteria

observed H_2 yields of 3.2 and 3.8 mol H_2/mol glucose for *T. maritima* and *T. neapolitana*, respectively. Similar H_2 yields were observed for both strains when using xylose and arabinose as substrate (2.7 and 3.2 $mol_{H2}/mol_{glucose\text{-}eq}$, respectively, for *T. maritima* and 3.4 and 3.8 $mol_{H2}/mol_{glucose\text{-}eq}$, respectively for *T. neapolitana*), which evidenced the affinity of these species for simple sugars. Belonging to the archaea domain, the species *Pyrococcus furiosus* can reach H_2 yields of 3.8 $mol_{H2}/mol_{glucose}$ consumed [82]. Despite the high hydrogen yields, the hydrogen productivities of both bacteria and archaea under thermophilic conditions are lower than those obtained by mesophilic microorganisms [36, 80, 83]. For this reason and considering the net energy gain, mesophilic temperatures are preferred for most applications. Regarding the mesophilic temperature range (25 °C–45 °C), it has been widely reported that the microorganisms related to high hydrogen yields belong to the *Clostridium* genus [84]. Several *Clostridium* species, such as *Clostridium butyricum, Clostridium pasterianum, Clostridium beijerinckii,* and *Clostridium bifermentas,* have been described as efficient HPB [84]. This is mainly because *Clostridium* sp. can ferment several types of carbohydrates via the PFOR pathway (Eq. 5.1) [71]. Besides, this strict anaerobic bacteria genus can form spores, allowing them to endure hostile environmental conditions for survival (e.g., extremely acidic or basic pH, dryness, and high/low temperatures) [72]. This particular feature is widely employed in practice to select *Clostridium* sp. in mixed cultures by applying different pre-treatments to the microbial inocula (see previous section) [19]). Furthermore, *Clostridium* species might also be influenced by environmental conditions (e.g., hydrogen partial pressure) [75]. For instance, hydrogen accumulation in the reactor headspace can lead to a shift from acetate to butyrate production at hydrogen partial pressures as low as 60.7 Pa [6]. At such values, the cell needs to re-oxidize the reducing equivalents by producing more reduced products, such as butyrate. This metabolic shift drops the maximum theoretical hydrogen to 2 mol of H_2/mol of glucose instead of 4 $mol_{H2}/mol_{glucose}$ (Eqs. 5.1 and 5.2, respectively). Additionally, VFAs accumulation could cause a pH drop, activating solventogenesis in some *Clostridium* species, such as *C. acetobutylicum* (Eq. 5.10 and 5.11, Table 5.1, for ethanol and butanol production, respectively) [85]. Herein, bacteria switch to a hydrogen-consuming metabolism giving. As a result, the production of acetone, butanol or ethanol reduces the hydrogen yield.

Besides bacteria belonging to the *Clostridium* genus, other non-spore forming strict anaerobic bacteria can produce hydrogen via the PFOR pathway (Eq. 5.1). Among these bacteria, *Acetanaerobacterium elongatum, Ethanoligenens harbinense,* and *Megasphaera* species have been reported as HPB in reactors treating different substrates, such as molasses, sucrose-based wastewaters and vinasses from the sugarcane industry [86–88]. In particular, *Ethanoligenens harbinense* was described as efficient HPB at low-pH values (pH between 3 and 4.5 and temperature between 35–37 °C), achieving yields of 3.4 ± 0.7 mol H_2/mol of sucrose consumed (~8.9 $mmol_{H2}/g_{COD}$) [87, 88]. Similarly, *Megasphaera* species was related to intermediate hydrogen yields (ranging from 0.8 $mmol_{H2}/g_{COD}$ to 2 $mmol_{H2}/g_{COD}$) [89]. In particular, *Megasphaera* species such as *Megasphaera elsdenii* were suspected to act as lactate-consuming bacteria producing hydrogen and butyrate [90].

Additionally, several facultative bacteria were also identified as HPB. Facultative bacteria can use the available oxygen either for their catabolism or for the fermentation of organic matter in an anaerobic environment. Facultative bacteria produce hydrogen via the PFL pathway, with a maximum theoretical yield of 2 $mol_{H2}/mol_{glucose}$ [71]. Bacteria belonging to the *Enterobacteriaceae* family, such as *Enterobacter* sp., *Escherichia* sp., *Klebsiella* sp., and *Citrobacter* sp., are known as facultative HPB [72]. The capability of producing hydrogen has been studied in several *Citrobacter* sp., attaining a maximum hydrogen yield of 1.2 $mol_{H2}/mol_{glucose}$ consumed [91, 92]. Following this trend, some bacteria belonging to the *Bacillus* genus (*Bacillus cereus, Bacillus amyloliquefaciens, Bacillus coagulans, Bacillus cloacae, Bacillus macerans, Bacillus licheniformis, and Bacillus polymyxa*) have been reported to produce hydrogen via the PFL pathway, achieving yields of around 2 mol H_2/mol glucose consumed [93, 94]. Despite the lower hydrogen yields obtained compared to those achieved by strictly anaerobic bacteria, especially *Clostridium* species, the advantage of facultative microorganisms is their ability to consume oxygen, creating an anaerobic environment that can eventually favour hydrogen production by strictly anaerobic microorganisms [93]. In addition, their rapid growth rate and hydrolytic capabilities make this group of bacteria interesting for process extension [71].

Non hydrogen-producing bacteria

Non-hydrogen producing bacteria can collaborate with HPB, mostly enhancing hydrogen production, or outcompete them for organic resources. Depending on the environmental conditions, some bacteria in the community play an ambiguous role in hydrogen production or consumption. Hydrolysis remains the limiting step in DF when working with complex substrates, such as lignin, starch, or different types of food waste, because hydrogen is produced from simple sugars [95]. Several microorganisms, such as *Bacillus* sp., *Bifidobacterium* sp., and some *Clostridium* sp., also described as HPB, were reported to release exo-enzymes, which contribute to hydrolyze the complex substrates [96]. Hence, the role of these bacteria is not only reduced to hydrogen production since they also participate in breaking down complex organic matter. On the other hand, some bacteria may outcompete the HPB for their substrate. As observed in Table 5.1, some of the reactions occurring during DF (Eq. 5.10 to 5.16) with mixed cultures do not yield hydrogen, and their final products cannot further be transformed into hydrogen. For instance, propionate producers, mostly belonging to the genera *Propionibacterium, Propionispira,* or *Schwartzia* sp., hinder hydrogen production (Eq. 5.13 and 5.14, Table 5.1) by competing with the HPB for substrate and reducing equivalents but also by consuming hydrogen as electron donor [71]. These microorganisms can be avoided by employing temperatures higher than 37 °C and HRT > 6–10h [56, 97].

In addition, ethanol can be generated from sugars without hydrogen production by bacteria belonging to the Enterobacteriales order (Eq. 5.15, Table 5.1) [98]. As already mentioned, the solventogenesis reaction, which is often correlated to some metabolic stress, such as a sudden pH decrease caused by VFAs accumulation, could use the reducing equivalents produced during glycolysis to form acetone, butanol, or ethanol in strictly anaerobic microorganisms [28, 99].

Finally, the lactic acid bacteria (LAB) exhibit an ambiguous role in hydrogen production via DF. LAB are ubiquitous bacteria that ferment sugars and produce principally lactic acid. These bacteria are divided into two types, (i) the homolactic bacteria, which produce only lactic acid as a fermentation product (Eq. 5.12, Table 5.1), and (ii) the heterolactic bacteria, which can produce lactic acid and acetate or lactic acid and ethanol as fermentation products (Eq. 5.10 and 5.11, Table 5.1, respectively) [78]. These bacteria are often related to process instability and are considered substrate competitors to HPB (according to Eq. 5.10–5.12 from Table 5.1) [75]. *Lactobacillus*, *Sporolactobacillus*, or *Streptococcus* sp. were reported to hinder hydrogen production from sugars, as the lactic acid pathway from glucose does not release hydrogen [100].

Species from the genus *Lactobacillus* sp. produce bacteriocins that may inhibit *Clostridium* sp., also contributing to process instability [53]. Lactic acid accumulation is often related to fermentation failure and could generate a harsh environment for the HPB (at pH values close to 3 or 4) [53]. Nonetheless, there is no final agreement regarding the detrimental role of LAB. In fact, some authors reported that the presence of LAB could benefit hydrogen production [89, 101]. On the one hand, the LAB are facultative bacteria that could consume oxygen, generating an anaerobic environment for the strictly anaerobic bacteria. Besides, it has been reported that LAB can produce hydrolytic enzymes that could break down complex organic matter, increasing substrate availability for HPB [102, 103].

Moreover, hydrogen can be produced from lactic acid via the acrylate and the PFOR pathways [104]. In some cases, acetate can be used as a supplementary electron acceptor during lactic acid fermentation, as the hydrogenases have a lower activity when fermenting lactic acid than glucose [105]. It was reported that some bacteria from the *Clostridium* genus could consume lactic acid and acetate to produce butyrate and hydrogen [106]. For instance, [107] found a positive correlation between the consumption of lactic acid and acetate and the increasing abundance of the species *Clostridium sp., C. beijerinckii* and *C. butyricum.*

Hydrogen consuming-microorganisms

Three principal groups of microorganisms can directly consume molecular hydrogen in DF reactors when working with mixed cultures: sulphate-reducing bacteria, hydrogenotrophic methanogenic archaea, and homoacetogenic bacteria. In Table 5.1, the reactions produced by these microorganisms are shown in Eqs. 5.17, 5.18, and 5.19, respectively.

Sulphate-reducing bacteria have a high affinity for hydrogen, being able to metabolize it at extremely low concentrations of 0.1 Pa. However, controlling the pH at values lower than 6 has been proven to avoid effectively hydrogen consumption by these bacteria [108].

Hydrogen consumption by methanogenic archaea has been widely reported in the literature. Hydrogenotrophic methanogenic archaea consume hydrogen and carbon dioxide to produce methane. Several operational strategies have been used to avoid hydrogen consumption by methanogens (e.g., pretreatments and low HRTs, see the previous Section). Even though these strategies are often employed in combination to avoid methane production, sometimes they are not completely effective, and methane can still be detected, probably because hydrogenotrophic methanogens are more resilient than acetoclastic methanogens [109, 110].

The third group of microorganisms that consume hydrogen and the most problematic is the homoacetogenic bacteria. These bacteria can produce acetate from hydrogen and carbon dioxide. The homoacetogenic bacteria is not a phylogenetic group. Hence several bacteria from different phyla have been identified as homoacetogenic [111]. For instance, some *Clostridium* species, namely *C. aceticum,*

Clostridium thermoautotrophicum, Clostridium thermoaceticum, Clostridium stercorarium, and *C. ljungdahlii* were identified as homoacetogens. Some of these homoacetogens can produce hydrogen or consume it due to the bi-directional hydrogenases that they own [112, 113]. Indeed, the homoacetogens identification is complex as it seems to be a reaction independent of the temperature, pH, or microbial inoculum and is even enhanced in high cell density reactors [42, 114].

Moreover, the produced metabolite is acetate, which is common to other DF reactions. Some theoretical equations have been proposed to identify homoacetogenesis activity [115, 116]. Nevertheless, the main cause of homoacetogenesis has been proposed to be the high hydrogen partial pressure in the reactors.

5.4 Conclusion

DF is a promising biological process to produce hydrogen due to its application versatility for carbohydrate-rich materials. However, this process suffers from low yields due to thermodynamic limitations and competitive metabolic pathways. Operational parameters such as pH, temperature, OLR, and HRT, as well as reactor configuration and the substrate selected, affect DF performance. Even though some trends were identified, the strong dependence on operational parameters makes experimental investigations of paramount importance to select working values. Factors such as hydrogen and carbon dioxide partial pressures and comparative studies dealing with substrate and inoculum pretreatments should be assessed to further improve the process efficiency. For this reason, DF optimization should include multifactor effects (operational conditions and microbiome) as well as economic, energy, and environmental aspects to ensure an economically-viable process at a larger scale.

Acknowledgements Jose Antonio Magdalena would like to thank the Complutense University of Madrid for the financing of his contract at LBE-INRAE (France), with funds from the Ministry of Universities for the requalification of the Spanish University System for 2021–2023 (Modality 1. Margarita Salas), coming from the European Union-Next generation EU funding. This work was carried out in the framework of the HyDS Project (part of the ERDF-REACT-EU- H2VERT project). Lucie Perat would like to thank ENGIE Lab CRIGEN and the Occitanie region (in the context of Défi Clé Hydrogène projects) for the funding of her PhD in LBE-INRAE (France) in the MELINBIOH project.

References

1. Ramírez-Morales JE, Tapia-Venegas E, Toledo-Alarcón J, Ruiz-Filippi G (2015) Simultaneous production and separation of biohydrogen in mixed culture systems by continuous dark fermentation. Water Sci Technol 71:1271–1285. https://doi.org/10.2166/wst.2015.104
2. Peters K, Sargent F (2023) Formate hydrogenlyase, formic acid translocation and hydrogen production: dynamic membrane biology during fermentation. Biochimica et Biophysica Acta (BBA)—Bioenergetics 1864:1–8. https://doi.org/10.1016/j.bbabio.2022.148919

3. Sokol KP, Robinson WE, Oliveira AR et al (2019) Reversible and selective interconversion of hydrogen and carbon dioxide into formate by a semiartificial formate hydrogenlyase mimic. J Am Chem Soc 141:17498–17502. https://doi.org/10.1021/jacs.9b09575
4. Morra S (2022) Fantastic [FeFe]-Hydrogenases and where to find them. Front Microbiol 13.https://doi.org/10.3389/fmicb.2022.853626
5. Peters JW, Schut GJ, Boyd ES et al (2015) [FeFe]- and [NiFe]-hydrogenase diversity, mechanism, and maturation. Biochimica et Biophysica Acta—Molecular Cell Res 1853:1350–1369. https://doi.org/10.1016/j.bbamcr.2014.11.021
6. Angenent LT, Karim K, Al-Dahhan MH et al (2004) Production of bioenergy and biochemicals from industrial and agricultural wastewater. Trends Biotechnol 22:477–485. https://doi.org/10.1016/j.tibtech.2004.07.001
7. Guo XM, Trably E, Latrille E et al (2010) Hydrogen production from agricultural waste by dark fermentation: a review. Int J Hydrogen Energy 35:10660–10673. https://doi.org/10.1016/j.ijhydene.2010.03.008
8. Monlau F, Barakat A, Trably E et al (2013) Lignocellulosic materials into biohydrogen and biomethane: impact of structural features and pretreatment. Crit Rev Environ Sci Technol 43:260–322. https://doi.org/10.1080/10643389.2011.604258
9. Guo X, Trably E, Latrille E et al (2014) Predictive and explicative models of fermentative hydrogen production from solid organic waste: role of butyrate and lactate pathways. Int J Hydrogen Energy 39:7476–7485. https://doi.org/10.1016/j.ijhydene.2013.08.079
10. Lay J-J, Fan K-S, Chang J-1, Ku C-H (2003) Influence of chemical nature of organic wastes on their conversion to hydrogen by heat-shock digested sludge. Int J Hydrogen Energy 28:1361–1367.https://doi.org/10.1016/S0360-3199(03)00027-2
11. Dong L, Zhenhong Y, Yongming S et al (2009) Hydrogen production characteristics of the organic fraction of municipal solid wastes by anaerobic mixed culture fermentation. Int J Hydrogen Energy 34:812–820. https://doi.org/10.1016/j.ijhydene.2008.11.031
12. Hanaki K, Matsuo T, Nagase M (1981) Mechanism of inhibition caused by long-chain fatty acids in anaerobic digestion process. Biotechnol Bioeng 23:1591–1610. https://doi.org/10.1002/bit.260230717
13. Reyna-Gómez LM, Molina-Guerrero CE, Alfaro JM et al (2019) Effect of carbon/nitrogen ratio, temperature, and inoculum source on hydrogen production from dark codigestion of fruit peels and sewage sludge. Sustainability 11:1–13. https://doi.org/10.3390/su11072139
14. Zhang Z, Zhang G, Li W et al (2016) Enhanced biogas production from sorghum stem by co-digestion with cow manure. Int J Hydrogen Energy 41:9153–9158. https://doi.org/10.1016/j.ijhydene.2016.02.042
15. Yasin NHM, Mumtaz T, Hassan MA, Abd Rahman N (2013) Food waste and food processing waste for biohydrogen production: a review. J Environ Manage 130:375–385. https://doi.org/10.1016/j.jenvman.2013.09.009
16. Bai MD, Cheng SS, Chao YC (2004) Effects of substrate components on hydrogen fermentation of multiple substrates. Water Sci Technol 50:209–216
17. Anzola-Rojas M del P, Gonçalves da Fonseca S, Canedo da Silva C, et al (2015) The use of the carbon/nitrogen ratio and specific organic loading rate as tools for improving biohydrogen production in fixed-bed reactors. Biotechnol. Reports 5:46–54.https://doi.org/10.1016/j.btre.2014.10.010
18. Argun H, Kargi F, Kapdan IK, Oztekin R (2008) Biohydrogen production by dark fermentation of wheat powder solution: effects of C/N and C/P ratio on hydrogen yield and formation rate. Int J Hydrogen Energy 33:1813–1819. https://doi.org/10.1016/j.ijhydene.2008.01.038
19. Bundhoo MAZ, Mohee R, Hassan MA (2015) Effects of pre-treatment technologies on dark fermentative biohydrogen production: a review. J Environ Manage 157:20–48. https://doi.org/10.1016/j.jenvman.2015.04.006
20. Rafieenia R, Lavagnolo MC, Pivato A (2018) Pre-treatment technologies for dark fermentative hydrogen production: current advances and future directions. Waste Manage 71:734–748. https://doi.org/10.1016/j.wasman.2017.05.024

21. Pagliaccia P, Gallipoli A, Gianico A et al (2016) Single stage anaerobic bioconversion of food waste in mono and co-digestion with olive husks: impact of thermal pretreatment on hydrogen and methane production. Int J Hydrogen Energy 41:905–915. https://doi.org/10.1016/j.ijhydene.2015.10.061
22. Cui M, Shen J (2012) Effects of acid and alkaline pretreatments on the biohydrogen production from grass by anaerobic dark fermentation. Int J Hydrogen Energy 37:1120–1124. https://doi.org/10.1016/j.ijhydene.2011.02.078
23. Monlau F, Sambusiti C, Barakat A et al (2014) Do furanic and phenolic compounds of lignocellulosic and algae biomass hydrolyzate inhibit anaerobic mixed cultures? A comprehensive review. Biotechnol Adv 32:934–951. https://doi.org/10.1016/j.biotechadv.2014.04.007
24. Quéméneur M, Hamelin J, Barakat A et al (2012) Inhibition of fermentative hydrogen production by lignocellulose-derived compounds in mixed cultures. Int J Hydrogen Energy 37:3150–3159. https://doi.org/10.1016/j.ijhydene.2011.11.033
25. Toledo-Alarcón J, Capson-Tojo G, Marone A et al (2018) Basics of bio-hydrogen production by dark fermentation. In: Liao Q, Chang J, Herrmann C, Xia A (eds) Bioreactors for microbial biomass and energy conversion. Springer, Singapore, pp 199–220
26. Khanal S, Chen, W-H, Li, L, Sung,S (2003) Biological hydrogen production: effects of pH and intermediate products. Int J Hydrogen Energy 1–9.https://doi.org/10.1016/j.ijhydene.2003.11.002
27. Serebryakova LT, Sheremetieva ME (2006) Characterization of catalytic properties of hydrogenase isolated from the unicellular cyanobacterium *Gloeocapsa alpicola* CALU 743. Biochem Mosc 71:1370–1376. https://doi.org/10.1134/S0006297906120133
28. Van Ginkel S, Logan BE (2005) Inhibition of biohydrogen production by undissociated acetic and butyric acids. Environ Sci Technol 39:9351–9356. https://doi.org/10.1021/es0510515
29. Liu D, Zeng RJ, Angelidaki I (2008) Effects of pH and hydraulic retention time on hydrogen production versus methanogenesis during anaerobic fermentation of organic household solid waste under extreme-thermophilic temperature (70 °C). Biotechnol Bioeng 100:1108–1114. https://doi.org/10.1002/bit.21834
30. Levin DB, Pitt L, Love M (2004) Biohydrogen production: prospects and limitations to practical application. Int J Hydrogen Energy 29:173–185. https://doi.org/10.1016/S0360-3199(03)00094-6
31. Azman NF, Abdeshahian P, Kadier A et al (2016) Biohydrogen production from de-oiled rice bran as sustainable feedstock in fermentative process. Int J Hydrogen Energy 41:145–156. https://doi.org/10.1016/j.ijhydene.2015.10.018
32. Koesnandar NN, Yamamoto A, Nagai S (1991) Enzymatic reduction of cystine into cysteine by cell-free extract of *Clostridium thermoaceticum.* J Ferment Bioeng 72:11–14. https://doi.org/10.1016/0922-338X(91)90138-7
33. Sattar A, Arslan C, Ji C et al (2016) Quantification of temperature effect on batch production of bio-hydrogen from rice crop wastes in an anaerobic bio reactor. Int J Hydrogen Energy 41:11050–11061. https://doi.org/10.1016/j.ijhydene.2016.04.087
34. De Gioannis G, Muntoni A, Polettini A, Pomi R (2013) A review of dark fermentative hydrogen production from biodegradable municipal waste fractions. Waste Manage 33:1345–1361. https://doi.org/10.1016/j.wasman.2013.02.019
35. Shin H-S, Youn J-H, Kim S-H (2004) Hydrogen production from food waste in anaerobic mesophilic and thermophilic acidogenesis. Int J Hydrogen Energy 29:1355–1363. https://doi.org/10.1016/j.ijhydene.2003.09.011
36. Ananthi V, Ramesh U, Balaji P, et al (2022) A review on the impact of various factors on biohydrogen production. Int J Hydrogen Energy 1–13.https://doi.org/10.1016/j.ijhydene.2022.08.046
37. Hawkes FR, Hussy I, Kyazze G et al (2007) Continuous dark fermentative hydrogen production by mesophilic microflora: principles and progress. Int J Hydrogen Energy 32:172–184. https://doi.org/10.1016/j.ijhydene.2006.08.014
38. Jung K-W, Kim D-H, Kim S-H, Shin H-S (2011) Bioreactor design for continuous dark fermentative hydrogen production. Biores Technol 102:8612–8620. https://doi.org/10.1016/j.biortech.2011.03.056

39. Chen Y, Yin Y, Wang J (2021) Recent advance in inhibition of dark fermentative hydrogen production. Int J Hydrogen Energy 46:5053–5073. https://doi.org/10.1016/j.ijhydene.2020.11.096
40. Hallenbeck PC (2005) Fundamentals of the fermentative production of hydrogen. Water Sci Technol 52:21–29. https://doi.org/10.2166/wst.2005.0494
41. Ghimire A, Frunzo L, Pirozzi F et al (2015) A review on dark fermentative biohydrogen production from organic biomass : process parameters and use of by-products. Appl Energy 144:73–95. https://doi.org/10.1016/j.apenergy.2015.01.045
42. Saady NMC (2013) Homoacetogenesis during hydrogen production by mixed cultures dark fermentation: Unresolved challenge. Int J Hydrogen Energy 38:13172–13191. https://doi.org/10.1016/j.ijhydene.2013.07.122
43. Chang S, Li J, Liu F, Yu Z (2012) Effect of different gas releasing methods on anaerobic fermentative hydrogen production in batch cultures. Front Environ Sci Eng 6:901–906. https://doi.org/10.1007/s11783-012-0403-1
44. Kisielewska M, Dębowski M, Zieliński M (2015) Improvement of biohydrogen production using a reduced pressure fermentation. Bioprocess Biosyst Eng 38:1925–1933. https://doi.org/10.1007/s00449-015-1434-3
45. Mizuno O, Dinsdale R, Hawkes FR et al (2000) Enhancement of hydrogen production from glucose by nitrogen gas sparging. Biores Technol 73:59–65. https://doi.org/10.1016/S0960-8524(99)00130-3
46. Willquist K, Claassen PAM, van Niel E (2009) Evaluation of the influence of CO_2 on hydrogen production by *Caldicellulosiruptor saccharolyticus*. Int J Hydrogen Energy 34:4718–4726. https://doi.org/10.1016/j.ijhydene.2009.03.056
47. Júnior ADNF, Pages C, Latrille E et al (2020) Biogas sequestration from the headspace of a fermentative system enhances hydrogen production rate and yield. Int J Hydrogen Energy 45:11011. https://doi.org/10.1016/j.ijhydene.2020.02.064
48. Siriwongrungson V, Zeng RJ, Angelidaki I (2007) Homoacetogenesis as the alternative pathway for H_2 sink during thermophilic anaerobic degradation of butyrate under suppressed methanogenesis. Water Res 41:4204–4210. https://doi.org/10.1016/j.watres.2007.05.037
49. Kraemer JT, Bagley DM (2006) Supersaturation of dissolved H_2 and CO_2 during fermentative hydrogen production with N_2 Sparging. Biotechnol Lett 28:1485–1491. https://doi.org/10.1007/s10529-006-9114-7
50. Palomo-Briones R, Razo-Flores E, Bernet N, Trably E (2017) Dark-fermentative biohydrogen pathways and microbial networks in continuous stirred tank reactors: Novel insights on their control. Appl Energy 198:77–87. https://doi.org/10.1016/j.apenergy.2017.04.051
51. Chang J-J, Wu J-H, Wen F-S et al (2008) Molecular monitoring of microbes in a continuous hydrogen-producing system with different hydraulic retention time. Int J Hydrogen Energy 33:1579–1585. https://doi.org/10.1016/j.ijhydene.2007.09.045
52. Ueno Y, Sasaki D, Fukui H et al (2006) Changes in bacterial community during fermentative hydrogen and acid production from organic waste by thermophilic anaerobic microflora. J Appl Microbiol 101:331–343. https://doi.org/10.1111/j.1365-2672.2006.02939.x
53. Elbeshbishy E, Dhar BR, Nakhla G, Lee H-S (2017) A critical review on inhibition of dark biohydrogen fermentation. Renew Sustain Energy Rev 79:656–668. https://doi.org/10.1016/j.rser.2017.05.075
54. Lin C-Y, Chang C-C, Hung C-H (2008) Fermentative hydrogen production from starch using natural mixed cultures. Int J Hydrogen Energy 33:2445–2453. https://doi.org/10.1016/j.ijhydene.2008.02.069
55. Elreedy A, Fujii M, Tawfik A (2019) Psychrophilic hydrogen production from petrochemical wastewater via anaerobic sequencing batch reactor: techno-economic assessment and kinetic modelling. Int J Hydrogen Energy 44:5189–5202. https://doi.org/10.1016/j.ijhydene.2018.09.091
56. Davila-Vazquez G, Cota-Navarro CB, Rosales-Colunga LM et al (2009) Continuous biohydrogen production using cheese whey: improving the hydrogen production rate. Int J Hydrogen Energy 34:4296–4304. https://doi.org/10.1016/j.ijhydene.2009.02.063

57. Kim S-H, Han S-K, Shin H-S (2006) Effect of substrate concentration on hydrogen production and 16S rDNA-based analysis of the microbial community in a continuous fermenter. Process Biochem 41:199–207. https://doi.org/10.1016/j.procbio.2005.06.013
58. Rincón B, Sánchez E, Raposo F et al (2008) Effect of the organic loading rate on the performance of anaerobic acidogenic fermentation of two-phase olive mill solid residue. Waste Manage 28:870–877. https://doi.org/10.1016/j.wasman.2007.02.030
59. Palomo-Briones R, Trably E, López-Lozano NE et al (2018) Hydrogen metabolic patterns driven by *Clostridium-Streptococcus* community shifts in a continuous stirred tank reactor. Appl Microbiol Biotechnol 102:2465–2475. https://doi.org/10.1007/s00253-018-8737-7
60. Show K-Y, Lee D-J, Chang J-S (2011) Bioreactor and process design for biohydrogen production. Biores Technol 102:8524–8533. https://doi.org/10.1016/j.biortech.2011.04.055
61. Kim S-H, Han S-K, Shin H-S (2008) Optimization of continuous hydrogen fermentation of food waste as a function of solids retention time independent of hydraulic retention time. Process Biochem 43:213–218. https://doi.org/10.1016/j.procbio.2007.11.007
62. Liu D, Liu D, Zeng RJ, Angelidaki I (2006) Hydrogen and methane production from household solid waste in the two-stage fermentation process. Water Res 40:2230–2236. https://doi.org/10.1016/j.watres.2006.03.029
63. Carolin Christopher F, Kumar PS, Vo D-VN, Joshiba GJ (2021) A review on critical assessment of advanced bioreactor options for sustainable hydrogen production. Int J Hydrogen Energy 46:7113–7136. https://doi.org/10.1016/j.ijhydene.2020.11.244
64. Han W, Wang B, Zhou Y et al (2012) Fermentative hydrogen production from molasses wastewater in a continuous mixed immobilized sludge reactor. Biores Technol 110:219–223. https://doi.org/10.1016/j.biortech.2012.01.057
65. Sinharoy A, Kumar M, Pakshirajan K (2020) Chapter 14—an overview of bioreactor configurations and operational strategies for dark fermentative biohydrogen production. In: Singh L, Yousuf A, Mahapatra DM (eds) Bioreactors. Elsevier, pp 249–288
66. Gavala HN, Skiadas IV, Ahring BK (2006) Biological hydrogen production in suspended and attached growth anaerobic reactor systems. Int J Hydrogen Energy 31:1164–1175. https://doi.org/10.1016/j.ijhydene.2005.09.009
67. Kumar G, Mudhoo A, Sivagurunathan P et al (2016) Recent insights into the cell immobilization technology applied for dark fermentative hydrogen production. Biores Technol 219:725–737. https://doi.org/10.1016/j.biortech.2016.08.065
68. Magrini FE, de Almeida GM, da Maia SD et al (2021) Effect of different heat treatments of inoculum on the production of hydrogen and volatile fatty acids by dark fermentation of sugarcane vinasse. Biomass Conv Bioref 11:2443–2456. https://doi.org/10.1007/s13399-020-00687-0
69. Luo L, Sriram S, Johnravindar D et al (2022) Effect of inoculum pretreatment on the microbial and metabolic dynamics of food waste dark fermentation. Biores Technol 358:1–9. https://doi.org/10.1016/j.biortech.2022.127404
70. Luo G, Karakashev D, Xie L et al (2011) Long-term effect of inoculum pretreatment on fermentative hydrogen production by repeated batch cultivations: homoacetogenesis and methanogenesis as competitors to hydrogen production. Biotechnol Bioeng 108:1816–1827. https://doi.org/10.1002/bit.23122
71. Cabrol L, Marone A, Tapia-Venegas E et al (2017) Microbial ecology of fermentative hydrogen producing bioprocesses: useful insights for driving the ecosystem function. FEMS Microbiol Rev 41:158–181. https://doi.org/10.1093/femsre/fuw043
72. Cao Y, Liu H, Liu W et al (2022) Debottlenecking the biological hydrogen production pathway of dark fermentation: insight into the impact of strain improvement. Microb Cell Fact 21:1–16. https://doi.org/10.1186/s12934-022-01893-3
73. Ergal İ, Fuchs W, Hasibar B et al (2018) The physiology and biotechnology of dark fermentative biohydrogen production. Biotechnol Adv 36:2165–2186. https://doi.org/10.1016/j.biotechadv.2018.10.005
74. Hovorukha V, Havryliuk O, Gladka G et al (2021) Hydrogen dark fermentation for degradation of solid and liquid food waste. Energies 14:1–12. https://doi.org/10.3390/en14071831

75. Castelló E, Nunes Ferraz-Junior AD, Andreani C et al (2020) Stability problems in the hydrogen production by dark fermentation: possible causes and solutions. Renew Sustain Energy Rev 119:1–16. https://doi.org/10.1016/j.rser.2019.109602
76. Tapia-Venegas E, Ramirez-Morales JE, Silva-Illanes F et al (2015) Biohydrogen production by dark fermentation: scaling-up and technologies integration for a sustainable system. Rev Environ Sci Biotechnol 14:761–785. https://doi.org/10.1007/s11157-015-9383-5
77. Wang Q, Li H, Feng K, Liu J (2020) Oriented fermentation of food waste towards high-value products: a review. Energies 13:1–29. https://doi.org/10.3390/en13215638
78. Madigan MT, Martinko JM, Bender KS, et al (2014) In: Brock biology of microorganisms, 14th edn. Pearson
79. Ferreira TB, Rego GC, Ramos LR et al (2018) Selection of metabolic pathways for continuous hydrogen production under thermophilic and mesophilic temperature conditions in anaerobic fluidized bed reactors. Int J Hydrogen Energy 43:18908–18917. https://doi.org/10.1016/j.ijhydene.2018.08.177
80. Wong YM, Wu TY, Juan JC (2014) A review of sustainable hydrogen production using seed sludge via dark fermentation. Renew Sustain Energy Rev 34:471–482. https://doi.org/10.1016/j.rser.2014.03.008
81. Eriksen NT, Riis ML, Holm NK, Iversen N (2011) H_2 synthesis from pentoses and biomass in *Thermotoga spp.* Biotech Lett 33:293–300. https://doi.org/10.1007/s10529-010-0439-x
82. Chou C-J, Shockley KR, Conners SB et al (2007) Impact of substrate glycoside linkage and elemental sulfur on bioenergetics of and hydrogen production by the hyperthermophilic archaeon *Pyrococcus furiosus*. Appl Environ Microbiol 73:6842–6853. https://doi.org/10.1128/AEM.00597-07
83. Pawar SS, Van Niel EWJ, Niel EWJV (2013) Thermophilic biohydrogen production: how far are we? Appl Microbiol Biotechnol 97:7999–8009. https://doi.org/10.1007/s00253-013-5141-1
84. Wang J, Yin Y (2021) *Clostridium* species for fermentative hydrogen production: an overview. Int J Hydrogen Energy 46:34599–34625. https://doi.org/10.1016/j.ijhydene.2021.08.052
85. Xue C, Cheng C (2019) Chapter two—butanol production by *Clostridium*. In: Li Y, Ge X (eds) Advances in bioenergy. Elsevier, pp 35–77
86. Ferraz Júnior ADN, Etchebehere C, Zaiat M (2015) Highs organic loading rate on thermophilic hydrogen production and metagenomic study at an anaerobic packed-bed reactor treating a residual liquid stream of a Brazilian biorefinery. Biores Technol 186:81–88. https://doi.org/10.1016/j.biortech.2015.03.035
87. Mota VT, Ferraz Júnior ADN, Trably E, Zaiat M (2018) Biohydrogen production at pH below 3.0: Is it possible? Water Res 128:350–361. https://doi.org/10.1016/j.watres.2017.10.060
88. Ren N, Xing D, Rittmann BE et al (2007) Microbial community structure of ethanol type fermentation in bio-hydrogen production. Environ Microbiol 9:1112–1125. https://doi.org/10.1111/j.1462-2920.2006.01234.x
89. Etchebehere C, Castelló E, Wenzel J et al (2016) Microbial communities from 20 different hydrogen-producing reactors studied by 454 pyrosequencing. Appl Microbiol Biotechnol 100:3371–3384. https://doi.org/10.1007/s00253-016-7325-y
90. Ohnishi A, Hasegawa Y, Fujimoto N, Suzuki M (2021) Biohydrogen production by Mixed Culture of *Megasphaera elsdenii* with lactic acid bacteria as lactate-driven dark fermentation. Biores Technol 343:1–9. https://doi.org/10.1016/j.biortech.2021.126076
91. Oh Y-K, Kim H-J, Park S et al (2008) Metabolic-flux analysis of hydrogen production pathway in *Citrobacter amalonaticus* Y19. Int J Hydrogen Energy 33:1471–1482. https://doi.org/10.1016/j.ijhydene.2007.09.032
92. Hamilton C, Hiligsmann S, Beckers L et al (2010) Optimization of culture conditions for biological hydrogen production by *Citrobacter freundii* CWBI952 in batch, sequenced-batch and semicontinuous operating mode. Int J Hydrogen Energy 35:1089–1098. https://doi.org/10.1016/j.ijhydene.2009.10.073
93. Łukajtis R, Holowacz I, Kucharska K et al (2018) Hydrogen production from biomass using dark fermentation. Renew Sustain Energy Rev 91:665–694. https://doi.org/10.1016/j.rser.2018.04.043

94. Kumar P, Patel SKS, Lee J-K, Kalia VC (2013) Extending the limits of *Bacillus* for novel biotechnological applications. Biotechnol Adv 31:1543–1561. https://doi.org/10.1016/j.biotechadv.2013.08.007
95. Mugnai G, Borruso L, Mimmo T et al (2021) Dynamics of bacterial communities and substrate conversion during olive-mill waste dark fermentation: prediction of the metabolic routes for hydrogen production. Biores Technol 319:1–12. https://doi.org/10.1016/j.biortech.2020.124157
96. Hung C-H, Chang Y-T, Chang Y-J (2011) Roles of microorganisms other than clostridium and enterobacter in anaerobic fermentative biohydrogen production systems–a review. Bioresour Technol 102:8437–8444. https://doi.org/10.1016/j.biortech.2011.02.084
97. Sivagurunathan P, Sen B, Lin CY (2014) Overcoming propionic acid inhibition of hydrogen fermentation by temperature shift strategy. Int J Hydrogen Energy 39:19232–19241. https://doi.org/10.1016/j.ijhydene.2014.03.260
98. Zhou M, Yan B, Wong JWC, Zhang Y (2018) Enhanced volatile fatty acids production from anaerobic fermentation of food waste: a mini-review focusing on acidogenic metabolic pathways. Biores Technol 248:68–78. https://doi.org/10.1016/j.biortech.2017.06.121
99. Dessì P, Porca E, Frunzo L et al (2018) Inoculum pretreatment differentially affects the active microbial community performing mesophilic and thermophilic dark fermentation of xylose. Int J Hydrogen Energy 43:9233–9245. https://doi.org/10.1016/j.ijhydene.2018.03.117
100. Chatellard L, Trably E, Carrère H (2016) The type of carbohydrates specifically selects microbial community structures and fermentation patterns. Biores Technol 221:541–549. https://doi.org/10.1016/j.biortech.2016.09.084
101. Sikora A, Błaszczyk M, Jurkowski M, et al (2013) Lactic acid bacteria in hydrogen-producing consortia: on purpose or by coincidence? In: Lactic acid bacteria—R & D for Food, health and livestock purposes. IntechOpen, pp 487–514
102. Chang J-J, Chou C-H, Ho C-Y et al (2008) Syntrophic co-culture of aerobic *Bacillus* and anaerobic *Clostridium* for bio-fuels and bio-hydrogen production. Int J Hydrogen Energy 33:5137–5146. https://doi.org/10.1016/j.ijhydene.2008.05.021
103. Li SL, Lin JS, Wang YH et al (2011) Strategy of controlling the volumetric loading rate to promote hydrogen-production performance in a mesophilic-kitchen-waste fermentor and the microbial ecology analyses. Biores Technol 102:8682–8687. https://doi.org/10.1016/j.biortech.2011.02.067
104. García-Depraect O, Castro-Muñoz R, Muñoz R, et al (2021) A review on the factors influencing biohydrogen production from lactate: the key to unlocking enhanced dark fermentative processes. Bioresource Technol 324:124595. https://doi.org/10.1016/j.biortech.2020.124595
105. Detman A, Mielecki D, Chojnacka A et al (2019) Cell factories converting lactate and acetate to butyrate: *Clostridium butyricum* and microbial communities from dark fermentation bioreactors. Microb Cell Fact 18:36. https://doi.org/10.1186/s12934-019-1085-1
106. García-Depraect O, Muñoz R, Rodríguez E et al (2021) Microbial ecology of a lactate-driven dark fermentation process producing hydrogen under carbohydrate-limiting conditions. Int J Hydrogen Energy 46:11284–11296. https://doi.org/10.1016/j.ijhydene.2020.08.209
107. García-Depraect O, Valdez-Vázquez I, Rene ER et al (2019) Lactate- and acetate-based biohydrogen production through dark co-fermentation of tequila vinasse and nixtamalization wastewater: metabolic and microbial community dynamics. Biores Technol 282:236–244. https://doi.org/10.1016/j.biortech.2019.02.100
108. Lin C, Chen H (2006) Sulfate effect on fermentative hydrogen production using anaerobic mixed microflora. Int J Hydrogen Energy 31:953–960. https://doi.org/10.1016/j.ijhydene.2005.07.009
109. Castelló E, García y Santos C, Iglesias T, et al (2009) Feasibility of biohydrogen production from cheese whey using a UASB reactor: links between microbial community and reactor performance. Int J Hydrogen Energy 34:5674–5682.https://doi.org/10.1016/j.ijhydene.2009.05.060
110. Demirel B, Scherer P (2008) The roles of acetotrophic and hydrogenotrophic methanogens during anaerobic conversion of biomass to methane: a review. Rev Environ Sci Biotechnol 7:173–190. https://doi.org/10.1007/s11157-008-9131-1

111. Schuchmann K, Müller V (2016) Energetics and application of heterotrophy in acetogenic bacteria. Appl Environ Microbiol 82:4056–4069. https://doi.org/10.1128/AEM.00882-16
112. Huang Y, Zong W, Yan X et al (2010) Succession of the bacterial community and dynamics of hydrogen producers in a hydrogen-producing bioreactor. Appl Environ Microbiol 76:3387–3390. https://doi.org/10.1128/AEM.02444-09
113. Westerholm M, Dolfing J, Schnürer A (2019) Growth characteristics and thermodynamics of syntrophic acetate oxidizers. Environ Sci Technol 53:5512–5520. https://doi.org/10.1021/acs.est.9b00288
114. Dinamarca C, Gañán M, Liu J, Bakke R (2011) H_2 consumption by anaerobic non-methanogenic mixed cultures. Water Sci Technol 63:1582–1589. https://doi.org/10.2166/wst.2011.214
115. Arooj MF, Han SK, Kim SH et al (2008) Effect of HRT on ASBR converting starch into biological hydrogen. Int J Hydrogen Energy 33:6509–6514. https://doi.org/10.1016/j.ijhydene.2008.06.077
116. Arooj MF, Han SK, Kim SH et al (2008) Continuous biohydrogen production in a CSTR using starch as a substrate. Int J Hydrogen Energy 33:3289–3294. https://doi.org/10.1016/j.ijhydene.2008.04.022

Chapter 6
Experiences of Biohydrogen Production from Various Feedstocks by Dark Fermentation at Laboratory Scale

José de Jesús Montoya-Rosales, Casandra Valencia-Ojeda, Lourdes B. Celis, and Elías Razo-Flores

Abstract Over the past two decades, the advantages of biohydrogen as a clean fuel have been increasingly reported. From the various biological methods for producing hydrogen, dark fermentation has consistently delivered the highest hydrogen yields in laboratory-scale experiments. In addition, dark fermentation can be integrated into biorefinery and waste-to- energy systems, boosting the potential of biohydrogen for the development of the circular economy. However, many challenges must be overcome to achieve competitive dark fermentation performance with other technologies, such as metabolites production and substrate utilization efficiency. This chapter presents an overview of biohydrogen production efficiency obtained from various feedstocks and reactor configurations. In addition to reviewing the role of microbial communities in dark fermentation processes. Overall, this chapter underscores the importance of developing and implementing methods and practices to enhance biohydrogen production at the laboratory scale that allows the scale-up of this technology.

Keywords Dark fermentation · Feedstock type · Hydrogen · Microbial community · Reactor configuration

J. de J. Montoya-Rosales · C. Valencia-Ojeda · L. B. Celis · E. Razo-Flores (✉)
División de Ciencias Ambientales, Instituto Potosino de Investigación Científica y Tecnológica A.C., Camino a la Presa San José 2055, Lomas 4a Sección, 78216 San Luis Potosí, SLP, México
e-mail: erazo@ipicyt.edu.mx

J. de J. Montoya-Rosales
e-mail: jose.montoya@ipicyt.edu.mx

C. Valencia-Ojeda
e-mail: casandra.valencia@ipicyt.edu.mx

L. B. Celis
e-mail: celis@ipicyt.edu.mx

E. Razo-Flores
Departamento de Procesos y Tecnología, Universidad Autónoma Metropolitana Cuajimalpa, Av. Vasco de Quiroga 4871, Colonia Santa Fe Cuajimalpa, 05300 Ciudad de México, México

V. Alcaraz Gonzalez et al. (eds.), *Wastewater Exploitation*, Springer Water,
https://doi.org/10.1007/978-3-031-57735-2_6

6.1 Model Substrates for Hydrogen Production

Using model substrates in dark fermentation is a practical way to evaluate the efficiency of hydrogen (H_2) production. Model substrates are often simple organic molecules that facilitate monitoring their fate, conducting mass balances, estimating the metabolite distribution and kinetics, and calculating accurate theoretical product yields. The typical model substrates used for H_2 production by dark fermentation at the laboratory scale are glucose (monosaccharide), sucrose (disaccharide), and, to a lesser extent, lactose (disaccharide). Sucrose is composed of glucose and fructose, and lactose of galactose and glucose. The preferred substrates for laboratory experiments are glucose and sucrose because they are easily biodegraded and produce high rates of hydrogen (> 20 $L_{H_2}/(L\,d)$). However, they are not economically profitable at full scale [1]. The reason for using lactose as substrate relies on its abundance in cheese whey, which is the main waste stream in the dairy industry, representing 80–95% of the total volume of processed milk. Thereby, cheese whey is an economical and valuable lactose source that can be dried to a powder and used as a substrate, even at pilot or full-scale. Cheese whey powder has several advantages, such as reduced volume, high lactose concentration (> 61%), long shelf life, and easy storage and transportation compared with liquid cheese whey [2].

6.1.1 *Batch Experiments for Hydrogen Production with Model Substrates*

Batch experiments allow the evaluation of operational factors such as the type of substrate, inoculum source, mineral medium, pH, and temperature, among others, in addition to defining optimal operation conditions. For example, in batch experiments with glucose (5 g/L) as the substrate at mesophilic conditions (37 °C), and initial pH of 7.5, the maximum hydrogen yield (HY) amounted to 1.46 $mol_{H_2}/mol_{glucose}$ consumed; this yield represents 36.5% of the maximum theoretical yield (4 $mol_{H_2}/mol_{glucose}$) [2]. In the same study, with cheese whey as the substrate, the maximum HY was 3.1 $mol_{H_2}/mol_{lactose}$ (45% of the theoretical maximum, 8 $mol_{H_2}/mol_{lactose}$ consumed) at pH 6.0 and 15 g/L. Hence, the main application of batch experiments relies upon analyzing the hydrogen production potential of the inoculum or the substrates, or testing new strategies to enhance H_2 production. Due to the various ways of assessing the hydrogen production potential, a standardized protocol was recently established [3]. The protocol was validated in several laboratories from different countries and included the manual and automatic modes. The results showed repeatability, and reproducibility, and allowed the comparison of H_2 production with different inoculum sources and substrates [3].

An example of batch mode experiments to evaluate new strategies to avoid product inhibition is the addition of humic substances as an external redox mediator. The volumetric hydrogen production rate (VHPR) of 1.16 $L_{H_2}/(L\,d)$ was increased 1.8-fold upon the addition of chemically reduced leonardite. In the control without leonardite, the VHPR was 0.57 $L_{H_2}/(L\,d)$. This result shows that adding humic substances can increase the production of H_2 and carboxylic acids by serving as an additional electron pool [4]. In the same way, batch experiments were used to evaluate several conditions that improve the VHPR and HY. For instance, adding silicone oil in the range of 2.5–30% (v/v) increased H_2 production and yield. The proportion of 10% (v/v) silicone oil in the reaction medium was sufficient to facilitate H_2 mass transfer from the liquid to the gas phase and to ensure its availability in a continuous hydrogen-producing two-phase bioreactor [5].

6.1.2 Continuous Hydrogen Production with Model Substrates

Bioreactors for H_2 production via dark fermentation are classified as suspended-biomass and fixed-biomass reactors. From suspended-biomass reactors, the continuous stirred tank reactor (CSTR) is the most reported configuration for hydrogen production due to its simplicity in operation and efficient mass transfer [6]. However, one hindrance of the CSTR that limits H_2 productivity is the relatively low biomass concentrations, generally between 2 g volatile suspended solid (VSS)/L and 5 g_{VSS}/L [7–9]. The low biomass concentration achieved in this type of reactor is due to the low specific growth rates of H_2-producing bacteria (equivalent $\mu_{max} = 0.12$ to $0.28\ h^{-1}$) and the solids retention time (SRT), which is equal to the hydraulic retention time (HRT), which means that operation at low HRT (< 5–3.5 h) is not possible. Alternatively, fixed-biomass reactors can maintain high biomass concentrations (between 5 and 27 g_{VSS}/L) and operate at low HRT (< 6 h). Also, reactors such as the up-flow anaerobic sludge blanket, anaerobic filter, trickling bed, packed bed, and fluidized bed have been evaluated for hydrogen production [10–13]. The packed bed reactor showed the maximum average VHPR among the fixed-biomass reactors, 5.82 ± 0.98 $L_{H_2}/(L\,d)$, at organic loading rates (OLR) of 100 g chemical oxygen demand (COD)/(L d). In contrast, the CSTR exhibited a VHPR of 6.01 $L_{H_2}/(L\,d)$, at 24 g_{COD}/(L d), showing a superior performance not only in the VHPR but also in the HY of 1.29 $mol_{H_2}/mol_{glucose}$ consumed, which was about 2 times higher compared to the packed bed reactor [13].

Since the CSTR configuration offers better performance for H_2 production compared with fixed biomass reactors, this type of reactor was further investigated to evaluate the influence of operating parameters such as HRT, stirrer rotation, and OLR on H_2 production using cheese whey as substrate. One of the pioneer works about continuous hydrogen production using cheese whey was conducted by Davila-Vazquez et al. [7]. In this work, it was shown that HRT of 6 h and stirring conditions of 300 rpm improved H_2 production by preventing biomass wash-out and improving mass transfer conditions, respectively. These optimal operating parameters, together with selected microorganisms, enabled the system to increase H_2 production. The selected microorganisms mainly enhanced the acetate and butyrate pathways and provided VHPR values of up to 25 $L_{H_2}/(L\,d)$ [6–8]. Palomo-Briones et al. [6] investigated the effect of different mass transfer conditions using a series of CSTR at stirring speeds ranging from 100 to 400 rpm. The results showed that the hydrogen production increased from 4.40 $L_{H_2}/(L\,d)$ to a maximum VHPR of 7.67 $L_{H_2}/(L\,d)$ and the HY from 0.61 $mol_{H_2}/mol_{glucose}$ to 1.1 $mol_{H_2}/mol_{glucose}$, with the increase of the stirring speed from 100 to 400 rpm. Nevertheless, a stirring speed of 300 rpm resulted in VPHR and HY of 7.67 $L_{H_2}/(L\,d)$ and 1.0 $mol_{H_2}/mol_{glucose}$, respectively, as was also observed at the maximum stirring speed of 400 rpm, but now with lower energy input. The improvement in the H_2 production was attributed to the enhancement in mass transfer conditions, which decreased the concentrations of dissolved H_2 in the aqueous phase and favored the dominance of H_2-producing bacteria (e.g., *Clostridium* sp.), causing the increase in H_2 production through acetate and butyrate pathways. Similarly, variation of HRT from 6 to a maximum of 24 h showed a decrease in H_2 production, leading to the establishment of microbial communities containing *Sporolactobacillaceae* and *Streptococcaceae*, which displace the H_2-producing bacteria [14]. On the other hand, the OLR showed a positive correlation with carboxylic acids production when the OLR was increased from 60 to 180 g total carbohydrates (TC) as lactose/(L d). However, within this interval, the VHPR and HY increased until the threshold value of 140 g_{TC}/(L d). After that, the positive correlation was lost, and H_2 production started to decrease, as shown in Fig. 6.1 [5, 7–9]. This limit in H_2 production is associated with the accumulation of final products (CO_2, H_2, and carboxylic acids), which can originate product inhibition, causing a change in the metabolic routes and microbial community composition.

Although the CSTR is the leading reactor configuration for H_2 production with model substrates, fixed biomass reactors were also used in the early stages of H_2 production research. For example, up-flow anaerobic sludge blanket reactors were used to evaluate pH decrease (5.6–4.5), increment of OLR (20–40 g_{COD}/(L d)), and repeated heat treatment of the inoculum as strategies to supress methanogens. These strategies showed that lowering the pH resulted in higher H_2 production but favored homoacetogenic microorganisms. Likewise, organic shock loads up to 30 g_{COD}/(L d) increased the H_2 production rate (172%) and decreased the methane production (75%), being the best strategy to cope with methanogens without stopping reactor operation. On the other hand, after observing methane production in the reactor, a second heat treatment of the biomass suppressed methanogenesis. However, the reactor must be stopped and inoculated again [10, 15], which is impractical for a full-scale application. The studies carried out in an anaerobic sequencing batch biofilm reactor showed the importance of pretreating the inoculum (drying at 90 °C for 10 min) to suppress the activity of methanogens from the beginning. This reactor was operated in long-term mode (over 201 days) and produced H_2 (from 6.5 to 14.8 $L_{H_2}/(L\,d)$) without the interference of methanogenesis [16].

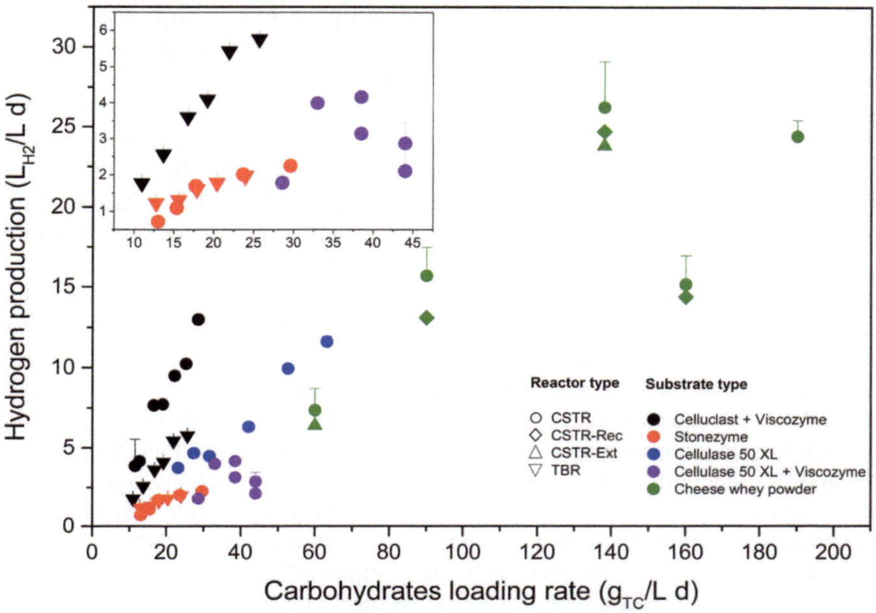

Fig. 6.1 Hydrogen production from cheese whey powder and different agave bagasse enzymatic hydrolysates using several continuous reactor configurations. CSTR: Continuous stirred tank reactor; CSTR-Rec: CSTR coupled to biomass recirculation; CSTR-Ext: CSTR coupled to CO_2 and H_2 extraction; TBR: Trickling bed reactor; TC: Total carbohydrates. Data was gathered from Davila-Vazquez et al. [7], Cota-Navarro et al. [8], Montoya-Rosales et al. [5, 9, 30], Valencia-Ojeda et al. [33]. The inset graph is a closer look of the hydrogen production results at the carbohydrates loading rate from 10 to 45 g_{TC}/L d

6.1.3 *Strategies and Perspectives*

Currently, the strategies to improve H_2 production through dark fermentation focus not only on increasing the VHPR but also on carboxylic acid production yields because the price of carboxylic acids (400–2500 USD/ton) is currently higher than that of H_2 (600–1800 USD/ton). Therefore, the generation, recovery, and commercialization of carboxylic acids have a positive economic impact on H_2 production technology from the biorefinery perspective [17].

In this regard, a continuous two-phase partitioning bioreactor was evaluated to decrease the concentration of H_2 and CO_2 in the aqueous phase and increase the concentration of carboxylic acids. This strategy developed in a conventional CSTR, fed with cheese whey, and supplemented with 10% (v/v) silicone oil as the organic phase, resulted in a 13% improvement in VHPR and 24% in carboxylic acids. The improvement was attributed to various factors: (i) decrease of H_2 and CO_2 concentrations in the aqueous phase, (ii) changes in microbial composition or metabolic pathways, or (iii) combination of both, which could move the scale toward more negative Gibbs free energy values. Therefore, the two-phase partitioning CSTR resulted in an efficient vehicle to increase H_2 and carboxylic acids production [9]. Another strategy seeking to cope with the low biomass concentration issue in the CSTR is biomass recirculation. Recirculated biomass increased the biomass concentration in the reactor by 200–300%, with the concomitant increase of the VHPR up to a maximum of 30.8 $L_{H_2}/(L\,d)$ and carboxylic acids production of 20 g/L at an OLR of 138 $g_{TC}/(L\ d)$. These results are attributed to the high biomass concentration achieved in the reactor and the enrichment of H_2-producing microorganisms such as *Clostridium* sp. [5]. These two strategies led to an increase in VHPR rates and carboxylic acid production by nearly 13–25% and 8–24%, respectively, compared to their CSTR controls, as shown in Fig. 6.1.

Managing carboxylic acids levels in the aqueous phase is essential to maintain non-inhibitory conditions for H_2 production. Inhibitory concentrations of carboxylic acids during dark fermentation can vary widely due to multiple factors. An adequate method to lower carboxylic acids concentrations in the aqueous phase is extractive fermentation, which entails their removal and recovery. Extractive fermentation is the process of in-situ extraction to remove the product (such as carboxylic acids, H_2, and CO_2) and other potential inhibitory compounds, hence increasing the HY. Nonetheless, there is a lack of knowledge regarding the behavior of continuous extractive fermentation systems under intensive H_2 production and the implications on microbial communities and metabolic pathways derived from their implementation. Strategies such as extraction of carboxylic acids by ionic exchange or electrodialysis could be applied because of their efficient recovery, low energy demand, and simple operation and design [18, 19]. Additionally, emerging technologies such as chain elongation and electro-fermentation have been explored as new strategies for dark fermentation revalorization [17, 20], which could be studied with the operational parameters previously defined in batch and continuous reactor operation.

6.2 Lignocellulosic Biomass for Hydrogen Production via Dark Fermentation: The Agave Bagasse Case Study

From the wide variety of biomass feedstocks for hydrogen production by dark fermentation, lignocellulosic biomass stands out due to its abundance and high polysaccharide content. The annual global production of lignocellulosic biomass in nature is estimated at 181.5 billion tons. Nowadays, it is estimated that around 8.2 billion tons (< 10%) are used for cooking, heating, construction material, and producing paper, cardboard, and textiles. Therefore, along with the development of technologies for biomass biorefineries, lignocellulosic biomass can be used to produce renewable biofuels [21].

Lignocellulosic biomass is mainly composed of lignin, cellulose, and hemicellulose. Cellulose and hemicellulose are considered suitable feedstocks for biofuel production. However, lignin affects the availability of carbohydrates during biomass hydrolysis and thus poses challenges for hydrogen production. A pretreatment step is often recommended to reduce the recalcitrance of the lignocellulosic biomass and obtain more accessible hemicellulose and cellulose fibers to ease the hydrolysis step. Most of the pretreatments aim to remove hemicellulose (e.g., dilute acid hydrolysis, steam explosion, ionic liquids) or/and lignin (e.g., alkaline hydrolysis, organosolv, oxidative hydrolysis) [22]. After pretreatment, the hydrolysis of cellulose and hemicellulose is required to increase the concentration of soluble carbohydrates for the dark fermentation step. In this regard, chemical (acid hydrolysis) and biological (enzymatic hydrolysis) are widely reported in the literature. In enzymatic hydrolysis, different enzymes are responsible for the breakdown of hemicellulose and cellulose into oligomers and monomers. Although chemical hydrolysis provides a higher sugar yield (20 g_{TC}/L) compared with enzymatic hydrolysis (8 g_{TC}/L), enzymatic hydrolysis has certain advantages in comparison with chemical hydrolysis, such as moderate operating conditions (30–50 °C) and higher HY (> $1.7\ mol_{H_2}/mol_{glucose}$ consumed). In addition, enzymatic hydrolysis does not generate dark fermentation inhibitors such as furfurals, hydroxy-methyl furfurals, vanillin, and other phenolic compounds [23, 24].

Currently, pretreatments and hydrolysis are hot research topics aiming to increase the use of lignocellulosic biomass as feedstock to produce biofuels at commercial scale. Nonetheless, a universal pretreatment scheme is challenging due to the diverse nature of lignocellulosic biomass. In general, batch dark fermentation with lignocellulosic biomass hydrolysates, such as oat straw and agave bagasse hydrolysates obtained from the combination of one or sequential pretreatments, followed by enzymatic hydrolysis, resulted in hydrogen production 1.5–2.5 times higher than the production obtained from hydrolysates of non-pretreated biomass [25, 26].

Agave bagasse is a lignocellulosic residue of local and regional importance in Mexico, which originates from the agave plants used as raw material to produce fibers and alcoholic beverages (tequila and mezcal). Due to its considerable availability (yearly production of about 1,537,000 tons) [27], high content of fermentable sugars (31–43% w/w cellulose, 11–22% w/w hemicellulose), and relatively low content of lignin (11–20% w/w) [28], agave bagasse can be used for H_2 production via dark fermentation. The laboratory-scale hydrogen production via dark fermentation using agave bagasse has been performed in batch and continuous reactors from various enzymatic hydrolysates.

6.2.1 *Effect of the Type of Enzyme in Batch Dark Fermentation Experiments*

In practice, for the hydrolysis of agave bagasse, there are two groups of commercial enzymes with cellulolytic and hemicellulolytic activities. The first group includes specific enzymes expressly designed for the saccharification of lignocellulosic materials. For example, Cellic® CTec2, Cellic® HTec2, Celluclast 1.5 L, and Viscozyme L (Novozymes, Denmark) [29–32]. The second group includes non-specific enzymes designed for other purposes and applications in industries other than biofuel production (such as those for textile and food markets); these enzymes are cheaper and more available on the Mexican market. For example, Cellulase 50 XL (Enziquim, Mexico) and Stonezyme (ENMEX, Mexico) [33, 34]. In this regard, the enzymatic hydrolysates obtained with Celluclast 1.5L reached higher hydrogen production ($1.8\,L_{H_2}/L$) in batch experiments than those hydrolysates obtained with Cellulase 50 XL ($1.15\,L_{H_2}/L$) and Stonezyme ($0.85\,L_{H_2}/L$). Nonetheless, the carbohydrates (e.g., glucose, xylose, fructose, mannose, cellobiose, etc.) released in the enzymatic treatment convey a certain quality to the hydrolysate, impacting hydrogen production. In this sense, achieving a similar hydrogen production would be possible using non-specialized enzymes. For instance, during enzymatic pretreatment and hydrolysis of lignocellulosic materials, a higher release of carbohydrate monomers (glucose, xylose) is desired for an efficient dark fermentative process.

6.2.2 *Effect of the Continuous Reactor Configurations on Hydrogen Production*

In two continuous reactor configurations (CSTR and trickling-bed reactor, TBR), H_2 production was investigated using different enzymatic hydrolysates from agave bagasse as substrates. The experimental data using the same agave bagasse enzymatic hydrolysates (Celluclast 1.5 L—Viscozyme) and similar OLR showed that higher VHPR were obtained in the CSTR than in the TBR (Fig. 6.1). For instance, with a CSTR, it was possible to obtain a maximum VHPR of $13.5\,L_{H_2}/(L\,d)$ at an OLR of 30 $g_{TC}/(L\ d)$, whereas the highest VHPR in the TBR was 2.3 times lower ($5.7\,L_{H_2}/(L\,d)$) at an OLR of 26.5 $g_{TC}/(L\ d)$ [30]. Additionally, with the CSTR, it was possible to achieve a higher carbohydrate loading rate (> 30 $g_{TC}/(L\ d)$). These experiments also showed that regardless of the type of hydrolysate, the way microbial biomass grows within the reactor is a crucial factor that modifies the H_2 production performance. Therefore, the CSTR offered more benefits, compared with the TBR, mainly because it allowed for better control of the environmental parameters such as HRT, SRT, fermentation broth temperature, pH, and consequently, the composition of microbial communities [14].

Overall, the production of hydrogen in CSTR systems was positively correlated to the carbohydrates loading rate using enzymatic agave bagasse hydrolysates obtained with different enzymes, as shown in Fig. 6.1. Interestingly, similar loading rates of carbohydrates evaluated under the same CSTR conditions but fed with different enzymatic hydrolysates resulted in remarkably different hydrogen production values. For example, at a carbohydrate loading rate of approximately 30 $g_{TC}/(L\ d)$, the hydrogen production ranged from 2 to $13.5\,L_{H_2}/(L\,d)$, as shown in Fig. 6.1. Considering that the concentration of carbohydrates is often referred to as one of the most determinant factors of dark fermentation, the variance in H_2 production at similar concentrations could be related to the type of carbohydrates present in the hydrolysates [33]. For instance, oligosaccharides (e.g., starch, glycogen, chitin) released during enzymatic hydrolysis may harm H_2 production since they are difficult to metabolize by H_2-producing bacteria due to their complexity and polymerization degree.

6.2.3 Perspectives on the Continuous Hydrogen Production from Lignocellulosic Biomass

The U.S. Department of Energy (DOE) set a 2022 target to achieve hydrogen production of 20 $L_{H_2}/(L\ d)$ from lignocellulosic substrates using biological methods. Specifically, studies conducted on CSTR using agave bagasse enzymatic hydrolysates with Celluclast 1.5 L—Viscozyme at an OLR of 30 $g_{TC}/(L\ d)$, [30], resulted in the highest VHPR (13.5 $L_{H_2}/(L\ d)$) achieved to date with agave bagasse hydrolysates. This result is 65% of the U.S. DOE target. Considering that the main bottleneck in continuous hydrogen production with agave bagasse is the type and concentration of carbohydrates released during enzymatic hydrolysis, the fine-tuning and application of pretreatment strategies is a priority. Using pretreated biomass in enzymatic hydrolysis would increase the concentration of carbohydrates released and strengthen the probability of reaching or surpassing the hydrogen production DOE target in continuous reactors. Currently, the pretreatments used for agave bagasse with promising results are hydrothermal [22], alkaline [35], dilute acid [36], organosolv [37], autohydrolysis [38], and ionic liquids [39].

6.3 Microbial Community Approach: Opening the Black Box

Most dark fermentation studies involve mixed cultures, making the efficiency of hydrogen productivity mainly dependent on the composition, diversity, and structure of the microbial community. In general, it has been widely reported that efficient dark fermentation is predominantly characterized by *Clostridium* species. *Clostridium*, a genus comprising spore-forming obligate anaerobes, is notably prevalent during periods of high H_2 production [40]. The H_2-producing members of *Clostridium* genus include a vast number of species: *C. acetobutylicum*, *C. butyricum*, *C. saccharobutylicum*, *C. tyrobutyricum*, *C. pasteurianum*, and *C. beijerinckii*, among others [41]. H_2-producing mixed cultures may also contain facultative anaerobes, for example, from the genera *Klebsiella* and *Kluyvera*. These microorganisms may be directly implicated in H_2 production and, nonetheless, may participate in the hydrolysis of oligomeric carbohydrates [40–42]. However, the theoretical HY

($0.8\ mol_{H_2}/mol_{glucose}$ consumed) reported for members of the *Klebsiella* genus are, in general, lower than those registered for *Clostridium* ($2.3\ mol_{H_2}/mol_{glucose}$ consumed) [43].

Lactic acid bacteria (LAB) are another group of microorganisms frequently detected in dark fermentation systems. Some genera belonging to LAB that were detected in dark fermentative reactors are *Sporolactobacillus*, *Lactobacillus*, *Lactococcus*, *Leuconostoc*, and *Streptococcus* [5, 9, 14, 41, 44, 45]. The first reports on LAB in dark fermentation reactors described these microorganisms as the main competitors of hydrogen-producing bacteria. Moreover, LAB have been documented with the capacity to reduce hydrogen production through substrate competition and the inhibitory effects arising from the significant production of lactic acid and antimicrobial peptides, commonly referred to as bacteriocins [41]. Nonetheless, recent reports have demonstrated that lactic acid produced by LAB can be used as an additional substrate by *Clostridium* species to produce butyrate and hydrogen. Members of the LAB bacterial group also consume oxygen in the medium to maintain anaerobic conditions [46, 47].

Overall, the microbial community results obtained from suspended-biomass and fixed-biomass reactors, independently of the reactor configuration and type of substrate, are shown in Fig. 6.2. The operating conditions modified the microbial community dominance in most systems. Interestingly, microbial community data in a CSTR fed with lignocellulosic (agave bagasse enzymatic hydrolysates) and non-lignocellulosic substrates (cheese whey and glucose) had similar Shannon index ($p > 0.05$): 1.09 (lignocellulosic) versus1.21 (non-lignocellulosic) despite the remarkable differences in hydrogen productivities as shown in Fig. 6.1 [5, 9, 14]. The reactors fed with cheese whey delivered the highest VHPR at an OLR of 138 g_{TC}/(L d), independently of the improvement strategy assessed. Under these conditions, *Clostridium* bacteria (relative abundance between 50 and 75%) and LAB (relative abundance of 15 and 45%) co-dominated the microbial community structure, as shown in Fig. 6.2 [5, 6, 9].

The microbial community data obtained from CSTR and TBR fed with agave bagasse enzymatic hydrolysates showed that simple microbial communities characterized the CSTR since the ecological niches were mainly dominated by *Clostridium* and *Sporolactobacillus* genera [14]. Conversely, in the TBR, the stratification of microbial biomass was linked to an increased abundance of ecological niches, resulting in higher microbial community diversity (1.65 ± 0.35). In this context, microorganisms of the genera *Caproiciproducens*, *Clostridium*, *Kluyvera*, *Sporolactobacillus*, *Lachnoclostridium*, and *Leuconostoc* prevailed and characterized the microbial communities of the TBR [14]. The fundamental link between community diversity and dark fermentation performance indicates that relatively simple microbial communities delivered high hydrogen production rates (Fig. 6.2). In summary, maintaining simple microbial communities over the long term requires precise control of operating conditions. For instance, employing short HRT (approximately 6 h) results in the development of clearly defined hydrogen-producing microbial communities co-dominated by LAB [44, 48]. Nonetheless, achieving the coupling of HRT with SRT is only feasible in CSTR systems. In contrast, in

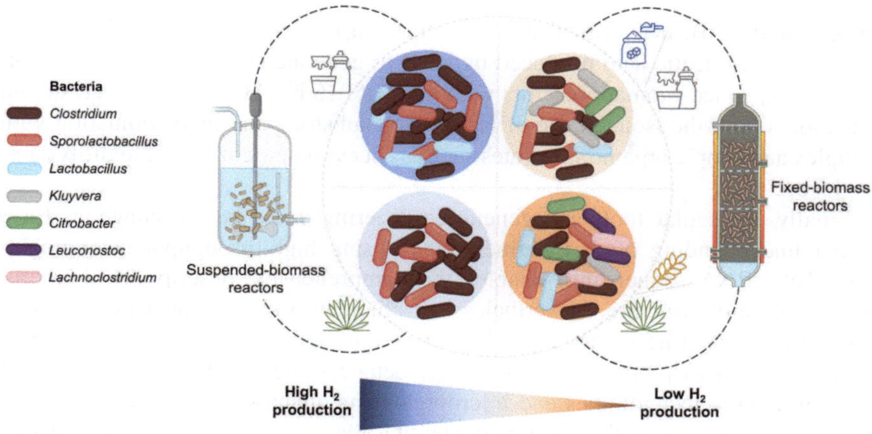

Fig. 6.2 Schematic composition of microbial communities in suspended-biomass reactors and fixed-biomass reactors fed with model substrates (glucose, cheese whey powder) and lignocellulosic substrates (agave bagasse hydrolysates and oat straw hydrolysates). Data was gathered from Davila-Vazquez et al. [7], Carrillo-Reyes et al. [15], Cisneros-Perez et al. [49], Palomo-Briones et al. [14, 44, 45], Montoya-Rosales et al. [5, 9]

systems with uncoupled HRT and SRT, a greater diversity of microorganisms tends to coexist, for example, in up-flow anaerobic sludge blanket reactors, TBR, and biofilm reactors, among others. This diversity can impede high hydrogen production rates and complicate the reactor control. All these observations underscore the crucial role of the microbial community diversity in influencing the performance of hydrogen-producing systems.

6.4 Conclusions and Final Remarks

Traditionally, dark fermentation research has concentrated on studying and optimizing operational parameters (pH, temperature, OLR, type of substrate) to enhance the H_2 productivities in terms of VHPR and HY. However, in this chapter, we reported that beyond the traditional operational parameters, factors such as the type and substrate concentration, by-product inhibition, and microbial community dynamics are bottlenecks currently facing biological hydrogen production by dark fermentation, which need to be studied in-depth. For instance, in this chapter, we presented some strategies such as pretreatment of lignocellulosic biomass, extraction of by-products, and improvement of mass transfer conditions that hold promise for the short- and long-term process development. Nevertheless, more studies and efforts are needed to improve hydrogen production by dark fermentation.

Derived from the observations that the highest hydrogen productivities reached in a CSTR were associated with a microbial structure co-dominated by H_2-producing bacteria and LAB, it is critical to concentrate on the hypothesis that *Clostridium* species can use lactic acid produced by LAB to generate butyrate and H_2, particularly in suspended-biomass reactors with low HRT (6 h). Also, the research should encompass both the isolated use of lactate as a substrate and its combination with complex and simple organic substrates such as cheese whey, enzymatic hydrolysates, or glucose.

Finally, molecular tools and genetic engineering could be advantageous for a deeper understanding of dark fermentation. Using high-throughput technologies (e.g., 16S rRNA sequencing) enables the comprehensive description of microbial community structure. Nevertheless, to elucidate the optimal dark fermentation pathway, it is necessary to include the omics sciences in the analysis (e.g., proteomics, transcriptomics, metabolomics, metagenomics) with the combination of high-throughput technologies and microorganisms quantification techniques (e.g., qPCR, flow cytometry). The challenging point is that omics science is still uncharted territory in the dark fermentation field, and transcriptomics and metagenomics have been applied very little in continuous reactors.

Acknowledgements This work was financially supported by SEP-CONACYT project A1-S-37174. E. Razo-Flores acknowledges the "*Cátedra Rodolfo Quintero-Ramírez*" granted by the Universidad Autónoma Metropolitana-Cuajimalpa.

References

1. Toledo-Alarcón J, Capson-Tojo G, Marone A, Paillet F, Júnior ADNF, Chatellard L, Bernet N, Trably E (2018) Basics of bio-hydrogen production by dark fermentation. In: Liao Q, Chang J-s, Herrmann C, Xia A (eds) Bioreactors for microbial biomass and energy Conversion. Springer, Singapore, pp 199–220. https://doi.org/10.1007/978-981-10-7677-0_6
2. Davila-Vazquez G, Alatriste-Mondragón F, de León-Rodríguez A, Razo-Flores E (2008) Fermentative hydrogen production in batch experiments using lactose, cheese whey and glucose: influence of initial substrate concentration and pH. Int J Hydrogen Energy 33(19):4989–4997. https://doi.org/10.1016/j.ijhydene.2008.06.065
3. Carrillo-Reyes J, Tapia-Rodríguez A, Buitrón G, Moreno-Andrade I, Palomo-Briones R, Razo-Flores E, Aguilar Juárez O, Arreola-Vargas J, Bernet N, Maluf Braga AF, Braga L, Castelló E, Chatellard L, Etchebehere C, Fuentes L, León-Becerril E, Méndez-Acosta HO, Ruiz-Filippi G, Tapia-Venegas E, Trably E, Wenzel J, Zaiat M (2019) A standardized biohydrogen potential protocol: an international round robin test approach. Int J Hydrogen Energy 44(48):26237–26247. https://doi.org/10.1016/j.ijhydene.2019.08.124
4. Núñez-Valenzuela P, Palomo-Briones R, Cervantes FJ, Razo-Flores E (2021) Humic substances improve the co-production of hydrogen and carboxylic acids by anaerobic mixed cultures. Int J Hydrogen Energy 46(65):32800–32808. https://doi.org/10.1016/j.ijhydene.2021.07.128
5. Montoya-Rosales JdeJ, Palomo-Briones R, Celis LB, Etchebehere C, Cházaro-Ruiz LF, Escobar-Barrios V, Razo-Flores E (2022) Coping with mass transfer constrains in dark fermentation using a two-phase partitioning bioreactor. Chem Eng J 445:136749. https://doi.org/10.1016/j.cej.2022.136749
6. Palomo-Briones R, Celis LB, Méndez-Acosta HO, Bernet N, Trably E, Razo-Flores E (2019) Enhancement of mass transfer conditions to increase the productivity and efficiency of dark

fermentation in continuous reactors. Fuel 254:115648. https://doi.org/10.1016/j.fuel.2019.115648
7. Davila-Vazquez G, Cota-Navarro CB, Rosales-Colunga LM, de León-Rodríguez A, Razo-Flores E (2009) Continuous biohydrogen production using cheese whey: improving the hydrogen production rate. Int J Hydrogen Energy 34(10):4296–4304. https://doi.org/10.1016/j.ijhydene.2009.02.063
8. Cota-Navarro CB, Carrillo-Reyes J, Davila-Vazquez G, Alatriste-Mondragon F, Razo-Flores E (2011) Continuous hydrogen and methane production in a two-stage cheese whey fermentation system. Water Sci Technol 64(2):367–374. https://doi.org/10.2166/wst.2011.631
9. Montoya-Rosales JdJ, Palomo-Briones R, Celis LB, Etchebehere C, Razo-Flores E (2020) Discontinuous biomass recycling as a successful strategy to enhance continuous hydrogen production at high organic loading rates. Int J Hydrogen Energy 45(35):17260–17269. https://doi.org/10.1016/j.ijhydene.2020.04.265
10. Carrillo-Reyes J, Celis LB, Alatriste-Mondragón F, Razo-Flores E (2014) Decreasing methane production in hydrogenogenic UASB reactors fed with cheese whey. Biomass Bioenerg 63:101–108. https://doi.org/10.1016/j.biombioe.2014.01.050
11. Carrillo-Reyes J, Celis LB, Alatriste-Mondragón F, Razo-Flores E (2012) Different start-up strategies to enhance biohydrogen production from cheese whey in UASB reactors. Int J Hydrogen Energy 37(7):5591–5601. https://doi.org/10.1016/j.ijhydene.2012.01.004
12. Cisneros-Pérez C, Etchebehere C, Celis LB, Carrillo-Reyes J, Alatriste-Mondragón F, Razo-Flores E (2017) Effect of inoculum pretreatment on the microbial community structure and its performance during dark fermentation using anaerobic fluidized-bed reactors. Int J Hydrogen Energy 42(15):9589–9599. https://doi.org/10.1016/j.ijhydene.2017.03.157
13. Carrillo-Reyes J, Cortés-Carmona MA, Bárcenas-Ruiz CD, Razo-Flores E (2016) Cell wash-out enrichment increases the stability and performance of biohydrogen producing packed-bed reactors and the community transition along the operation time. Renew Energy 97:266–273. https://doi.org/10.1016/j.renene.2016.05.082
14. Palomo-Briones R, Montoya-Rosales JdJ, Razo-Flores E (2021) Advances towards the understanding of microbial communities in dark fermentation of enzymatic hydrolysates: diversity, structure and hydrogen production performance. Int J Hydrogen Energy 46(54):27459–27472. https://doi.org/10.1016/j.ijhydene.2021.06.016
15. Carrillo-Reyes J, Celis LB, Alatriste-Mondragón F, Montoya L, Razo-Flores E (2014) Strategies to cope with methanogens in hydrogen producing UASB reactors: community dynamics. Int J Hydrogen Energy 39(22):11423–11432. https://doi.org/10.1016/j.ijhydene.2014.05.099
16. Carrillo-Reyes J, Trably E, Bernet N, Latrille E, Razo-Flores E (2016) High robustness of a simplified microbial consortium producing hydrogen in long term operation of a biofilm fermentative reactor. Int J Hydrogen Energy 41(4):2367–2376. https://doi.org/10.1016/j.ijhydene.2015.11.131
17. de Sousa e Silva A, Sales Morais NW, Maciel Holanda Coelho M, Lopes Pereira E, Bezerra dos Santos A (2020) Potentialities of biotechnological recovery of methane, hydrogen and carboxylic acids from agro-industrial wastewaters. Bioresour Technol Rep 10. https://doi.org/10.1016/j.biteb.2020.10040618
18. Redwood MD, Orozco RL, Majewski AJ, Macaskie LE (2012) Electro-extractive fermentation for efficient biohydrogen production. Biores Technol 107:166–174. https://doi.org/10.1016/j.biortech.2011.11.026
19. Gu J, Li Z, Xie G, Yang Y, Liu B, Ren ZJ, Xing D (2022) Electro-fermentation enhances H_2 and ethanol co-production by regulating electron transfer and substrate transmembrane transport. Chem Eng J 429:132223. https://doi.org/10.1016/j.cej.2021.132223
20. Toledo-Alarcón J, Moscoviz R, Trably E, Bernet N (2019) Glucose electro-fermentation as main driver for efficient H_2-producing bacteria selection in mixed cultures. Int J Hydrogen Energy 44(4):2230–2238. https://doi.org/10.1016/j.ijhydene.2018.07.091
21. Ashokkumar V, Venkatkarthick R, Jayashree S, Chuetor S, Dharmaraj S, Kumar G, Chen WH, Ngamcharussrivichai C (2022) Recent advances in lignocellulosic biomass for biofuels and value-added bioproducts–a critical review. Bioresour Technol 344:126195. https://doi.org/10.1016/j.biortech.2021.126195

22. Shiva, Rodríguez-Jasso RM, López-Sandin I, Aguilar MA, López-Badillo CM, Ruiz HA (2023) Intensification of enzymatic saccharification at high solid loading of pretreated agave bagasse at bioreactor scale. J Environ Chem Eng 11:109257. https://doi.org/10.1016/j.jece.2022.109257
23. Arreola-Vargas J, Alatriste-Mondragón F, Celis LB, Razo-Flores E, López-López A, Méndez-Acosta HO (2015) Continuous hydrogen production in a trickling bed reactor by using triticale silage as inoculum: effect of simple and complex substrates. J Chem Technol Biotechnol 90(6):1062–1069. https://doi.org/10.1002/jctb.4410
24. Arreola-Vargas J, Celis LB, Buitrón G, Razo-Flores E, Alatriste-Mondragón F (2013) Hydrogen production from acid and enzymatic oat straw hydrolysates in an anaerobic sequencing batch reactor: performance and microbial population analysis. Int J Hydrogen Energy 38(32):13884–13894. https://doi.org/10.1016/j.ijhydene.2013.08.065
25. Arreola-Vargas J, Razo-Flores E, Celis LB, Alatriste-Mondragón F (2015) Sequential hydrolysis of oat straw and hydrogen production from hydrolysates: role of hydrolysates constituents. Int J Hydrogen Energy 40(34):10756–10765. https://doi.org/10.1016/j.ijhydene.2015.05.200
26. Galindo-Hernández KL, Tapia-Rodríguez A, Alatriste-Mondragón F, Celis LB, Arreola-Vargas J, Razo-Flores E (2018) Enhancing saccharification of Agave Tequilana bagasse by oxidative delignification and enzymatic synergism for the production of hydrogen and methane. Int J Hydrogen Energy 43(49):22116–22125. https://doi.org/10.1016/j.ijhydene.2018.10.071
27. Consejo Regulador del Tequila (2022) https://www.crt.org.mx/EstadisticasCRTweb/. Accessed 30 Jan 2023
28. Palomo-Briones R, López-Gutiérrez I, Islas-Lugo F, Galindo-Hernández KL, Munguía-Aguilar D, Rincón-Pérez JA, Cortés-Carmona MA, Alatriste-Mondragón F, Razo-Flores E (2017) Agave bagasse biorefinery: processing and perspectives. Clean Technol Environ Policy 20(7):1423–1441. https://doi.org/10.1007/s10098-017-1421-2
29. López-Gutiérrez I, Razo-Flores E, Méndez-Acosta HO, Amaya-Delgado L, Alatriste-Mondragón F (2021) Optimization by response surface methodology of the enzymatic hydrolysis of non-pretreated agave bagasse with binary mixtures of commercial enzymatic preparations. Biomass Convers Biorefinery 11(6):2923–2935. https://doi.org/10.1007/s13399-020-00698-x
30. Montoya-Rosales JdeJ, Olmos-Hernandez DK, Palomo-Briones R, Montiel-Corona V, Mari AG, Razo-Flores E (2019) Improvement of continuous hydrogen production using individual and binary enzymatic hydrolysates of agave bagasse in suspended-culture and biofilm reactors. Bioresour Technol 283:251–260. https://doi.org/10.1016/j.biortech.2019.03.072
31. Contreras-Dávila CA, Méndez-Acosta HO, Arellano-García L, Alatriste-Mondragón F, Razo-Flores E (2017) Continuous hydrogen production from enzymatic hydrolysate of Agave Tequilana bagasse: effect of the organic loading rate and reactor configuration. Chem Eng J 313:671–679. https://doi.org/10.1016/j.cej.2016.12.084
32. Montiel Corona V, Razo-Flores E (2018) Continuous hydrogen and methane production from Agave Tequilana bagasse hydrolysate by sequential process to maximize energy recovery efficiency. Bioresour Technol 249:334–341. https://doi.org/10.1016/j.biortech.2017.10.032
33. Valencia-Ojeda C, Montoya-Rosales JdeJ, Palomo-Briones R, Montiel-Corona V, Celis LB, Razo-Flores E (2021) Saccharification of agave bagasse with Cellulase 50 XL is an effective alternative to highly specialized lignocellulosic enzymes for continuous hydrogen production. J Environ Chem Eng 9:105448. https://doi.org/10.1016/j.jece.2021.105448
34. Tapia-Rodríguez A, Ibarra-Faz E, Razo-Flores E (2019) Hydrogen and methane production potential of agave bagasse enzymatic hydrolysates and comparative technoeconomic feasibility implications. Int J Hydrogen Energy 44(33):17792–17801. https://doi.org/10.1016/j.ijhydene.2019.05.087
35. Perez-Pimienta JA, Poggi-Varaldo HM, Ponce-Noyola T, Ramos-Valdivia AC, Chavez-Carvayar JA, Stavila V, Simmons BA (2016) Fractional pretreatment of raw and calcium oxalate-extracted agave bagasse using ionic liquid and alkaline hydrogen peroxide. Biomass Bioenerg 91:48–55. https://doi.org/10.1016/j.biombioe.2016.05.001
36. Rios-Del Toro EE, Arreola-Vargas J, Cardenas-López RL, Valdez-Guzmán BE, Toledo-Cervantes A, González-Álvarez V, Méndez-Acosta HO (2021) Two-stage semi-continuous

hydrogen and methane production from undetoxified and detoxified acid hydrolysates of agave bagasse. Biomass Bioenerg 150:106130. https://doi.org/10.1016/j.biombioe.2021.106130
37. Pérez-Pimienta JA, Vargas-Tah A, López-Ortega KM, Medina-López YN, Mendoza-Pérez JA, Avila S, Singh S, Simmons BA, Loaces I, Martinez A (2017) Sequential enzymatic saccharification and fermentation of ionic liquid and organosolv pretreated agave bagasse for ethanol production. Biores Technol 225:191–198. https://doi.org/10.1016/j.biortech.2016.11.064
38. Rios-González LJ, Morales-Martínez TK, Rodríguez-Flores MF, Rodríguez-De la Garza JA, Castillo-Quiroz D, Castro-Montoya AJ, Martinez A (2017) Autohydrolysis pretreatment assessment in ethanol production from agave bagasse. Biores Technol 242:184–190. https://doi.org/10.1016/j.biortech.2017.03.039
39. Pérez-Pimienta JA, García-López RM, Méndez-Acosta HO, González-Álvarez V, Simmons BA, Méndoza-Pérez JA, Arreola-Vargas J (2021) Ionic liquid-water mixtures enhance pretreatment and anaerobic digestion of agave bagasse. Ind Crops Prod 171:113924. https://doi.org/10.1016/j.indcrop.2021.113924
40. Etchebehere C, Castello E, Wenzel J, Anzola-Rojas MdP, Borzacconi L, Buitron G, Cabrol L, Carminato VM, Carrillo-Reyes J, Cisneros-Perez C, Fuentes L, Moreno-Andrade I, Razo-Flores E, Filippi GR, Tapia-Venegas E, Toledo-Alarcon J, Zaiat M (2016) Microbial communities from 20 different hydrogen-producing reactors studied by 454 pyrosequencing. Appl Microbiol Biotechnol 100(7):3371–3384. https://doi.org/10.1007/s00253-016-7325-y
41. Castelló E, Nunes Ferraz-Junior AD, Andreani C, Anzola-Rojas MdP, Borzacconi L, Buitrón G, Carrillo-Reyes J, Gomes SD, Maintinguer SI, Moreno-Andrade I, Palomo-Briones R, Razo-Flores E, Schiappacasse-Dasati M, Tapia-Venegas E, Valdez-Vázquez I, Vesga-Baron A, Zaiat M, Etchebehere C (2020) Stability problems in the hydrogen production by dark fermentation: possible causes and solutions. Renew Sustain Energy Rev 119:109602. https://doi.org/10.1016/j.rser.2019.109602
42. Cabrol L, Marone A, Tapia-Venegas E, Steyer J-P, Ruiz-Filippi G, Trably E (2017) Microbial ecology of fermentative hydrogen producing bioprocesses: useful insights for driving the ecosystem function. FEMS Microbiol Rev 41(2):158–181. https://doi.org/10.1093/femsre/fuw043
43. Wong YM, Wu TY, Juan JC (2014) A review of sustainable hydrogen production using seed sludge via dark fermentation. Renew Sustain Energy Rev 34:471–482. https://doi.org/10.1016/j.rser.2014.03.008
44. Palomo-Briones R, Razo-Flores E, Bernet N, Trably E (2017) Dark-fermentative biohydrogen pathways and microbial networks in continuous stirred tank reactors: novel insights on their control. Appl Energy 198:77–87. https://doi.org/10.1016/j.apenergy.2017.04.051
45. Palomo-Briones R, Trably E, Lopez-Lozano NE, Celis LB, Mendez-Acosta HO, Bernet N, Razo-Flores E (2018) Hydrogen metabolic patterns driven by Clostridium-Streptococcus community shifts in a continuous stirred tank reactor. Appl Microbiol Biotechnol 102(5):2465–2475. https://doi.org/10.1007/s00253-018-8737-7
46. García-Depraect O, Martínez-Mendoza LJ, Diaz I, Muñoz R (2022) Two-stage anaerobic digestion of food waste: enhanced bioenergy production rate by steering lactate-type fermentation during hydrolysis-acidogenesis. Biores Technol 358:127358. https://doi.org/10.1016/j.biortech.2022.127358
47. Martínez-Mendoza LJ, Lebrero R, Muñoz R, García-Depraect O (2022) Influence of key operational parameters on biohydrogen production from fruit and vegetable waste via lactate-driven dark fermentation. Biores Technol 364:128070. https://doi.org/10.1016/j.biortech.2022.128070
48. García-Depraect O, Muñoz R, Rodríguez E, Rene ER, León-Becerril E (2021) Microbial ecology of a lactate-driven dark fermentation process producing hydrogen under carbohydrate-limiting conditions. Int J Hydrogen Energy 46(20):11284–11296. https://doi.org/10.1016/j.ijhydene.2020.08.209
49. Cisneros-Pérez C, Carrillo-Reyes J, Celis LB, Alatriste-Mondragón F, Etchebehere C, Razo-Flores E (2015) Inoculum pretreatment promotes differences in hydrogen production performance in EGSB reactors. Int J Hydrogen Energy 40(19):6329–6339. https://doi.org/10.1016/j.ijhydene.2015.03.048

Chapter 7
Microbial Communities in Dark Fermentation, Analytical Tools to Elucidate Key Microorganisms and Metabolic Profiles

Julián Carrillo-Reyes, Idania Valdez-Vazquez, Miguel Vital-Jácome, Alejandro Vargas, Marcelo Navarro-Díaz, Jonathan Cortez-Cervantes, and Ana P. Chango-Cañola

Abstract Hydrogen production from dark fermentation (DF) is a viable option for valorizing organic waste and agro-industrial effluents. Fermentative bacteria carry out this process through complex ecological interactions, like cooperation, syntrophy, and competition. Therefore, understanding such complex microbial dynamics is key in developing the process since performance and stability depend on it. In this sense, studying the DF microbiome helps understand possible interactions, metabolic pathways, and the microorganisms' role. This chapter discusses the need to analyze microbial communities from an ecological perspective and reviews existing tools for their characterization. A review of statistical methods and metabolism modeling

J. Carrillo-Reyes (✉) · I. Valdez-Vazquez · M. Vital-Jácome · A. Vargas · J. Cortez-Cervantes · A. P. Chango-Cañola
Laboratory for Research On Advanced Processes for Water Treatment, Unidad Académica Juriquilla, Instituto de Ingeniería, Universidad Nacional Autónoma de México, Blvd. Juriquilla 3001, 76230 Querétaro, Mexico
e-mail: JCarrilloR@iingen.unam.mx

I. Valdez-Vazquez
e-mail: IValdezV@iingen.unam.mx

M. Vital-Jácome
e-mail: MVitalj@iingen.unam.mx

A. Vargas
e-mail: AVargasc@iingen.unam.mx

J. Cortez-Cervantes
e-mail: JCortezC@iingen.unam.mx

A. P. Chango-Cañola
e-mail: AChangoC@iingen.unam.mx

M. Navarro-Díaz
Cluster of Excellence "Controlling Microbes to Fight Infections", University of Tübingen, Auf Der Morgenstelle 28, 72074 Tübingen, Germany

V. Alcaraz Gonzalez et al. (eds.), *Wastewater Exploitation*, Springer Water,
https://doi.org/10.1007/978-3-031-57735-2_7

are integrated to take advantage of the microbial characterization and elucidate key microorganisms, their interactions, and their correlation to operational parameters, favoring the efficiency of DF.

Keywords Microbial ecology · Molecular network analysis · Multivariate statistical analysis · Functional profiles

7.1 Introduction

Molecular hydrogen (H_2) is a promising alternative clean energy source and carrier, with the highest rate of gravimetric combustion energy, 2.75 times higher than other fuels, without greenhouse gasses emissions during its use [1]. To produce renewable H_2, it is sought that the raw materials are always available and do not imply a decrease in natural resources. The use of biomass, agro-industrial and industrial wastewater discharges, and urban organic solid wastes, among others, are of interest due to their richness in carbohydrates, organic acids, and proteins and, above all, for being abundant and inexpensive [2, 3]. Among different processes that produce biohydrogen, dark fermentation (DF) is considered the most economical [4]. In the last decade, the research focused on DF has increased dramatically using different feedstocks, where the major obstacle is the low H_2 yield compared to the theoretical maximum and its low level of technological readiness [5].

Using different wastes as a substrate to produce hydrogen depends on their content of carbohydrates or organic acids and, consequently, the potential yield. Recent advances show that dark fermentative hydrogen production is driven by two main metabolic pathways: carbohydrates and lactate dependents [6]. The first pathway involves the conversion of glucose into pyruvate, which is converted to acetyl CoA and reduces ferredoxin or formate, where hydrogen synthesis occurs through the oxidation of ferredoxin [7]. In the lactate pathway, the Flavin Adenine Dinucleotide-dependent lactate dehydrogenase forms stable complexes with the flavoprotein that converts pyruvate into a coenzyme (acetyl-CoA) to later form butyrate from a molecule of acetoacetyl-CoA reduced to butyryl-CoA [8].

Given the different residues available and their variable composition, several specialized microorganisms are attributed to performing the DF process. In such a process, a wide microbial diversity is present, where the contribution of different microorganisms is unclear and can contribute to or decrease the process performance [9]. The study of microbial communities should go beyond the "black box" approach to contribute to understanding the interactions and to know the effect of these interactions on hydrogen production [9, 10]. In this sense, previously have been reported around 22 common genera in DF reactors, such as *Clostridium, Enterobacter, Bacillus*, and *Lactobacillus,* among others, where the type of substrate and operating conditions can define each community, their function, and interactions [5, 11]. Sequencing *16S rRNA* gene amplicons with high-throughput technology is common for studying microbial communities in DF, and its application with other omics sciences has been widely reviewed [12–14]. Sequencing holds excellent potential for identifying common and uncommon taxonomic groups within communities. However, the results interpretation, the community correlation to process performance, and the ability to identify key microorganisms depend on further analysis. This chapter reviews the purpose of analytical tools like statistical multivariate, co-occurrence networks, metabolic prediction, and modeling. These tools offer a means to uncover microbial communities' ecological interactions and specific contributions to hydrogen production. This information can inform operational decisions to enhance the performance of DF processes.

7.2 Understanding Communities from an Ecological Perspective

Microbial communities show highly complex and dynamic properties that might be understood using an ecological focus [15]. Hydrogen production via DF occurs thanks to the cooperation between microbial community members (including bacteria, fungi, and probably viruses and protists). Despite their complexity, a systematic ecological research approach should [10]: (1) identify the microbial species present in a community and (2) their function; (3) describe ecological interactions; (4) analyze community-level properties; (5) understand the effect of the physicochemical environment in microbial dynamics and (6) identify broad patterns applicable to several types of communities.

The description of the taxonomical identity and abundance of microbial diversity in an ecosystem is often used to infer function in community contexts directly, but many factors alter this connection. It is important to remark that the structure of community results from the interplay between bacterial metabolism, environmental conditions (e. g. pH, nutrient availability), and historical and randomly occurring (i.e., stochastic) eco-evolutionary processes such as migration, adaptation, and biotic interactions [16]. These processes influence the assembly of microbial communities and the different states in which a community achieves stability in composition and function [17]. In the case of DF consortia, successful attempts have been made by modifying culture parameters that effectively increase the presence of hydrogen-producing bacteria. For example, Santiago et al. [18] explored how the hydraulic retention time (HRT) changed the assemblage in sequencing batch reactors, affecting the abundance of *Enterobacter*, *Clostridum*, and *Lactobacillus* bacteria.

The relationship between the taxonomical composition and the array of functions of microbial communities is complex. Microbial diversity alone cannot explain the metabolic capacities of bacterial communities. The disconnection between taxonomical and functional diversity can result from the phenotypic variation inside taxonomical groups (i.e., species, genera, families, etc.) [19]. For instance, the *Clostridium* genus is the most studied and used bacteria in hydrogen fermentation bioreactors. Notably, the *Clostridium* genus is highly polyphyletic (i.e., the members of the genus are not evolutionarily related but, instead, they were classified based only on phenotypic characteristics like spore formation or anaerobic growth) and, thus, exhibits wide functional diversity and pathogenetic capabilities [20]. In this same line, Wang et al. reviewed the high variability of multiple strains of *Clostridium* species in terms of hydrogen production and substrate utilization [21]. This intra-group functional diversity might be widespread in the bacteria that form part of hydrogen-producing consortia and must be studied individually.

In addition, bacterial metabolism includes several mechanisms that regulate the expression of functions [22]. These mechanisms are influenced by abiotic environmental factors (such as nutrient availability) and biotic factors, such as other organisms [23]. In this same line, some functions have been shown to arise from biotic interactions and cannot be predicted from the individual characteristics of populations [24]. As an example of how function (and interactions) can be affected by functional intra-group diversity, Perez-Rangel et al. [25] found that *Lactobacillus* species (which are typically associated with decreased hydrogen production) do not always inhibit hydrogen production.

Biotic interactions are among the most influencing mechanisms underlying microbial communities' composition and function. Despite their apparent simplicity, bacteria have been shown to engage in complex social behaviors. Hence, recent work on microbial ecology has focused on determining the types and effects of biotic interactions between microorganisms. Microbial interactions can be highly variable and occur within a continuous axis that ranges from strongly negative to strongly positive interactions, driven by the need to access (or limit the access of other organisms) to nutritional resources [24]. Hence, they strongly influence the growth of different populations and, thus, the function of whole microbial systems. Complexity increases because bacterial interactions change with the environment (temperature, moisture, pH, nutrients, and other organisms). The importance of detailing these context-dependent interactions in systems like bioreactors lies in the fact that changes in environmental conditions can have unexpected effects on populations responsible for the desired functions that are not necessarily related to the metabolism of the focus bacteria, but rather on their interactions with other bacteria [26]. In DF consortia, several types of interactions have been described. For example, it is known that species of the genus *Bacillus* participate in the hydrolysis of complex substrates that make simple sugars available to hydrogen producers [9]. On the other hand, species of the genus *Klebsiella* consume oxygen, allowing the growth of strictly anaerobic hydrogen producers (e.g., *Clostridium* species). Interactions that decrease hydrogen production have been identified, such as the production of bacteriocins (a type of competition for interference) or the consumption of hydrogen (a type of commensalism) [27].

Lastly, it is essential to remark that interactions are usually analyzed between species pairs for methodological (e.g., experimental, mathematical, or computational) simplicity. More realistically, interactions occur between groups of species. These interactions can lead to simultaneous but differential effects on multiple populations. Some of those effects include chain effects (where the effects of one interaction propagate to other populations), rock-paper-scissors mechanics, or high-order interactions (in which two or more species simultaneously affect a third focal species) [28]. Recent research on the topic promises to improve our predicting power on the properties of microbial communities and how they respond against perturbations. For instance, Sanchez-Gorostiaga et al. [29] analyzed how the robustness of a simple function (amylolytic activity) was altered by the interactions in a synthetic consortium composed of several species of *Bacillus* and a *Paenibacillus polymyxa* strain. In DF, few have focused on studying interactions between groups of bacteria. Recently, Fuentes-Santiago et al. [30] revealed positive interactions driving the fermentative hydrogen production from cheese whey and negative interactions affecting the process when the community used winery vinasses as substrate. Comprehending the microbial communities that drive DF holds significant potential and utility. However, integrating analytical tools that can explain such characteristics is not typically included in such studies.

7.3 Microbial Interactions and Tools to Elucidate Them

Microbial interactions can be classified as positive, negative, or neutral. Mutualism is a type of microbial interaction where both partners of varied species are beneficiated. Biofilms are a classic example of mutualism since they are widely recognized as a beneficial strategy that microbial communities use to protect themselves from external factors. Biofilm bioreactors have been widely used to improve hydrogen production in continuous reactors, conferring protection to the H_2-producing populations against inhibitory substances [31], increasing the organic loading rate [32], and avoiding the washed out of cells from the reactor [33]. Cross-feeding and synergism are other beneficial interactions between two populations. In cross-feeding, one species uses products excreted by another without affecting it. This beneficial interaction improved hydrogen production from glucose by a factor of 2.5 in co-cultures of *Clostridium acetobutylicum* with *Desulfovibrio vulgaris* [34]. In this co-culture, *C. acetobutylicum* grew on glucose to produce hydrogen, providing nutrients to *D. vulgaris*. This interaction triggered changes in the metabolic fluxes of *C. acetobutylicum*, resulting in improved hydrogen production. Synergism between H_2 producers was reported for *Enterococcus* spp. isolated from wheat straw [35]. A three-species culture of *Enterococcus* produced hydrogen from a consolidated bioprocess of xylan, producing more hydrogen than the sum of the hydrogen production from the single-species cultures. The negative interactions include competition (a loss-loss interaction), antagonism (a win-loss interaction), and amensalism (a loss-neutral interaction) [36]. Competition refers to when two populations have a similar niche and exclude each other [37]. In fermentative hydrogen production, *Lactobacillus* (or more generally, lactic acid bacteria, LAB) has been identified to establish negative interactions with *Clostridium* [38], a negative relationship that could be classified as competition if both populations lose and antagonism if only *Clostridium* loss. Amensalism refers to when one population harms another without benefit in return, and neutralism is when two populations do not interact. Sometimes, it is challenging to categorize competition, antagonism, and amensalism since the experimental approaches used to study these ecological interactions are unidirectional. Understanding these microbial interactions and the factors that influence them is particularly interesting nowadays.

There are different experimental approaches to mapping microbial interactions, which include culture-dependent methods and culture-independent methods. These approaches aim to describe at least one of the four attributes of a microbial interaction: reciprocity, strength, the mode of action, and the time scale [37]. The culture-dependent methods typically use co-culturing microorganisms to characterize and measure interactions between microorganisms. Under this approach, Wang et al. [39] demonstrated a mutualist relationship between *Bacillus cereus* A1 and *Brevundimonas naejangsanensis* B1. Both species cooperate for hydrolyzing starch, and the products of one species support the growth of the other. Also, culture-dependent methods map microbial interactions without co-culturing microorganisms; instead, one species is cultured in the cell-free spent medium of another species. Pérez-Rangel et al. [40] used this approach to demonstrate that the supernatant of LAB *Pediococcus acidilactici* inhibited the growth of the hydrogen producers *Enterococcus* due to the excretion of bacteriocin-like inhibitory substances. Despite its usefulness, the culture-dependent methods are insufficient to categorize microbial interactions in complex environments where all change in time: species diversity, fermentation products, and physicochemical parameters.

The culture-independent methods are based on high-throughput sequencing to determine species abundance. Once species abundance is obtained, a network serves to infer patterns of co-presence and mutual exclusion between taxa [36]. The theory of biological diversity states that if two species compete, they cannot coexist stably in an environment [41]. Therefore, two species have a positive interaction when they show a similar abundance pattern over multiple samples, and vice versa; two species have a negative interaction when they show a mutual exclusion. Under this approach, it is possible to categorize positive or negative interactions; however, it is impossible to map the reciprocity, the strength, and the time scale. Construction of microbial networks from abundance data requires specialized approaches: briefly, (i) species abundance matrix across relatively large numbers of samples or time points, (ii) compute pairwise similarity scores between two species using Pearson or Spearman correlations, (iii) determine the significance of the similarity score, which requires a null model (typically permutations) to assign p-values to each score, and (iv) only significant pairwise relationships are then combined to construct a network [36].

A *network* is a graphical visualization showing different species (nodes) and their connections (edges) at a specific time and space. Once the network is constructed, its analysis includes characterization of the network topology, module detection, and identification of hubs. Network topology refers to the network arrangement of nodes and edges (Fig. 7.1). Some metrics that describe the structural properties of networks are degree distribution, average path, and centrality metrics (degree centrality, closeness centrality, betweenness centrality, and eigenvector centrality) [42]. These network metrics describe the relative importance of each node inside

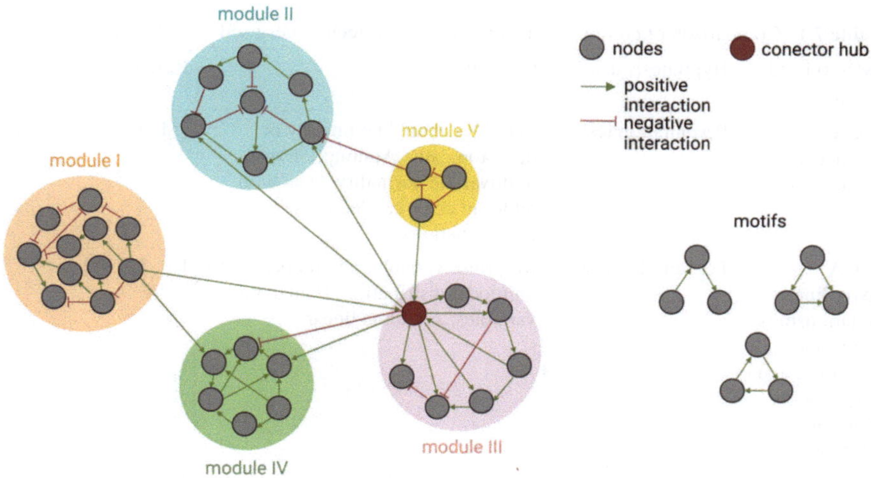

Fig. 7.1 Typical components that make up a co-occurrence network analysis

the network and its possible contribution to the ecosystem functionally. Ranking nodes by their degree (for example) could talk about their biological importance. Networks are often decomposed into smaller units called modules and motifs, nodes highly connected between them. At the local level, some motifs represent a particular ecological role, such as mutual predation between two species, a simple food chain, and a predator preying on two species [43]. Motifs are aggregated into modules connected to each other through a small number of linker nodes. Depending on how each node is connected inside the networks, it can be classified as module hubs and non-hubs [44]. Hubs are highly connected nodes representing potential keystone species, significantly affecting all the other ecosystem species [45]. Microbial networks represent an opportunity to elucidate the complex microbial interactions established in hydrogen reactors and identify keystone species in converting complex substrates into hydrogen.

7.4 Multivariate Statistical Analysis to Elucidate Key Microorganisms

Multivariate statistical methods are valuable tools to discern determinant factors for the efficiency of DF systems, either operational parameters or microbial community structure. There are various statistical methods available to analyze microbes and identify patterns related to important variables (e.g., VFA concentration, H_2 production, or inhibitory compound concentration) or taxonomic groups linked to the system performance (e.g., high/low H_2 yield, stability/instability periods, high/low organic loading rates, or sampling days). These analyses bring valuable information for

Table 7.1 Comparison of common multivariate analysis techniques used in dark fermentation

Method and purpose	Hypothesis test	Limitations	References
PCA. Dimensional reduction	Permutation test	Variables should be linearly or monotonically. Assumption of multivariate normality. Data with multiple zeros can lead to errors in the ordinations	[51, 52, 58, 75, 76]
CCA. Non-linear relationships between response and explanatory variables (X^2 distance)	Permutation test	Response variables should belong to unimodal distribution. Explanatory variables should be linear	[61, 77, 78]
RDA. Linear relationships between response and explanatory variables (Euclidean distance)	Permutation test, F-test, ANOVA or MANOVA	Not all the variance in the original response matrix is presented. Response data is a distance or (dis)similarity matrix, which can be considered distance-based RDA. Delete the influence of a set of explanatory variables (e.g., experimental block) before RDA	[53, 79, 80]
NMDS. Ordination of samples based on any distance or dissimilarity matrix	ANOSIM, PERMANOVA or PERMDISP	Avoid relating environmental variables (or other explanatory variables) to the distances in an NMDS ordination	[56–58]

optimizing the process, focusing on key microorganisms that have the most significant impact on the efficiency of DF [46, 47]. In this sense, multivariate statistical methods summarize high dimensional data to identify important patterns, relationships, or correlations among microbial communities and operational variables [48]. The most common multivariate analyses are principal components analysis (PCA), redundancy analysis (RDA), Non-metric multidimensional scaling (NMDS), and Canonical correspondence analysis (CCA) (Table 7.1).

7.4.1 Principal Components Analysis (PCA)

The PCA reduces the dimensionality of variables, converting original variables into a new set of principal components (PC) [48, 49]. The PC captures the data variance (> 70% or < 40% suggests a high or a low percentage of variance captured, respectively), and its significance is assessed by a permutation test [47, 50]. PCA plot allows the observation of trends from key variables and clusters based on categorical variables (e.g., HRT, OLR, etc.) [48, 49]. The PCA interpretation can consider scaling I and II. The scaling I (distance biplot) uses the variance–covariance matrix (usually microbial abundance data) to provide insights about relationships between the categorical variables using Euclidean distance [48]. The scaling II (correlation biplot) is focused on a correlation matrix (usually standardized data) to determine the correlations among the variables (e.g., microbial abundance data vs. operational conditions) [49]. The PCA has identified variables and microorganisms that can contribute during the various stages of DF. For example, during the DF of olive-mill wastes, PCA showed that lactate production was related to the higher abundance of *Turicibacter,* considered as lactic acid bacteria (LAB); meanwhile, butyrate concentration was related to H_2 producer bacteria (HPB) associated with the *Clostridium* genus and to lactic acid consumption [51]. Similarly, a PCA revealed a positive correlation between the abundance of *Clostridium*, H_2, acetate, and butyrate production during the fermenting food wastes [51].

7.4.2 Redundancy Analysis (RDA)

The RDA is a technique used to model linear relationships between two sets of variables: response/dependent variables (e.g., microbial abundance data) and explanatory/independent variables (e.g., operational/environmental conditions) [47]. The multivariate linear regression tests the response variable constrained to an explanatory variable, generating a matrix of the fitted values. In other words, the regression analysis seeks to understand the relationship between multiple explanatory variables and how they collectively influence the response variables [49]. The RDA method produces two ordinations; one in the space is performed using the fitted values matrix as the response variables and another in the space defined by the explanatory variables as the predictors [47]. Like PCA, the RDA results can be interpreted by the scaling chosen (scaling I and II) [47]. Also, the significance of the RDA model can be assessed using a permutation test, F-test, ANOVA, or MANOVA [47, 49]. Besides, redundancy statistic (R^2) is a measure of the strength of the linear relationship between response and explanatory variables, whose values are between 0 (poor fit) and 1 (good fit). The R^2 can be adjusted considering the number of explanatory variables [49]. The RDA has been used to find key microorganisms to DF linked with adding trace elements that improve the H_2 production. The Fe^{2+} addition was positively correlated with H_2 production, increasing *Enterobacteriaceae* and *Acinetobacter* abundance, taxon related to hydrogen production, and hydrolytic bacteria, respectively [53].

7.4.3 Non-metric Multidimensional Scaling (NMDS)

NMDS is a method for visually representing the association among a set of samples in a lower dimensional space using any distance or dissimilarity matrix [47, 49]. NMDS creates an ordination adjusting the positions of the samples until a minimum stress value is achieved [49]. It has been suggested that stress values of < 0.05 are excellent, < 0.10 are good, < 0.20 are usable, and > 0.20 are not acceptable [54]. To determine the statistical significance of ordination measures between groups, either PERMANOVA, analysis of group similarities (ANOSIM), or homogeneity of group dispersions (PERMDISP) can be used [55]. NMDS using Bray–Curtis dissimilarity matrix have commonly been used to assess variables that affect microbial community structure during DF. For instance, the effect of low N and P concentrations [56], changes in pH and temperature [57], and the impact of the type of hydrolysate and reactor configuration on hydrogen production [58].

7.4.4 Canonical Correspondence Analysis (CCA)

Similarly to RDA, the CCA method performs a direct comparison to find relationships between explanatory variables (e.g., operational/environmental conditions) and response variables (e.g., microbial abundance data) using multivariate linear regression to be represented on biplot [59]. The difference lies in that the CCA preserves the chi-square distance instead of the Euclidean distance (using RDA) in the matrix of the fitted values [59]. By preserving the chi-square distance, CCA can better capture the non-linear relationships between response and explanatory variables, leading to a more accurate representation of the data; however, rare microorganisms may disproportionately influence the analysis [48, 59]. The interpretation of the CCA plot depends on the scaling chosen, as with other statistical approaches. The scaling I are relationships between categorical variables. As a result, categorical variables serve as the centroids of the response variables, and their separation between categorical variables represents their chi-square distances. Dots close to centroids are more likely to achieve that condition. The scaling II focuses on the relationships among response variables. The interpretation is identical to scaling I, except that the term categorical variables are replaced with response variables [47]. Also, CCA offers hypothesis testing via a permutation test [60].

The CCA has identified microorganisms associated with DF performance. For instance, a high H_2 production increased with the abundance of *Clostridium* (HPB), *Azotobacter* (hydrolytic bacteria), and *Lactobacillus* (LAB), along with positive correlations with acetic and butyric acids [61].

Comparing the results of multiple multivariate analyses can provide a more comprehensive understanding of the relationships between environmental factors and microbial community composition. It can help identify robust and reliable patterns consistent across different methods [47, 62]. In multivariate analysis, PCA can frequently serve as a useful beginning point by enabling you to find trends, groupings, important variables, and potential outliers [47]. The RDA can assess the statistical significance of the relationships between explanatory and response variables, providing valuable insights into the key factors that shape microbial community structure [48]. Although Euclidean distance is a commonly used metric for analyzing raw species abundance data, it may not always be appropriate. This is particularly true when the data contains null abundance or non-linear relationships; in such cases, CCA can be used as an alternative [59]. Meanwhile, NMDS can be used to assess categorical variables clustering with similar microbial compositions, examining their distances and infer the degree of dissimilarity and potential ecological processes driving the observed patterns [47].

7.5 Prediction of Functional Profiles

After characterizing microbial communities in biohydrogen reactors either by the *16S rRNA* marker gene or by metagenomics—the so-called taxonomic profile—a natural question arises: what is the function of the microorganisms found? Understanding the microbial communities' functions is crucial to predicting their response to disturbances and environmental changes. The so-called functional profile describes a microbial community's metabolic capacity based on the presence or absence of genes involved in metabolic pathways, i.e., it is a picture of a community's metabolism based on the metagenome. Recent bioinformatic advances allowed the development of functional prediction tools based on DNA metabarcoding or metagenomics. Functional prediction based on *16S rRNA* is preferred due to the higher cost of shotgun sequencing and metagenomics, among which PICRUSt and Tax4Fun are the most popular tools [63]. The most recent versions of PICRUSt2 and Tax4Fun2, allow using the *16S rRNA* amplicon sequences as input data, expands the genome reference databases, and include new valuable features such as predictions from *18S rRNA* of the fungal kingdom and calculations of functional redundancy [64, 65]. These tools first compare the input data, which can be taxonomy or *16S rRNA* amplicon sequences, with reference genomes available in the tools' databases (e.g., SILVA, RDP, or Greengenes). Then, inferred metagenomes are constructed with the matches found between the input data and the reference genomes and are mapped to specialized databases containing known microbial genes, enzymes, or metabolic pathways (e.g., KEGG, COG, MetaCyC, CAZy, among others). Finally, a gene-function-pathway association is constructed. The abundance of these functions is determined based on the taxonomic abundance in the inferred metagenomes.

Functional prediction based on metagenomics is less reported but is possible thanks to specialized tools such as eggNOG-mapper, GhostKOALA, MG-RAST, and PANNZER2, among others [66]. Most of these tools are based on comparing predicted gene sequences with known genes in databases such as KEGG and COG, using different BLAST variants [67]. The main advantage of metagenome-based prediction is that it shows a more realistic picture of the microbial community's functional capacity because functions are inferred based on the actual genetic content. Besides, other phenomena, such as lateral gene transfer and gene loss, can be studied [63].

Functional profiles prediction in H_2 production systems by DF is very recent (Table 7.2). Most of these investigations have used batch fermentations and pure substrates as models, but the tendency is towards using laboratory-scale bioreactors and agro-industrial effluents. The primary tools used to predict functional profiles are based on *16S rRNA* and the first version of PICRUSt. However, the most recent versions of PICRUSt2, Tax4Fun2, and even prediction based on metagenomics have already started to be used. The main interest has been predicting genes related to fermentative pathways associated with H_2 production, considering 2–40 genes. The purpose of predicting functional profiles in the different studies is diverse and very specific; however, the vast majority have used this tool to infer the abundance level of genes related to H_2 production due to substrate composition, the use of H_2 production enhancing strategies, and changes in reactor operation variables.

Functional prediction tools applied to DF have found evidence about the contribution of specific metabolic pathways, especially those associated with H_2 production from lactate and acetate, which has been the study object of several investigations [51, 68–70]. The interest in lactate-based metabolic pathways for H_2 production has increased in recent years, and it is expected that these pathways will be critical to develop H_2 production systems treating agro-industrial effluents [6]. In this sense, predicting functional profiles will be essential for studying and exploiting these metabolisms. Functional prediction can be used to study the environmental factors triggering the induction of lactate-based metabolic pathways or genes related to the cellular communication mechanisms regulating the balance between H_2-producing bacteria and lactic acid bacteria.

Regarding the main limitations of functional prediction, a study recently revealed that PICRUSt, PICRUSt2, and Tax4Fun are probably limited in predicting functions when dealing with non-human samples [71]. Metagenome-based prediction might also be limited because prediction accuracy may vary depending on the functional categories, with better performances for genes related to "housekeeping" functions. The authors argued that one of the main reasons for these limitations could be the bias of genome databases toward human-related microorganisms and that developing specific tools for particular environments using appropriate reference databases could improve the prediction of functional profiles [71]. An excellent example of specific tools is CowPI, a functional inference tool specific to the rumen microbiome, which performed better than PICRUSt using a specialized database for this environment [72]. For this reason, it is essential to highlight and support recent achievements, such as specialized genetic catalogs for studying microbial communities in wastewater treatment plants and anaerobic digestion reactors [73, 74]. However, as the technology for H_2 production by DF evolves, efforts should be directed toward developing specific databases for this process and specific tools for predicting functional profiles.

Table 7.2 The use of tools for functional profile prediction in studying H_2 production by dark fermentation

Reactor Configuration	Substrate	Purpose of the functional analysis	Reference
16S rRNA + PICRUSt			
Batch	Glucose	To study the metabolic profile of co-cultures under different mixing ratios	[68]
		To examine the effect of inoculum pretreatment using ionizing radiation	[81]
		To study the effect of $CaFe_2O_4$ nanoparticles	[82]
	Pretreated sewage sludge	To study the effect of Fe^{2+} on the abundance of genes related to H_2 production	[83]
	Olive mill waste	To study substrate conversion routes of olive mill waste	[51]
Dynamic membrane bioreactor	Glucose	To compare the metabolic potential with and without biomass immobilization	[84]
		To predict gene abundance related to lactic acid and H_2 production	[69]
16S rRNA + Tax4Fun2			
Fluidized bed reactor	Cheese whey and glycerol	To infer genes related to different organic loading rates	[85]
Batch	Cheese whey and winery vinasses	To study the metabolic profile of different substrates composition	[30]
Metagenomics + GhostKOALA			
Batch	Molasses, lactate, acetate	To identify the presence of genes encoding fermentation	[70]
Packet-bed reactors	Sugar-beet molasses	To compare the metabolic potential of microbial communities	[86]

7.6 Modelling Tools to Drive Communities and Metabolisms

A model is a useful abstract representation of a physical system. Usually, models aim to represent only the relationship between causes and effects. However, sometimes, they also want to look more deeply into the system to see the internal workings of the process. In the case of bioprocesses, the systems are generally highly complex, with internal relationships and intricate networks that may need to be considered. Nevertheless, the mathematical models of anaerobic processes range from straightforward algebraic relations to non-linear partial differential equations. The models allow us to propose better strategies for operating these processes, especially when the systems are viewed from the system dynamics perspective. This section explores two perspectives for modeling DF: from a macroscopic perspective using differential equations and a microscopic perspective using metabolic models on a genomic scale.

7.6.1 Dynamic Models for Dark Fermentation

A crucial milestone for Anaerobic digestion (AD) dynamic modeling was the publication of the Anaerobic Digestion Model 1 (ADM1) [87]. Ever since, 504 publications have followed, reporting applications, modifications, new process inclusions, and the adaptation of ADM1 to model DF. The first two attempts to adapt the ADM1 to DF were: (i) suppressing the acetogenic and methanogenic steps; (ii) imposing operational conditions that hinder these bioreactions (e.g., low pH values that inhibit methanogenesis or low HRT values to washout the methanogens) [88]. Another approach used the ADM1 as a basis and simplified the model for conditions pertaining to DF, keeping only some bioreactions modifying them, and adding other processes [89–92]. An interesting approach assumes the thermodynamic limitations during DF and considers variable stoichiometry [93]. Recent proposals use dedicated software concentrating on the disintegration of complex particulates and the hydrolysis steps with fixed stoichiometry, fitting the parameters to experimental conditions [94, 95].

Models for DF based on ADM1 have thus been used successfully to fit the experimental data obtained from biohydrogen production from several residues. For example, when using cheese whey as a substrate, [96] used a variable stoichiometry approach to predict biohydrogen production at different HRT values. The same substrate was studied using an ADM1-based model to explain the role of lactate and acetate in biohydrogen production [97], while [98] used a simpler model to propose a feedback control strategy for maximizing hydrogen production from acid cheese whey. Other substrates used to fit ADM1-based models have been sugarcane vinasse [99], cocoa and coffee mucilage and swine manure [100], and macroalgae [101].

Other proposals for dynamic modeling of DF do not explicitly use the ADM1 as a basis but instead build the model explicitly. For example, [102] proposed a straightforward model that considers only two bioreactions and a single biomass to model hydrogen production from molasses, [103] have proposed a model that is suitable for two-phase AD systems, whose first stage is DF, and [104] incorporated a clarifier in the model.

The models reviewed above comprise mechanistic models, i.e., based on performing mass and energy balances on the system to find the underlying differential equations. This construction method is opposed to empirical or statistical models that build the relationship between causes and effects (*inputs* and *outputs*, respectively, in dynamical systems theory). Nevertheless, the kinetics that define the different bioreaction rates involved in these mechanistic models have a certain degree of empiricism, as most of these relations come from experimental observations and/or experience. The models (especially those based on ADM1) are also usually classified as *unstructured* because the different microorganisms that perform the biochemical reactions are not distinguished individually. A group or guild of microorganisms is assumed to be lumped as a single entity performing a metabolic task. In this sense, a single microbial species, for example, may belong simultaneously to the acetate consumers but also be involved in the hydrolysis of carbohydrates.

Furthermore, in *unstructured* models, the biomass of the different microbial groups is not differentiated; they act as entities that may grow, utilize substrates for conversion into other metabolites, or die due to decay. We do not "see" the inner parts of the cells and the mechanisms that provoke these metabolic changes. In contrast, *structured* models distinguish the inner metabolic processes inside the cells. They differentiate between the different aspects comprising a group of microorganisms or even a single microbial species: cell count, ATP content, amount of pyruvate, etc. A simple structured model may distinguish between the biomass's active fraction and a storage product.

7.6.2 Genome-Scale Metabolic Models for Dark Fermentation

The first publications date from 2009 for genome-scale metabolic models [105], and the interest in this modeling tool is increasing. Genome-scale metabolic models (GEMs) are stoichiometry-based models that describe an organism or a community's entire metabolic reactions and pathways using gene-protein-reaction associations formulated based on genome annotation and experimental data [106, 107]. GEMs are *structured* models because they contain detailed information about intracellular processes. Nowadays, GEMs can be reconstructed using several tools [108]. At the same time, the metabolic fluxes through the metabolic network can be predicted using optimization techniques such as flux balance analysis [109], for which there are also several software tools [110].

Typically, GEMs have been used to predict targets for gene manipulation of bacteria and eukaryotes to enhance the microbial production of several chemicals [107]. In recent years, there has been a growing interest in using GEMs to study the complexity of natural and engineered microbial communities for bioprocesses. In particular, the GEMs applied to microbial communities are expected to have some of the following applications: (i) to elucidate microbial interactions, exchange of metabolites, and the relationship between the presence and function of microbial species [15]; (ii) to design optimizing strategies for the production of chemicals and materials of interest, based on the simulation of different environmental scenarios [111]; (iii) to integrate the data provided by omic sciences such as metagenomics, metatranscriptomics, metaproteomics, or metabolomics [112]; (iv) as tools for the design of synthetic microbial communities with many biotechnological applications [113].

The use of GEMs in DF processes can be divided into three stages in time. The first stage corresponds to the *in-silico* reconstruction and modeling of specific microorganisms, using flux balance analysis to guide genetic modification through metabolic engineering [114]. GEMs have been developed for several H_2-producing bacteria, such as *Clostridium butyricum*, *Escherichia coli*, *Citrobacter amalonaticus*, and *Thermotoga maritima* [115, 116]. The second stage comprises the reconstruction of simplified GEMs for mixed cultures based on the concept of the universal bacterium, i.e., an approach where the microbial community is considered a single entity and all metabolic reactions and metabolites from the different species or guilds are combined into a single model [111]. This approach was used with several aims in the study of DF, for example, to elucidate the acetogenic consumption of H_2, to study the effect of inoculum pretreatment strategies, to study the effect of pH, and to identify genetic modification targets [117–119]. The third stage, the present stage, is characterized by the availability of several microbiology and molecular tools that facilitate the study of microbial communities to produce H_2. These tools are expected to encourage using GEMs in DF systems [13]. In a recent study, GEMs were used to simulate the exchange of metabolites in co-cultures of *Clostridium beijerinckii* and *Yokenella regensburgei*, predicting an increase of biomass and H_2 based on the cross-feeding of lactate to *C. beijerinckii* [120]. Another recent study proposed incorporating electron bifurcation reactions to improve the prediction capability of current GEMs for DF [121].

7.7 Learnings from Microbial Communities, Engineering Perspectives

Understanding microbial communities from an ecological perspective involves a holistic view of the composition and interactions between microorganisms and their environment. By examining these relationships, it is possible to gain insights into the complex microbial ecosystems critical to many ecosystems' functioning. Further advances in this research area might make it possible to precisely understand why some microbial species are dominant, how new ecological niches emerge, or why the communities that inhabit bioreactors show shifts in their metabolic pathways. The understanding that leads to controlling and predicting the composition and function of microbial communities is based on deciphering the complex multivariate processes that occur within them. At the same time, the management of bioreactors can greatly benefit from understanding the ecological mechanisms that rule their behavior. This kind of property represents the current frontier in microbial ecology research.

As engineering selected consortia, DF usually has easily defined functions and reduced diversity, which can help simplify the measurable characteristics of microbial communities. Also, these systems have defined physical boundaries, and sometimes it is easier to control the species composition and cellular density or simulate ecological processes (e.g., invasion). For instance, bioaugmentation has introduced one or more strains to an existing consortium to improve hydrogen production (e.g. by increasing substrate degradation capabilities). Ecologically, bioaugmentation can be seen as a form of invasion in which a foreign species has to establish in a new community. For this, the parameters like the substrate and inoculation ratio must be carefully considered. For instance, Valdez-Vazquez et al. bioaugmented the native microbial communities of several complex substrates with hydrolytic and fermentative *Clostridium* species [122]. In the latter example, the authors identified that the bioaugmentation was only successful on certain substrates and inoculation ratios. Invasion parameters and conditions should be studied to improve bioaugmentation or to prevent the establishment of bacterial species detrimental to hydrogen production.

According to several authors, the future of H_2 production by DF is in designing synthetic microbial communities designed to maximize productivity and take full advantage of the composition of complex substrates [13, 123, 124]. The extended use of functional profiles could help us to reach this understanding and to overcome some of the most critical current limitations for the process, including the low H_2 yield, production of inhibitory by-products, presence of hydrogen-consuming bacteria, and substrate complexity [125]. Functional prediction can be used to study the effects on the metabolism of recent technological advances and strategies to improve H_2 production, for example, biomass pretreatments, biomass immobilization, additives, bioaugmentation, and metabolic engineering tools [123, 125]. However, these applications will depend on our ability to validate the prediction of functions in DF because many genes predicted from metagenomes may not be expressed under a given condition. For this reason, the analysis of functional profiles can be combined with the information provided by metatranscriptomics and metaproteomics to provide better insights into functional interactions in microbial communities [126, 127]. Research articles and reviews on dark fermentation have been published for at least 21 years and have increased significantly in the last five years [128], so vast information about the process and microbial communities has been generated. Soon, the standardization of protocols for analyzing microorganisms and metabolisms in DF should be considered to have easily comparable data sets, as has been done with the protocols for determining biohydrogen potential [129].

Multivariate statistical analysis is a powerful tool for determining patterns and relationships between variables. Another potential tool is differential analysis, which identifies biologically relevant features that differ significantly in abundance between groups, using only read count data [130]. In this sense, such analysis has been applied, for instance, to evaluate differential transcriptomic profiles in fermentation systems [131]. However, it also can be applied to reveal key microorganisms capable of distinguishing different groups of samples or conditions, as has been used in AD which may serve as potential biomarkers of different operational conditions and performance [132]. This information can be used to understand the underlying mechanisms or factors driving the observed differences. Additionally, biomarkers can be used as indicators of microbial activity or diagnostic tools to detect and monitor microbial communities in different contexts. Despite several studies applying the approach of finding biomarkers using differential analysis in an analogous system like anaerobic digestion [133–135], such tools are underused for DF.

Regarding the future of GEMs in DF, it is worth observing their use in other bioprocesses like AD, with interesting applications already found in the literature;

for example, [136] reconstructed 836 GEMs to represent the functional capabilities of bacteria and archaea in AD, managing to predict many microbial interactions and exchange of metabolites. In another example, [137] used reported GEMs for two methanogenic microorganisms, *Methanosarcina barkeri*, and *Methanococcus maripaludis*, to simulate methanogenesis in the ADM1 model, combining these two modeling tools with dynamic flux balance analysis to predict intracellular fluxes, and taking a step forward to improve the modeling of AD. Even more promising are, for example, developments based on GEMs to simulate microbial ecosystems in time and space with application to biofilms [138] or the combined use of machine learning and GEMs to harness multi-omics information [139]. It is only a matter of time before these approaches can be applied to DF, so soon, we could have powerful tools that will allow us to understand microbial interactions better and drive microbial metabolism towards more efficient DF processes.

7.8 Conclusions

Including an ecological perspective during the microbial community characterization and analysis is a highly valuable tool to understand dark fermentative systems performance and to propose operational strategies to increase hydrogen or carboxylic acids production. The main challenge is to go from a simple microbial characterization to a deep ecological perspective analysis, meaning understanding the interaction between community members and the environment. To such aim, combining new sequencing technologies based on omics sciences with network analysis, multivariate statistical analysis, functional profiles, and their mathematical modeling are available tools. Such kind of understanding of fermentative microbial communities is on the edge of knowledge and can help to take full advantage of their metabolic flexibility.

Acknowledgements This work was supported by the project "Grupos Interdisciplinarios de Investigación (GII)", Instituto de Ingeniería, UNAM; and the DGAPA-UNAM PAPIIT projects (grants no. IN102721, IN105423, IN103722, and TA101822).

References

1. Saravanan A, Senthil Kumar P, Khoo KS et al (2021) Biohydrogen from organic wastes as a clean and environment-friendly energy source: Production pathways, feedstock types, and future prospects. Bioresour Technol 342.https://doi.org/10.1016/J.BIORTECH.2021.126021
2. Camacho CI, Estévez S, Conde JJ et al (2022) Dark fermentation as an environmentally sustainable WIN-WIN solution for bioenergy production. J Clean Prod 374.https://doi.org/10.1016/J.JCLEPRO.2022.134026
3. Boodhun BSF, Mudhoo A, Kumar G et al (2017) Research perspectives on constraints, prospects and opportunities in biohydrogen production. Int J Hydrogen Energy 42:27471–27481. https://doi.org/10.1016/J.IJHYDENE.2017.04.077

4. Aydin MI, Karaca AE, Qureshy AMMI, Dincer I (2021) A comparative review on clean hydrogen production from wastewaters. J Environ Manage 279:111793. https://doi.org/10.1016/J.JENVMAN.2020.111793
5. Lopez-Hidalgo AM, Smoliński A, Sanchez A (2022) A meta-analysis of research trends on hydrogen production via dark fermentation. Int J Hydrogen Energy 47:13300–13339. https://doi.org/10.1016/J.IJHYDENE.2022.02.106
6. García-Depraect O, Castro-Muñoz R, Muñoz R et al (2021) A review on the factors influencing biohydrogen production from lactate: the key to unlocking enhanced dark fermentative processes. Bioresour Technol 324:124595. https://doi.org/10.1016/J.BIORTECH.2020.124595
7. Rajesh Banu J, Kavitha S, Yukesh Kannah R et al (2020) Industrial wastewater to biohydrogen: possibilities towards successful biorefinery route. Bioresour Technol 298.https://doi.org/10.1016/J.BIORTECH.2019.122378
8. Detman A, Mielecki D, Chojnacka A et al (2019) Cell factories converting lactate and acetate to butyrate: clostridium butyricum and microbial communities from dark fermentation bioreactors. Microb Cell Fact 18:1–12. https://doi.org/10.1186/S12934-019-1085-1/FIGURES/5
9. Cabrol L, Marone A, Tapia-Venegas E et al (2017) Microbial ecology of fermentative hydrogen producing bioprocesses: useful insights for driving the ecosystem function. FEMS Microbiol Rev 41:158–181. https://doi.org/10.1093/femsre/fuw043
10. Navarro-Díaz M, Martinez-Sanchez ME, Valdez-Vazquez I, Escalante AE (2020) A framework for integrating functional and microbial data: the case of dark fermentation H_2 production. Int J Hydrogen Energy 45:31706–31718. https://doi.org/10.1016/J.IJHYDENE.2020.08.189
11. Castelló E, Braga L, Fuentes L, Etchebehere C (2018) Possible causes for the instability in the H_2 production from cheese whey in a CSTR. Int J Hydrogen Energy 43:2654–2665. https://doi.org/10.1016/j.ijhydene.2017.12.104
12. Franzosa EA, Hsu T, Sirota-Madi A et al (2015) Sequencing and beyond: integrating molecular "omics" for microbial community profiling. Nat Rev Microbiol 13:360–372. https://doi.org/10.1038/nrmicro3451
13. Dzulkarnain ELN, Audu JO, Wan Dagang WRZ, Abdul-Wahab MF (2022) Microbiomes of biohydrogen production from dark fermentation of industrial wastes: current trends, advanced tools and future outlook. Bioresources Bioprocess 9:1–25. https://doi.org/10.1186/S40643-022-00504-8
14. Verma S, Pandey AK (2020) Omics tools: approaches for microbiomes analysis to enhance bioenergy production. Waste Energy Prospects Appl 207–234.https://doi.org/10.1007/978-981-33-4347-4_9/FIGURES/2
15. Zuñiga C, Zaramela L, Zengler K (2017) Elucidation of complexity and prediction of interactions in microbial communities. Microb Biotechnol 10:1500–1522. https://doi.org/10.1111/1751-7915.12855
16. Ladau J, Eloe-Fadrosh EA (2019) Spatial, temporal, and phylogenetic scales of microbial ecology. Trends Microbiol 27:662–669. https://doi.org/10.1016/j.tim.2019.03.003
17. Cordero OX, Datta MS (2016) Microbial interactions and community assembly at microscales. Curr Opin Microbiol 31:227–234. https://doi.org/10.1016/J.MIB.2016.03.015
18. Santiago SG, Trably E, Latrille E et al (2019) The hydraulic retention time influences the abundance of Enterobacter, Clostridium and Lactobacillus during the hydrogen production from food waste. Lett Appl Microbiol 69:138–147. https://doi.org/10.1111/LAM.13191
19. Maistrenko OM, Mende DR, Luetge M, et al (2020) Disentangling the impact of environmental and phylogenetic constraints on prokaryotic within-species diversity. ISME J 14:5 14:1247–1259. https://doi.org/10.1038/s41396-020-0600-z
20. Parks DH, Chuvochina M, Waite DW et al (2018) A standardized bacterial taxonomy based on genome phylogeny substantially revises the tree of life. Nat Biotechnol 36:10 36:996–1004. https://doi.org/10.1038/nbt.4229

21. Wang J, Yin Y (2021) Clostridium species for fermentative hydrogen production: an overview. Int J Hydrogen Energy 46:34599–34625. https://doi.org/10.1016/J.IJHYDENE.2021.08.052
22. Niehaus L, Boland I, Liu M et al (2019) Microbial coexistence through chemical-mediated interactions. Nat Commun 10:1 10:1–12. https://doi.org/10.1038/s41467-019-10062-x
23. Pacheco AR, Segrè D (2019) A multidimensional perspective on microbial interactions. FEMS Microbiol Lett 366:125. https://doi.org/10.1093/FEMSLE/FNZ125
24. Madsen JS, Sørensen SJ, Burmølle M (2017) Bacterial social interactions and the emergence of community-intrinsic properties. Curr Opin Microbiol 42:104–109. https://doi.org/10.1016/J.MIB.2017.11.018
25. Pérez-Rangel M, Barboza-Corona JE, Navarro-Díaz M et al (2021) The duo Clostridium and Lactobacillus linked to hydrogen production from a lignocellulosic substrate. Water Sci Technol 83:3033–3040.https://doi.org/10.2166/WST.2021.186
26. Bittleston LS, Gralka M, Leventhal GE et al (2020) Context-dependent dynamics lead to the assembly of functionally distinct microbial communities. Nat Commun 2020 11:1 11:1–10. https://doi.org/10.1038/s41467-020-15169-0
27. Castelló E, Nunes Ferraz-Junior AD, Andreani C et al (2020) Stability problems in the hydrogen production by dark fermentation: possible causes and solutions. Renew Sustain Energy Rev 119.https://doi.org/10.1016/j.rser.2019.109602
28. Gibbs T, Levin SA, Levine JM (2022) Coexistence in diverse communities with higher-order interactions. Proc Natl Acad Sci U S A 119:e2205063119. https://doi.org/10.1073/PNAS.2205063119/SUPPL_FILE/PNAS.2205063119.SAPP.PDF
29. Sanchez-Gorostiaga A, Bajić D, Osborne ML et al (2019) High-order interactions distort the functional landscape of microbial consortia. PLoS Biol 17:e3000550. https://doi.org/10.1371/JOURNAL.PBIO.3000550
30. Fuentes-Santiago V, Valdez-Vazquez I, Vital-Jácome M et al (2023) Carbohydrates/acid ratios drives microbial communities and metabolic pathways during biohydrogen production from fermented agro-industrial wastewater. J Environ Chem Eng 11:110302. https://doi.org/10.1016/J.JECE.2023.110302
31. Muñoz-Páez KM, Alvarado-Michi EL, Moreno-Andrade I et al (2020) Comparison of suspended and granular cell anaerobic bioreactors for hydrogen production from acid agave bagasse hydrolyzates. Int J Hydrogen Energy 45:275–285. https://doi.org/10.1016/J.IJHYDENE.2019.10.232
32. Carrillo-Reyes J, Trably E, Bernet N et al (2016) High robustness of a simplified microbial consortium producing hydrogen in long term operation of a biofilm fermentative reactor. Int J Hydrogen Energy 41.https://doi.org/10.1016/j.ijhydene.2015.11.131
33. Hung CH, Lee KS, Cheng LH et al (2007) Quantitative analysis of a high-rate hydrogen-producing microbial community in anaerobic agitated granular sludge bed bioreactors using glucose as substrate. Appl Microbiol Biotechnol 75:693–701. https://doi.org/10.1007/S00253-007-0854-7/FIGURES/6
34. Benomar S, Ranava D, Cárdenas ML et al (2015) Nutritional stress induces exchange of cell material and energetic coupling between bacterial species. Nat Commun 6:1 6:1–10. https://doi.org/10.1038/ncomms7283
35. Valdez-Vazquez I, Pérez-Rangel M, Tapia A et al (2015) Hydrogen and butanol production from native wheat straw by synthetic microbial consortia integrated by species of Enterococcus and Clostridium. Fuel 159:214–222. https://doi.org/10.1016/J.FUEL.2015.06.052
36. Faust K, Raes J (2012) Microbial interactions: from networks to models. Nat Rev Microbiol 10:8 10:538–550. https://doi.org/10.1038/nrmicro2832
37. Gupta G, Ndiaye A, Filteau M (2021) Leveraging experimental strategies to capture different dimensions of microbial interactions. Front Microbiol 12:700752. https://doi.org/10.3389/FMICB.2021.700752/BIBTEX
38. Etchebehere C, Castelló E, Wenzel J et al (2016) Microbial communities from 20 different hydrogen-producing reactors studied by 454 pyrosequencing. Appl Microbiol Biotechnol 100.https://doi.org/10.1007/s00253-016-7325-y

39. Wang S, Tang H, Peng F et al (2019) Metabolite-based mutualism enhances hydrogen production in a two-species microbial consortium. Commun Biol 2:1 2:1–11. https://doi.org/10.1038/s42003-019-0331-8
40. Pérez-Rangel M, Valdez-Vazquez I, Martínez-Zavala SA et al (2022) Evaluation of inhibitory compounds produced by bacteria isolated from a hydrogen-producing bioreactor during the self-fermentation of wheat straw. J Appl Microbiol 133:1989–2001. https://doi.org/10.1111/JAM.15708
41. Saha M, Mukherjee G, Basu A, Sil AK (2021) Study of Potential Interrelationship Criteria of Microorganisms for Sustainable Diversity. Microbes in Microbial Communities 71–90.https://doi.org/10.1007/978-981-16-5617-0_3/FIGURES/5
42. Winterbach W, Mieghem P Van, Reinders M et al (2013) Topology of molecular interaction networks. BMC Syst Biol 7:1 7:1–15. https://doi.org/10.1186/1752-0509-7-90
43. Klaise J, Johnson S (2017) The origin of motif families in food webs. Sci Rep 7:1 7:1–11. https://doi.org/10.1038/s41598-017-15496-1
44. Guimerà R, Amaral LAN (2005) Functional cartography of complex metabolic networks. Nature 433:7028 433:895–900. https://doi.org/10.1038/nature03288
45. Paine RT (1969) A note on trophic complexity and community stability. 103:91–93. https://doi.org/10.1086/282586
46. Li L, Peng X, Wang X, Wu D (2018) Anaerobic digestion of food waste: a review focusing on process stability. Bioresour Technol 248:20–28. https://doi.org/10.1016/j.biortech.2017.07.012
47. Buttigieg PL, Ramette A (2014) A guide to statistical analysis in microbial ecology: a community-focused, living review of multivariate data analyses. FEMS Microbiol Ecol 90:543–550. https://doi.org/10.1111/1574-6941.12437
48. Ramette A (2007) Multivariate analyses in microbial ecology. FEMS Microbiol Ecol 62:142–160
49. Legendre P (2012) Numerical ecology
50. Camargo A (2022) PCAtest: testing the statistical significance of principal component analysis in R. PeerJ 10.https://doi.org/10.7717/peerj.12967
51. Mugnai G, Borruso L, Mimmo T et al (2021) Dynamics of bacterial communities and substrate conversion during olive-mill waste dark fermentation: prediction of the metabolic routes for hydrogen production. Bioresour Technol 319:124157. https://doi.org/10.1016/J.BIORTECH.2020.124157
52. Luo L, Sriram S, Johnravindar D et al (2022) Effect of inoculum pretreatment on the microbial and metabolic dynamics of food waste dark fermentation. Bioresour Technol 358:127404. https://doi.org/10.1016/J.BIORTECH.2022.127404
53. Yin Y, Chen Y, Wang J (2021) Co-fermentation of sewage sludge and algae and Fe^{2+} addition for enhancing hydrogen production. Int J Hydrogen Energy 46:8950–8960. https://doi.org/10.1016/J.IJHYDENE.2021.01.009
54. Clarke KR (1993) Non-parametric multivariate analyses of changes in community structure. Aust J Ecol 18:117–143. https://doi.org/10.1111/j.1442-9993.1993.tb00438.x
55. Chong J, Liu P, Zhou G, Xia J (2020) Using MicrobiomeAnalyst for comprehensive statistical, functional, and meta-analysis of microbiome data. Nat Protoc 15:799–821. https://doi.org/10.1038/s41596-019-0264-1
56. Yang F, Wang S, Li H et al (2023) Differences in responses of activated sludge to nutrients-poor wastewater: Function, stability, and microbial community. Chem Eng J 457:141247. https://doi.org/10.1016/J.CEJ.2022.141247
57. Eng F, Fuess LT, Bovio-Winkler P et al (2022) Optimization of volatile fatty acid production by sugarcane vinasse dark fermentation using a response surface methodology. Links between performance and microbial community composition. Sustain Energy Technol Assessments 53:102764. https://doi.org/10.1016/J.SETA.2022.102764
58. Palomo-Briones R, de Jesus Montoya-Rosales J, Razo-Flores E (2021) Advances towards the understanding of microbial communities in dark fermentation of enzymatic hydrolysates: Diversity, structure and hydrogen production performance. Int J Hydrogen Energy 46:27459–27472.https://doi.org/10.1016/J.IJHYDENE.2021.06.016

59. Legendre P, Gallagher ED (2001) Ecologically meaningful transformations for ordination of species data. Oecologia 129:271–280. https://doi.org/10.1007/s004420100716
60. Gelfand A, Fuentes M, Hoeting JA, Smith RL (2017) Handbook of environmental and ecological statistics, 1st edition. CRC Press
61. Pérez-Rangel M, Barboza-Corona JE, Navarro-Díaz M et al (2021) The duo Clostridium and Lactobacillus linked to hydrogen production from a lignocellulosic substrate. Water Sci Technol 83.https://doi.org/10.2166/wst.2021.186
62. Carballa M, Regueiro L, Lema JM (2015) Microbial management of anaerobic digestion: exploiting the microbiome-functionality nexus. Curr Opin Biotechnol 33:103–111
63. Djemiel C, Maron PA, Terrat S et al (2022) Inferring microbiota functions from taxonomic genes: a review. Gigascience 11:1–30. https://doi.org/10.1093/gigascience/giab090
64. Douglas GM, Maffei VJ, Zaneveld JR et al (2020) PICRUSt2 for prediction of metagenome functions. Nat Biotechnol 38:685–688. https://doi.org/10.1038/s41587-020-0548-6
65. Wemheuer F, Taylor JA, Daniel R et al (2020) Tax4Fun2: prediction of habitat-specific functional profiles and functional redundancy based on 16S rRNA gene sequences. Environmental Microbiomes 15:1–12. https://doi.org/10.1186/S40793-020-00358-7/FIGURES/5
66. Yang C, Chowdhury D, Zhang Z et al (2021) A review of computational tools for generating metagenome-assembled genomes from metagenomic sequencing data. Comput Struct Biotechnol J 19:6301–6314. https://doi.org/10.1016/J.CSBJ.2021.11.028
67. Boratyn GM, Camacho C, Cooper PS et al (2013) BLAST: a more efficient report with usability improvements. Nucleic Acids Res 41:W29–W33. https://doi.org/10.1093/NAR/GKT282
68. Park JH, Kim DH, Baik JH et al (2021) Improvement in H2 production from Clostridium butyricum by co-culture with Sporolactobacillus vineae. Fuel 285:119051. https://doi.org/10.1016/j.fuel.2020.119051
69. Jung JH, Sim YB, Park JH et al (2021) Novel dynamic membrane, metabolic flux balance and PICRUSt analysis for high-rate biohydrogen production at various substrate concentrations. Chem Eng J 420:127685. https://doi.org/10.1016/j.cej.2020.127685
70. Detman A, Laubitz D, Chojnacka A et al (2021) Dynamics of dark fermentation microbial communities in the light of lactate and butyrate production. Microbiome 9:1–21. https://doi.org/10.1186/s40168-021-01105-x
71. Sun S, Jones RB, Fodor AA (2020) Inference-based accuracy of metagenome prediction tools varies across sample types and functional categories. Microbiome 8:1–9. https://doi.org/10.1186/S40168-020-00815-Y/FIGURES/5
72. Wilkinson TJ, Huws SA, Edwards JE et al (2018) CowPI: A rumen microbiome focussed version of the PICRUSt functional inference software. Front Microbiol 9:1095. https://doi.org/10.3389/FMICB.2018.01095/BIBTEX
73. Dueholm MKD, Nierychlo M, Andersen KS et al (2022) MiDAS 4: a global catalogue of full-length 16S rRNA gene sequences and taxonomy for studies of bacterial communities in wastewater treatment plants. Nat Commun 13:1 13:1–15. https://doi.org/10.1038/s41467-022-29438-7
74. Ma S, Jiang F, Huang Y et al (2021) A microbial gene catalog of anaerobic digestion from full-scale biogas plants. Gigascience 10:1–10. https://doi.org/10.1093/GIGASCIENCE/GIAA164
75. Fuess LT, Ferraz ADN, Machado CB, Zaiat M (2018) Temporal dynamics and metabolic correlation between lactate-producing and hydrogen-producing bacteria in sugarcane vinasse dark fermentation: the key role of lactate. Bioresour Technol 247:426–433. https://doi.org/10.1016/J.BIORTECH.2017.09.121
76. Carrillo-Reyes J, Trably E, Bernet N et al (2016) High robustness of a simplified microbial consortium producing hydrogen in long term operation of a biofilm fermentative reactor. Int J Hydrogen Energy 41:2367–2376. https://doi.org/10.1016/J.IJHYDENE.2015.11.131
77. Etchebehere C, Castelló E, Wenzel J et al (2016) Microbial communities from 20 different hydrogen-producing reactors studied by 454 pyrosequencing. Appl Microbiol Biotechnol. https://doi.org/10.1007/s00253-016-7325-y

78. Hernández C, Alamilla-Ortiz ZL, Escalante AE et al (2019) Heat-shock treatment applied to inocula for H2 production decreases microbial diversities, interspecific interactions and performance using cellulose as substrate. Int J Hydrogen Energy 44:13126–13134. https://doi.org/10.1016/J.IJHYDENE.2019.03.124
79. Jia X, Wang Y, Ren L et al (2019) Early warning indicators and microbial community dynamics during unstable stages of continuous hydrogen production from food wastes by thermophilic dark fermentation. Int J Hydrogen Energy 44:30000–30013. https://doi.org/10.1016/j.ijhydene.2019.08.082
80. Huang J, Pan Y, Liu L et al (2022) High salinity slowed organic acid production from acidogenic fermentation of kitchen wastewater by shaping functional bacterial community. J Environ Manage 310:114765. https://doi.org/10.1016/J.JENVMAN.2022.114765
81. Yin Y, Wang J (2021) Predictive functional profiling of microbial communities in fermentative hydrogen production system using PICRUSt. Int J Hydrogen Energy 46:3716–3725. https://doi.org/10.1016/J.IJHYDENE.2020.10.246
82. Zhang J, Zhang H, Zhang J et al (2022) Improved biohydrogen evolution through calcium ferrite nanoparticles assisted dark fermentation. Bioresour Technol 361:127676. https://doi.org/10.1016/J.BIORTECH.2022.127676
83. Yin Y, Wang J (2021) Mechanisms of enhanced hydrogen production from sewage sludge by ferrous ion: Insights into functional genes and metabolic pathways. Bioresour Technol 321:124435. https://doi.org/10.1016/J.BIORTECH.2020.124435
84. Park JH, Park JH, Lee SH et al (2020) Metabolic flux and functional potential of microbial community in an acidogenic dynamic membrane bioreactor. Bioresour Technol 305:123060. https://doi.org/10.1016/J.BIORTECH.2020.123060
85. de Souza Almeida P, de Menezes CA, Camargo FP et al (2022) Producing hydrogen from the fermentation of cheese whey and glycerol as cosubstrates in an anaerobic fluidized bed reactor. Int J Hydrogen Energy 47:14243–14256.https://doi.org/10.1016/J.IJHYDENE.2022.02.176
86. Detman A, Laubitz D, Chojnacka A et al (2021) Dynamics and complexity of dark fermentation microbial communities producing hydrogen from sugar beet molasses in continuously operating packed bed reactors. Front Microbiol 11:3303. https://doi.org/10.3389/FMICB.2020.612344/BIBTEX
87. Batstone DJ, Keller J, Angelidaki I et al (2002) The IWA Anaerobic Digestion Model No 1 (ADM1). Water Sci Technol 45:65–73. https://doi.org/10.2166/WST.2002.0292
88. Gadhamshetty V, Arudchelvam Y, Nirmalakhandan N, Johnson DC (2010) Modeling dark fermentation for biohydrogen production: ADM1-based model vs gompertz model. Int J Hydrogen Energy 35:479–490. https://doi.org/10.1016/J.IJHYDENE.2009.11.007
89. Ntaikou I, Gavala HN, Lyberatos G (2010) Application of a modified Anaerobic Digestion Model 1 version for fermentative hydrogen production from sweet sorghum extract by Ruminococcus albus. Int J Hydrogen Energy 35:3423–3432. https://doi.org/10.1016/J.IJHYDENE.2010.01.118
90. Antonopoulou G, Gavala HN, Skiadas IV, Lyberatos G (2012) Modeling of fermentative hydrogen production from sweet sorghum extract based on modified ADM1. Int J Hydrogen Energy 37:191–208. https://doi.org/10.1016/J.IJHYDENE.2011.09.081
91. Pradhan N, Dipasquale L, D'Ippolito G et al (2016) Kinetic modeling of fermentative hydrogen production by Thermotoga neapolitana. Int J Hydrogen Energy 41:4931–4940. https://doi.org/10.1016/J.IJHYDENE.2016.01.107
92. Guellout Z, Clion V, Benguerba Y et al (2018) Study of the dark fermentative hydrogen production using modified ADM1 models. Biochem Eng J 132:9–19. https://doi.org/10.1016/J.BEJ.2017.12.015
93. Penumathsa BKV, Premier GC, Kyazze G et al (2008) ADM1 can be applied to continuous bio-hydrogen production using a variable stoichiometry approach. Water Res 42:4379–4385. https://doi.org/10.1016/J.WATRES.2008.07.030
94. Alexandropoulou M, Antonopoulou G, Lyberatos G (2018) A novel approach of modeling continuous dark hydrogen fermentation. Bioresour Technol 250:784–792. https://doi.org/10.1016/J.BIORTECH.2017.12.005

95. Alexandropoulou M, Antonopoulou G, Lyberatos G (2022) Modeling of continuous dark fermentative hydrogen production in an anaerobic up-flow column bioreactor. Chemosphere 293:133527. https://doi.org/10.1016/J.CHEMOSPHERE.2022.133527
96. Montecchio D, Yuan Y, Malpei F (2018) Hydrogen production dynamic during cheese whey Dark Fermentation: new insights from modelization. Int J Hydrogen Energy 43:17588–17601. https://doi.org/10.1016/J.IJHYDENE.2018.07.146
97. Blanco VMC, Oliveira GHD, Zaiat M (2019) Dark fermentative biohydrogen production from synthetic cheese whey in an anaerobic structured-bed reactor: performance evaluation and kinetic modeling. Renew Energy 139:1310–1319. https://doi.org/10.1016/j.renene.2019.03.029
98. Muñoz-Páez KM, Vargas A, Buitrón G (2023) Feedback control-based strategy applied for biohydrogen production from acid cheese whey. Waste Biomass Valorization 14:447–460. https://doi.org/10.1007/S12649-022-01865-Z/TABLES/3
99. Couto PT, Eng F, Naessens W et al (2020) Modelling sugarcane vinasse processing in an acidogenic reactor to produce hydrogen with an ADM1-based model. Int J Hydrogen Energy 45:6217–6230. https://doi.org/10.1016/J.IJHYDENE.2019.12.206
100. Amado M, Barca C, Hernández MA, Ferrasse JH (2021) Evaluation of energy recovery potential by anaerobic digestion and dark fermentation of residual biomass in Colombia. Front Energy Res 9:321. https://doi.org/10.3389/FENRG.2021.690161/BIBTEX
101. Kim B, Jeong J, Kim J et al (2022) Mathematical modeling of dark fermentation of macroalgae for hydrogen and volatile fatty acids production. Bioresour Technol 354:127193. https://doi.org/10.1016/J.BIORTECH.2022.127193
102. Aceves-Lara CA, Latrille E, Bernet N et al (2008) A pseudo-stoichiometric dynamic model of anaerobic hydrogen production from molasses. Water Res 42:2539–2550. https://doi.org/10.1016/J.WATRES.2008.02.018
103. Borisov M, Dimitrova N, Simeonov I (2020) Mathematical modeling and stability analysis of a two-phase biosystem. Processes 8:791.https://doi.org/10.3390/PR8070791
104. Hafez H, El NMH, Nakhla G (2010) Steady-state and dynamic modeling of biohydrogen production in an integrated biohydrogen reactor clarifier system. Int J Hydrogen Energy 35:6634–6645. https://doi.org/10.1016/J.IJHYDENE.2010.04.012
105. Rodriguez J, Premier GC, Guwy AJ et al (2009) Metabolic models to investigate energy limited anaerobic ecosystems. Water Sci Technol 60:1669–1675. https://doi.org/10.2166/WST.2009.224
106. Kim WJ, Kim HU, Lee SY (2017) Current state and applications of microbial genome-scale metabolic models. Curr Opin Syst Biol 2:10–18. https://doi.org/10.1016/j.coisb.2017.03.001
107. Gu C, Kim GB, Kim WJ et al (2019) Current status and applications of genome-scale metabolic models. Genome Biol 20:1–18. https://doi.org/10.1186/s13059-019-1730-3
108. Mendoza SN, Olivier BG, Molenaar D, Teusink B (2019) A systematic assessment of current genome-scale metabolic reconstruction tools. Genome Biol 20:158. https://doi.org/10.1186/s13059-019-1769-1
109. Orth JD, Thiele I, Palsson BO (2010) What is flux balance analysis? Nat Biotechnol 28:245–248. https://doi.org/10.1038/nbt.1614
110. Lakshmanan M, Koh G, Chung BKS, Lee DY (2014) Software applications for flux balance analysis. Brief Bioinform 15:108–122. https://doi.org/10.1093/bib/bbs069
111. Perez-Garcia O, Lear G, Singhal N (2016) Metabolic network modeling of microbial interactions in natural and engineered environmental systems. Front Microbiol 7.https://doi.org/10.3389/fmicb.2016.00673
112. Dahal S, Yurkovich JT, Xu H et al (2020) Synthesizing systems biology knowledge from omics using genome-scale models. Proteomics 20:1900282. https://doi.org/10.1002/PMIC.201900282
113. Lawson CE, Harcombe WR, Hatzenpichler R et al (2019) Common principles and best practices for engineering microbiomes. Nat Rev Microbiol 17:725–741
114. Cai G, Jin B, Monis P, Saint C (2011) Metabolic flux network and analysis of fermentative hydrogen production. Biotechnol Adv 29:375–387. https://doi.org/10.1016/j.biotechadv.2011.02.001

115. Oh YK, Raj SM, Jung GY, Park S (2011) Current status of the metabolic engineering of microorganisms for biohydrogen production. Bioresour Technol 102:8357–8367. https://doi.org/10.1016/J.BIORTECH.2011.04.054
116. Nogales J, Gudmundsson S, Thiele I (2012) An in silico re-design of the metabolism in Thermotoga maritima for increased biohydrogen production. Int J Hydrogen Energy 37:12205–12218. https://doi.org/10.1016/J.IJHYDENE.2012.06.032
117. Rafieenia R, Pivato A, Schievano A, Lavagnolo MC (2018) Dark fermentation metabolic models to study strategies for hydrogen consumers inhibition. Bioresour Technol 267:445–457. https://doi.org/10.1016/j.biortech.2018.07.054
118. Chaganti SR, Kim DH, Lalman JA (2011) Flux balance analysis of mixed anaerobic microbial communities: effects of linoleic acid (LA) and pH on biohydrogen production. Int J Hydrogen Energy 36:14141–14152. https://doi.org/10.1016/J.IJHYDENE.2011.04.161
119. Gonzalez-Garcia RA, Aispuro-Castro R, Salgado-Manjarrez E et al (2017) Metabolic pathway and flux analysis of H_2 production by an anaerobic mixed culture. Int J Hydrogen Energy 42:4069–4082. https://doi.org/10.1016/J.IJHYDENE.2017.01.043
120. Schwalm ND, Mojadedi W, Gerlach ES et al (2019) Developing a microbial consortium for enhanced metabolite production from simulated food waste. Fermentation 5.https://doi.org/10.3390/fermentation5040098
121. Mostafa A, Im S, Kim J et al (2022) Electron bifurcation reactions in dark fermentation: An overview for better understanding and improvement. Bioresour Technol 344:126327. https://doi.org/10.1016/J.BIORTECH.2021.126327
122. Valdez-Vazquez I, Castillo-Rubio LG, Pérez-Rangel M et al (2019) Enhanced hydrogen production from lignocellulosic substrates via bioaugmentation with Clostridium strains. Ind Crops Prod 137:105–111. https://doi.org/10.1016/J.INDCROP.2019.05.023
123. Villanueva-Galindo E, Vital-Jácome M, Moreno-Andrade I (2022) Dark fermentation for H_2 production from food waste and novel strategies for its enhancement. Int J Hydrogen Energy. https://doi.org/10.1016/j.ijhydene.2022.11.339
124. Wang J, Yin Y (2019) Progress in microbiology for fermentative hydrogen production from organic wastes. Crit Rev Environ Sci Technol 49:825–865. https://doi.org/10.1080/10643389.2018.1487226
125. Morya R, Raj T, Lee Y et al (2022) Recent updates in biohydrogen production strategies and life–cycle assessment for sustainable future. Bioresour Technol 366:128159. https://doi.org/10.1016/J.BIORTECH.2022.128159
126. Ding J, Wei D, An Z et al (2020) Succession of the bacterial community structure and functional prediction in two composting systems viewed through metatranscriptomics. Bioresour Technol 313:123688. https://doi.org/10.1016/J.BIORTECH.2020.123688
127. Salvato F, Hettich RL, Kleiner M (2021) Five key aspects of metaproteomics as a tool to understand functional interactions in host-associated microbiomes. PLoS Pathog 17:e1009245. https://doi.org/10.1371/JOURNAL.PPAT.1009245
128. Sillero L, Sganzerla WG, Forster-Carneiro T et al (2022) A bibliometric analysis of the hydrogen production from dark fermentation. Int J Hydrogen Energy 47:27397–27420. https://doi.org/10.1016/J.IJHYDENE.2022.06.083
129. Carrillo-Reyes J, Tapia-Rodríguez A, Buitrón G et al (2019) A standardized biohydrogen potential protocol: an international round robin test approach. Int J Hydrogen Energy 44:26237–26247. https://doi.org/10.1016/J.IJHYDENE.2019.08.124
130. Nearing JT, Douglas GM, Hayes MG et al (2022) Microbiome differential abundance methods produce different results across 38 datasets. Nat Commun 13.https://doi.org/10.1038/s41467-022-28034-z
131. Munir RI, Spicer V, Krokhin OV et al (2016) Transcriptomic and proteomic analyses of core metabolism in Clostridium termitidis CT1112 during growth on α-cellulose, xylan, cellobiose and xylose. BMC Microbiol 16:1–21. https://doi.org/10.1186/S12866-016-0711-X/TABLES/4
132. Tzun-Wen Shaw G, Liu AC, Weng CY et al (2017) Inferring microbial interactions in thermophilic and mesophilic anaerobic digestion of hog waste. PLoS ONE 12:e0181395. https://doi.org/10.1371/JOURNAL.PONE.0181395

133. Zhang Y, Guo B, Zhang L, Liu Y (2020) Key syntrophic partnerships identified in a granular activated carbon amended UASB treating municipal sewage under low temperature conditions. Bioresour Technol 312:123556. https://doi.org/10.1016/J.BIORTECH.2020.123556
134. Xu R, Yang ZH, Zheng Y et al (2018) Organic loading rate and hydraulic retention time shape distinct ecological networks of anaerobic digestion related microbiome. Bioresour Technol 262:184–193. https://doi.org/10.1016/J.BIORTECH.2018.04.083
135. Huang Q, Liu Y, Dhar BR (2023) Boosting resilience of microbial electrolysis cell-assisted anaerobic digestion of blackwater with granular activated carbon amendment. Bioresour Technol 381:129136. https://doi.org/10.1016/J.BIORTECH.2023.129136
136. Basile A, Campanaro S, Kovalovszki A et al (2020) Revealing metabolic mechanisms of interaction in the anaerobic digestion microbiome by flux balance analysis. Metab Eng 62:138–149. https://doi.org/10.1016/j.ymben.2020.08.013
137. Weinrich S, Koch S, Bonk F et al (2019) Augmenting biogas process modeling by resolving intracellular metabolic activity. Front Microbiol 10:1–14. https://doi.org/10.3389/fmicb.2019.01095
138. Dukovski I, Bajić D, Chacón JM et al (2021) A metabolic modeling platform for the computation of microbial ecosystems in time and space (COMETS). Nat Protoc 16:11 16:5030–5082. https://doi.org/10.1038/s41596-021-00593-3
139. Antonakoudis A, Barbosa R, Kotidis P, Kontoravdi C (2020) The era of big data: genome-scale modelling meets machine learning. Comput Struct Biotechnol J 18:3287–3300. https://doi.org/10.1016/j.csbj.2020.10.011

Part IV
Microbial Fuel Cells

Chapter 8
Microbial Fuel Cell Systems for Wastewater Treatment and Energy Generation from Organic Carbon and Nitrogen: Fundamentals, Optimization and Novel Processes

Vitor Cano, Gabriel Santiago de Arruda, Julio Cano, Victor Alcaraz-Gonzalez, René Alejandro Flores-Estrella, and Theo Syrto Octavio de Souza

Abstract Multiple types of wastewater contain abundant nutrients and chemical energy within organic and inorganic compounds, as well as reusable water. Thus, based on a circular economy perspective, wastewater treatment plants could be designed as biorefineries and potential energy producers, reducing the demand for fossil fuels and, consequently, CO_2 emissions. The microbial fuel cell (MFC) is a bioelectrochemical system aligned with this systematic perspective. This novel technology, relying on interactions between electrochemically active bacteria

V. Cano (✉) · T. S. O. de Souza
Department of Hydraulic and Environmental Engineering, Polytechnic School, University of São Paulo, Av. Prof. Almeida Prado, 83, trav. 2, Cidade Universitária, 05508-900 São Paulo, São Paulo, Brazil
e-mail: vitorc@usp.br

G. S. de Arruda
Institute of Energy and Environment, University of São Paulo, Av. Prof. Luciano Gualberto, 1289, Cidade Universitária, 05508-900 São Paulo, São Paulo, Brazil
e-mail: gsarruda@usp.br

J. Cano
School of Arts, Sciences and Humanities, University of São Paulo, Rua Arlindo Béttio, 1000 - Ermelino Matarazzo, 03828-000 São Paulo, São Paulo, Brazil
e-mail: julio.cano@usp.br

V. Alcaraz-Gonzalez
Universidad de Guadalajara—CUCEI, Blvd. Marcelino Garcia Barragán 1451, Col Olímpica, 44430 Guadalajara, Jalisco, México
e-mail: victor.agonzalez@academicos.udg.mx

R. A. Flores-Estrella
Departamento Ciencias de La Sustentabilidad, El Colegio de La Frontera Sur, Carretera Antiguo Aeropuerto Km. 2.5, C.P. 30700 Tapachula, Chiapas, México
e-mail: rene.flores@ecosur.mx

V. Alcaraz Gonzalez et al. (eds.), *Wastewater Exploitation*, Springer Water,
https://doi.org/10.1007/978-3-031-57735-2_8

and solid-state electrodes (anode and cathode), can treat wastewater and directly convert biodegradable compounds into electricity. MFCs are mostly considered for converting organic carbon into electricity, but recent findings demonstrated their ability to convert nitrogen compounds as well. New possibilities include the oxidation of ammonia nitrogen on the anode coupled with current generation and utilization of oxidized nitrogen (nitrite and nitrate) as electron acceptors on the cathode. This is particularly important for industrial and agro-industrial effluents, which commonly presents high concentration of organic carbon and nitrogen species. Despite its remarkable advantages, large-scale implementation of MFC technology is still limited by relatively low power densities. In this regard, the proper selection of electrode materials can optimize the MFC performance in terms of power output. Tridimensional carbonaceous electrodes offer unique properties regarding morphology, surface, adhesion of the biofilm, electron transfer, and oxidation of the substrate. Furthermore, the capacitive properties of high specific surface electrodes can be utilized in charge and discharge cycles to increase power output by combining faradaic and non-faradaic currents. Operating parameters such as temperature and resistance of the external circuit can affect the level of the current generation by modulating the activity and predominance of electroactive bacteria in the biofilm of MFCs. In this chapter, MFC systems are reviewed. Fundamentals of the process occurring within MFCs are presented, and main contributions in the field of MFC focused on carbon and nitrogen conversions, tridimensional electrodes, and capacitive properties are discussed. Then, relevant insights towards their potential applications in energy generation from wastewater are presented.

Keywords Bioelectricity · Capacitive electrodes · Microbial fuel cell · Nitrogen · Wastewater treatment

8.1 Introduction

Water resources, as well as environmental sanitation and energy sectors, are complex and interdisciplinary, and are closely related to the economic development, public services, and human well-being. Hence the interconnection between the sanitation and energy sectors is very relevant regarding sustainability. Integrated actions aimed at protecting water bodies and diversifying the energy matrix with renewable sources are relevant to the public interest and are sustainable development goals [1].

The impacts on environment and public health resulting from the release of wastewater into natural water bodies have been known for a long time. Therefore, the development of wastewater treatment technologies is crucial for the promotion of sustainable water use and protection, and public health protection [2]. Biological and physicochemical processes are commonly explored in conventional wastewater treatment technologies. These technologies include aerobic activated sludge system, membrane separation, adsorption, chemical precipitation, coagulation, and electrochemical treatment [3]. Currently, high treatment efficiencies can be achieved by these systems. However, their implementation has been oriented by a linear approach model focused on the removal of organic matter and nutrients instead of its utilization as a resource. Moreover, high sludge generation and energy consumption are typical problems in conventional high-rate wastewater treatment plants (WWTP) [4, 5].

Alternatively, several studies have been carried out in the exploration of new sources of water, energy and nutrients. In this sense, the utilization of wastewater to obtain resources, such as energy, presents great relevance, since this matrix contains abundant nutrients and chemical energy within inorganic and organic matter, as well as reusable water [6]. In this sense, wastewater treatment plants could be designed as biorefineries, not only consuming energy, but also as potential producers, reducing the demand for fossil fuels and CO_2 emissions [7]. This concept guides the development towards a cyclic perspective, in which wastes are valuable resources, in contrast to the current linear model.

Currently, the anaerobic digestion aiming at methane or hydrogen production for energy generation has been considered as an alternative approach aligned with the circular economy perspective. However, in order to generate electricity from the biogas, a number of conversion processes are necessary, including the conversion of the organic matter into the biogas, and further conversion into electric current, which limits the overall electricity production efficiency [8].

Another technology aligned with a systematic perspective that can overcome the conversion efficiency limitation, is the microbial fuel cell (MFC). It is a novel technology based on bioelectrochemical processes that can treat wastewater and directly generate electricity simultaneously, relying on electrochemically active bacteria [9, 10]. The literature reports that MFC has the ability to directly generate energy from various biodegradable compounds, including pure compounds and complex substrates [11]. In this sense, energy generation in MFC was demonstrated for the treatment of municipal wastewater [12], industrial effluents [13], landfill leachate [14] among others.

Despite the advantage of directly converting organic and inorganic substrates into electricity, the MFC technology still has important limitations for full-scale application due to high-cost materials, such as catalysts, and a lack of knowledge about optimal operation conditions. Thus, in order to increase the energy generation yield and reduce costs, improvements are necessary [15, 16]. In this context, this chapter seeks to contribute to the development of microbial fuel cell technology for potential future applications in energy generation from wastewater treatment. The specific objectives are to present fundamentals of the processes occurring within MFCs and main contributions in the field of MFC optimization, considering carbon and nitrogen conversions, electrode materials, biocapacitive properties, as well as main parameters controlling the system's efficiency in treating wastewater.

8.2 Fundamentals of the Microbial Fuel Cell

The microbial fuel cell (MFC) is a bioelectrochemical system (BES) that relies on the bio-catalytic capabilities of electrochemically-active microorganisms or enzymes to oxidize an electron donor, converting the chemical energy within biodegradable compounds into clean electricity [11, 17, 18]. Electroactive microorganisms developed mechanisms to transfer electrons to electron acceptors, including Fe(III) and Mn(IV) oxides, through their outer membrane. This process, that is inherent to electroactive microorganisms and is the biological machinery of the MFC, is referred to as extracellular electron transfer (EET) [19]. Hence the MFC operation is based on interactions between an electron acceptor and the microbial metabolism by combining the oxidation of an electron donor with the EET ability of electroactive microorganisms [20].

In MFC systems, biochemical reactions are catalyzed by microorganisms' enzymes responsible for the degradation of biodegradable substrates on the surface of a solid electron acceptor, named anode, releasing electrons and protons [15]. Typically, MFCs consist of an anode oxidizing organic matter under anaerobic condition and a cathode exposed to oxygen. To maintain different conditions in the anode and cathode chambers, a proton transfer system is utilized as a separator [9]. Electrons released during the oxidation reaction on the anode migrate through an external circuit to the cathode. The electron flow in the external circuit is the electricity generated by the MFC system. The protons released in the same oxidation reaction on the anode migrate to the cathode through the proton transfer system (usually a proton exchange membrane, PEM). On the cathode, electrons and protons are combined with oxygen in a reduction reaction, forming H_2O [21, 22]. So, the potential difference between the microorganism's enzymes responsible for the EET and the reduction reaction of the electron acceptor in the cathode chamber generate the voltage [15]. Figure 8.1 illustrates the main components of a conventional MFC and the synergy between electroactive microorganisms and the anode.

Considering its operation fundamentals, the MFC offers advantages over other technologies used in the wastewater treatment and energy sectors [15, 23, 24]: (i) theoretical high energy conversion efficiency based on the direct conversion of organic substrate into electricity in a single step; (ii) operation at different temperature ranges, including mesophilic and thermophilic conditions; (iii) biogas treatment is not necessary, since CO_2 is the final gaseous product; (iv) direct utilization of oxygen for organic matter degradation is not necessary, avoiding costs related to an aeration system; and (v) relatively low sludge production rate.

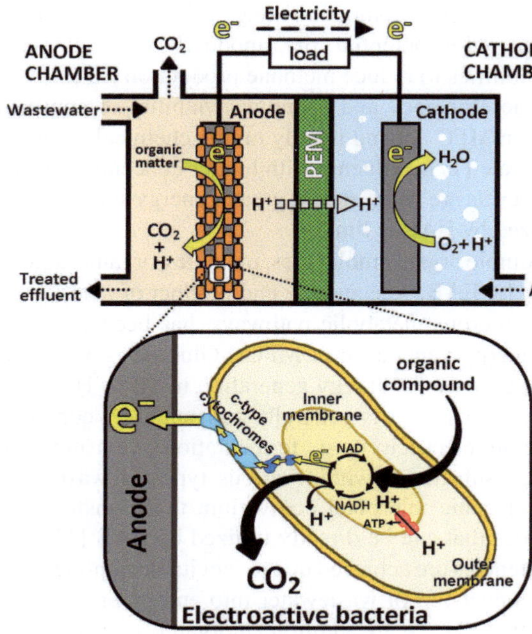

Fig. 8.1 Illustration of a conventional MFC with two separated chambers, schematically showing the electroactive metabolism of bacteria on the anode

8.2.1 Organic Matter Oxidation on the Anode

The MFC generates energy relying on the activity of electroactive microorganisms consuming organic matter. The most studied electrochemically-active microorganisms are bacteria, named electroactive bacteria (EAB) or exoelectrogens [9, 19, 25].

Although EAB metabolism occurs in the absence of oxygen, it differs from the metabolism of microorganisms usually present in strictly anaerobic environments. Unlike methanogenesis, the final products of the reactions catalyzed by EAB are H_2O and CO_2. Thus, MFC does not present considerable contribution to global warming [18]. Considering glucose as electron donor and oxygen as electron acceptor, the redox reactions that characterize the MFC technology are presented as follows [26]:

$$C_6H_{12}O_6 + 6H_2O \rightarrow 6CO_2 + 24H^+ + 24e^- \ E^{\circ\prime} = -0.43 \text{ V vs. SHE at pH 7} \quad (8.1)$$

$$O_2 + 4H^+ + 4e^- \rightarrow 2H_2O \ E^{\circ\prime} = 0.82 \text{ V vs. SHE at pH 7} \quad (8.2)$$

In MFCs, competition for electron donors (biodegradable organic substrate) commonly occurs between EAB, fermentative and methanogenic organisms [22]. Therefore, the bacterial consortium leading to methanogenesis reduces the number of electrons available for current generation by EAB. Thus, to avoid the loss of coulombic efficiency (recovery of electrons from consumed organic matter) in the

MFC, inhibitor compounds (e.g., 2-bromoethanesulfonate), oxygen pulses, control of the redox potential, pH, anode chamber temperature, and anode potential are possibilities to reduce methane production [27].

The efficiency and economic viability of converting organic waste into bioenergy in MFC depend mainly on the chemical composition and concentration of the substrate [17]. Effluents with higher concentrations of organic matter at low oxidation levels are theoretically greater energy sources; however the range of compounds utilized by EAB is limited.

Simple organic molecules, including organic acids, are preferred for energy generation by EAB. Acetate, the end product of complex organic molecules degradation by numerous metabolic pathways, has been widely accepted as an ideal substrate for energy generation in MFCs. Glucose is also commonly considered a compatible substrate for energy generation in MFC. However, it has lower conversion efficiency because it is generally consumed by competing metabolisms that does not generate electricity (e.g., fermentation). Consumption of more complex and recalcitrant substrates, within various types of wastewater, depends on developing of a syntrophic microbial consortium that transforms complex molecules in carbon sources that can be directly utilized by EAB [17, 28, 29]. While lower conversion efficiencies are achieved due to inevitable non-electrogenic competitive metabolism, the conversion of wastewater into energy presents advantages in terms of sustainability [30–32]. In addition, inorganic compounds commonly found in wastewater, such as nitrogen, have been recently demonstrated to be an energy source in bioelectrochemical systems [33–35].

8.2.2 Final Electron Acceptor Reduction on the Cathode

Oxygen is widely used in MFCs as electron acceptor in the cathode due to its availability in the atmosphere. Also, oxygen has high redox potential, is not toxic, and does not generate greenhouse gases as a product of the reduction reaction [15, 36]. Under typical operating conditions a high potential of + 0.82 V versus Standard Hydrogen Electrode (SHE) can be achieved in the cathode by oxygen reduction [9].

However, feasible electron acceptors in the cathodic chamber are not limited to oxygen, and other substances can be considered depending on the MFC configuration and objective. The necessary conditions for an electron acceptor to be used in a MFC are: having high redox potential, fast reaction kinetics, being economically feasible and widely available [37]. In addition to oxygen, the use of other electron acceptors is reported in the literature, such as Fe(III), Mn(IV), ferricyanide, permanganate, dichromate, perchlorate, sulfate, and nitrate [38]. Table 8.1 shows some electron acceptors and the possible half-reaction potentials in the MFC.

In the EAB metabolism, organic substrates are consumed to produce energy that is stored intracellularly as NADH [9, 39]. The EAB species grow relying on the difference between the energy level of NADH and the final electron acceptor [18]. Thus, their respiration depends on the cathode reaction. The electrode must have a potential

Table 8.1 Reactions of interest regarding the cathode of MFCs

Reaction	$E^{\circ\prime}$ V versus SHE (pH 7)	Reaction	$E^{\circ\prime}$ V versus SHE (pH 7)
Oxygen/H_2O	+ 0.82	$Cr_2O_7^{2-}/Cr^{3+}$	+ 0.365
ClO_3^-/Cl^-	+ 0.81	NO_3^-/NH_4^+	+ 0.36
ClO_4/Cl^-	+ 0.81	$[Fe(CN)_6]^{3-}/[Fe(CN)_6]^{4-}$	+ 0.36
Fe_3^+/Fe_2^+	+ 0.77	NO_2^-/NO	+ 0.35
NO_3^-/N_2	+ 0.75	NO_2^-/NH_4^+	+ 0.34
$C_2H_4Cl_2/C_2H_4$	+ 0.739	O_2/H_2O_2	+ 0.260
C_2Cl_4/C_2HCl_3	+ 0.574	SeO_4^{2-}/Se	+ 0.322
C_2HCl_3/*cis*-$C_2H_2Cl_2$	+ 0.550	$HSeO_3^-/Se$	+ 0.26
C_2H_3Cl/C_2H_4	+ 0.45		

Source Adapted from [36]

(determined by the reduction reaction) greater than the NADH oxidation to allow the electrons flux from the respiratory chain to the electrode. Considering the redox potential of oxygen (+ 0.82 V vs. SHE) and the redox potential of NADH (− 0.32 V vs. SHE), a maximum voltage of 1.14 V is expected for the MFC, independently of the oxidized substrate [9, 18].

8.2.3 Extracellular Electron Transfer (EET)

The current generation in MFC depends on the occurrence of EET, transporting electrons from the respiratory metabolism of bacteria to the anode. The EAB relies on two types of mechanisms to perform EET: (i) direct and (ii) mediated electron transfer. Figure 8.2 schematically illustrates the EET mechanisms.

The direct electron transfer mechanism depends on direct contact between the microbial cell and a solid-state electron acceptor, which is established by redox proteins and/or cell appendages on the external membrane [40]. The direct EET mechanism best known so far belongs to *Geobacter sulfurreducens*, which relies on a series of c-type cytochromes (periplasmic and in the outer membrane), with OmcZ as the most important one, to transfer electrons from its inner membrane to the outer cell surface. These bacteria also developed a mechanism to transfer electrons over micrometer-long distances using type IV pili, known as nanowires [10, 20]. Biofilms formed by bacteria relying on nanowires are not limited by the electrode surface area. This results in the formation of a solid conductive matrix allowing bacteria in outer part of the biofilm to access the electrode. Hence, high current density can be achieved since in this condition EET is not solely dependent on electron shuttles or direct contact between bacteria and electrode [40].

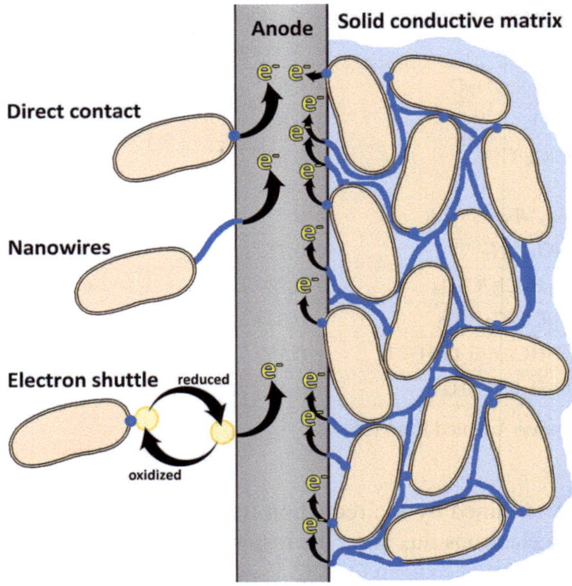

Fig. 8.2 Schematic overview of the EET mechanisms for the use of a solid-state electron acceptor, including direct and mediated EET

Regarding the mediated EET, EAB relies on the utilization of electron shuttles (redox mediators). The electron shuttles can access the respiratory mechanism of EAB, supporting the production of ATP while being reduced. The electron shuttle carries the electrons, obtained in respiration, outside bacteria, to a solid electron acceptor (i.e. anode). On the anode, the electron shuttles are oxidized, transferring electrons, and becoming available to carry more electrons [39]. Electron shuttles are typically organic or inorganic compounds that can be reversibly reduced. Phenazine derivatives, flavins, and quinones are examples of mediators produced by the electrogens *Pseudomonas aeruginosa, Shewanella oneidensis,* and *Lactococcus lactis*, respectively [20, 41]. An important implication of EET depending on self-secreted mediators is that the rate of the process is limited by the production, diffusion, and availability of electron shuttles. In addition, mediators can also be consumed by microorganisms that do not contribute to the current generation. In contrast, it means that current generation can be potentially increased with the addition of artificial electron shuttles, such as methylene blue (MB), methyl viologen (MV), anthraquinone-2,6-disulfonate (AQDS), 2-hydroxy-1,4-naphthoquinone, and resazurin [20]. Furthermore, some electron-shuttles are naturally available in wastewater, such as humic acids and polyphenolic compounds [42].

A great number of microorganisms were reported as capable of achieving EET. However, for most of them the molecular mechanism responsible for transporting electrons out of their cells remains uncertain. Furthermore, how bacteria interact with each other in mixed culture, how external factors affect them, and how they can couple oxidation of different compounds with EET under different conditions needs to be better comprehended [20]. Considering that significantly more electrogen organisms are expected to be discovered with MFC advances, more studies are necessary to comprehend their role in energy generation and how operation parameters affect their performance [43].

8.2.4 Diversity and Activity of Electroactive Bacteria

Electroactive microorganisms are a group of organisms whose respiration is connected to a mechanism that transport electrons out of the cell toward insoluble or solid-state electron acceptors [9]. Diverse microorganisms have been reported as electroactive, including bacteria from different phyla, archaea, and even eukaryotes such as *Saccharomyces cerevisiae* [44].

Electroactive bacteria (EAB) are generally the main group found in MFC. One of the most studied EAB is *Geobacter sulfurreducens* because it often dominates in the anode biofilm under conditions considered optimal for energy generation. *Shewanella oneidensis* is less predominant in the biofilm of bioelectrochemical systems but has also been considered as a model EAB which relies on the production of redox mediators and, thus, is found in suspension without direct contact with the anode. In addition, EAB include *Bacillus, Klebsiella, Rhodopseudomonas, Desulfuromonas, Geomonas, and Geothrix,* among others. However, the diversity of EAB has still to be completely discovered since new bacteria are continually identified as capable of performing EET [19, 44].

Even though many studies have focused on pure culture to test the electrogenic ability of certain isolated species, it has been systematically demonstrated that mixed cultures achieve higher energy output. In addition, mixed culture applied for wastewater treatment commonly have lower operating costs and can be used for a wider range of substrates with higher adaptive capacity to external conditions [45].

Several factors can affect the structure and composition of microbial communities in MFC. For example, external resistance (R_{ext}) is an important operation parameter affecting the microbial community in MFC since it regulates the electron flow from the anode to the cathode [43, 46]. The reduction in resistance increases the availability of the anode as an electron acceptor, which is directly related to EET [47]. Thus, changes in R_{ext} affect voltage, anode potential, current, biofilm structure and morphology, microbial metabolism, organic matter removal, and coulombic efficiency (CE) of the MFC [20, 48].

Suzuki et al. [49] reported that a MFC with R_{ext} of 1000 Ω led to the enrichment of highly electroactive bacteria in the anode biofilm compared to 10 Ω. In their study, *Geobacter metallireducens* was observed with both types of R_{ext} but with more negative onset potential and higher performance at 1000 Ω. Katuri et al. [50] found that MFCs operated at lower R_{ext} achieved higher current generation with lower microbial diversity. In this regard, the R_{ext} -influenced anodic potential may intensify competition between members of the EAB and non-EAB community [51].

The competition between EAB and non-EAB metabolism can occur indirectly, as it influences the rate of substrate consumption and the associated proton production rate on the anode. Hence, R_{ext} can affect the pH in the anodic biofilm [36, 52]. The acidified anolyte can decrease the electrogenic activity and, consequently, affect the performance of EAB and non-EAB [9, 36, 53]. Regarding the direct effects of R_{ext} on competition between EAB and non-EAB, a lower MFC voltage created by reducing the R_{ext} is considered to naturally select EAB that can still meet their metabolic energy needs with a low potential gradient between the redox potential of their electron donor and the anode [51].

Besides energy generation performance, the lower R_{ext} decreases the time required for start-up, suggesting that it controls the growth rate and adaptive behavior of the anodic community [54]. Hence, understanding the development and dynamics of microbial communities under the influence of R_{ext} is essential to modulate the reactions inside the MFC, aiming to improve the system's performance [51].

Temperature is another crucial factor that controls bacteria growth rate and kinetics, influencing the initial biofilm formation process [11]. EAB were reported to be active in the temperature range of 5–90 °C, while as the temperature rises, species richness decreases and equitability and power output increase [16, 43, 45].

Although low energy generation has been achieved in lower temperatures, psychrophilic populations were found in anodic biofilm with electron transfer ability [55]. Regarding higher temperatures, it was reported by Mei et al. [16] changes in the microbial community structure as a function of temperature. At 10 °C, the most abundant genera were *Acidovorax* (16.35%), *Zoogloea* (14.87%), *Simplicispira* (11.76%), and *Geothrix* (6.47%). In comparison, at 20 °C most of the bacteria belonged to the genus *Pelobacter* (47.46%), and at 30 °C, *Geobacter* (11.77%), *Azonexus* (8.66%), *Bacteroidetes* (7.81%) e *Thauera* (6.85%) dominated the biofilm, resulting in higher energy generation.

The studies have mainly focused on mesophilic MFC. Hence, the knowledge concerning EAB in thermophilic conditions is limited. The genera *Thermincola* and *Caloramator* have been frequently reported with high relative abundance in thermophilic MFCs. *Thermincola ferriacetica* is a Gram-positive bacterium commonly found in thermophilic MFCs and was demonstrated to perform EET [56, 57]. The genus *Caloramator*, a known thermophilic, strict anaerobic and chemorganoheterotroph, was identified as an EAB with a direct EET mechanism [57]. Other bacteria generating current without exogenous mediators at thermophilic conditions were also reported, such as *Thermincola potens* and *Calditerrivibrio nitroreducens*. Still, the bacterial diversity in thermophilic MFC has not yet been well elucidated, and many EAB are expected to be identified in MFC at temperatures higher than 45 °C [43, 45, 57].

Most studies with MFC focus on the biofilm attached to the anode. However, some microorganisms can use electrons generated on the anode, catalyzing the reduction reaction on the cathode [37, 58]. This group of electroactive bacteria is named electrotroph. These bacteria can be used as biocatalysts for oxygen reduction, substituting expensive chemical catalysts, such as platinum. Furthermore, biocathodes can use other electron acceptors, such as nitrate and nitrite, expanding the applications and opportunities of the technology [59–61]. Regarding of bacterial community, biocathodes show high diversity, including members from Proteobacteria, Bacteroidetes, Actinobacteria, Firmicutes, Aquificae, and Planctomycetes. A key factor controlling the microbial community structure on the cathode is the type of electron acceptor available in the cathode chamber [58, 61].

8.3 Processes Involving Nitrogen in MFC

The release of nitrogen into water bodies is a major concern due to eutrophication and toxic effects. Considering the high concentration of ammonia nitrogen, this is particularly relevant for wastewater from the dewatering of digested biosolids [62], landfill leachate [63], swine manure [64], and vinasse from ethanol production [65].

Biological nitrogen removal is usually based on nitrification, followed by denitrification. Nitrification only occurs when oxygen is present at concentrations high enough to support the growth of strictly aerobic nitrifying bacteria that oxidize ammonia to nitrite and nitrate with a total demand of 4.57 g O_2/g NH_4^+–N [66]. The aeration in a conventional WWTP (i.e., activated sludge process) is responsible for electricity consumption of about 0.6 kWh/m^3, which represents 50–90% of the total electricity consumed in a WWTP, and from 15 to 49% of total costs within a plant [4, 67, 68]. Moreover, denitrification requires a source of organic carbon as an electron donor to reduce nitrate and nitrite to dinitrogen (N_2). Thus, in WWTP, external carbon is commonly added to support denitrification. This results in higher costs for the treatment and can lead to residual organic matter in the final effluent [15, 69].

Regarding nitrogen in MFCs, both oxidation and reduction reactions have been reported in MFCs. These processes can potentially change the energy balance of nitrogen treatment in WWTP [35, 70].

8.3.1 Reactions on the Anode

Ammonia oxidation in MFC is fundamentally characterized by reduced (or null) oxygen consumption, so aeration is unnecessary (Fig. 8.3, anode chamber). Furthermore, the process can generate electricity. The first evidence of the utilization of ammonia as an electron donor by EAB on the anode was reported by He et al. [71]. The process is currently under lab-scale investigation, and energy savings are projected for its implementation in WWTP [35].

He et al. [71] observed current generation when adding ammonium chloride, ammonium sulfate, or ammonium phosphate in an MFC with a rotating disk cathode. Their MFC achieved mean ammonium removal efficiency up to 67%, with AOB bacteria found in the anode and nitrite as the main product. These results suggested

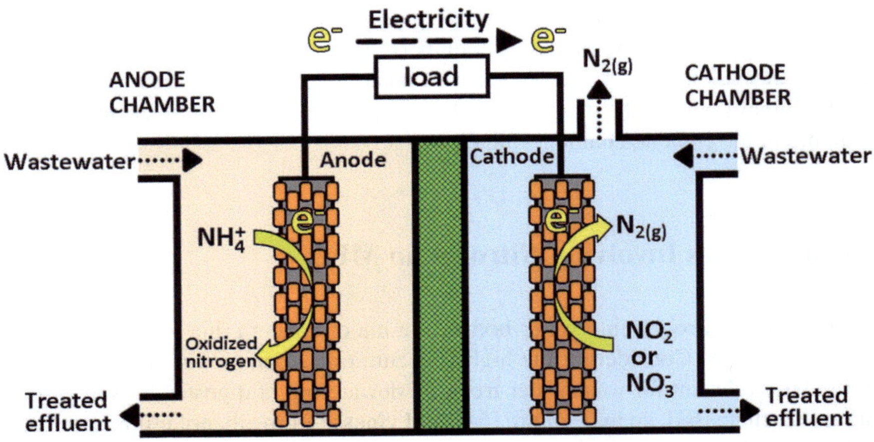

Fig. 8.3 Diagram of the processes involving nitrogen oxidation and reduction in the bioanode and biocathode of MFC

ammonium was directly used as the anodic fuel or indirectly as substrates for nitrifiers to produce organic compounds for heterotroph EAB. Similarly, Qu et al. [72] achieved ammonium oxidation and proposed the anode as the electron acceptor for a microbial community dominated by *Nitrosomonas europaea* in a dual chamber microbial electrolysis cell (MEC). However, nitrate was reported as the main product of ammonium oxidation instead of nitrite.

A significant contribution to understanding the pathway of ammonia oxidation with current generation was described by Vilajeliu-Pons et al. [34]. In their study, complete anaerobic conversion of ammonium by *Nitrosomonas,* an ammonia-oxidizing genus, to dinitrogen gas, with low production of NO_2^-, NO_3^-, N_2O, was demonstrated in a continuous flow BES. In addition, Shaw et al. [73] reported current generation with successful ammonia oxidation in absence of nitrite by *Candidatus* Brocadia and *Candidatus* Scalindua, which are anaerobic ammonia oxidation (anammox) bacteria. Finally, it was also demonstrated that a pure culture of *Acidimicrobiaceae* sp. A6 was capable of using the anode as an electron acceptor during the oxidation of ammonium [74].

Most of the studies focused directly on adapting mixed culture sludge from different origins (such as WWTP, soil, marine sediments, laboratory bioreactor, etc.) to use ammonia as an electron donor. An alternative approach was reported by Tang et al. [75] by switching acetate medium to ammonium medium after the formation of stable electrogenic acetate-oxidizing biofilms. The results showed ammonia removal efficiency of 82% with a dominance of *Ignavibacteriaceae*, *Geobacteraceae,* and *Nitrosomonadaceae* families.

External energy is usually supplied to the bioelectrochemical system to control the anode potential in ammonia oxidation. However, a recent study demonstrated that energy can be generated from biological ammonia oxidation in MFC without energy input. Regarding the structure and composition of the microbial community on the anode, the power density level substantially changed depending on the dominant functional group. This indicates that the rate and efficiency of electron transfer to the anode also depends on the type of bacteria oxidizing ammonia [35].

8.3.2 Reactions on the Cathode

Regarding the second step of nitrogen treatment, it was experimentally demonstrated that electroactive bacteria on the cathode could catalyze the nitrate reduction using the cathode as an electron donor in a process named bioelectrochemical denitrification [76]. Several studies focusing on this process have been published since then [21, 24, 33, 37, 77]. In this process, nitrate, instead of oxygen, is used as an electron acceptor in a biocathode, allowing the system to remove organic matter (in the anodic chamber) and nitrogen (in the cathode chamber) simultaneously without external carbon addition [21].

So, this kind of MFC is a complete bioelectrochemical system in which electroactive microorganisms catalyze both anodic and cathodic reactions. It is a promising solution for wastewater treatment with high nitrogen content [11]. At the same time, it is also a low-cost alternative to catalyze the cathodic reactions compared to chemical catalyzers, such as platinum, and avoids costs related to oxygen supply [78].

Gregory et al. [76] reported, for the first time, that a graphite electrode connected to a potentiostat could be used as an electron donor for nitrate reduction by microorganisms on the cathode, with a consistent stoichiometric ratio between electrons consumption and nitrate reduction. However, the bioelectrochemical denitrification coupled with organic matter oxidation on the anode, generating energy, was demonstrated only in 2007 with a maximum power density of 8 W/m^3 [33]. Later, organic matter oxidation, nitrification, and cathodic denitrification were achieved simultaneously in a MFC with two cathodes, resulting in ammonia nitrogen removal between 84 and 97% and total nitrogen removal up to 90% [21]. Also focusing on simultaneous reactions, Virdis et al. [77] achieved nitrification and denitrification in the same cathodic chamber, with dissolved oxygen at 4 mg L^{-1}, resulting in nitrogen removal up to 94%.

8.4 Microbial Fuel Cell Applications and Optimization

In the last decades, MFC technology has been tested for a number of different objectives. Applications often reported include biosensing, soil bioremediation, energy generation through marine sediments, desalination, and wastewater treatment [17, 24, 79]. As the system degrades several compounds found in wastewater, its application for wastewater treatment combined with energy generation has received special attention [18, 80].

8.4.1 Wastewater Treatment

Shizas and Bagley [81] estimated that, in a domestic wastewater treatment plant in Toronto, Canada, there was 9.3 times more energy in the wastewater than what was used to treat it. The current level of energy generation of MFC and its costs are considered a barrier for its implementation in wastewater treatment, especially compared to anaerobic digestion technologies. However, continuous advances in MFC performance over the last decades have drawn attention to a possible future implementation in wastewater treatment plants [8, 24].

The first studies regarding MFC application for wastewater treatment were published in the 1990s [9]. In 2004, it was demonstrated that domestic sewage could be treated at high levels simultaneously with energy generation, with a power density up to 26 mW/m^2 [82]. Since then, numerous studies have reported improvements in energy output, treatment efficiency, and costs of MFCs treating wastewater [11, 22, 37, 80, 82]. For example, Feng et al. [83], using a stacked 250 L MFC, obtained a maximum power density of 0.47 W/m^3, with a COD removal of 79%. According to the authors, the estimated initial investment cost of the system was US$ 2500, while the energy cost of operation was 0.5 kWh/m^3, approximately 50% lower than conventional aerobic treatment.

Several urban, industrial, and agro-industrial wastes have already been studied in the MFC field. Wastewater generated in the industrial and agro-industrial sectors commonly presents a high concentration of organic matter and nutrients, such as nitrogen, potassium, and phosphorus [5]. Depending on the source and industrial process, chemical oxygen demand (COD) can reach from 2000 to > 30,000 mg/L. In addition, the characteristics of certain types of effluents are advantageous for the use of MFC, such as the high concentration of simple organic molecules, the presence of redox mediators, such as lignin and phenolic compounds, and high conductivity due to the elevated concentration of nitrogen, potassium, sodium and sulfate, and elevated temperature, which help the charge transfer inside the MFC, reducing the ohmic resistance [22, 84–87]. For example, Sakdaronnarong et al. [22] reported the treatment of vinasse from sugar cane by a MFC with a total volume of 4 L. Using lignin (0.1 mol/L) as a redox mediator, with an external resistance of 560 Ω, a power density of up to 93 W/m^2 and COD removal of 81% was achieved in 6 days.

In the past, attempts to scale up the MFC technology for domestic wastewater treatment were not successful because of the low conductivity of the sewage and not suitable reactor configurations. Currently, the possibility of developing a self-sufficient MFC for the treatment of domestic sewage is conjectured based on novel configurations and approaches [18, 88]. In this regard, the system configuration has been reported as crucial for the performance of pilot-scale MFCs. Using liquid electrolytes in the cathode tends to result in higher treatment performance but low power output and undesirable energy consumption for the aeration of the catholyte. On the contrary, the utilization of MFCs with cathodes exposed to air (air-cathodes) presents improved power density but requires post-treatment to meet the typical threshold for discharge [88]. Furthermore, the scaling-up of MFC has been considered towards modularity and stacking of small MFC units [89, 90]. Since 2014, numerous studies have reported the application of large MFCs systems (volume > 100 L) with promising results [6, 83, 88, 91]. A recent study by Rossi et al. [92] assessed a pilot scale MFC (850 L) consisting of 17 anode modules treating domestic wastewater and reported a power density of 0.043 W/m^2, with COD removal of 49 ± 15%.

8.4.2 Operational Conditions and Optimization

The current performance of MFC is explained in part by thermodynamic limitations and high energy losses associated with the structural components of the MFC and the microorganisms. The MFC has thermodynamic barriers due to the nature of biochemical reactions. In addition, part of the energy is consumed by the microbial metabolism, and lost to ohmic resistances associated with electron transfer from the cell to the electrode, due to mass transport in the biofilm. In addition, resistances associated with the anode and cathode, electrolyte and membrane, hinder electron transport, reducing energy recovery and conversion efficiency [18, 93].

Optimization of the system's performance can be achieved from the knowledge concerning bioelectrochemical processes within the MFC and the determination of the operational parameters influence over the efficiency of treatment and energy generation [15]. Among the operational characteristics considered relevant to improve the economic feasibility of the MFC, the electrode material and operating conditions, such as the temperature, are highlighted [16, 32].

8.4.2.1 Temperature

Temperature is an operation variable whose effect on the performance of traditional anaerobic systems is well known. However, it is still unclear how it affects MFC performance [16]. It is assumed that temperature influences kinetics, mass transfer, thermodynamics, and microbial community structure and, thus, can catalyze reactions on the anode and cathode of MFCs [11, 94]. In addition, the relation between temperature and the reactor material is important. As the temperature increases, it can lead to higher electrical conductivity [94] but reduces the treatment performance of adsorptive electrode materials [11].

Lower temperature resulted in losses in power density and organic matter removal rate, as it decreased biofilm growth rate and enzymatic activity. In temperatures lower than 15 °C, more time was required for a start-up with a power density of 709 mW/m^2, which was 55% lower in relation to operation at 30 °C [95]. However, it has been speculated that MFC biofilms may have a self-heating effect that could reduce the impact of low temperatures on their performance [96].

Regarding temperatures > 20 °C, positive effects in energy generation and treatment efficiency have been reported. In an air–cathode MFC fed with acetate as an electron donor, a power density 15% higher was achieved at 30 °C, compared to 20 °C, while high COD removal (> 90%) was observed independently of the temperature [16]. Similarly, Tee et al. [11] reported higher power density, 74 mW/m^3, at 35 °C in comparison to 15 °C, 20 °C and 30 °C. It was proposed that electrogen activity is intensified at temperatures > 20 °C, but enzymes are denatured at temperatures > 40 °C, decreasing the external electron transfer to the anode.

Jong et al. [97] were among the first researchers to demonstrate the ability of a thermophilic community to generate energy in a MFC operated at 55 °C, with a power density of 1030 mW/m^2 and CE of 80%. Later, energy generation in a thermophilic MFC (55 °C) was confirmed with 37 mW/m^2 and CE of 89% [98]. Another study achieved a current density between 209 and 254 mA/m^2 at 60 °C, which was tenfold higher compared to 22 °C [56].

In a membrane-less MFC fed with synthetic wastewater, thermophilic temperature (55 °C) was also favorable since it reduced the non-EAB biomass attached to the cathode, but mean COD removal efficiency, between 40 and 59%, was lower in comparison to mesophilic condition (averages ranging from 56 to 64%)[86].

For Fu et al. [57] the thermophilic microbial community at 55 °C performed better with an average electrical voltage generation of 74.8 mV when compared to the mesophilic unit (25 °C), with a generation of 24.9 mV. This increase in energy generation in thermophilic units is related to the superior ionic conductivity compared to mesophilic ones, due to the greater mobility of ions in the solution, decreasing the internal resistance and increasing the charge transfer [57, 99].

While the temperature was demonstrated to improve the performance of the anode, it can adversely affect the cathode reactions. It was shown that the internal resistance associated with the cathode was substantially higher at 55 °C in comparison with 25 °C and 35 °C. Oxygen solubility in the catholyte decreases with temperature rise, reducing the availability of dissolved oxygen in the cathode [32].

8.4.2.2 External Resistance

The applied external resistance (R_{ext}) is another important factor recently found to affect the growth rate of electro-active bacteria. However, the extent of the effects within a MFC is still controversial [49, 54]. While Suzuki et al. [49] observed power density five times higher for a MFC with 1000 Ω in relation to 10 Ω, Rismani-Yazdi et al. [46] reported higher power density, between 1.14 and 2.4 times, with a low R_{ext} of 20 Ω against 249, 249, 480 and 1000 Ω.

As an electric power source, the maximum power of a MFC is achieved when the R_{ext} and internal resistance (R_{int}) are equal. Thus, if R_{ext} is higher or lower than the R_{int}, losses in power output are expected due to changes in voltage and current [100, 101]. By applying lower R_{ext}, larger current densities are expected. These increased current densities should be accompanied by improved coulombic efficiency and power [100], and increased substrate removal capabilities [102, 103].

It is relevant to point out that R_{int} is subject to change over time, for example, due to changing operating conditions and biofilm characteristics [104, 105]. In this sense, the dynamic control of the R_{ext} to equal the R_{int} in a MFC can increase its power output [106].

However, the optimal R_{ext} of the system may reflect some dependence on the anolyte composition, as indicated by Yi et al. [107]. Córdova-Bautista et al. [108] demonstrated R_{ext} effects in conjunction with other relevant process variables in an MFC, for example, the pH of the anolyte. Furthermore, R_{ext} could have an irreversible effect on anodic biofilm composition [109]. R_{ext} can influence the anode potential and the electrical current, determining the amount of energy available for EET metabolism. This affects EAB growth and can potentially change the balance between EAB and competing microorganisms in the electrically conductive biofilm. Thus, start-up and transition to a steady-state condition in MFC may be controlled by R_{ext}. [50, 54, 101, 110, 111]. On the other hand, forming a thicker biofilm can influence the diffusion of the substrate to the inner layers of the biofilm, limiting the energy generation potential of an MFC [112, 113].

In addition to the impact on the anode chamber, the cathode potential can also be affected by R_{ext}; as the resistance of the circuit is changed, the transference of positive charges can be enhanced to the cathode. This affects the rate of reactions at the cathode, which is relevant, especially for MFCs limited by cathodic reactions [114].

8.4.2.3 Electrode Material

The composition, morphology, and surface characteristics of the electrode material affect several key factors of the MFC system. These factors include the adhesion of the biofilm, kinetics of electron transfer, and oxidation of the substrate [93, 115]. Thus, one of the main challenges in developing MFCs concerns identifying materials that optimize energy generation, increase coulomb efficiency, and minimize the costs [9, 116].

The electrode polarization losses, which reduce the MFC performance, are associated with activation, ohmic, and mass transport losses. Activation losses are related to the beginning of the oxidation or reduction reaction on the electrode and electron transfer from the bacterial cells to the electrode surface. Ohmic losses arise due to the electrical resistance of the MFC components, electrolyte and through the PEM. Mass transport losses are caused by limitations in the mass transfer of the reactants to the electrode and the products out of the electrode. Hence, mass transport losses are relevant at high current, especially when a thick non-conductive biofilm is developed on the electrodes [9, 117].

The activation losses related to the anode can be minimized by increasing the surface area and improving the electron transfer between the bacteria and the anode. Concerning the cathode, it can also be minimized by increasing the area but also by using electron acceptors whose reaction presents faster kinetic, by adding chemical catalyzers, or by using microorganisms as catalyzers configuring a biocathode.

Ohmic losses can be minimized by improving the conductivity of the MFC components, for example reducing the spacing between the electrodes and avoiding materials with high resistivity. Finally, three-dimensional electrodes can allow a more efficient transport of reactants and products associated with reactions on the electrodes, reducing the mass transport resistance [3, 118].

Thus, the characteristics required for the electrode material are high conductivity, biocompatibility, chemical and physical stability, high specific surface, high porosity, low cost, and easy manufacturing. In addition to these characteristics, it is important to consider how the material used affects the bacteria's ability to transfer electrons to its surface [9].

Metallic materials are usually considered for electrodes confection since they present high conductivities concerning other materials. However, as it is necessary to be non-corrosive, fewer options are compatible with MFCs [93, 116]. Stainless steel has favorable mechanical properties for long-term operation, and, thus, is considered a promising option for large scale applications. In addition, it is manufactured with different compositions and morphologies [115, 119]. In the study of Pocaznoi et al. [115], stainless steel showed higher current densities compared to graphite under the same inoculum and polarization conditions. Current densities obtained for stainless steel, under polarization of $-$ 0.2 and $+$ 0.1 V vs. Saturated Calomel Electrode (SCE), were 20.6 A/m^2 and 35 A/m^2, respectively, while the graphite electrode showed values of 9.5 and 11 A/m^2.

However, stainless steel has disadvantages, such as its smooth surface and the presence of Cr, culminating in low bacterial adhesion [116]. Because of this, studies have been carried out to modify the surface of stainless steel through various methodologies, such as flame oxidation, flame deposition, and nanocarbon coating [119]. The flame oxidation leads to the formation of iron oxide nanoparticles on the stainless steel surface, increasing biocompatibility [116]. However, these kinds of modifications decrease the chemical stability of the material and can lead to corrosion when exposed to wastewater (especially with high salinity), increase internal resistance, or result in unstable currents [120].

Electrode made of carbonaceous materials is another promising option due to its properties: chemical stability; high conductivity; and good biocompatibility. This kind of electrode is available in various configurations that favor the adhesion of microorganisms (plane, fabric, brush, granules, etc.) [116, 119]. The following carbonaceous materials are commonly utilized in MFCs: carbon (paper, mesh, felt, fabric, foam), graphite (stem, granules, leaves, brush), and reticulated vitreous carbon [9, 93, 115]. However, currently, the high cost associated with these materials, especially when they are modified with catalysts or nanoparticles, mays result in economic unfeasibility [120].

Nonetheless, the volumetric power density of MFCs can be maximized utilizing high-performance electrodes based on three-dimensional materials that allow biofilm formation on the external and internal surfaces [93, 121]. In a comparative study, the higher current density achieved by a carbon cloth electrode concerning stainless steel and graphite was attributed to its three-dimensional configuration, resulting in an additional surface area available for biofilm colonization [115]. In this sense, granular activated carbon (GAC) has a high surface area for bacteria adhesion, resulting in more efficient electron transfer and increased energy generation. Due to its high specific surface area, it has been proposed as an efficient substitute for chemical catalyzers in the oxygen reduction reaction.

By combining the MFC processes with the use of GAC, greater efficiency of wastewater treatment has been reported due to the adsorption capacity of the GAC. For example, Tee et al. [11] evaluated the effect of GAC as an anode in a single chamber MFC (working volume of 5.65 L) with a cathode exposed to air, applied to treat palm oil mill effluent. Using a ceramic cylinder as a proton exchange system and an external resistance of 50 Ω, they obtained an average COD removal of up to 93.5%. The authors attributed the high efficiency to the GAC adsorption characteristics and high surface area. The maximum energy density observed was 74 mW/m^3, with a CE of 10.6%.

Several studies reported the use of GAC as anode and/or cathode with different configurations but most achieved power densities as low as 6 W/m^3 [11, 122–126]. On the other hand, other studies reported power densities up to 50 W/m^3, suggesting that this material can lead to efficient MFCs [91, 127]. Hence, the combination of low-cost materials with high conductivity and specific surface areas, such as stainless steel and granular activated carbon, represents an opportunity to achieve high energy generation and treatment efficiency without increasing the overall costs.

8.4.2.4 Capacitive Electrodes

A possible approach to increase power generation is the application of capacitive materials as electrodes [128]. This strategy is based on increasing the bioelectrode conversion rate through three-dimensional electrodes with a high surface area available for microbial growth [121]. These materials usually have porous structures (i.e., granular activated carbon) that increase the number of catalytic sites (and, therefore, the electron transfer rate). Furthermore, they also allow for charge storage when electrons produced by bacteria (known as faradaic current) are stored in the form of an electrical double layer (EDL). Electrochemical systems are defined within a boundary between conventional batteries and capacitors based on certain characteristics controlled by the EDL (i.e., power density, current generation, and charge/discharge cycle life) [129]. EDL occurs at any electrode/solution interface, but it is intensified in capacitive electrodes (Fig. 8.4).

The system must be operated intermittently for electricity storage purposes, which means that it must undergo a charging and discharging process (Fig. 8.5). The charging process forms the EDL, in the capacitive material, and then is released during discharge producing a non-faradaic current (a current not attributed to any redox process) called capacitive current (Icap). In a capacitive bioanode, charging results from the chemical oxidation of the substrate by bacteria, generating a faradaic current (Ifar). As microorganisms keeps this oxidation throughout discharging process, a sum of Icap and Ifar will be released [128].

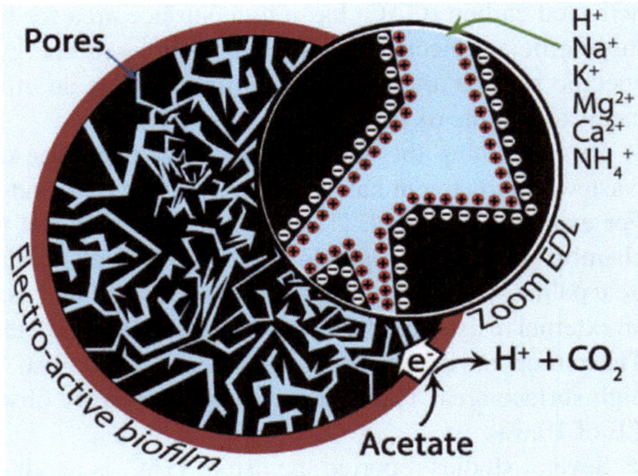

Fig. 8.4 Illustration of the structure of capacitive granules with an electrochemically active biofilm. *Source* [121]

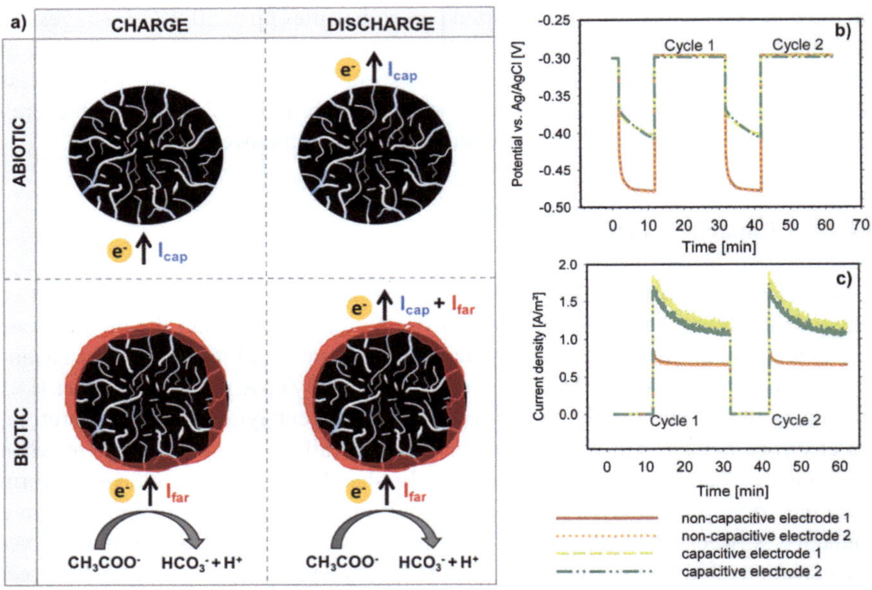

Fig. 8.5 **a** Illustration of the charging and discharging process of capacitive granules with and without microorganisms; **b** potential and **c** current density behavior for two cycles of a charge – discharge experiment with 10 min of charging and 20 min of discharging. Source: [128, 130]. Deeka et al. [130] Reprinted (adapted) with permission from Deeke A, Sleutels THJA, Hamelers HVM, Buisman CJN (2012) Capacitive bioanodes enable renewable energy storage in microbial fuel cells. Environ Sci Technol 46:3554–3560. Copyright 2012 American Chemical Society

A study points out that 40% more current can be obtained from a capacitive electrode during intermittent operation compared to continuous operation of a non-capacitive electrode [131] which proves to be a promising solution given the thermodynamics limitation of MFC's.

8.5 Conclusions

The MFC technology has gained attention in the last decades due to its promising application in which organic matter from wastewater can be directly converted into electricity by biological activity. Such ability was expected to quickly result in significant contributions to the sustainable development goals with structural and integrated advances in the sanitation and energy sectors toward a circular economy approach. However, despite significant advances in terms of energy generation and wastewater treatment performance, the MFC system is still under low technology readiness level and outcompeted by other longer-studied technologies. Currently, factors limiting the power output of MFCs include energy losses by microbial metabolism and materials, which impede the implementation of the technology at large-scale. These barriers have oriented the studies in the field of bioelectrochemistry to investigate complex interdisciplinary phenomena involving molecular biology, materials science, and chemical engineering. Henceforth novel findings have improved the comprehension of the energy generation process in MFC. This resulted in advanced configurations and operation strategies (i.e., capacitive electrodes with charge and discharge cycles), novel applications from processes involving inorganic compounds from wastewater (bioelectrochemical nitrification and denitrification), and culminated in recent new pilot-scale studies. The continuous improvements in MFC highlight a promising future for this technology, comprising not only the primary objective of generating renewable energy but also novel applications focused on advanced wastewater treatment processes.

Acknowledgements The authors are grateful to FAPESP [Process number: 2021/07925-6 and 2019/27180-5] and Coordenação de Aperfeiçoamento de Pessoal de Nível Superior - Brazil (CAPES) [Finance Code 001].

References

1. United Nations (2015) Transforming our world: the 2030 agenda for sustainable development. Available via United Nations' digital library. https://sustainabledevelopment.un.org/post2015/transformingourworld/publication. Transforming our world: The 2030 agenda for sustainable development A/RES/70/1 Accessed 08 Jul 2023
2. Heller L, de Nascimento NO (2005) Pesquisa e desenvolvimento na área de saneamento no Brasil: necessidades e tendências. Eng San Amb 10:24–35https://doi.org/10.1590/S1413-41522005000100004

3. Palanisamy G, Jung H-Y, Sadhasivam T et al (2019) A comprehensive review on microbial fuel cell technologies: Processes, utilization, and advanced developments in electrodes and membranes. J Clean Prod 221:598–621. https://doi.org/10.1016/j.jclepro.2019.02.172
4. Gude VG (2015) Energy and water autarky of wastewater treatment and power generation systems. Renew Sust Energ Rev 45:52–68. https://doi.org/10.1016/j.rser.2015.01.055
5. Hossain MI, Cheng L, Cord-Ruwisch R (2019) Energy efficient COD and N-removal from high-strength wastewater by a passively aerated GAO dominated biofilm. Bioresour Technol 283:148–158. https://doi.org/10.1016/j.biortech.2019.03.056
6. Tota-Maharaj K, Paul P (2015) Performance of pilot-scale microbial fuel cells treating wastewater with associated bioenergy production in the Caribbean context. Int J Energy Environ Eng 6:213–220. https://doi.org/10.1007/s40095-015-0169-x
7. Meneses-Jácome A, Diaz-Chavez R, Velásquez-Arredondo HI et al (2016) Sustainable Energy from agro-industrial wastewaters in Latin-America. Renew Sust Energ Rev 56:1249–1262. https://doi.org/10.1016/j.rser.2015.12.036
8. McCarty PL, Bae J, Kim J (2011) Domestic wastewater treatment as a net energy producer–can this be achieved? Environ Sci Technol 45:7100–7106. https://doi.org/10.1021/es2014264
9. Logan BE (2008) Microbial fuel cells, 1st edn. Wiley-Interscience, Hoboken
10. Slate AJ, Whitehead KA, Brownson DAC, Banks CE (2019) Microbial fuel cells: an overview of current technology. Renew Sust Energy Rev 101:60–81. https://doi.org/10.1016/j.rser.2018.09.044
11. Tee P-F, Abdullah MO, Tan IAW et al (2017) Effects of temperature on wastewater treatment in an affordable microbial fuel cell-adsorption hybrid system. J Environ Chem Eng 5:178–188. https://doi.org/10.1016/j.jece.2016.11.040
12. Commault AS, Lear G, Weld RJ (2015) Maintenance of geobacter-dominated biofilms in microbial fuel cells treating synthetic wastewater. Bioelectrochemistry 106:150–158. https://doi.org/10.1016/j.bioelechem.2015.04.011
13. Abbasi U, Jin W, Pervez A et al (2016) Anaerobic microbial fuel cell treating combined industrial wastewater: correlation of electricity generation with pollutants. Bioresour Technol 200:1–7. https://doi.org/10.1016/j.biortech.2015.09.088
14. Vázquez-Larios AL, Solorza-Feria O, Poggi-Varaldo HM et al (2014) Bioelectricity production from municipal leachate in a microbial fuel cell: Effect of two cathodic catalysts. Int J Hydrog Energy 39:16667–16675. https://doi.org/10.1016/j.ijhydene.2014.05.178
15. Al-Mamun A, Baawain MS (2015) Accumulation of intermediate denitrifying compounds inhibiting biological denitrification on cathode in Microbial Fuel Cell. J Environ Health Sci Eng. https://doi.org/10.1186/s40201-015-0236-5
16. Mei X, Xing D, Yang Y et al (2017) Adaptation of microbial community of the anode biofilm in microbial fuel cells to temperature. Bioelectrochemistry 117:29–33. https://doi.org/10.1016/j.bioelechem.2017.04.005
17. Pant D, Van Bogaert G, Diels L, Vanbroekhoven K (2010) A review of the substrates used in microbial fuel cells (MFCs) for sustainable energy production. Bioresour Technol 101:1533–1543. https://doi.org/10.1016/j.biortech.2009.10.017
18. Sun M, Zhai L-F, Li W-W, Yu H-Q (2016) Harvest and utilization of chemical energy in wastes by microbial fuel cells. Chem Soc Rev 45:2847–2870. https://doi.org/10.1039/C5CS00903K
19. Lovley DR, Holmes DE (2022) Electromicrobiology: the ecophysiology of phylogenetically diverse electroactive microorganisms. Nat Rev Microbiol 20:5–19. https://doi.org/10.1038/s41579-021-00597-6
20. Philips J, Verbeeck K, Rabaey K, Arends J (2015) Electron transfer mechanisms in biofilms. In: Microbial electrochemical and fuel cells: fundamentals and applications, 1st edn. Elsevier. Woodhead Publishing, Sawston, pp 67–113
21. Zhang F, He Z (2012) Integrated organic and nitrogen removal with electricity generation in a tubular dual-cathode microbial fuel cell. Process Biochem 47:2146–2151. https://doi.org/10.1016/j.procbio.2012.08.002

22. Sakdaronnarong C, Ittitanakam A, Tanubumrungsuk W et al (2015) Potential of lignin as a mediator in combined systems for biomethane and electricity production from ethanol stillage wastewater. Renew Energy 76:242–248. https://doi.org/10.1016/j.renene.2014.11.009
23. Rabaey K, Verstraete W (2005) Microbial fuel cells: novel biotechnology for energy generation. Trends Biotechnol 23:291–298. https://doi.org/10.1016/j.tibtech.2005.04.008
24. Sun H, Xu S, Zhuang G, Zhuang X (2016) Performance and recent improvement in microbial fuel cells for simultaneous carbon and nitrogen removal: A review. J Environ Sci 39:242–248. https://doi.org/10.1016/j.jes.2015.12.006
25. Koch C, Harnisch F (2016) What is the essence of microbial electroactivity? Front Microbiol.https://doi.org/10.3389/fmicb.2016.01890
26. Bajracharya S, Sharma M, Mohanakrishna G et al (2016) An overview on emerging bioelectrochemical systems (BESs): technology for sustainable electricity, waste remediation, resource recovery, chemical production and beyond. Renew Energy 98:153–170. https://doi.org/10.1016/j.renene.2016.03.002
27. Kaur A, Boghani HC, Michie I et al (2014) Inhibition of methane production in microbial fuel cells: Operating strategies which select electrogens over methanogens. Bioresour Technol 173:75–81. https://doi.org/10.1016/j.biortech.2014.09.091
28. Gezginci M (2016) The effect of different substrate sources used in microbial fuel cells on microbial community. JSM Environ Sci Ecol. https://doi.org/10.47739/2333-7141/1035
29. Zhao Y-G, Zhang Y, She Z et al (2017) Effect of substrate conversion on performance of microbial fuel cells and anodic microbial communities. Environ Eng Sci 34:666–674. https://doi.org/10.1089/ees.2016.0604
30. Toczyłowska-Mamińska R, Pielech-Przybylska K, Sekrecka-Belniak A, Dziekońska-Kubczak U (2020) Stimulation of electricity production in microbial fuel cells via regulation of syntrophic consortium development. Appl Energy. https://doi.org/10.1016/j.apenergy.2020.115184
31. Rodrigues ICB, Leão VA (2020) Producing electrical energy in microbial fuel cells based on sulphate reduction: a review. Environ Sci Pollut Res 27:36075–36084. https://doi.org/10.1007/s11356-020-09728-7
32. Cano V, Cano J, Nunes SC, Nolasco MA (2020) Electricity generation influenced by nitrogen transformations in a microbial fuel cell: assessment of temperature and external resistance. Renew Sust Energy Rev. https://doi.org/10.1016/j.rser.2020.110590
33. Clauwaert P, Rabaey K, Aelterman P et al (2007) Biological denitrification in microbial fuel cells. Environ Sci Technol 41:3354–3360. https://doi.org/10.1021/es062580r
34. Vilajeliu-Pons A, Koch C, Balaguer MD et al (2018) Microbial electricity driven anoxic ammonium removal. Water Res 130:168–175. https://doi.org/10.1016/j.watres.2017.11.059
35. Cano V, Nolasco MA, Kurt H et al (2023) Comparative assessment of energy generation from ammonia oxidation by different functional bacterial communities. Sci Total Environ. https://doi.org/10.1016/j.scitotenv.2023.161688
36. Bajracharya S, ElMekawy A, Srikanth S, Pant D (2016) 6 - Cathodes for microbial fuel cells. In: Scott K, Yu EH (eds) Microbial electrochemical and fuel cells, 1st edn. Woodhead Publishing, Boston, pp 179–213
37. He C-S, Mu Z-X, Yang H-Y et al (2015) Electron acceptors for energy generation in microbial fuel cells fed with wastewaters: a mini-review. Chemosphere 140:12–17. https://doi.org/10.1016/j.chemosphere.2015.03.059
38. Ucar D, Zhang Y, Angelidaki I (2017) An overview of electron acceptors in microbial fuel cells. Frontiers Microbiol. https://doi.org/10.3389/fmicb.2017.00643
39. Kumar R, Singh L, Zularisam AW (2016) Exoelectrogens: Recent advances in molecular drivers involved in extracellular electron transfer and strategies used to improve it for microbial fuel cell applications. Renew Sust Energy Rev 56:1322–1336. https://doi.org/10.1016/j.rser.2015.12.029
40. Torres CI, Marcus AK, Lee H-S et al (2010) A kinetic perspective on extracellular electron transfer by anode-respiring bacteria. FEMS Microbiol Rev 34:3–17. https://doi.org/10.1111/j.1574-6976.2009.00191.x

41. Hodgson DM, Smith A, Dahale S et al (2016) Segregation of the anodic microbial communities in a microbial fuel cell cascade. Front Microbiol. https://doi.org/10.3389/fmicb.2016.00699
42. Hsueh C-C, Wu C-C, Chen B-Y (2019) Polyphenolic compounds as electron shuttles for sustainable energy utilization. Biotechnol Biofuels. https://doi.org/10.1186/s13068-019-1602-9
43. Li M, Zhou M, Tian X et al (2018) Microbial fuel cell (MFC) power performance improvement through enhanced microbial electrogenicity. Biotechnol Adv 36:1316–1327. https://doi.org/10.1016/j.biotechadv.2018.04.010
44. Logan BE, Rossi R, Ragab A, Saikaly PE (2019) Electroactive microorganisms in bioelectrochemical systems. Nat Rev Microbiol 17:307–319. https://doi.org/10.1038/s41579-019-0173-x
45. Dai K, Wen J-L, Zhang F et al (2017) Electricity production and microbial characterization of thermophilic microbial fuel cells. Bioresour Technol 243:512–519. https://doi.org/10.1016/j.biortech.2017.06.167
46. Rismani-Yazdi H, Christy AD, Carver SM et al (2011) Effect of external resistance on bacterial diversity and metabolism in cellulose-fed microbial fuel cells. Bioresour Technol 102:278–283. https://doi.org/10.1016/j.biortech.2010.05.012
47. Liu T, Yu Y, Li D et al (2016) The effect of external resistance on biofilm formation and internal resistance in Shewanella inoculated microbial fuel cells. RSC Adv 6:20317–20323. https://doi.org/10.1039/C5RA26125B
48. Kondaveeti SK, Seelam JS, Mohanakrishna G (2018) Anodic electron transfer mechanism in bioelectrochemical systems. In: Das D (ed) Microbial fuel cell: a bioelectrochemical system that converts waste to watts. 1st edn. Springer International Publishing, Cham, pp 87–100. https://doi.org/10.1007/978-3-319-66793-5_5
49. Suzuki K, Kato Y, Yui A et al (2018) Bacterial communities adapted to higher external resistance can reduce the onset potential of anode in microbial fuel cells. J Biosci Bioeng 125:565–571. https://doi.org/10.1016/j.jbiosc.2017.12.018
50. Katuri KP, Scott K, Head IM et al (2011) Microbial fuel cells meet with external resistance. Bioresour Technol 102:2758–2766. https://doi.org/10.1016/j.biortech.2010.10.147
51. Jung S, Regan JM (2011) Influence of external resistance on electrogenesis, methanogenesis, and anode prokaryotic communities in microbial fuel cells. Appl Environ Microbiol 77:564–571. https://doi.org/10.1128/AEM.01392-10
52. Franks AE, Nevin KP, Jia H et al (2008) Novel strategy for three-dimensional real-time imaging of microbial fuel cell communities: monitoring the inhibitory effects of proton accumulation within the anode biofilm. Energy Environl Sci 2:113–119. https://doi.org/10.1039/B816445B
53. Zhuang L, Zhou S, Li Y, Yuan Y (2010) Enhanced performance of air-cathode two-chamber microbial fuel cells with high-pH anode and low-pH cathode. Bioresour Technol 101:3514–3519. https://doi.org/10.1016/j.biortech.2009.12.105
54. Buitrón G, López-Prieto I, Zúñiga IT, Vargas A (2017) Reduction of start-up time in a microbial fuel cell through the variation of external resistance. Energy Procedia 142:694–699. https://doi.org/10.1016/j.egypro.2017.12.114
55. Lu L, Ren N, Zhao X et al (2011) Hydrogen production, methanogen inhibition and microbial community structures in psychrophilic single-chamber microbial electrolysis cells. Energy Environ Sci 4:1329. https://doi.org/10.1039/c0ee00588f
56. Mathis BJ, Marshall CW, Milliken CE et al (2008) Electricity generation by thermophilic microorganisms from marine sediment. Appl Microbiol Biotechnol 78:147–155. https://doi.org/10.1007/s00253-007-1266-4
57. Fu Q, Kobayashi H, Kawaguchi H et al (2013) Electrochemical and phylogenetic analyses of current-generating microorganisms in a thermophilic microbial fuel cell. J Biosci Bioeng 115:268–271. https://doi.org/10.1016/j.jbiosc.2012.10.007
58. Zhang G, Feng S, Jiao Y et al (2017) Cathodic reducing bacteria of dual-chambered microbial fuel cell. Int J Hydrog Energy 42:27607–27617. https://doi.org/10.1016/j.ijhydene.2017.06.095

59. Huang L, Regan JM, Quan X (2011) Electron transfer mechanisms, new applications, and performance of biocathode microbial fuel cells. Bioresour Technol 102:316–323. https://doi.org/10.1016/j.biortech.2010.06.096
60. Virdis B, Rabaey K, Yuan Z, Keller J (2008) Microbial fuel cells for simultaneous carbon and nitrogen removal. Water Res 42:3013–3024. https://doi.org/10.1016/j.watres.2008.03.017
61. Puig S, Coma M, Desloover J et al (2012) Autotrophic denitrification in microbial fuel cells treating low ionic strength waters. Environ Sci Technol 46:2309–2315. https://doi.org/10.1021/es2030609
62. Johnson DB, Schideman LC, Canam T, Hudson RJM (2018) Pilot-scale demonstration of efficient ammonia removal from a high-strength municipal wastewater treatment sidestream by algal-bacterial biofilms affixed to rotating contactors. Algal Res 34:143–153. https://doi.org/10.1016/j.algal.2018.07.009
63. Cano V, Vich DV, Rousseau DPL et al (2019) Influence of recirculation over COD and N–NH_4 removals from landfill leachate by horizontal flow constructed treatment wetland. Int J Phytoremediation 21:998–1004. https://doi.org/10.1080/15226514.2019.1594681
64. Xu S, Zhu J, Meng Z et al (2019) Hydrogen and methane production by co-digesting liquid swine manure and brewery wastewater in a two-phase system. Bioresour Technol. https://doi.org/10.1016/j.biortech.2019.122041
65. España-Gamboa E, Mijangos-Cortes J, Barahona-Perez L et al (2011) Vinasses: characterization and treatments. Waste Manag Res 29:1235–1250. https://doi.org/10.1177/0734242X10387313
66. Daims H, Lücker S, Wagner M (2016) A new perspective on microbes formerly known as nitrite-oxidizing bacteria. Trends Microbiol 24:699–712. https://doi.org/10.1016/j.tim.2016.05.004
67. Drewnowski J, Remiszewska-Skwarek A, Duda S, Łagód G (2019) Aeration process in bioreactors as the main energy consumer in a wastewater treatment plant. Review of solutions and methods of process optimization. Processes. https://doi.org/10.3390/pr7050311
68. Gu Y, Li Y, Li X et al (2017) The feasibility and challenges of energy self-sufficient wastewater treatment plants. Appl Energy 204:1463–1475. https://doi.org/10.1016/j.apenergy.2017.02.069
69. Zhu T, Zhang Y, Quan X, Li H (2015) Effects of an electric field and iron electrode on anaerobic denitrification at low C/N ratios. Chem Eng J 266:241–248. https://doi.org/10.1016/j.cej.2014.12.082
70. Nolasco MA, Cano V, Cano J (2022) Sistema bioeletroquímico sequencial empilhável para geração de eletricidade. BR Patent BR102021005323A2 Mar 2021
71. He Z, Kan J, Wang Y et al (2009) Electricity production coupled to ammonium in a microbial fuel cell. Environ Sci Technol 43:3391–3397. https://doi.org/10.1021/es803492c
72. Qu B, Fan B, Zhu S, Zheng Y (2014) Anaerobic ammonium oxidation with an anode as the electron acceptor: anaerobic ammonium oxidation with electrode reduction. Environ Microbio Rep 6:100–105. https://doi.org/10.1111/1758-2229.12113
73. Shaw DR, Ali M, Katuri KP et al (2020) Extracellular electron transfer-dependent anaerobic oxidation of ammonium by anammox bacteria. Nat Commun. https://doi.org/10.1038/s41467-020-16016-y
74. Ruiz-Urigüen M, Steingart D, Jaffé PR (2019) Oxidation of ammonium by Feammox *Acidimicrobiaceae* sp. A6 in anaerobic microbial electrolysis cells. Environ Sci Water Res Technol 5:1582–1592. https://doi.org/10.1039/C9EW00366E
75. Tang J, Chen S, Huang L et al (2017) Acceleration of electroactive anammox (electroammox) start-up by switching acetate pre-acclimated biofilms to electroammox biofilms. Bioresour Technol 243:1257–1261. https://doi.org/10.1016/j.biortech.2017.08.033
76. Gregory KB, Bond DR, Lovley DR (2004) Graphite electrodes as electron donors for anaerobic respiration. Environ Microbiol 6:596–604. https://doi.org/10.1111/j.1462-2920.2004.00593.x
77. Virdis B, Rabaey K, Rozendal RA et al (2010) Simultaneous nitrification, denitrification and carbon removal in microbial fuel cells. Water Res 44:2970–2980. https://doi.org/10.1016/j.watres.2010.02.022

78. Al-Mamun A, Lefebvre O, Baawain MS, Ng HY (2016) A sandwiched denitrifying biocathode in a microbial fuel cell for electricity generation and waste minimization. Int J Environ Sci Technol 13:1055–1064. https://doi.org/10.1007/s13762-016-0943-1
79. Moqsud MA, Omine K, Yasufuku N et al (2013) Microbial fuel cell (MFC) for bioelectricity generation from organic wastes. Waste Manag 33:2465–2469. https://doi.org/10.1016/j.wasman.2013.07.026
80. AlSayed A, Soliman M, Eldyasti A (2020) Microbial fuel cells for municipal wastewater treatment: from technology fundamentals to full-scale development. Renew Sust Energy Rev. https://doi.org/10.1016/j.rser.2020.110367
81. Shizas I, Bagley DM (2004) Experimental determination of energy content of unknown organics in municipal wastewater streams. J Energy Eng 130:45–53. https://doi.org/10.1061/(ASCE)0733-9402(2004)130:2(45)
82. Liu H, Ramnarayanan R, Logan BE (2004) Production of electricity during wastewater treatment using a single chamber microbial fuel cell. Environ Sci Technol 38:2281–2285. https://doi.org/10.1021/es034923g
83. Feng Y, He W, Liu J et al (2014) A horizontal plug flow and stackable pilot microbial fuel cell for municipal wastewater treatment. Bioresour Technol 156:132–138. https://doi.org/10.1016/j.biortech.2013.12.104
84. Sierra-Alvarez R, Lettinga G (2007) The methanogenic toxicity of wastewater lignins and lignin related compounds. J Chem Technol Biotechnol 50:443–455. https://doi.org/10.1002/jctb.280500403
85. Pandey P, Shinde VN, Deopurkar RL et al (2016) Recent advances in the use of different substrates in microbial fuel cells toward wastewater treatment and simultaneous energy recovery. Appl Energy 168:706–723. https://doi.org/10.1016/j.apenergy.2016.01.056
86. Penteado ED, Fernandez-Marchante CM, Zaiat M et al (2017) Influence of carbon electrode material on energy recovery from winery wastewater using a dual-chamber microbial fuel cell. Environ Technol 38:1333–1341. https://doi.org/10.1080/09593330.2016.1226961
87. Cano V (2020) Energy generation in a novel microbial fuel cell: characterization and dynamics of microbial communities using organic matter and ammonia as electron donors. Dissertation, University of Sao Paulo
88. Rossi R, Logan BE (2022) Impact of reactor configuration on pilot-scale microbial fuel cell performance. Water Res. https://doi.org/10.1016/j.watres.2022.119179
89. Greenman J, Ieropoulos IA (2017) Allometric scaling of microbial fuel cells and stacks: the lifeform case for scale-up. J Power Sources 356:365–370. https://doi.org/10.1016/j.jpowsour.2017.04.033
90. Jadhav DA, Mungray AK, Arkatkar A, Kumar SS (2021) Recent advancement in scaling-up applications of microbial fuel cells: From reality to practicability. Sust Energy Technol Assess. https://doi.org/10.1016/j.seta.2021.101226
91. Wu S, Li H, Zhou X et al (2016) A novel pilot-scale stacked microbial fuel cell for efficient electricity generation and wastewater treatment. Water Res 98:396–403. https://doi.org/10.1016/j.watres.2016.04.043
92. Rossi R, Hur AY, Page MA et al (2022) Pilot scale microbial fuel cells using air cathodes for producing electricity while treating wastewater. Water Res. https://doi.org/10.1016/j.watres.2022.118208
93. Dumitru A, Scott K (2016) 4—Anode materials for microbial fuel cells. In: Scott K, Yu EH (eds) Microbial electrochemical and fuel cells, 1st edn. Woodhead Publishing, Boston, pp 117–152
94. Penteado ED (2017) Tratamento de águas residuárias em células a combustível microbianas e geração de energia elétrica direta: fundamentos e aplicação. Dissertation, University of Sao Paulo
95. Cheng S, Xing D, Logan BE (2011) Electricity generation of single-chamber microbial fuel cells at low temperatures. Biosens Bioelectron 26:1913–1917. https://doi.org/10.1016/j.bios.2010.05.016

96. Heidrich ES, Dolfing J, Wade MJ et al (2018) Temperature, inocula and substrate: Contrasting electroactive consortia, diversity and performance in microbial fuel cells. Bioelectrochemistry 119:43–50. https://doi.org/10.1016/j.bioelechem.2017.07.006
97. Jong BC, Kim BH, Chang IS et al (2006) Enrichment, performance, and microbial diversity of a thermophilic mediatorless microbial fuel cell. Environ Sci Technol 40:6449–6454. https://doi.org/10.1021/es0613512
98. Wrighton KC, Agbo P, Warnecke F et al (2008) A novel ecological role of the Firmicutes identified in thermophilic microbial fuel cells. Mult J Microb Ecol 2:1146–1156. https://doi.org/10.1038/ismej.2008.48
99. Carballa M, Smits M, Etchebehere C et al (2011) Correlations between molecular and operational parameters in continuous lab-scale anaerobic reactors. Appl Microbiol and Biotechnol 89:303–314. https://doi.org/10.1007/s00253-010-2858-y
100. Aelterman P, Versichele M, Marzorati M et al (2008) Loading rate and external resistance control the electricity generation of microbial fuel cells with different three-dimensional anodes. Bioresour Technol 99:8895–8902. https://doi.org/10.1016/j.biortech.2008.04.061
101. Pinto RP, Srinivasan B, Guiot SR, Tartakovsky B (2011) The effect of real-time external resistance optimization on microbial fuel cell performance. Water Res 45:1571–1578. https://doi.org/10.1016/j.watres.2010.11.033
102. Song T-S, Yan Z-S, Zhao Z-W, Jiang H-L (2010) Removal of organic matter in freshwater sediment by microbial fuel cells at various external resistances. J Chem Technol Biotechnol 85:1489–1493. https://doi.org/10.1002/jctb.2454
103. Chen S, Patil SA, Brown RK, Schröder U (2019) Strategies for optimizing the power output of microbial fuel cells: transitioning from fundamental studies to practical implementation. Appl Energy 233:15–28. https://doi.org/10.1016/j.apenergy.2018.10.015
104. Woodward L, Perrier M, Srinivasan B et al (2010) Comparison of real-time methods for maximizing power output in microbial fuel cells. AIChE J 56:2742–2750. https://doi.org/10.1002/aic.12157
105. Grondin F, Perrier M, Tartakovsky B (2012) Microbial fuel cell operation with intermittent connection of the electrical load. J Power Sources 208:18–23. https://doi.org/10.1016/j.jpowsour.2012.02.010
106. Premier GC, Kim JR, Michie I et al (2011) Automatic control of load increases power and efficiency in a microbial fuel cell. J Power Sources 196:2013–2019. https://doi.org/10.1016/j.jpowsour.2010.09.071
107. Yi Y, Xie B, Zhao T et al (2019) Effect of external resistance on the sensitivity of microbial fuel cell biosensor for detection of different types of pollutants. Bioelectrochemistry 125:71–78. https://doi.org/10.1016/j.bioelechem.2018.09.003
108. Córdova-Bautista Y, Paraguay-Delgado F, Pérez Hernández B et al (2018) Influence of external resistance and anodic pH on power density in microbial fuel cell operated with B. Subtilis BSC-2 strain. Appl Ecol Environ Res 16:1983–1997. https://doi.org/10.15666/aeer/1602_19831997
109. Pasternak G, Greenman J, Ieropoulos I (2018) Dynamic evolution of anodic biofilm when maturing under different external resistive loads in microbial fuel cells electrochemical perspective. J Power Sources 400:392–401. https://doi.org/10.1016/j.jpowsour.2018.08.031
110. Chae K-J, Choi M-J, Kim K-Y et al (2010) Methanogenesis control by employing various environmental stress conditions in two-chambered microbial fuel cells. Bioresour Technol 101:5350–5357. https://doi.org/10.1016/j.biortech.2010.02.035
111. Gustave W, Yuan Z-F, Sekar R et al (2019) The change in biotic and abiotic soil components influenced by paddy soil microbial fuel cells loaded with various resistances. J Soils Sediments 19:106–115. https://doi.org/10.1007/s11368-018-2024-1
112. Karamanev DG, Nikolov LN (1996) Application of inverse fluidization in wastewater treatment: from laboratory to full-scale bioreactors. Environ Prog 15:194–196. https://doi.org/10.1002/ep.670150319
113. Gatti MN, Milocco RH (2017) A biofilm model of microbial fuel cells for engineering applications. Int J Energy Environ Eng 8:303–315. https://doi.org/10.1007/s40095-017-0249-1

114. Su J-C, Tang S-C, Su P-J, Su J-J (2019) Real-time monitoring of micro-electricity generation through the voltage across a storage capacitor charged by a simple microbial fuel cell reactor with fast fourier transform. Energies. https://doi.org/10.3390/en12132610
115. Pocaznoi D, Calmet A, Etcheverry L et al (2012) Stainless steel is a promising electrode material for anodes of microbial fuel cells. Energy Environ Sci 5:9645–9652. https://doi.org/10.1039/c2ee22429a
116. Yamashita T, Ishida M, Asakawa S et al (2016) Enhanced electrical power generation using flame-oxidized stainless steel anode in microbial fuel cells and the anodic community structure. Biotechnol Biofuels.https://doi.org/10.1186/s13068-016-0480-7
117. Linardi M (2010) Introdução a ciência e tecnologia de células a combustível, 1st edn. Artliber, São Paulo
118. Lu M, Li SFY (2012) Cathode reactions and applications in microbial fuel cells: a review. Crit Revi Environ Sci Technol 42:2504–2525. https://doi.org/10.1080/10643389.2011.592744
119. Peng L, Zhang X-T, Yin J et al (2016) Geobacter sulfurreducens adapts to low electrode potential for extracellular electron transfer. Electrochim Acta 191:743–749. https://doi.org/10.1016/j.electacta.2016.01.033
120. Ledezma P, Donose BC, Freguia S, Keller J (2015) Oxidised stainless steel: a very effective electrode material for microbial fuel cell bioanodes but at high risk of corrosion. Electrochim Acta 158:356–360. https://doi.org/10.1016/j.electacta.2015.01.175
121. Borsje C, Liu D, Sleutels THJA et al (2016) Performance of single carbon granules as perspective for larger scale capacitive bioanodes. J Power Sources 325:690–696. https://doi.org/10.1016/j.jpowsour.2016.06.092
122. Karra U, Muto E, Umaz R et al (2014) Performance evaluation of activated carbon-based electrodes with novel power management system for long-term benthic microbial fuel cells. Int J Hydrog Energy 39:21847–21856. https://doi.org/10.1016/j.ijhydene.2014.06.095
123. Jiang D, Li B (2009) Granular activated carbon single-chamber microbial fuel cells (GAC-SCMFCs): a design suitable for large-scale wastewater treatment processes. Biochem Eng J 47:31–37. https://doi.org/10.1016/j.bej.2009.06.013
124. Kalathil S, Lee J, Cho MH (2011) Granular activated carbon based microbial fuel cell for simultaneous decolorization of real dye wastewater and electricity generation. New Biotechnol 29:32–37. https://doi.org/10.1016/j.nbt.2011.04.014
125. Jin Y (2014) Reaction mechanism on anode filled with activated carbon in microbial fuel cell. J Chem Pharm Res 6(5):333–339
126. Nam J-Y, Kim H-W, Shin H-S (2010) Ammonia inhibition of electricity generation in single-chambered microbial fuel cells. J Power Sources 195:6428–6433. https://doi.org/10.1016/j.jpowsour.2010.03.091
127. He Z, Wagner N, Minteer SD, Angenent LT (2006) An upflow microbial fuel cell with an interior cathode: assessment of the internal resistance by impedance spectroscopy. Environ Sci Technol 40:5212–5217. https://doi.org/10.1021/es060394f
128. Caizán-Juanarena L, Sleutels T, Borsje C, ter Heijne A (2020) Considerations for application of granular activated carbon as capacitive bioanode in bioelectrochemical systems. Renew Energy 157:782–792. https://doi.org/10.1016/j.renene.2020.05.049
129. Caizán-Juanarena L, Servin-Balderas I, Chen X et al (2019) Electrochemical and microbiological characterization of single carbon granules in a multi-anode microbial fuel cell. J Power Sources. https://doi.org/10.1016/j.jpowsour.2019.04.042
130. Deeke A, Sleutels THJA, Hamelers HVM, Buisman CJN (2012) Capacitive bioanodes enable renewable energy storage in microbial fuel cells. Environ Sci Technol 46:3554–3560. https://doi.org/10.1021/es204126r
131. Deeke A, Sleutels THJA, Heijne AT et al (2013) Influence of the thickness of the capacitive layer on the performance of bioanodes in microbial fuel cells. J Power Sources 243:611–616. https://doi.org/10.1016/j.jpowsour.2013.05.195

Part V
Microbial Electrolysis Cells

Chapter 9
Online Optimization of Microbial Electrolysis Cells

Ixbalank Torres-Zúñiga, José de Jesús Colín-Robles, Glenda Cea-Barcia, and Victor Alcaraz-Gonzalez

Abstract In this chapter, two Extremum Seeking Control (ESC) strategies are presented to maximize the hydrogen production rate of a Microbial Electrolysis Cell (MEC). First, a model-free ESC strategy based on a simple sliding mode controller, with the input-output map gradient as a sliding variable, is proposed to online maximize the MEC productivity by using the dilution rate as a manipulated variable. Next, the model of the MEC is considered to calculate the input-output gradient to offline solve an optimization problem. The maximum HPR computed is then used as a reference by a Super-Twisting controller to compute the optimum dilution rate to track the MEC productivity to its maximum value. Both algorithms are proposed in discrete time to be easily implemented in digital systems such as microcontrollers, DSP, or FPGA. The results of both gradient-based optimization strategies are compared by simulations considering an uncertainty parameter framework in the MEC model.

Keywords Microbial electrolysis cell · Extremum seeking control · Gradient-based optimization · Finite differences · Sliding mode control

I. Torres-Zúñiga (✉)
C. A. Telemática, Department of Electronics Engineering, Universidad de Guanajuato, Carr. Salamanca-Valle de Santiago Km. 3.5+1.8, Comunidad de Palo Blanco, C. P. 36885 Salamanca, Mexico
e-mail: ixbalank_torres@ifac-mail.org

J. de Jesús Colín-Robles
Department of Automotive Systems, Tecnológico Nacional de México / ITS de Purísima del Rincón, Blvd. del Valle # 2301 Guardarrayas, C. P. 36425 Purísima del Rincón, Mexico
e-mail: jesus.cr@purisima.tecnm.mx

G. Cea-Barcia
Department of Environmental Engineering, Universidad de Guanajuato, Ex Hacienda El Copal Km. 9, Carr. Irapuato-Silao, C. P. 36500 Irapuato, Mexico
e-mail: glendacea@ugto.mx

V. Alcaraz-Gonzalez
Departamento de Ingeniería Química, Universidad de Guadalajara, Blvd. Marcelino García Barragan 1421, Esq. Calzada Olímpica, C.P. 44430 Guadalajara, Mexico
e-mail: victor.agonzalez@academicos.udg.mx

V. Alcaraz Gonzalez et al. (eds.), *Wastewater Exploitation*, Springer Water,
https://doi.org/10.1007/978-3-031-57735-2_9

9.1 Introduction

The total worldwide energy demand is constantly increasing, resulting in an energy crisis and environmental pollution. Based on the International Energy Agency, in 2030, the total worldwide energy demand will be doubled [1]. So far, most of this enormous energy demand is supplied by fossil fuels. The utilization of fossil fuels generates greenhouse gases and toxic and hazardous substances, occasioning global warming and human and ecosystem damage. Therefore, the need for ecofriendly and renewable alternative energy sources has risen [1].

In this context, globally public and private research institutions work on the development of new technologies that can utilize renewable sources for generating clean energy. Among the clean energies, hydrogen is one of the fuels that arouses the greatest interest because it is a renewable energy carrier and the most abundant element in the universe. The hydrogen gas has unique energy properties, such as high energy density (120 MJ/Kg) and low volumetric energy density (8 MJ/L). In addition, hydrogen is clean burning characteristics, since its primary combustion product is water. However, it is not available directly on the planet, it must be obtained from renewable and non-renewable sources such as biomass, water, seawater, wastewater, and fossil fuels [1–3].

According to an international agreement, hydrogen production is classified into different colors. The blue hydrogen is produced from the steam reforming of natural gas and the CO_2 produced is captured and stored underground using industrial techniques. The gray hydrogen is produced from non-renewable fossil fuels such as natural gas or coal by steam reforming/auto-thermal reforming processes, but, in contrast to blue hydrogen, the produced CO_2 is not captured. Brown hydrogen is produced from hydrocarbon feedstock such as coal or methane via the gasification process, although with very high CO_2 emissions. The black hydrogen is produced from coal gasification, producing syngas and then, hydrogen is separated from the other gases using absorbers or special membranes, while the remaining gases are released into the atmosphere. Green hydrogen is produced from either fresh water, seawater, or wastewater, and electricity by an electrolysis process. In this process, water is transformed into hydrogen (H_2) and oxygen (O_2) with zero carbon emissions. In green hydrogen production, the electricity used for its production comes from renewable energy or low-carbon energy. However, for a practical application of this technology, the hydrogen produced must be stored or used in a hydrogen fuel cell to produce electricity, water, and heat. Nevertheless, the main challenges of hydrogen production are the development of integrated, efficient, and low-cost processes. [1, 2, 4].

The water electrolysis process was discovered by the English scientists William Nicholson and Anthony Carlisle in 1800, which discovered that applying electric current to water produced hydrogen and oxygen gas. Then, in 1960, General Electric developed hydrogen fuel cells for generating electricity in the Apollo and Gemini space missions. Nowadays, water electrolysis and fuel cell technology are used for electric vehicles, for energy production by burn, for the production of specialty chemicals, or various other small-scale applications [1, 4]. Regarding hydrogen sources, in 2020, most of the hydrogen production was obtained from fossil fuels, emitting 830 million tons/year of CO_2, instead, less than 5% of total hydrogen production was obtained from renewable resources. Hence, under this scenario, the main challenge is to obtain hydrogen from renewable sources, efficiently and economically [1].

On the other hand, by combining the principles of classical electrochemistry with bioreactor engineering, a new generation of electrochemical bio-based technologies capable of recovering energy and metabolites from waste organic matter emerged at the beginning of this century. These are the so-called bioelectrochemical systems (BES), which can be broadly classified into microbial fuel cells (MFCs) if they operate in galvanic mode, or microbial electrolysis cells (MEC) if they operate in electrolytic mode. With the participation of microorganisms growing at both the anode and cathode of an electrochemical cell, a wide variety of biochemical reactions can take place, both in the presence of oxygen or in its absence. In most BES, such biochemical reactions are carried out at the anode by microorganisms called anodophilic respiring bacteria wherein the organic matter is oxidized, and then, a release of protons into the bulk phase, as well as electrons into the anode [5] is carried out. Protons can migrate to the cathode wherein they are reduced in a variety of other chemical and biochemical reactions. At both anode and cathode, reactions can be spontaneous, in which case an electric current flow (i.e., electricity) occurs, or non-spontaneous, in which the flow of electrons must be induced by the application of a potential differential. This is the case of MEC, in which hydrogen is produced at the cathode by applying a small voltage (> 0.112 V in theory, > 0.2 V in practice) [6].

In recent years, several control strategies have been developed to improve the performance of microbial electrolysis cells. For instance, in [7] a method for controlling the applied voltage in a MEC is applied at laboratory-scale MECs fed with acetate or synthetic wastewater to maximize the rate of hydrogen production. In [8] an integrated approach, involving process modeling, optimization, and neural network-based and Adaptive-PID controllers, has been proposed to deal with the nonlinearity and complexity of the microbial interactions to produce biohydrogen under optimal conditions in a MEC. In [9] two nonlinear adaptive control laws are proposed and applied in a MEC model, the main control objective is to improve the electrical current generated and thus, the production of biohydrogen gas in the MEC, using the dilution rate and the applied potential as individual control inputs. In [10] a fuzzy logic control scheme on a MEC to track the output hydrogen flow rate to a set point value into an operating region was implemented. A very complete review on modeling, control and optimization of MEC systems can be found in [11–13].

In order to online optimize the performance of bioprocesses at steady state, different real-time optimization (RTO) strategies have been developed, see for example [14–22]. Such optimization approaches basically consist of maximizing (or minimizing) some objective function or measurable criterion, while certain economic, safety or quality constraints hold, improving in such a way the system performance. In general, the performance of RTO techniques depends heavily on the accuracy of the plant model, which has led to the development of adapted RTO methods that take into account possible mismatches. For instance, direct input adaptation approaches take advantage of the steady state plant invariance properties to transform the optimization problem into a feedback control problem. Among such methods, the Extremum Seeking Control (ESC) is RTO strategy that allows to lead the system towards the extreme of a measurable concave (convex) function corresponding to optimal operating conditions at steady state [19]. ESC is particularly useful when there exist local minima or maxima that appear commonly in nonlinear functions that may be present either in the model, in the control law, or in the cost function of an optimization problem [14].

The better the system model structure is known, the greater the controller efficiency is achieved, but even under a minimum knowledge scenario, ESC offers great robustness. In this sense, ESC can be classified into two main categories [19]:

- Model-based strategies: Whether the model is well known, its structure can be exploited directly by the controller, but it can also be used to estimate unmeasured variables as well as unknown or uncertain parameters by taking advantage of some of its robustness properties. In addition, the knowledge of the model structure allows for solving the optimization problem offline, while an output feedback controller leads the process toward the optimal region.
- Model-free strategies: In this category, the gradient of the measurable objective function is estimated directly from the online information, and the output feedback controller works only with both the available online information and the estimated gradient. Actually, prior knowledge about the model structure is not required, allowing the optimum to be achieved simply by driving the process in the direction in which the approximated objective function gradient vanishes.

On the other hand, effluent treatment systems suffer from a lack of online suitable sensors for measuring important variables (e.g., biomass concentration) necessary for use with optimization and control techniques. In addition to the highly non-linear nature of this type of system, this lack of sensors strongly increases the degree of unknowledge and uncertainty of the system. In order to deal with these uncertainties together, methods based on the theory of sliding modes have been proposed for use in conjunction with ESC. For example, in [23] this combination of ESC and sliding modes is used to maximize methane production in anaerobic digesters. In [24], the concentration of biomass in a bioreactor with uncertain kinetics is optimized even when the target function is not measured and needs to be estimated. In [25], a bank of weighted Super-Twisting observers is used to estimate a virtual output which in turn is used as input to an output-feedback ESC to control and optimize the substrate concentration while biomass production is also enhanced in a fed-batch bioreactor. In [26] the authors propose a gradient-based extremum-seeking control strategy, in which, the optimization algorithm and the gradient estimation are performed by the Super-Twisting algorithm. Such ESC strategy is applied to a MEC for maximizing the hydrogen production rate.

In this chapter, two gradient-based extremum-seeking control strategies are proposed to maximize the hydrogen production rate of a microbial electrolysis cell.

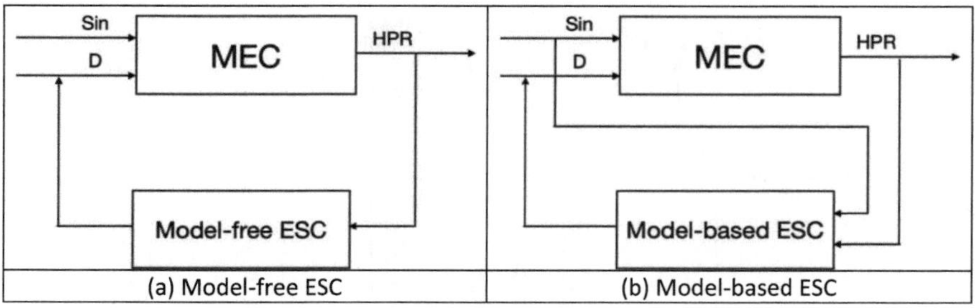

Fig. 9.1 ESC strategies

The first one is a model-free ESC strategy, in which only the measured output is used to compute the optimum dilution rate that maximizes the hydrogen production rate (HPR) of the MEC at steady state (see Fig. 9.1a). The second one is a model-based ESC strategy, in which the maximum HPR is calculated offline using the available model of the MEC, therefore the inlet substrate is used, as well, to compute the HPR at the steady state. Then, a nonlinear PI controller is used to online track such maximum value by computing the dilution rate that maximizes the productivity of the MEC at steady state (see Fig. 9.1b). This way, the chapter is organized as follows: in Sect. 9.2 the model of the MEC is introduced. In Sect. 9.3 the optimization problem to solve is stated and the optimization conditions are discussed. In Sect. 9.4 the gradient-based ESC strategy proposed is presented, first, a model-free ESC is developed, then a model-based ESC is developed. Finally, in Sect. 9.5 a comparison of the results of both ESC strategies is discussed and final remarks are stated.

9.2 Mathematical Model of the MEC

Due to the involvement of microbial ecosystems, the production of biohydrogen in a MEC is a quite complex process. In addition, models describing MEC dynamics are highly nonlinear. Both factors make these processes very difficult to operate and control. Nevertheless, these drawbacks can be palliated by applying an integrated engineering approach, which consists of developing and carrying out modeling, process control, and optimization simultaneously [8].

The MEC system considered in this chapter is a continuous isothermal bioreactor composed of an anodic chamber and a cathodic chamber communicating with each other through a cationic membrane (see Fig. 9.2). The model used to describe the dynamic behavior of this MEC system is the simplest found in the literature and consists of a set of nonlinear ordinary differential equations (ODE), which considers two microbial populations in the anodic chamber.

The dynamic mass balance equations are defined as:

$$\dot{S}(t) = D(t)(S_{in}(t) - S(t)) - k_a \mu_a X_a(t) - k_m \mu_m X_m(t) \tag{9.1a}$$

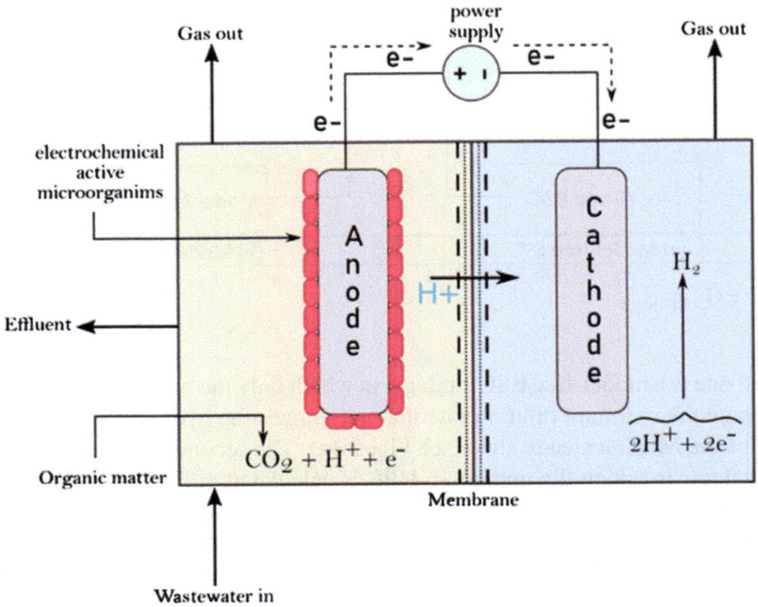

Fig. 9.2 Schematic diagram of the MEC

$$\dot{X}_a(t) = \mu_a X_a(t) - k_{da} X_a(t) - \alpha_a D(t) X_a(t) \tag{9.1b}$$

$$\dot{X}_m(t) = \mu_m X_m(t) - k_{dm} X_m(t) - \alpha_m D(t) X_m(t) \tag{9.1c}$$

where S is the substrate concentration (mg/L), X_a and X_m are the concentration of the anodophilic and acetoclastic methanogenic microorganisms, respectively (mg/L). S_{in} is the inlet substrate concentration (mg/L). D is the dilution rate: $D = Q_{in}/V_r$ $(1/d)$, where Q_{in} is the input flow rate (L/d) and V_r is the reactor volume (L). k_a and k_m are the yield coefficients for substrate consumption in the anode by the anodophilic (mg[S]/mg[X_a]) and the acetoclastic methanogenic (mg[S]/mg[X_m]) bacteria, respectively. μ_a and μ_m are the growth rates $(1/d)$ for anodophilic and acetoclastic methanogenic microorganisms, respectively. k_{da} and k_{dm} are the microbial decay rates $(1/d)$. Similarly, α_a and α_m are the dimensionless biofilm retention constants for anodophilic and acetoclastic methanogenic microorganisms, respectively.

Current intensity generated by the anodophilic microorganisms $I_{MEC}(A)$ may be calculated as:

$$I_{MEC}(t) = \left(\gamma_S k_a \mu_a L_f \left(1 - f_S^0\right) + \gamma_X b L_f\right) A_a X_a(t), \tag{9.2}$$

where γ_S and γ_S (mF/MW$_S$ and mF/MW$_X$) are the yield coefficients related to the number of coulombs that is possible to obtain from the substrate and the biomass, respectively (F is the Faraday constant, while MW_S and MW_X are the substrate and biomass molecular weight, respectively. f_S^0 is the dimensionless fraction of electrons used for the cell synthesis. b is the endogenous decay coefficient $(1/d)$, L_f is the biofilm thickness (m) and A_a is the anode area (m^2). The kinetic equations are given as

$$\mu_a = \mu_{\max a} \frac{S}{K_{Sa} + S} \frac{1}{1 + e^{\frac{-F\eta}{RT}}}, \tag{9.3a}$$

$$\mu_m = \mu_{\max m} \frac{S}{K_{Sm} + S}, \tag{9.3b}$$

where $\mu_{\max}$ is the maximum growth rate, K is the half-rate Monod constant, F is the Faraday constant $\left(\text{Ad/mol e}^-\right)$, R is the gas ideal constant ((J/mol K)), $\eta = E_{anode} - E_{Ka}$ is the local potential, with $E_{anode}(V)$ is the anode potential and $E_{Ka}(V)$ is the half-maximum-rate anodic electron acceptor potential, i.e., the potential that occurs when $S = K_{Sa}$ and the rate is 1/2 the maximum rate [28].

The hydrogen production rate HPR ($L[H_2]/Ld$) is defined as:

$$HPR(t) = \frac{1}{V_r} Q_{H_2}(t) = \frac{Y_{H_2} A_a RT}{mFPV_r} I_{MEC}, \tag{9.4}$$

where Q_{H2} is the hydrogen flow rate produced by the MEC (L/d), Y_{H_2} is the dimensionless cathode efficiency, T is the temperature (K), P is the pressure inside the cathodic chamber (Atm), F is the Faraday constant and R is the gas ideal constant (Atm L/mol K). Table 9.1 shows the parameters of Model (9.1)–(9.4) [27].

9.3 Problem Formulation

Let us consider the model (9.1)–(9.4) in a compact form as

$$\frac{dx}{dt} = f(x, u, w);\ x(0) = x_0 \tag{9.5a}$$

$$y(t) = J(x), \tag{9.5b}$$

where $x = [SX_aX_m]^T \in \mathcal{D} \subseteq \mathbb{R}^3$, is the state vector, $u = D \in \mathcal{N} \subseteq \mathbb{R}$ is the control input, $\text{w} = S_{in} \in \mathcal{M} \subseteq \mathbb{R}$ is an uncontrolled input and $y = HPR \in \mathbb{R}$ is the performance output. Besides, the functions $f : \mathcal{D} \times \mathcal{N} \times \mathcal{M} \to \mathbb{R}^3$ and $J : \mathcal{D} \to \mathbb{R}$ are smooth on $\mathcal{D}$. First, let us consider the following assumptions [29, 30]:

Table 9.1 Parameters of the MEC model

Parameter	Value	Unit
α_a	0.62	
α_b	0.5	
K_{da}	0.05 $\mu_{\max a}$	1/d
K_{dm}	0.05 $\mu_{\max m}$	1/d
k_a	0.667	mg[S]/mg[X_a]
k_m	4.7067	mg[S]/mg[X_m]
f_S^0	0.3	
γ_S	950	mF/MW$_S$
γ_X	800	mF/MW$_X$
b	0.05	1/d
L_f	25×10^{-6}	m
A_a	0.01	m^2
K_{Sa}	20	mg[S]/L
$\mu_{\max a}$	1.97	1/d
K_{Sm}	80	mg[S]/L
$\mu_{\max m}$	0.03	1/d
Y_{H_2}	0.8	
m	2	mol e$^-$/mol M
V_r	1	L
R	0.08205	Atm L/mol K
F	1.1167	Ad/mol e$^-$
P	1	Atm
T	198.15	K
η	0.3	Volts

Assumption 9.1 There exists a smooth function $l : \mathcal{N} \times \mathcal{M} \rightarrow \mathcal{D}$ such that $f(x^*, u, w) = 0$ if and only if $x^* = l(u, w)$.

Assumption 9.2 The operating point $x^* = l(u, w)$ of the system (9.5) is locally exponentially stable for each $u \in \mathcal{N}$ and for each $w \in \mathcal{M}$.

Assumption 9.3 The input–output map at steady state.

$$y = J(l(u, w)) \tag{9.6}$$

is concave and unimodal in the operating region $\mathcal{N}$.

Assumption 9.4 For each $w \in \mathcal{M}$, there exists $u^* \in \mathcal{N}$ such that:

$$\frac{\partial}{\partial u}J\big(l(u^*, w)\big) = 0,$$
$$\frac{\partial^2}{\partial u^2}J\big(l(u^*, w)\big) < 0.$$

Using these assumptions, we reduce the n-dimensional optimization problem of maximizing the performance function (9.5b) to the one-dimensional optimization problem of maximizing (9.6) at steady state. In addition, we assure that $J(l(u^*, w))$ has a unique maximizer $u^* \in \mathcal{N}$, for each $w \in \mathcal{M}$.

As mentioned before, in this chapter we are interested in maximizing the hydrogen production rate of the MEC at steady state by optimizing the dilution rate. Such an optimization problem can be stated as:

$$\begin{aligned} &\max_u J(l(u, w)) \\ &\text{such that :} \\ &\tfrac{\mathrm{d}x}{\mathrm{d}t} = f(x, u, w) \\ &y(t) = J(x). \end{aligned} \tag{9.7}$$

As stated in Assumption 9.2, the MEC must reach an exponentially stable operating point x* to achieve the optimal dilution ratio that maximizes the hydrogen production rate. This operating point is obtained by solving the system of nonlinear algebraic equations that results from equalizing the right-hand side of Eq. (9.1) to zero. Under normal operating conditions, the concentration of acetoclastic methanogenic biomass in the MEC is zero at steady state. Thus, the rate of hydrogen production as a function of the input substrate concentration and the dilution rate, both at steady state, is given by:

$$HPR(D, S_{in}) = J(l(u, w)) = \frac{Y_{H_2}RT}{mFPV_r}I^*_{MEC}(u, w), \tag{9.8}$$

where:

$$I^*_{MEC}(u, w) = \left(\frac{\gamma_S k_a \mu_{\max a} S^*(u)\big(1 - f_S^0\big)}{(\mathrm{K_{Sa}} + \mathrm{S^*(u)})\Big(1 + e^{\frac{-F\eta}{RT}}\Big)} + \gamma_X b\right) L_f A_a X_a^*(u, w),$$

$$X_a^*(u, w) = \frac{1 + e^{\frac{-F\eta}{RT}}}{k_a \mu_{\max a}}\left(\frac{K_{Sa} w}{S^*(u)}u + (w - K_{Sa})u - S^*(u)u\right),$$

$$S^*(u) = \frac{K_{Sa}K_{da} + \alpha_a K_{Sa} u}{\frac{\mu_{\max a}}{1+e^{\frac{-F\eta}{RT}}} - K_{da} - \alpha_a u}.$$

Figure 9.3 shows the function $HPR(D, S_{in})$ in Eq. (9.8). As it can be seen, the input–output map $HPR(D, S_{in})$ is concave and unimodal in the operating region $\mathcal{N} = [1, 3]$.

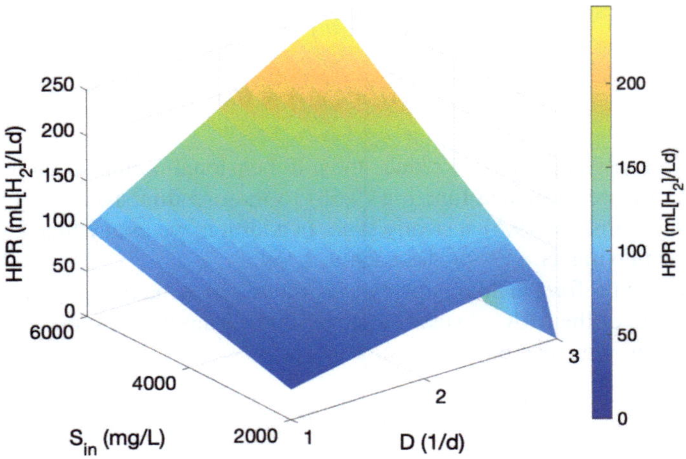

Fig. 9.3 Hydrogen production rate as function of the dilution rate and the inlet substrate

Let us now consider the linearized model of the MEC given by:

$$\dot{\overline{x}}(t) = A\overline{x}(t) + B_u\overline{u}(t) + B_w\overline{w}(t) \tag{9.9}$$

where $\overline{x} = x - x^*$, $\overline{u} = u - u^*$ and $\overline{w} = w - w^*$, A is the Jacobian matrix of $f(x, u, w)$ with respect to x, while B_u and B_w are vectors defined as the derivatives of $f(x, u, w)$ with respect to u and w, respectively. This way, matrix A is defined as

$$A = \left.\frac{\partial f(x, u, w)}{\partial x}\right|_{(x^*, u^*, w^*)}$$

Figure 9.4 shows the eigenvalues of matrix A. As can be regarded, λ_1, λ_2 and λ_3 are real and negative in the operating region defined for $\mathcal{N} = [1, 3]$ and $\mathcal{M} = [2000, 6000]$. It must be pointed out that the most the dilution rate approaches to $3(1/d)$, the eigenvalues λ_1 and λ_2 approach to zero. This way, matrix A is Hurwitz in the operating region, and therefore, the operating point x^* of the MEC modeled by the state space model of Eq. (9.5) is locally exponentially stable for each $u \in \mathcal{N}$ and for each $w \in \mathcal{M}$.

In the following section, two gradient-based ESC strategies are developed to solve the optimization problem (9.7). Both strategies use sliding mode controllers to maximize the hydrogen production rate of the MEC at steady state, and are proposed in discrete time to be easily implemented in digital systems.

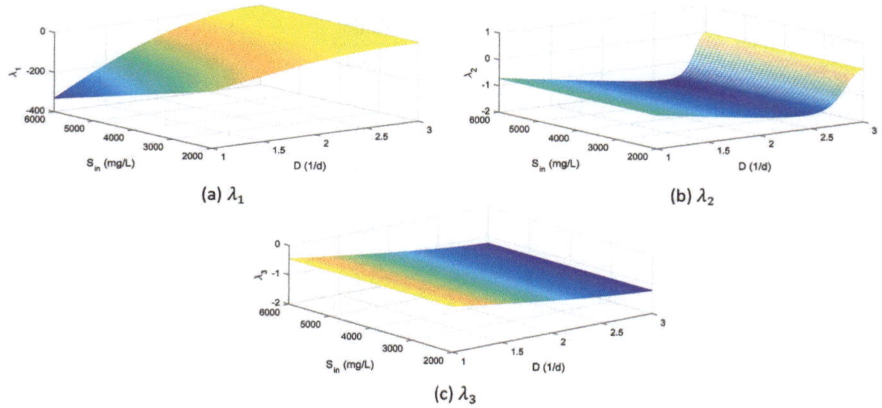

Fig. 9.4 Eigenvalues of matrix A

9.4 Gradient-Based Optimization

In the last sixty years a large variety of algorithms for unconstrained optimization of smooth functions has been developed. There exist two fundamental strategies to solve such kind of optimization problems. On the one hand, the line search algorithms choose a direction to search, along this direction, for the solution. On the other hand, in the trust region algorithms, the information gathered about the objective function is used to construct a model function whose behavior near the solution is similar to that of the actual objective function.

The steepest ascent direction $\partial J/\partial u$ is the most obvious choice for search direction for line search methods. It is intuitive, among all the directions we could move, it is the one along which J increases most rapidly. This way, gradient-based algorithms are line search methods that moves along $\partial J/\partial u$ at every step [31].

To compute the input u that maximizes the performance output $y = J(l(u, w)) = J(u, w)$ for each w, starting at $u(0)$, gradient-based optimization algorithms compute a sequence of iterates $u(k)$, for $k \in \mathbb{Z}^+$ ending when either it seems that a solution value has been approximated with sufficient accuracy or when no more progress can be achieved. For doing this, the algorithm takes advantage of the gradient $\partial J/\partial u$ for deciding how to move from one iterate $u(k)$ to the next $u(k+1)$ which yields a larger value of J than the foregoing one produced by $u(k)$.

Thus, the new iterate $u(k+1)$ can then be computed as:

$$u(k+1) = u(k) + \alpha_d \frac{\partial J}{\partial u}; \; u(0) = u_0, \alpha_d > 0, \tag{9.10}$$

where α_d is the distance to move along the gradient.

Table 9.2 Uncertain parameters of the MEC model

Parameter	Value	Unit
α_a	0.6944	
α_m	0.5250	
K_{da}	0.1034	1/d
K_{dm}	0.0061	1/d
k_a	0.7671	$mg[S]/mg[X_a]$
k_m	3.7654	$mg[S]/mg[X_m]$
K_{Sa}	23	$mg[S]/L$
$\mu_{\max a}$	2.5216	1/d
K_{Sm}	68	$mg[S]/L$
$\mu_{\max a}$	0.3420	1/d
T	238.52	K
η	0.33	V

In the following sections, the gradient-based optimization algorithm (9.10) will be used to maximize the productivity of the MEC. Two ESC strategies will be considered. The first one is a model-free ESC strategy, in which, only the information of the control input and the output performance will be used to online compute the gradient $\partial J/\partial u$ that is used in Eq. (9.10). In the second one, the model of the MEC will be used to offline compute the gradient $\partial J/\partial u$ that is used in Eq. (9.10). The optimum control input will be then used to offline compute the maximum productivity of the MEC at steady state by using Eq. (9.8). Next, such maximum productivity will be considered as reference by a Super-Twisting controller to online track the productivity of the MEC to this maximum value.

In order to simulate the MEC model, parametric uncertainties will be considered up to 30% of each parameter nominal value. Table 9.2 shows the uncertain parameters and their value used in simulations.

Simulations will be performed in MATLAB, for a simulation period of 180 days. The optimization will start at day 10 from the beginning of the MEC operation considering an optimization period of 8 h. Finally, the model of the MEC (1)–(4) will be solved using the stiff solver ode15s.

9.4.1 Model-Free Extremum Seeking Control

Let us approximate the gradient $\partial J/\partial u$ by finite differences as

$$\frac{\partial J}{\partial u} \approx \frac{\Delta J}{\Delta u}.$$

By replacing it in (9.10) it is obtained

$$u(k+1) = u(k) + \alpha_d \frac{\Delta J}{\Delta u}. \tag{9.11}$$

This approximation certainly adds uncertainty to the algorithm, which may seem like a disadvantage. However, the sign of the gradient provides valuable information that can be used to palliate this inconvenience. Indeed, the sign of the gradient in Eq. (9.11) indicates the direction in which the objective function grows faster, and then, it can be used lonely instead of the full gradient. Thus, Eq. (9.11) can be rewritten as:

$$u(k+1) = u(k) + \alpha_d \text{sign}\left(\frac{\Delta J}{\Delta u}\right); \, u(0) = u_0, \alpha_d > 0, \tag{9.12}$$

with the sign function defined as:

$$\text{sign}\left(\frac{\Delta J}{\Delta u}\right) = \begin{cases} 1; \frac{\Delta J}{\Delta u} > 0 \\ -1; \frac{\Delta J}{\Delta u} < 0 \end{cases}$$

and $sign(0) \in [-1, 1]$ [28]. Equation (9.12) induces a quasi-sliding mode, with sliding surface defined by the finite difference $\Delta J/\Delta u$ [33].

Let us consider the ESC strategy in Eq. (9.12) with $\alpha_d = 0.03$. Figure 9.5 shows the inlet substrate considered in this section.

Figure 9.6 shows the simulation results. Figure 6a shows the dilution rate computed by Eq. (9.12). Please notice that the dilution rate oscillates around the optimum value presenting a chattering behavior, this is due to the sliding variable

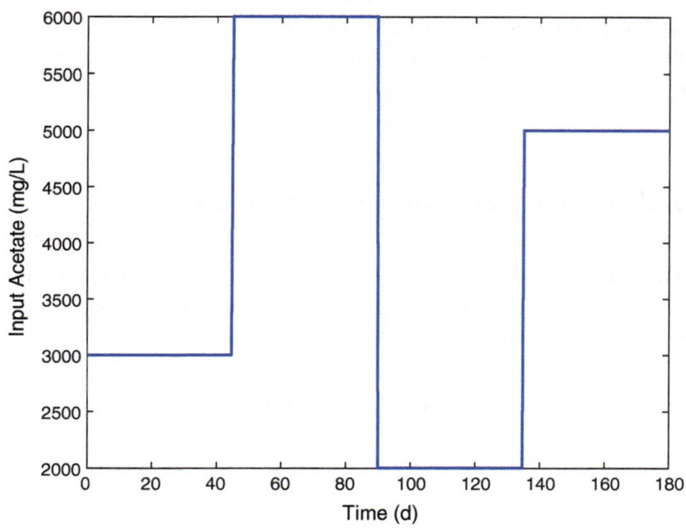

Fig. 9.5 Inlet substrate

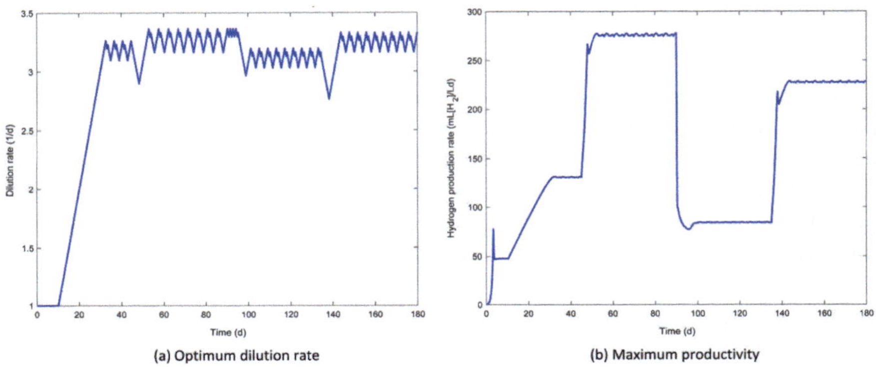

Fig. 9.6 Results of the model-free ESC strategy

dynamics once the sliding surface has been reached. Figure 6b shows the hydrogen production rate obtained by injecting the optimum dilution rate to the MEC, from day 31 to day 45 a HPR of $131mL[H_2]/Ld$ has been reached, from day 52 to day 90 the HPR reached is 277 mL$[H_2]/Ld$, from day 98 to day 134 a HPR of 84 $mL[H_2]/Ld$ is reached, and finally, from day 142 to day 180 the HPR reached is $227mL[H_2]/Ld$. Please notice that the maximum hydrogen production rate changes according to the inlet substrate fed. As it can be seen, the maximum productivity is reached after 21 days from the beginning of the optimization. Then, the maximum productivity is again reached after around 10 days once the inlet substrate has changed.

Figure 9.7 shows the state variables of the MEC. As it can be regarded, the acetogenic biomass and the current intensity follow the inlet substrate behavior. On the other hand, the methanogenic biomass rapidly decays to zero, while the substrate is around 90% consumed all along the simulation, indicating the correct performance of the MEC.

9.4.2 Model-Based Extremum Seeking Control

The optimum dilution rate can be calculated by differentiating the function $J(u, w)$ of Eq. (9.8) with respect to u and equating the result to zero (first-order optimality condition). This way, the gradient $\partial J/\partial u$ is defined as:

$$\frac{\partial J}{\partial u} = \left(\frac{Y_{H_2} A_a RT}{mFPV_r}\right)\frac{\partial I^*_{MEC}}{\partial u}, \tag{9.13}$$

where:

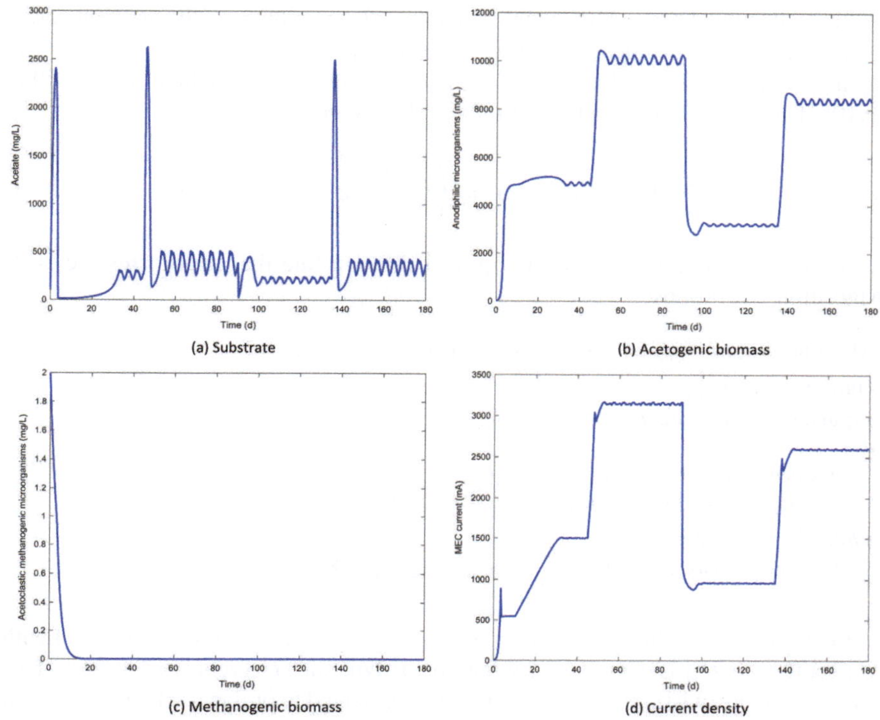

Fig. 9.7 State variables of the MEC

$$\frac{\partial I_{MEC}^*}{\partial u} = \left[\frac{(K_{Sa} + S^*(u))\left(1 + e^{\frac{-F\eta}{RT}}\right)\left(\gamma_S k_a \mu_{\max a} \frac{\partial S^*}{\partial u}(1 - f_s^0)\right)}{\left((K_{Sa} + S^*(u))\left(1 + e^{\frac{-F\eta}{RT}}\right)\right)^2} - \frac{\gamma_S k_a \mu_{\max a} S^*(u)(1 - f_s^0)\frac{\partial S^*}{\partial u}\left(1 + e^{\frac{-F\eta}{RT}}\right)}{\left((K_{Sa} + S^*(u))\left(1 + e^{\frac{-F\eta}{RT}}\right)\right)^2} \right] \left[L_f A_a X_a^*(u, w)\right] + \left[\frac{\gamma_S k_a \mu_{\max a} S^*(u)(1 - f_s^0)}{(K_{Sa} + S^*(u))\left(1 + e^{\frac{-F\eta}{RT}}\right)} + \gamma_X b \right] \left[L_f A_a \frac{\partial X_a^*}{\partial u} \right],$$

$$\frac{\partial S^*}{\partial u} = \frac{\left(\frac{\mu_{\max a}}{1 + e^{\frac{-F\eta}{RT}}} - K_{da} - \alpha_a u\right)\alpha_a K_{Sa} + (K_{Sa} K_{da} + \alpha_a K_{Sa} u)\alpha_a}{\left(\frac{\mu_{\max a}}{1 + e^{\frac{-F\eta}{RT}}} - K_{da} - \alpha_a u\right)^2},$$

$$\frac{\partial X_a^*}{\partial u} = \frac{1 + e^{\frac{-F\eta}{RT}}}{k_a \mu_{\max a}} \left[\frac{S^*(u) K_{Sa} w - K_{Sa} w u \frac{\partial S^*}{\partial u}}{(S^*(u))^2} + w - K_{Sa} - S^*(u) - u \frac{\partial S^*}{\partial u} \right]$$

Since the model of the MEC includes parametric uncertainties, the gradient calculated in Eq. (9.13) includes parametric uncertainties as well. In order to bypass such parametric uncertainties, let us only consider the sign of the gradient in Algorithm (9.10). This way, a robust optimization algorithm can be proposed as

$$u(k+1) = u(k) + \alpha_d \text{sign}\left(\frac{\partial J}{\partial u}\right); \alpha_d > 0, u(0) = u_0. \tag{9.14}$$

In order to offline compute the optimal dilution rate that maximizes the HPR Algorithm 1 is proposed.

Algorithm 1: Model-based optimization algorithm

Outputs: u_{opt}, y_{max}
Inputs: $u_0, w, \alpha_d, tolerance$
do
  To calculate the gradient $\frac{\partial J}{\partial u}$ using Eq. (9.13)
  To compute the optimal dilution rate u_{opt} using Eq. (9.14)
while $\left|\frac{\partial J}{\partial u}\right| \geq tolerance$
To calculate the maximum HPR $y_{max} = u_{opt}$ using Eq. (9.8)

Let us define the sliding variable $\sigma(t) = y(t) - y_{\max}$. By differentiating $\sigma(t)$ with respect to time t, it is easy to verify that the σ dynamics has relative degree 1. This way, the Super-Twisting algorithm can be implemented to online track the maximum hydrogen production rate $HPR_{\max}$ offline computed by Algorithm 1 [32].

Let us consider the discrete-time anti-windup Super-Twisting controller [33]:

$$\begin{aligned} u(k+1) &= -\lambda\sqrt[2]{|\sigma(k)|}\,\text{sign}(\sigma(k)) + u_1(k+1), \\ u_1(k+1) &= \begin{cases} u_1(k) - \alpha\,\text{sign}(\sigma(k))\Delta t;\ u_1(0) = u_{opt}, u_{\min} \leq u \leq u_{\max}, \\ u_1(k) - G_{aw}(u(k) - u_{\min})\Delta t;\ u < u_{\min}, \\ u_1(k) - G_{aw}(u(k) - u_{\max})\Delta t;\ u > u_{\max}, \end{cases} \end{aligned} \tag{9.15}$$

where u_1 is a nominal input, λ and α are constant gains, G_{aw} is the anti-windup scheme gain, $u_{\min}$ is the minimum allowed control input, $u_{\max}$ is the maximum allowed control input, and Δt is the sample period. The correct selection of gains λ and α makes possible to render the sliding surface $\sigma(k)$ into a quasi-sliding regime. Besides, the control input enters the segment $(u_{\min}, u_{\max})$ in finite time and remains there provided that for $k = 0$ the initial value u_{opt} is inside the segment as well [32, 33].

Let us consider the ESC strategy of Algorithm 1 and the Super-Twisting controller in Eq. (9.15), with parameters of Table 9.3, applied considering the inlet substrate of Fig. 9.5.

Figure 9.8 shows the simulation results, in red dashed lines the results offline computed by the gradient-based optimization Algorithm 1, while in blue lines the results generated by the Super-Twisting controller (9.15). Figure 8a shows the

Table 9.3 Parameters of the model-based ESC strategy

Parameter	Value	Units
α_d	0.001	
λ	0.1	
α	0.01	
G_{aw}	30.0	
$u_{\min}$	1.0	1/d
$u_{\max}$	3.0	1/d
Δt	0.5	h

optimum dilution rate. As it can be seen, the dilution rate computed by the Super-Twisting controller is different to the optimum dilution rate offline computed by the gradient-based optimization algorithm. This is due to the parametric uncertainties used to simulate the model of the MEC. Figure 8b shows the maximum productivity, from day 16 to day 45 a HPR of 117 mL[H_2]/Ld has been reached, from day 47 to day 90 the HPR reached is 245 mL[H_2]/Ld, from day 92 to day 134 a HPR of 75 mL[H_2]/Ld is reached, and finally, from day 139 to day 180 the HPR reached is 202 mL[H_2]/Ld. Please notice that the maximum hydrogen production rate changes according to the inlet substrate fed. Nevertheless, in this case, the chattering is much less than the chattering present in the model-free ESC results. This is due to the Super-Twisting controller. As it can be seen, the maximum productivity is reached after 7 days from the beginning of the optimization. Then, the maximum productivity is again reached in less than 3 days once the inlet substrate has changed.

Figure 9.9 shows the state variables of the MEC. As it can be seen, the acetogenic biomass and the current intensity follow the inlet substrate behavior. On the other hand, the methanogenic biomass rapidly decays to zero, while the substrate is almost completely consumed all along the simulation, indicating the correct performance of the MEC.

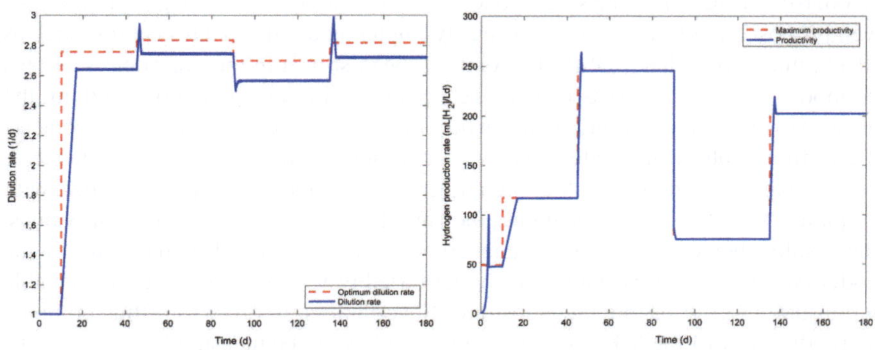

(a) Optimum dilution rate (b) Maximum productivity

Fig. 9.8 Results of Model-based ESC strategy

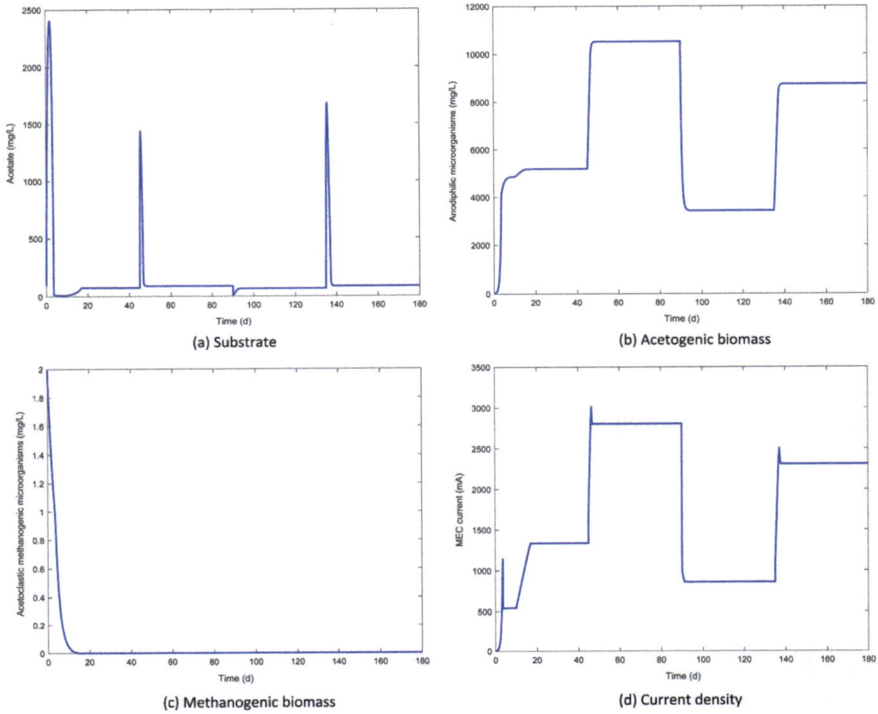

(a) Substrate (b) Acetogenic biomass (c) Methanogenic biomass (d) Current density

Fig. 9.9 State variables of the MEC

9.5 Conclusion

In this chapter, two extremum seeking control strategies have been proposed to online maximize the hydrogen production rate of a microbial electrolysis cell. The first one is a model-free ESC, while the second one is a model-based ESC. Both algorithms have been proposed in discrete time to be easily implemented in digital systems, such as FPGA, microcontrollers, DSP, etc. As it has been shown in the precedent sections, the model-free ESC consists of just one equation with one degree of freedom, the distance to move along the gradient. On the other hand, the model-based ESC consists of an offline optimization algorithm with two degrees of freedom, the distance to move along the gradient and the tolerance, and the Super-Twisting controller, with six parameters, the controller gains, the sample period and the saturation values. The results showed both, shorter convergence time and smaller chattering of the model-based ESC. Furthermore, the model-based ESC generates a dilution rate able of consuming most of the inlet substrate. Nevertheless, in order to offline compute the productivity of the MEC, the inlet substrate have to be measured or at least estimated. Therefore, from a computational point of view, the model-free ESC is cheaper because it computes the gradient by finite differences from both, the performance output measured and the control input (the decision variable), and the optimization algorithm consists of just one difference equation with one parameter.

As it can be regarded in the result figures, the maximum productivity obtained by the model-free ESC is 277 mL[H_2]/Ld against 245 mL[H_2]/Ld obtained by the model-based ESC, for an inlet substrate concentration of 6 g/L. Such difference is due to the fact that, while the model-free ESC optimizes the process directly, the model-based ESC optimizes an uncertain model of the process. It must be pointed out that the maximum productivities obtained in this work are similar to those reported in literature [21, 26].

Acknowledgements This chapter has been written with the support of the National System of Researchers (SNII) program, CONAHCyT Mexico.

References

1. Shiva KS, Lim H (2022) An overview of water electrolysis technologies for green hydrogen production. Energy Rep 8:13793–13813
2. Ciniviz M, Köse H (2012) Hydrogen use in internal combustion engine: A review. Int J Automot Eng Technol 1:1–15
3. He H, Huang Y, Nakadomari A, Masrur H, Krishnan N, Mikhaylov A, Hemeida AM, Senjyu T (2023) Potential and economic viability of green hydrogen production from seawater electrolysis using renewable energy in remote Japanese islands. Renew Energy 202:1436–1447
4. Peng L, Wei Z (2020) Catalyst engineering for electrochemical energy conversion from water to water: water electrolysis and the hydrogen fuel cell. Engineering 6(6):653–679
5. Escapa A, Mateos R, Martínez EJ, Blanes J (2016) Microbial electrolysis cells: an emerging technology for wastewater treatment and energy recovery: from laboratory to pilot plant and beyond. Renew Sustain Energy Rev 55:942–956
6. Logan BE, Call D, Cheng S, Hamelers HVM, Sleutels THJA, Jeremiasse AW, Rozendal RA (2008) Microbial electrolysis cells for high yield hydrogen gas production from organic matter. Environ Sci Tech 42(23):8630–8640
7. Tartakovsky B, Mehta P, Santoyo G, Guiot SR (2011) Maximizing hydrogen production in a microbial electrolysis cell by real-time optimization of applied voltage. Int J Hydrogen Energy 36(17):10557–10564
8. Azwar MY, Hussain MA, Abdul Wahab AK, Zanil MF (2015) A comparative study between neural networks NN-based and adaptive-PID controllers for the optimal bio-hydrogen gas production in microbial electrolysis cell reactor. In: Proceedings of the 12th international symposium on process systems engineering and 25th European symposium on computer aided process engineering, vol 37 of Computer Aided Chemical Engineering, Elsevier, pp 1529–1534
9. Alcaraz-González V, Rodriguez-Valenzuela G, Gómez-Martínez JJ, Luiz DG, Flores-Estrella RA (2021) Hydrogen production automatic control in continuous microbial electrolysis cells reactors used in wastewater treatment. J Environ Manage 281:111869
10. Mun Hong GK, Azlan Hussain M, Abdul Wahab AK (2021) Fuzzy logic controller implementation on a microbial electrolysis cell for biohydrogen production and storage. Chinese J Chem Eng 40:149–159
11. Recio-Garrido D, Perrier M, Tartakovsky B (2016) Modeling, optimization and control of bioelectrochemical systems. Chem Eng J 289:180–190
12. Mohd Asrul MA, Farid Atan M, Halim Yun HA, Hui Lai JC (2021) Mathematical model of biohydrogen production in microbial electrolysis cell: a review. Int J Hydrogen Energy 46(75):37174–37191

13. Mohd Asrul MA, Atan MF, Halim Yun HA, Hui Lai JC (2022) A review of advanced optimization strategies for fermentative biohydrogen production processes. Int J Hydrogen Energy 47(38):16785–16804
14. Ariyur KB, Krstic M (2003) Real-time optimization by extremum-seeking control. Wiley-Interscience, New Jersey
15. Dewasme L, Feudjio Letchindjio CG, Torres I, Vande Wouwer A (2017) Micro-algae productivity optimization using extremum-seeking control. In: Proceedings of the 25th mediterranean conference on control and automation, pp 672–677
16. Torres Zúñiga I, Villa-Leyva A, Vargas A, Buitrón G (2018) Experimental validation of online monitoring and optimization strategies applied to a biohydrogen production dark fermenter. Chem Eng Sci 190:48–59
17. Feudjio Letchindjio CG, Dewasme L, Deschenes JS, Vande Wouwer A (2019) An extremum seeking strategy based on block-oriented models: application to biomass productivity maximization in microalgae cultures. Ind Eng Chem Res 58:13481–13494
18. Lara-Cisneros G, Dochain D, Álvarez-Ramírez J (2019) Model based extremum seeking controller via modelling-error compensation approach. J Process Control 80:193–201
19. Dewasme L, Vande Wouwer A (2020) Model-free extremum seeking control of bioprocesses: a review with a worked example. Processes 8(10):1209–1239
20. Dewasme L, Vande Wouwer A, Feudjio Letchindjio CG, Ahmad A, Engell S (2021) Maximum-likelihood extremum seeking control of microalgae cultures. IFAC-PapersOnLine 54(3):336–341
21. Colín-Robles de J, Torres-Zúñiga I, Ibarra-Manzano MA, Alcaraz-González V (2021) FPGA-based implementation of an optimization algorithm to maximize the productivity of a microbial electrolysis cell. Processes 9(7):1111
22. López-Caamal F, Hernández-Escoto H, Torres I (2022) A gradient-based extremum seeking control of a substrate and product inhibited alcoholic fermentation process. IFAC-PapersOnLine 55(7):638–642
23. Lara-Cisneros G, Aguilar-Lopez R, Femat R (2015) On the dynamic optimization of methane production in anaerobic digestion via extremum-seeking control approach. Comput Chem Eng 75:49–59
24. Angulo MT (2015) Nonlinear extremum seeking inspired on second order sliding modes. Automatica 57:51–55
25. Vargas A, Moreno JA, Vande Wouwer A (2015) Super-twisting estimation of a virtual output for extremum-seeking output feedback control of bioreactors. J Process Control 35:41–49
26. Torres-Zúñiga I, López-Caamal F, Hernández-Escoto H, Alcaraz-González V (2021) Extremum seeking control and gradient estimation based on the Super-Twisting algorithm. J Process Control 105:223–235
27. Flores-Estrella RA, Rodríguez-Valenzuela G, Ramírez-Landeros JR, Alcaraz-González V, González-Álvarez V (2020) A simple microbial electrochemical cell model and dynamic analysis towards control design. Chem Eng Commun 207(4):493–505
28. Marcus AK, Torres CI, Rittmann BI (2007) Conduction-based modeling of the biofilm anode of a microbial fuel cell. Biotechnol Bioeng 98(6):1171–1182
29. Krstic M, Wang H-H (2000) Stability of extremum seeking feedback for general nonlinear dynamic systems. Automatica 36:595–601
30. Zhang C, Ordonez R (2012) Extremum-seeking control and applications: a numerical optimization-based approach. Springer, London
31. Nocedal J, Wright SJ (2000) Numerical optimization, 2nd edn. Springer, New York
32. Shtessel Y, Edwards C, Fridman L, Levant A (2014) Sliding mode control and observation. Springer, New York
33. Salgado I, Kamal S, Bandyopadhyay B, Chairez I, Fridman L (2016) Control of discrete time systems based on recurrent super-twisting-like algorithm. ISA Trans 64:47–55

Part VI
Bioethanol and Butanol Systems

Chapter 10
Optimizing Bioethanol Production via Extremum Seeking Control in a Continuous Stirred Tank Bioreactor

Fernando López-Caamal, Glenda Cea-Barcia, Héctor Hernández-Escoto, and Ixbalank Torres-Zúñiga

Abstract This chapter addresses bioethanol production in a continuous stirred tank bioreactor including inhibitory processes that might be present due to large substrate, biomass, and bioethanol concentrations. We aim to maximize the online productivity of the reactor despite the lack of knowledge of the mathematical model and external perturbations. To this end, as a reference, we determine the capability of the process to be optimized; thus, we use a mathematical model of the bioethanol production process. This mathematical model accounts for inhibition by modifying the reaction rates that are commonly considered. We fit the parameters of such a model with data from a typical bioethanol production scenario. In addition, we perform a sensitivity analysis to determine the key parameters regarding variations of the steady-state. By considering the dilution rate as the CSTR's input and the productivity as output, we show the convexity of the input–output map. Such convexity shows that the bioethanol production process is suitable for optimizing its productivity using the dilution rate. A suitable control strategy to steer the CSTR to an optimal state is Extremum Seeking Control since it does not require the knowledge of the location of the optimal productivity nor the dynamical model. In order to show the applicability of the methodology, we use such an Extremum Seeking Control technique in two different ranges of species concentrations. The first one considers

F. López-Caamal · H. Hernández-Escoto
División de Ciencias Naturales y Exactas del Campus Guanajuato, Departamento de Ingeniería Química, Universidad de Guanajuato, Guanajuato, México
e-mail: fernando.lopez@ugto.mx

H. Hernández-Escoto
e-mail: hhee@ugto.mx

G. Cea-Barcia
División de Ciencias de La Vida del Campus Irapuato-Salamanca, Departamento de Ciencias Ambientales, Universidad de Guanajuato, Irapuato, México
e-mail: glendacea@ugto.mx

I. Torres-Zúñiga (✉)
C.A. Telemática, División de Ingenierías del Campus Irapuato-Salamanca. Departamento de Ingeniería Electrónica, Universidad de Guanajuato, Salamanca, México
e-mail: ixbalank_torres@ifac-mail.org

V. Alcaraz Gonzalez et al. (eds.), *Wastewater Exploitation*, Springer Water,
https://doi.org/10.1007/978-3-031-57735-2_10

a large substrate concentration, typical in industrial applications, and the second one considers wastewater from tequila production as raw material for bioethanol production. Via numerical simulations, we show the performance of the extremum seeking control for both scenarios.

Keywords Bioethanol · Continuous bioreactor · Extremum seeking control · Optimization

10.1 Introduction

Since several decades ago, R&D about biofuels has been increasing because they represent renewable and environmentally friendly energy sources; and particularly speaking, bioethanol is at the door of large-scale production and consumption: worldwide production may exceed 100 billion liters by 2022 [1]. Some reasons are that its raw material is abundant worldwide, the process steps in its production have mature technical feasibility, and it can be used as a fuel, solvent, and several industrial applications [2]. The use of bioethanol as a fuel dates back to 1908 when Henry Ford designed his car that run just as well on alcohol fuels as on traditional gasoline and set up an ethanol facility. However, the use of ethanol continued till 1940, when fossil fuels were available at lower prices [3]. Bioethanol is a high-octane number biofuel that is produced from the fermentation of reducing sugars (sugar beet, sugar cane, and sweet sorghum), grain (corn, wheat, barley, and rye), potatoes, fruits, and vegetable waste and lignocellulosic residues. In purest form, bioethanol is a colorless clear liquid that boils at 78 °C and freezes at − 112 °C. Greenhouse gas emissions from bioethanol fuel are very low, which makes it more attractive as a renewable energy source [2, 3]. Bioethanol is typically used in transport as ethanol fuel mixtures with gasoline (Ethanol Blend: E5 to E100) or octane increaser as Ethyl Tertiary Butyl Ether (ETBE). ETBE (from 45% bioethanol and 55% isobutylene) is starting to be used in many countries instead of Methyl Tertiary Butyl Ether (MTBE) due to its human toxicity effect, increasing the demand for bioethanol [3]. The bioethanol fuel production chain comprises several important steps: biomass pretreatment, saccharification, fermentation, recovery and concentration of ethanol, bioethanol purification, and wastewater and waste handling [4]. However, this production chain has taken diverse process configurations, depending on, at first instance, the type of the raw material [5]. In early research, first-generation feedstocks such as reducing sugars from sugar cane or corn were considered, and second-generation feedstocks such as lignocellulosic material because they are agro-industrial residues. Lately, diverse sources that have not hit human food consumption have been considered [6].

In this light, wastewater from the spirits industry may be deemed as raw material for bioethanol production. Tequila production in Mexico was 374 million liters in 2020, as a result, between 3740 and 4488 million liters of tequila vinasses were generated [7]. Tequila production generates industrial wastewater such as tequila vinasses and process waters. Tequila vinasses remain in the bottom of the distiller after the distillation of the must of fermented agave (*Agave tequilana Weber* var. azul). Often these vinasses are discharged into water bodies, causing damage to the ecosystem due to their low pH (3.4–4.5) and high COD content (60,000–100,000 mg/L). Vinasses are dark red-brown since they contain melanoidins and phenolic compounds that are some of the products formed in the Maillard reaction [8]. These chemical characteristics make tequila vinasse a difficult treat waste. Traditional wastewater treatments are not effective for treating tequila vinasses being necessary advanced physic-chemical and biological combined treatments for efficiently treating these effluents, which involve high treatment costs [8]. An alternative to reduce the costs of the process and the consumption of water in the production chains is the circular economy approach. A circular economy is a production and consumption model involving sharing, leasing, reusing, repairing, refurbishing, and recycling existing materials and products as long as possible. In this sense, reusing tequila vinasses to produce more value-added products and a source of water and nutrients is of great interest to reduce water stress, the cost of processes, and waste generation. This chapter focuses on a fermentation process with tequila vinasse as a raw material. However, there exist alternative approaches for vinasse treatment. For instance, [9] describes a three-stage process for producing lactate, biohydrogen, and methane.

The processes that appear in the chain production of bioethanol are diverse, depending on the raw material and substances to carry out the transformation of the raw material up to reducing sugars. Next, the common process is fermentation, in which the reducing sugars are converted to ethanol by yeast. The process is carried out in a tank bioreactor, in which a batch or continuous operation can be performed. There are numerous works addressing the modeling, design, optimization, and control of this class of processes [10, 11], aimed to obtain high performance of the fermentation, e.g., ethanol production rate, also called productivity, given by the flowrate multiplied by the ethanol concentration in the fermentation broth.

Real-time optimization (RTO) of steady-state plants aims to improve process performance by optimizing a measurable criterion or objective function under economic, safety, or quality constraints. One of the most critical limitations of earlier RTO methods lies in model accuracy. In order to bypass model mismatches, direct input adaptation methods transform the RTO problem into a feedback control problem, exploiting the invariance properties of the steady-state plant. These latter methods include Extremum Seeking Control (ESC), a model-free feedback control strategy which drives the system towards optimal operating conditions corresponding to the extremum of a measurable convex objective function [12]. ESC is applicable in situations where there is a nonlinearity in the control problem, and the nonlinearity has a local minimum or maximum. The nonlinearity may be in the system as a physical nonlinearity or in the control objective, added to the system through a cost functional of an optimization problem [13]. Given that no knowledge of the model is required to implement ESC, it is widely applicable in biochemical systems whose mathematical models are often imperfect. Among many other works, an overview, a particular application, and a review with a worked example of ESC in bioreactors can be found in [12, 14, 15], respectively. Due to its finite-time convergence and robustness capabilities, sliding mode approaches have been used in ESC, achieving a more robust and shorter optimization time. See for instance [16–18].

This chapter presents a model that accounts for inhibition due to large substrate, biomass, and product concentrations. As one may find in Sect. 10.2, such a model is a nonlinear, ordinary differential equation with ten parameters whose numerical values we fit via nonlinear programming techniques described in Sect. 10.2.3. In Sect. 10.2.4 we perform a sensitivity analysis to determine which parameters have a more significant impact on the location of the steady-state. In Sect. 10.3, to maximize a performance index, we avail of our extremum seeking approach in [18], which depends on an accurate computation of the input–output gradient to steer the state of the fermenter to its maximum productivity. Such an approach uses the Super-Twisting Algorithm [19–21] both for estimating the input–output gradient and to update the dilution rate of the fermenter. In Sect. 10.4, we present two case studies. The first refers to the industrial production of bioethanol. In contrast, the second refers to the fermentation of tequila vinasse to attain extra ethanol production and remove the remaining substrate.

10.2 Model

10.2.1 Fermentation Modeling

Several attempts have been made to develop mathematical models that reproduce the behavior of substrate S, biomass X, and product P in a fermentation process. In general, the mathematical models account for the following reaction:

$$nS + X \xrightarrow{v_1} 2X + mP, \tag{10.1}$$

which represents biomass growth using the consumption of n units of substrate. As a result of biomass metabolism, m units of the product are generated. Regarding ethanol production, biomass, substrate, and product might be *Saccharomyces Cerevisiae*, glucose, and ethanol. Although describing the yield coefficients and the species involved, Reaction (10.1) does not show the rate at which the reactants become products. To fulfill such purpose, several reaction rates have been selected. For instance, a Monod [22] reaction rate

$$v = \mu_{\max} \frac{[S][X]}{K + [S]},$$

where $[Z]$ denotes the concentration of species Z, is commonly used to represent biomass growth due to its almost linear behavior for small $[S]$, which saturates at larger values of the substrate concentration. Among many others, works like [23–26] have used such a reaction rate in ranges of $[S]$ that do not hinder biomass growth.

However, depending on the behavior of the biomass, for larger concentrations of substrate the reaction rate may decrease as the substrate concentration increases. In such cases a Haldane reaction rate [27] is preferred and has been used, for instance, in works like [28–30], in order to account for inhibition due to substrate. By considering $[X]$ constant, this reaction rate grows almost linearly as $[S]$ increases before a threshold. After such a threshold, the reaction rate decreases asymptotically, thus considering biomass growth inhibition via large substrate concentrations. Such a reaction rate may take the form:

$$v_1 = K_1 \frac{[S][X]}{[S]^2 + K_2[S] + K_3}.$$

Now, to account for the effect of the inhibition in reaction (10.1) due to large concentrations of biomass, authors like [31, 32] have modified the Monod and Haldane reaction rates by multiplying them by the factor

$$v^* = \left(1 - \frac{[X]}{K_1}\right)^{K_2};$$

such a term equals one when $[X] = 0$ and decreases to zero as $[X]$ tends to K_1.

Another reaction rate of interest is a modified version of that presented in [33], which we will use to represent biomass degradation due to large concentrations of product, that is $X + P \xrightarrow{V_2} P$. Such a reaction rate takes the form

$$v_2 = K_1 \frac{[P]}{K_2[X] + K_3[P]^2 + K_4}[X].$$

In such a reaction rate, the factor of $[X]$ is small for small $[P]$; however, as $[P]$ increases, the rate v_2 increases. Conversely, when $[X]$ is small the factor of $[X]$ is large, as compared as when it grows. This latter mechanism represents the reduction of the susceptibility of the biomass colonies to the lethal effect of high alcohol concentrations, for instance, due to the creation of colonies of films. If such an effect is not markedly present in the fermentation process under consideration, one may set $K_2 = 0$.

For a more extensive review of mathematical models for bioethanol production, we refer the interested reader to works like [11, 34] or [35] for a control theory perspective on bioreactors. In the following section, a fermentation in a CSTR is presented.

10.2.2 Bioethanol Production Model in a CSTR

This section produces the dynamic model of a constant temperature, continuous stirred tank reactor, CSTR, in which fermentation occurs. Such a reactor is subject to an influx and efflux characterized by a time-varying dilution rate $D(t)$. As described in the previous section, the fermentation process is subject to substrate, biomass, and product inhibition [32, 36–38] and has natural biomass dead. Thus, we consider the following set of reactions:

$$nS + X \to^{v_1} 2X + mP, \quad X + P \to^{v_2} P, \quad X \to^{v_3} 0. \tag{10.2}$$

As in the previous section S, X, and P denote the substrate, biomass, and product, respectively; in turn, the constants $n, m \ \in \mathbb{R} > 0$ denote the yield coefficients in the biomass growth process. Let the species' concentration vector be

$$\boldsymbol{c}(t) = ([S][X][P])^T. \tag{10.3}$$

To account for biomass inhibition due to high concentrations of substrate, we avail of a Haldane reaction rate [27] with an inhibitory factor due to large concentrations of biomass [32],

$$v_1(c_1, c_2) = k_{11} \frac{c_1 c_2}{c_1^2 + k_{12} c_1 + k_{13}} \left(1 - \frac{c_2}{k_{14}}\right)^{k_{15}}. \tag{10.4a}$$

To account for biomass inhibition due to large product concentrations, we consider the following reaction rate:

$$v_2(c_2, c_3) = k_{21} \frac{c_3}{k_{22} c_2 + k_{23} c_3^2 + k_{24}} c_2. \tag{10.4b}$$

Finally, we use a mass action kinetics for the biomass natural death,

$$v_3(c_2) = k_3 c_2. \tag{10.4c}$$

Thus, in compact notation, the model for this CSTR is

$$\frac{\mathrm{d}}{\mathrm{d}t} \boldsymbol{c}(t) = \boldsymbol{N} \boldsymbol{v}(\boldsymbol{c}) + D(t)(\boldsymbol{c}_{in} - \boldsymbol{c}), \tag{10.5a}$$

where the matrix $\boldsymbol{N}$ is the stoichiometric matrix from the reaction network in (10.2), which takes the form

$$\boldsymbol{N} = \begin{pmatrix} -n & 0 & 0 \\ 1 & -\theta_1 & -\theta_2 \\ m & 0 & 0 \end{pmatrix}. \tag{10.5b}$$

Here $n = 30.303$ and $m = 7$. Now, θ_1 and θ_2 equal 0 or 1, whether one requires to account for inhibition by product or biomass death in the conditions of the reactor, respectively. For instance, for fermentation conditions with low substrate and product concentrations, one may set $\theta_1 = \theta_2 = 0$, in order to solely consider the process of biomass growth. In addition, the reaction rate vector is

$$v(c) = \begin{pmatrix} k_{11}\frac{c_1c_2}{c_1^2+k_{12}c_1+k_{13}}\left(1-\frac{c_2}{k_{14}}\right)^{k_{15}} \\ k_{21}\frac{c_3}{k_{22}c_2+k_{23}c_3^2+k_{24}}c_2 \\ k_3c_2 \end{pmatrix}. \tag{10.5c}$$

To model different scenarios involving the CSTR, we allow the influx to be endowed with concentrations of all the species and gather them in the following vector:

$$c_{in}(t) = ([S]_{in}\ [X]_{in}\ [P]_{in})^T.$$

For instance, in our case study in Sect. 10.4, the influx to the reactor has concentrations of substrate and biomass since we consider that the raw material for the fermentation process comes from wastewater from tequila production industries. Now, in certain setups in which the efflux of the CSTR is partially feedback, the influx is then endowed with a concentration of ethanol.

10.2.3 Parameter Fitting

This section, considers the species concentration in time, as presented in [32]. Please notice that such a model does not account for biomass inhibition via high concentrations of product presented in the previous section. However, we use the data from [32] to show that model (10.5) can perform a certain behavior. For this purpose, we define the following cost function, which describes how much the training data differs from that obtained with our model:

$$f(k) = \varphi + \sum_{i=1}^{3} w_i \left(\sum_{\forall j} (c_i^*[t_j] - c_i^*[t_j])^2 \right)^{1/2}, \tag{10.6}$$

where $c_i[t_j]$ represent the value of species c_i at the time instant t_j, which results from simulating (10.5) with a particular $\boldsymbol{k}$, defined as

$$\boldsymbol{k} = (k_{11}\ k_{12}\ k_{13}\ k_{14}\ k_{15}\ k_{21}\ k_{22}\ k_{23}\ k_{24}\ k_3)^T, \tag{10.7}$$

In turn, the values $c_i^*[t_j]$ are those of the simulation of [32]. Both simulations were performed with an initial concentration in g/L of

$$\boldsymbol{c}_0 = (50\ 2\ 0)^T, \tag{10.8a}$$

and with the following concentrations in the influx

$$c_{in} = (60.518\ 30\ 10)^T, \tag{10.8b}$$

along with a dilution rate of $D = 0.5[1/h]$. Furthermore, in Eq. (10.6), w_i is a constant weight different for each species in $\boldsymbol{c}$. In addition, the constant φ penalizes the presence of negative values in $\boldsymbol{k}$. That is, if any entry of $\boldsymbol{k}$ is negative, φ takes a value of 1000 in order to discard negative values for the kinetic parameters of the reaction network. Thus, the parameter fitting problem may be posed as an optimization problem in which a vector $\boldsymbol{k}$ is to be found such that (10.6) is minimized. That is

$$\begin{aligned} &\min_{\boldsymbol{k}} f(\boldsymbol{k}) \\ &\text{subject to} \begin{cases} \frac{\mathrm{d}}{\mathrm{d}t}\boldsymbol{c}(t) = \boldsymbol{N}\boldsymbol{v}(\boldsymbol{c}) + D(t)(\boldsymbol{c}_{in} - \boldsymbol{c}) \\ \boldsymbol{y}[t_j] = \boldsymbol{c}[t_j] \end{cases}. \end{aligned} \tag{10.9}$$

Here $\boldsymbol{c}[t_j]$ denotes the discrete measurement of $\boldsymbol{c}(t)$ in the instant t_j. To solve this optimization problem, we make use of Matlab's *fminsearch* [39] and *fmincon* [40, 41]. For the latter one we chose the option "Active-Set Optimization." Such functions require an initial guess $\boldsymbol{k}$, which we obtain via a stochastic search, i.e., by repeating 10^4 random samplings of uniformly distributed vectors $\boldsymbol{k}$ and evaluating the objective function in (10.6), we chose the set of parameters $\boldsymbol{k}$ which exhibits the smallest value as an initial guess for the nonlinear optimization routines *fminsearch* and *fmincon*. The parameters obtained with such optimization algorithms are in Table 10.1. In turn, Fig. 10.1 shows the solution of (10.5) with the fitted parameters (blue, continuous line) versus the training data (red, empty markers). It is important to mention that although model (10.5) considers different processes, it reproduces the behavior of model in [32].

Table 10.1 Parameters obtained from the optimization problem in (10.9)

Parameter	Value	Unit	Reference
k_{11}	922.8228	g/(Lh)	This chapter
k_{12}	1.8947×10^3	g/L	This chapter
k_{13}	114.4452	$(g/L)^2$	This chapter
k_{14}	100	g/L	[32]
k_{15}	0.9	1	[32]
k_{21}	1.0184×10^3	g/(Lh)	This chapter
k_{22}	46.1869	g/L	This chapter
k_{23}	48.5117	1	This chapter
k_{24}	254.8337	$(g/L)^2$	This chapter
k_3	3	1/h	This chapter

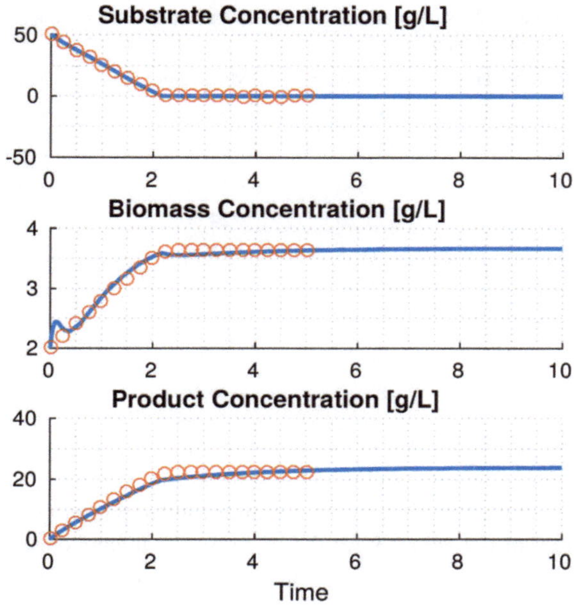

Fig. 10.1 Assessment of the solution of (10.5) with the parameters in Table 10.1 versus data from the model in [32]. The continuous line is the solution of model (10.5), whilst the empty markers denote training data

10.2.4 Parameter Sensitivity

In order to evaluate the range in which each model parameter can vary, a sensitivity analysis on Model (10.5) is performed by considering a Monte Carlo methodology. The parameter vector $\boldsymbol{k}$ in (10.7) influences the fermentation process dynamics and the steady-state. By considering the nominal value of the vector $\boldsymbol{k}$ of Table 10.1, the steady-state of the vector $\boldsymbol{c}$ is computed. Then, a set of vectors $\boldsymbol{k}_{LH}$ is constructed by using the Latin Hypercube sampling method [42]. Values between $-$ 30% and 30% of the elements of the vector $\boldsymbol{k}$ stratified over 10 levels with 10 samples randomly distributed into each interval to generate 100 simulations were considered. Such a Latin Hypercube was constructed using the Matlab function *lhdesign*. Each element of the vector $\boldsymbol{c}$ is then computed at steady-state for each vector $\boldsymbol{k}_{LH}$. Therefore, the absolute error due to the parameter uncertainties is determined for each state variable as $\varepsilon_S = |[S]_{ss}(\boldsymbol{k}) - [S]_{ss}(\boldsymbol{k_{LH}})|$, $\varepsilon_X = |[X]_{ss}(\boldsymbol{k}) - [X]_{ss}(\boldsymbol{k_{LH}})|$, and $\varepsilon_P = |[P]_{ss}(\boldsymbol{k}) - [P]_{ss}(\boldsymbol{k_{LH}})|$. An absolute error $\varepsilon = [\varepsilon_S\ \varepsilon_X\ \varepsilon_P] = [0.004\ 0.183\ 1.194]$, corresponding to 5% of the state vector $\boldsymbol{c}$ at steady-state, is selected as the maximum allowed.

From the subset of correctly simulated vector $\boldsymbol{k}_{LH}$ the minimal and the maximum values of every parameter are picked out to define its variation range. Table 10.2 shows both minimum and maximum values determined for each parameter of the vector $\boldsymbol{k}$ for the substrate. As it can be regarded, the parameter k_{22} is the most sensitive since the steady-state of the substrate is affected when k_{22} varies 15.30%. On the other hand, the rest of the parameters have similar sensitivity over $\pm$ 20%.

Table 10.2 Variation range in absolute and percentage values for each parameter that influence the substrate dynamics

Parameter	Min. value	Min. value (%)	Max. value	Max. value (%)
k_{11}	659.82	−28.50	1196.90	29.70
k_{12}	1331.97	−29.70	2366.48	24.90
k_{13}	90.76	−20.70	145.69	27.30
k_{14}	75.10	−24.90	127.90	27.90
k_{15}	0.65	−27.90	1.14	26.70
k_{21}	715.94	−29.70	1314.75	29.10
k_{22}	35.52	−23.10	53.25	15.30
k_{23}	36.43	−24.90	59.14	21.90
k_{24}	188.32	−26.10	315.23	23.70
k_3	2.18	−27.30	3.87	29.10

Table 10.3 shows both minimum and maximum values determined for each parameter of the vector $\boldsymbol{k}$ for the biomass. As it can be regarded, parameter k_3 is the most sensitive parameter, whereas the other parameters have similar sensitivity. Nevertheless, nothing about the sensitivity of the vector $\boldsymbol{k}$ on the product dynamics could be concluded by considering the Latin Hypercube constructed

In order to analyze the sensitivity of the vector $\boldsymbol{k}$ on the product dynamics, each element of the vector $\boldsymbol{k}$ is modified as $k_j = k_j \pm 5\% k_j$, for $j = 1, 2, \ldots, 10$, to generate a set of 20 elements to simulate. Table 10.4 shows the simulation results. Clearly, the steady-state of $[P]$ is not considerably affected by the modified vector $\boldsymbol{k}$. As it can be observed, the parameter k_{11} is the most sensitive parameter because by modifying it in 5% the product at steady-state is modified in 0.36%, while the rest of the parameters vary between 0.34% and 0.35%.

Table 10.3 Variation range in absolute and percentage values for each parameter that influence the biomass dynamics

Parameter	Min value	Min value (%)	Max value	Max value (%)
k_{11}	648.74	−29.70	1191.36	29.10
k_{12}	1366.08	−27.90	2434.69	28.50
k_{13}	81.83	−28.50	147.75	29.10
k_{14}	70.30	−29.70	127.90	27.90
k_{15}	0.68	−24.90	1.16	29.10
k_{21}	722.05	−29.10	1314.75	29.10
k_{22}	33.02	−28.50	59.90	29.70
k_{23}	35.27	−27.30	62.63	29.10
k_{24}	192.91	−24.30	330.52	29.70
k_3	2.28	−24.30	3.46	15.30

Table 10.4 Simulation results of product dynamics corresponding to the vector $\boldsymbol{k}$ modified one at a time

k_{11}	k_{12}	k_{13}	k_{14}	k_{15}	k_{21}	k_{22}	k_{23}	k_{24}	k_3	Variation of $[P]_{SS}$ (%)
968.96	1894.70	114.45	100.00	0.90	1018.40	46.19	48.51	254.83	3.00	0.3618
876.68	1894.70	114.45	100.00	0.90	1018.40	46.19	48.51	254.83	3.00	0.3427
922.82	1989.44	114.45	100.00	0.90	1018.40	46.19	48.51	254.83	3.00	0.3475
922.82	1799.97	114.45	100.00	0.90	1018.40	46.19	48.51	254.83	3.00	0.3585
922.82	1894.70	120.17	100.00	0.90	1018.40	46.19	48.51	254.83	3.00	0.3494
922.82	1894.70	108.72	100.00	0.90	1018.40	46.19	48.51	254.83	3.00	0.3573
922.82	1894.70	114.45	105.00	0.90	1018.40	46.19	48.51	254.83	3.00	0.3537
922.82	1894.70	114.45	95.0	0.90	1018.40	46.19	48.51	254.83	3.00	0.3530
922.82	1894.70	114.45	100.00	0.95	1018.40	46.19	48.51	254.83	3.00	0.3530
922.82	1894.70	114.45	100.00	0.85	1018.40	46.19	48.51	254.83	3.00	0.3537
922.82	1894.70	114.45	100.00	0.90	1069.32	46.19	48.51	254.83	3.00	0.3515
922.82	1894.70	114.45	100.00	0.90	967.48	46.19	48.51	254.83	3.00	0.3551
922.82	1894.70	114.45	100.00	0.90	1018.40	48.50	48.51	254.83	3.00	0.3534
922.82	1894.70	114.45	100.00	0.90	1018.40	43.88	48.51	254.83	3.00	0.3533
922.82	1894.70	114.45	100.00	0.90	1018.40	46.19	50.94	254.83	3.00	0.3550
922.82	1894.70	114.45	100.00	0.90	1018.40	46.19	46.09	254.83	3.00	0.3515
922.82	1894.70	114.45	100.00	0.90	1018.40	46.19	48.51	267.58	3.00	0.3534
922.82	1894.70	114.45	100.00	0.90	1018.40	46.19	48.51	242.09	3.00	0.3533
922.82	1894.70	114.45	100.00	0.90	1018.40	46.19	48.51	254.83	3.15	0.3467
922.82	1894.70	114.45	100.00	0.90	1018.40	46.19	48.51	254.83	2.85	0.3594

10.2.5 Inhibition Effects on the Steady-State

In this section, we show the effect of increasing or decreasing reaction rates responsible for the fermentation and its inhibition. We vary selected kinetic parameters representing a fermentation mechanism to assess such an effect.

The nominal parameters may be found in Table 10.1. As can be seen in Eq. (10.4a), when we vary k_{11} we alter the maximum reaction rate of the conversion of substrate to product. This is shown in Fig. 10.2, which depicts the location of steady-states as a function of both k_{11} and the dilution rate; the latter one is presented in a logarithmic scale. The red marker denotes the nominal condition in which the parameter fitting of Sect. 2.3 was performed. One may notice that if k_{11} is decreased below a threshold the fermentation process is hindered, whereas for this particular scenario an increase of k_{11} does not longer increase the product concentration. Regarding the dilution rate, one may notice that for large values, ca. $100[1/h]$, the reactor is in a washout condition.

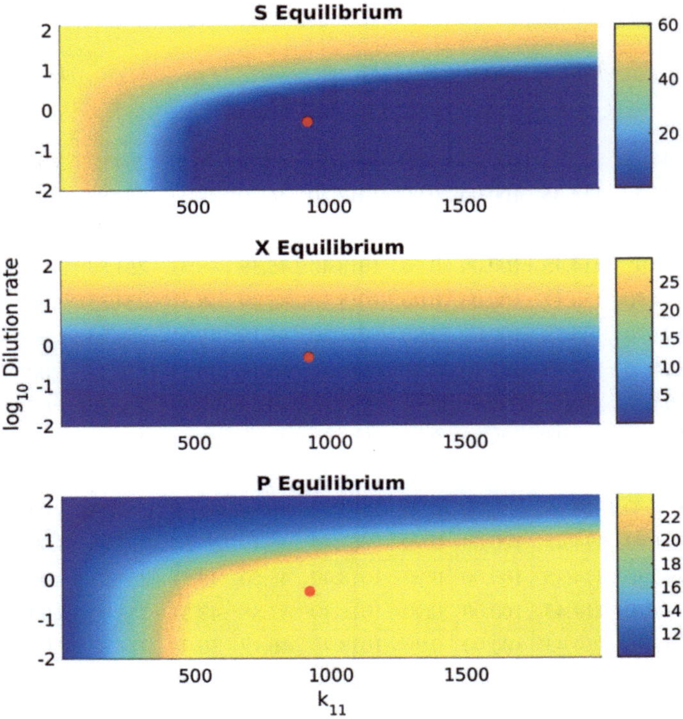

Fig. 10.2 Location of steady-state with variations of k_{11}

Now, regarding the inhibitory effect of the Haldane and Levenspiel reaction rate, an increase of either k_{12} or k_{15} decrease the steady-state concentration of the ethanol. This effect may be seen in Figs. 10.3 and 10.4, respectively. In particular, a large increase of the k_{15} is required to completely knock out the product concentration. Variations of k_{13} and k_{14} exert no significant effect on the steady-state location.

With respect to reaction v_2, an increase of k_{21} diminishes the steady-state concentration of ethanol, cf. Figure 10.5; in turn, by decreasing k_{23} below a threshold a significant decrease of ethanol concentration may be appreciated (Fig. 10.6).

To finalize this section, Fig. 10.7 shows the effect of variations of k_3. Notice that larger values of k_3 reduce the steady-state concentration of ethanol. Notably, for all figures shown in this section, the biomass concentration remains constant as the kinetic parameters are varied, except for variations of k_3. In the following section, we determine the capability of Model (10.5) to be optimized and perform a maximization of bioethanol production via an ESC approach.

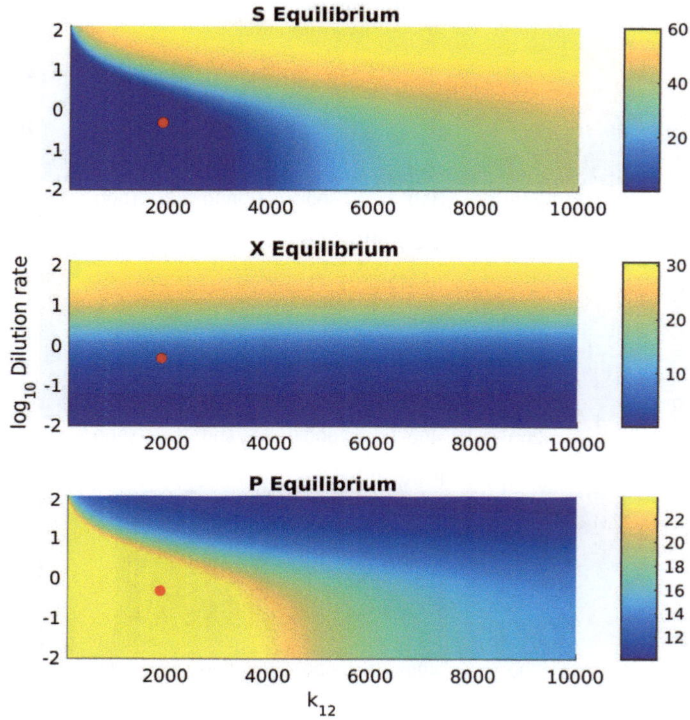

Fig. 10.3 Location of steady-state with variations of k_{12}

10.3 Extremum Seeking Control

This section provides the methodology used in the case studies section. We begin by defining how one may determine whether the model under consideration is suitable for optimization, and then, the strategy to achieve such optimal conditions is presented.

10.3.1 Input–Output Map Convexity

To determine whether this model is suitable for optimization, we use the following procedure to show the convexity of the input–output map. Let

$$u(t) = D(t); \tag{10.10}$$

that is to say, the dilution rate D is considered as the controlled input of the fermentation process. Different performance indexes may be chosen depending on the desired

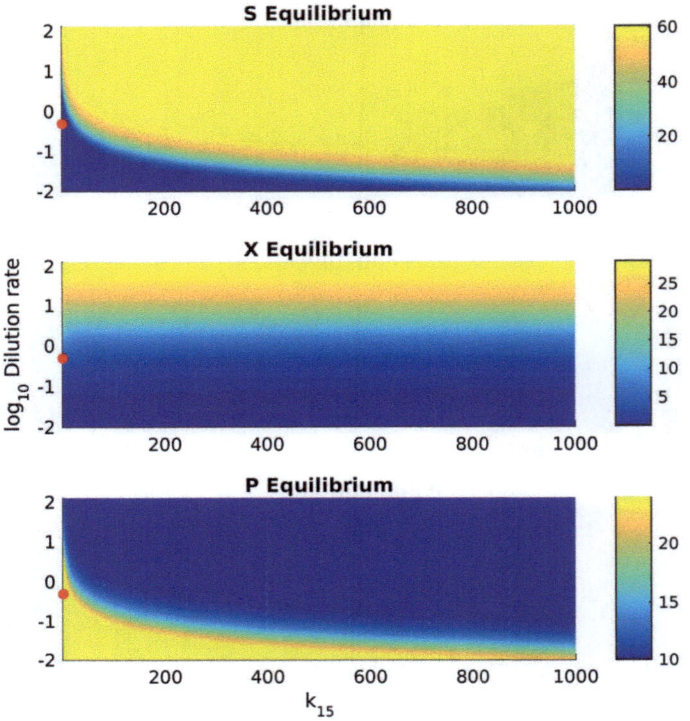

Fig. 10.4 Location of steady-state with variations of k_{15}

application. However, such a performance index is a function of the state and control inputs:

$$J(t) = f(c, u); \tag{10.11}$$

It is assumed that we have the measurements of $u(t)$ and $J(t)$ online. In order to show the concavity of the input–output map in equilibrium, one has to determine the steady-states of (10.5) as a function of $u(t)$.

10.3.2 Extremum Seeking Control

In the previous section, we postulate that if the steady-state map from dilution rate to the steady-state of productivity is concave, then the fermenter is suitable for ESC. As described earlier, our goal is to maximize a productivity index of the fermenter in a steady-state via an online update of the dilution rate. Such optimization problem may be stated as:

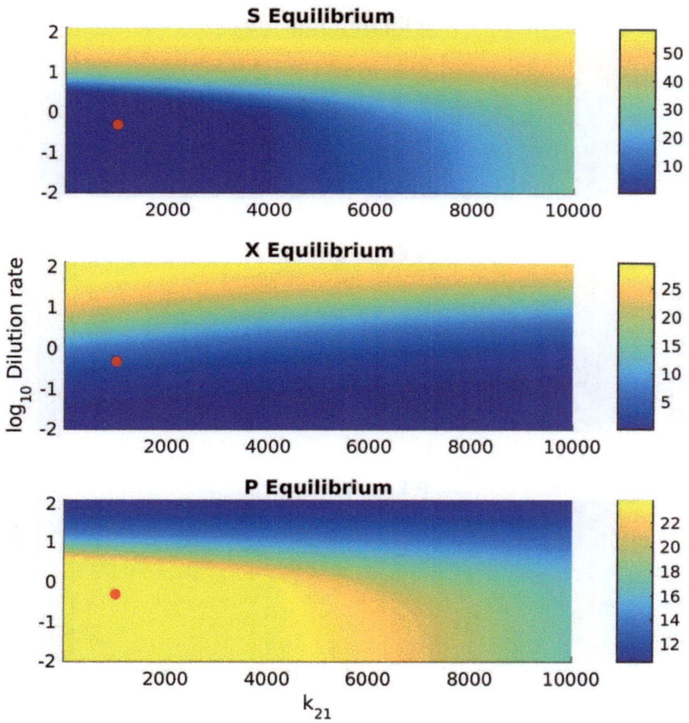

Fig. 10.5 Location of steady-state with variations of k_{21}

$$\max_{u\in\mathbb{R}_+} J(u)$$
$$\text{subject to}\begin{cases} \frac{\mathrm{d}}{\mathrm{d}t}\boldsymbol{c}(t) = \boldsymbol{N}\boldsymbol{v}(\boldsymbol{c}) + u(\boldsymbol{c}_{in} - \boldsymbol{c}) \\ J(t) = f(\boldsymbol{c}, u) \end{cases}. \tag{10.12}$$

where $J(\boldsymbol{c}, u)$ is the unknown input–output map at steady-state. In order to online solve the previous optimization problem, we use our Super-Twisting (STA) based approach [18], which we have used for different models and scenarios in works like [43–45]. Such a STA has the form:

$$u(t) = -\lambda|\sigma|^{\frac{1}{2}}sign(\sigma) + u_1(t), \quad \dot{u}_1(t) = -\alpha sign(\sigma), \tag{10.13}$$

here $\alpha, \lambda > 0$, and σ is a sliding variable [46]. By considering $\sigma = -\mathrm{d}J/\mathrm{d}u$, the Super-Twisting-based Extremum Seeking Controller (10.13) generates the control input $u(t)$ that maximizes the objective function $J(u)$ at steady-state [18]. Such an approach requires the input–output gradient, which we compute via the differentiator in [21]. To this end, let

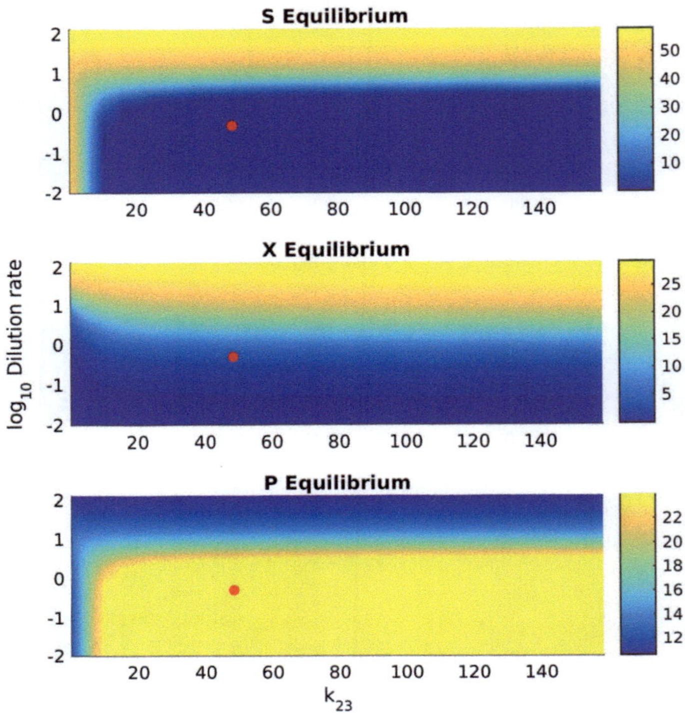

Fig. 10.6 Location of steady-state with variations of k_{23}

$$\theta := (y\ u)^T, \tag{10.14}$$

with first time derivative

$$\omega := (\dot{y}\ \dot{u})^T. \tag{10.15}$$

A finite-time estimate $\omega(t)$ may be obtained via

$$\dot{\hat{\theta}}(t) = -k_1\phi_1\big(\hat{\theta} - \theta\big) + \hat{\omega}(t), \quad \dot{\hat{\omega}}(t) = -k_2\phi_2\big(\hat{\theta} - \theta\big), \tag{10.16}$$

which is another form of the STA. There $\hat{\theta}(\hat{\omega}, resp.)$ denotes the estimation of $\theta(\omega, resp.)$, and

$$\phi_1(\boldsymbol{x}) := \Big(\eta \boldsymbol{x}_2^{-p} + \beta + \gamma \boldsymbol{x}_2^{q}\Big)\boldsymbol{x}, \quad \phi_1(0) := 0, \tag{10.17a}$$

$$\phi_2(\boldsymbol{x}) := \Big(\eta(1-p)\boldsymbol{x}_2^{-p} + \beta + \gamma(1+q)\boldsymbol{x}_2^{q}\Big)\phi_1(\boldsymbol{x}), \tag{10.17b}$$

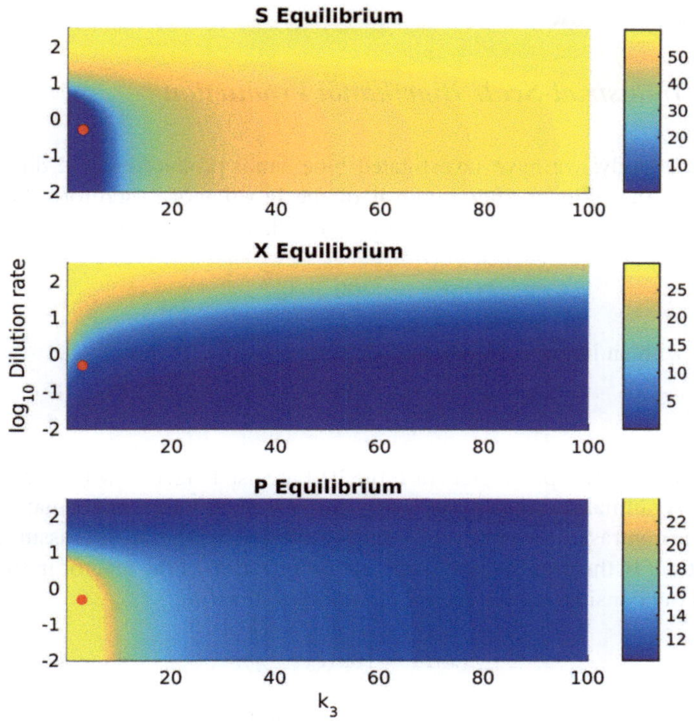

Fig. 10.7 Location of steady-state with variations of k_3

where $\boldsymbol{x}(t)_2 := \sqrt{\boldsymbol{x}^T(t)\boldsymbol{x}(t)}$. In addition, $\eta, \beta, \gamma > 0$, $\frac{1}{2} \geq p > 0$, and $q > 0$; and k_1 and k_2 are such that the matrix

$$\boldsymbol{A} = \begin{pmatrix} -k_1 & 1 \\ -k_2 & 0 \end{pmatrix} \tag{10.18}$$

is Hurwitz. Thus, the estimate of the input–output gradient is approximated as follows:

$$\frac{\mathrm{d}J}{\mathrm{d}u} := \frac{\hat{\omega}_1}{\hat{\omega}_2}. \tag{10.19}$$

In the following section, the optimization scheme applied to Model (10.5) is used for different scenarios.

10.4 Case Studies

10.4.1 Industrial-Scale Bioethanol Production

In this case study, we have investigated bioethanol production in a digester with feedback to the effluent after removal of the bioethanol. Therefore, the influent concentration c_{in} contains a concentration of both substrate and biomass:

$$\boldsymbol{c}_{in} = (200\ 30\ 0)^T,$$

Along with an initial concentration of biomass

$$\boldsymbol{c}_0 = (0\ 30\ 0)^T.$$

Here, the reaction rate parameters in Table 10.1 are used together with model reported in (10.5). A similar setup was studied in [32] with a different mathematical model; however, in contrast to the cited publication, a constant temperature is assumed during fermentation. In the present case study, an optimization of the reactor in steady state is carried out considering the following objective function:

$$J(t) = 10u(\overline{c}_3 - \chi). \tag{10.20}$$

Here the term $u\overline{c}_3$ represents steady-state ethanol productivity, whereas the term $u\chi$ penalizes large dilution rates. The objective function in (10.20) is proportional to the reactor's productivity. Please notice that for computing J, one requires the steady-state concentrations; in this case, we computed them via a simulation during a large enough period. It is important to note that when $\chi = 0$ the objective function is proportional to the reactor productivity, however, in such a case $J(t)$ is not a convex function of the dilution rate (data not shown). In contrast, for a range of values of χ such an objective function is concave, as shown in Fig. 10.8. Panel (a) shows Eq. (10.20) as a function of the dilution rate for values of χ ranging from 4 up to 22; whereas panel (b) shows the optimal productivity and steady-state concentration of bioethanol as a function of the parameter χ. Thus, by setting a value for χ, one may choose a desired productivity or bioethanol concentration to comply with the requirements of a particular application. For our case study, we choose a value of $\chi = 16$, the red marker in Fig. 10.8 shows the optimal conditions with such a value of χ, which, in turn, leads to a dilution rate of $u \approx 0.6310[1/h]$, with a maximum objective function of $J \approx 42.2$ along with bioethanol productivity around $14.3[g/(Lh)]$ and the following steady-state concentrations in the reactor: $\overline{c}_1 \approx 101.7628[g/L]$, $\overline{c}_2 \approx 4.6231[g/L]$, $\overline{c}_3 \approx 22.6951[g/L]$.

The optimum conditions are tracked via the ESC described in Sect. 10.3. To this end, the following parameters for the control input (10.13) are used:

$$\{\lambda, \alpha\} = \left\{70\text{x}10^{-6}, 700\text{x}10^{-6}\right\}.$$

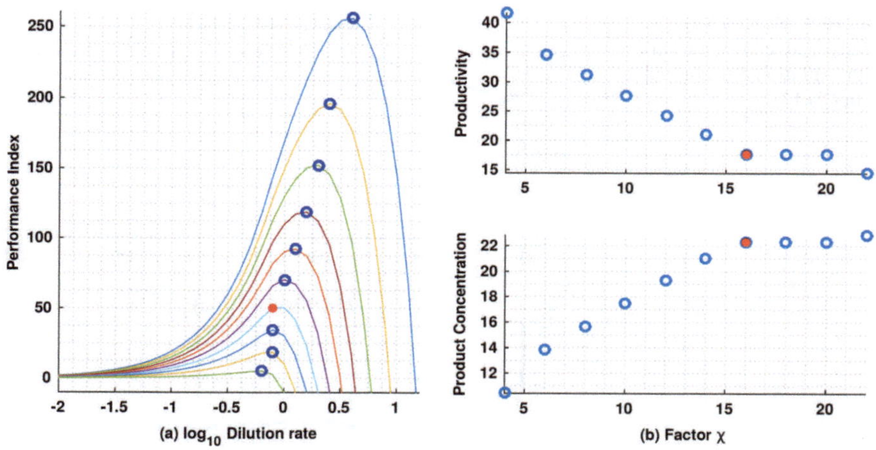

Fig. 10.8 Objective function for steady-state optimization of model (10.5) and the corresponding productivity and bioethanol concentration in the optimal conditions. The red markers in all panels denote the optimal condition selected for the optimization

Such control input is updated every $8[h]$ and kept constant during such an interval. However, the estimation of the input–output gradient is continuously computed via (10.16) and (10.19) along with the parameter values $\{\eta, \beta, \gamma\} = \{0.15, 0.15, 0.15\}$, $\{p, q\} = \{0.35, 0.30\}$, and $\{k_1, k_2\} = \{1, 1\}$. Figure 10.9 shows the closed-loop dynamics for the reactor concentrations with such parameters. In turn, Fig. 10.10 depicts the computed dilution rate, the value of the objective function, and the estimated input–output gradient. In both figures, the red, discontinuous line denotes the location of conditions that lead to the maximum J.

Please notice, in the lower panel of Fig. 10.10 that the gradient ripples around zero, showing that the maximum value of J has been achieved.

10.4.2 Bioethanol Production from Tequila Vinasse

In this section, the results from the previous section are used to design a control function *u(t)* that maximizes the productivity of the reactor used in (10.11). For the calculation, wastewater from tequila production is used for our fermentation process. This provides an additional amount of ethanol when cleaning the wastewater from residual substrate.

Tequila vinasse was used for the experiment. Tequila vinasse was sampled from a local tequila distillery and directly stored at 4 °C before analysis. The biomass concentration in the vinasse was determined by the Volatile Suspended Solids (VSS) method as reported in [47]. Additionally, the Colony Forming Unit Assay (CFU) was performed in fresh vinasse to measure viable cell numbers in CFU/mL. This

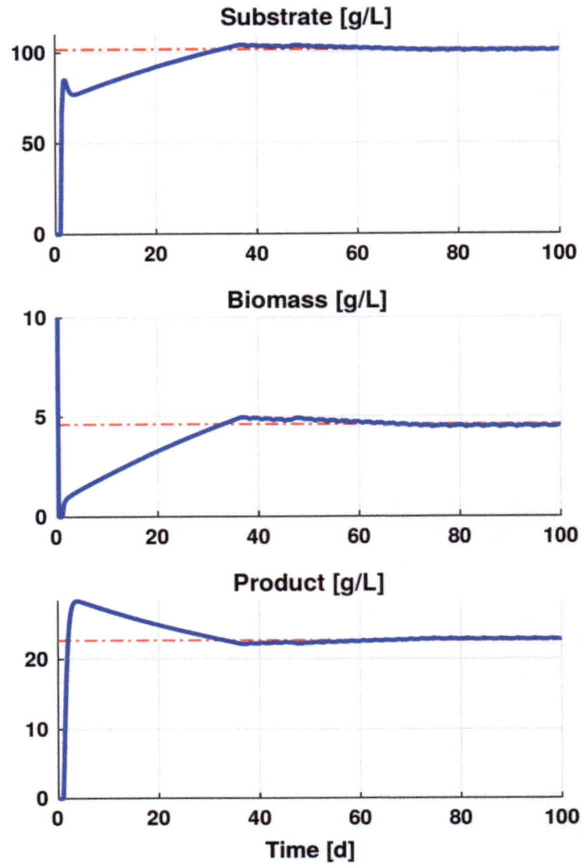

Fig. 10.9 Reactor's steady-state concentrations with a dilution rate computed as in (10.13)

is an indication of the number of cells that remain viable enough to proliferate and form small colonies [48]. For chemical characterization, tequila vinasse was settled, and the supernatant was filtered at $1.2\mu m$ with glass fiber filters (Whatman, USA). The filtered vinasse had a pH of 4.0. Total reducing sugars of filtered vinasse were determined by the Miller method (DNS) at 540 nm [49], and Chemical Oxygen Demand (COD) was determined by the 8000 Reactor Digestion Method, using a DRB 200 Reactor (Hach Company, CO, USA). In turn, the concentration of ethanol is assumed to be below a measurable threshold, as reported in [8]. Thereby we consider it to be zero.

The vinasse is fed into a CSTR containing substrate and biomass. However, substrate and biomass are present in very low concentrations, so that the parameters θi (see also 5b) were set to zero, thus also the initial ethanol concentration could be neglected, which would otherwise have caused an inhibition or death of the biomass in the reactor. With these initial simplifications (see also 5), the model results in:

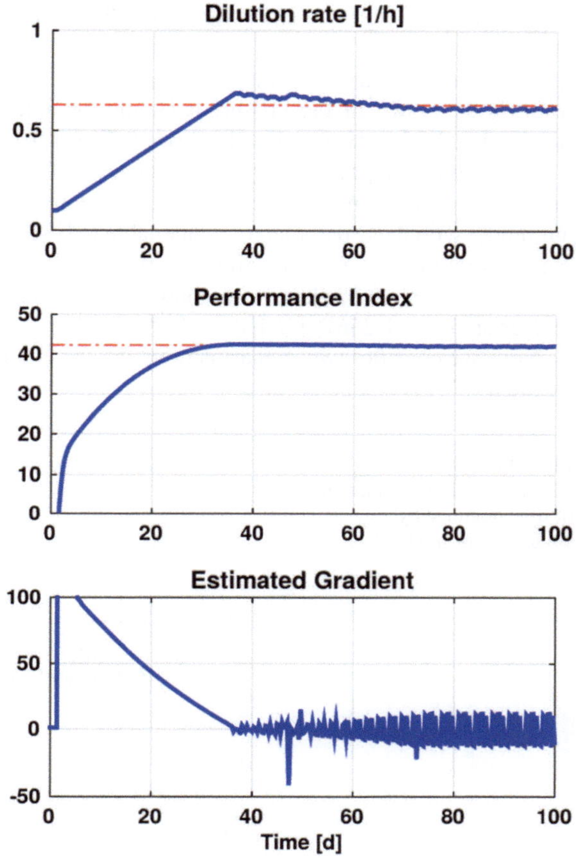

Fig. 10.10 Dilution rate computed with (10.13) along with the objective function J in (10.20)

$$\frac{\mathrm{d}}{\mathrm{d}t}\begin{pmatrix} c_1(t) \\ c_2(t) \\ c_3(t) \end{pmatrix} = \begin{pmatrix} -n \\ 1 \\ m \end{pmatrix} k_{11}\frac{c_1 c_2}{c_1^2 + k_{12}c_1 + k_{13}}\left(1 - \frac{c_2}{k_{14}}\right)^{k_{15}} + u(t)(\boldsymbol{c}_{in} - \boldsymbol{c}), \tag{10.21a}$$

$$J(t) = u(t)c_3(t), \tag{10.21b}$$

where the parameters k_{1i} are as in Table 10.1 and $\{m, n\} = \{7, 30.303\}$.

Please be wary that we have chosen bioethanol productivity, uc_3, as the performance index to optimize.

The analysis of the tequila vinasse, described above, resulted in a concentration of 4.995[g/L] for substrate (reducing sugars), 0.4728[g/L] of biomass. According to [8] we assume that the vinasse does not contain a measurable concentration of ethanol. For the simulation, the initial concentrations in the reactor as well as the concentrations in the feed were used:

$$c_{in} = c_0 = (4.995\ 0.4728\ 0)^T.$$

Taking into account these parameters and a dilution rate of 0.5 [1/h], the simulation result is as shown in Fig. 10.11. It should be noted that the productivity in the steady-state in this case is slightly above 0.5 [g/(L h)].

The simulated productivity as a function of the dilution rate is depicted in Fig. 10.12. There, one may see that the productivity has a global maximum; thus, our model with these parameters is amenable to ESC approaches. The location of the maximum point occurs at the dilution rate of $\overline{u} \approx 2.2[1/h]$, yielding a productivity of $J_{max} \approx 1.93[g/(Lh)]$. In such conditions, the steady-state concentrations of the reactor are $\overline{c}_1 \approx 1.2661[g/L]$, $\overline{c}_2 \approx 0.5962[g/L]$, $\overline{c}_3 \approx 0.8614[g/L]$.

Now, we consider two scenarios in which the control law (13) and the estimated gradient (10.19) can be used.

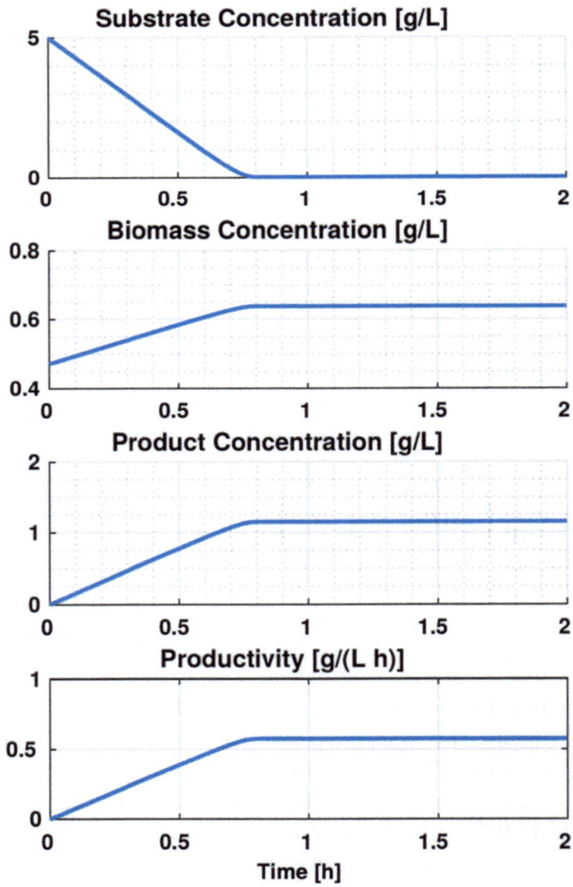

Fig. 10.11 Species concentrations as a function of time obtained by simulation of model (21)

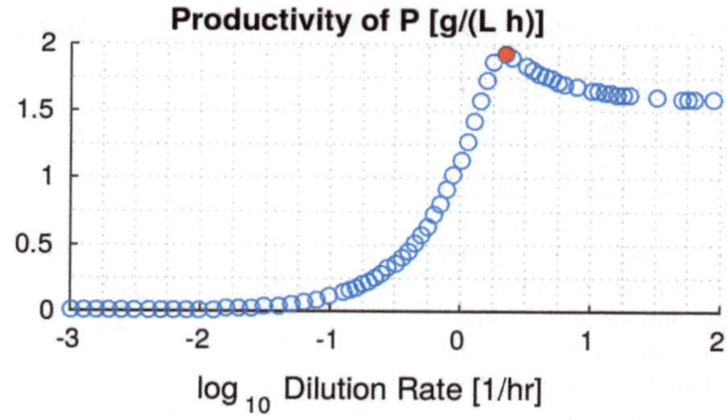

Fig. 10.12 Ethanol productivity as a function of the dilution rate

Scenario 1

The computed dilution rate $D(t)$ is updated in discrete time steps and is kept constant between updates. This approach is inspired by work done by [18, 43, 44] for different models and scenarios. It is worth mentioning here that this is the common framework of traditional ESC schemes. These schemes allow the plant to first reach a steady-state generated by an input value before such a value is updated.

Scenario 2

Here, the dilution rate is continuously updated based on the continuous estimation of the gradient. Note, if the gradient estimate has converged against the actual one, then for signals with bounded second derivative, we can consider both to be the same for any future time; we thus allow an initial period for the gradient estimate to converge before continuously updating the dilution rate. The update of the dilution rate is stopped when its absolute value is below a certain threshold. This stopping criterion prevents the denominator of the gradient estimate in Eq. (10.19) from being in the range of the integration noise and thus avoids an inaccurate gradient estimate.

10.4.2.1 Results for Scenario 1

In this scenario, the dilution rate, $D(t)$, is computed each $4[h]$ according to (10.13) with $\{\lambda, \alpha\} = \{0.05, 0.008\}$. The computed dilution rate is kept constant between update events. The estimation of the gradient, however, is continuously computed via (10.19) and (10.16) with the following parameter values: $\{\eta, \beta, \gamma\} = \{0.02, 0.02, 0.35\}$, $\{p, q\} = \{0.30, 0.30\}$, and $\{k_1, k_2\} = \{1, 1\}$.

Although being continuously computed, the gradient estimation is only required when the control input is updated. Under these conditions, Fig. 10.13 presents the controlled performance of Model (10.21) with the control input (10.13). The continuous line represents the concentrations as a function of time, whereas the discontinuous one represents the location of the steady-state concentrations when productivity maximization occurs. In turn, Fig. 10.14 shows in its upper panel the dilution rate that leads to optimum productivity, as shown in the figure's middle panel. The lower panel

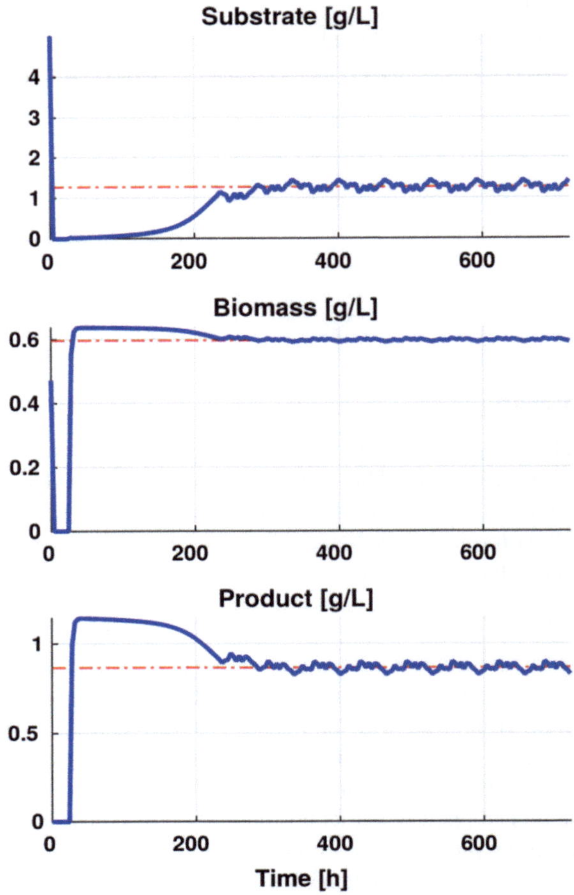

Fig. 10.13 Closed loop dynamics for model (10.21) with control (10.13). The control input is kept constant during a period of $4[h]$ at the end of which is updated. In all panels, the continuous line represents the species concentration, and the discontinuous one is the location of the steady-state when the maximum productivity occurs

shows the gradient estimation, which is used to compute the control input; given that such gradient ripples around 0, we can ensure that the productivity surrounds its optimal value. From Figs. 10.13 and 10.14, one may see that the maximum productivity is attained at around $230[h]$. Now, by comparing the productivity attained with a constant dilution rate of $0.5[1/h]$, the productivity shown in the middle panel of Fig. 10.14 is around 3 or 4 times larger with this ESC scheme; it takes, however, around 10 days to achieve the optimal productivity.

10.4.2.2 Results for Scenario 2

In this scenario, the dilution rate was continuously updated according to (10.13) with the following parameters $\{\lambda, \alpha\} = \{0.8, 0.6\}$. The control algorithm (10.13) is continuously updated until the estimate of the dilution rate $u(t)$ satisfies $|\hat{\omega}_2| <$

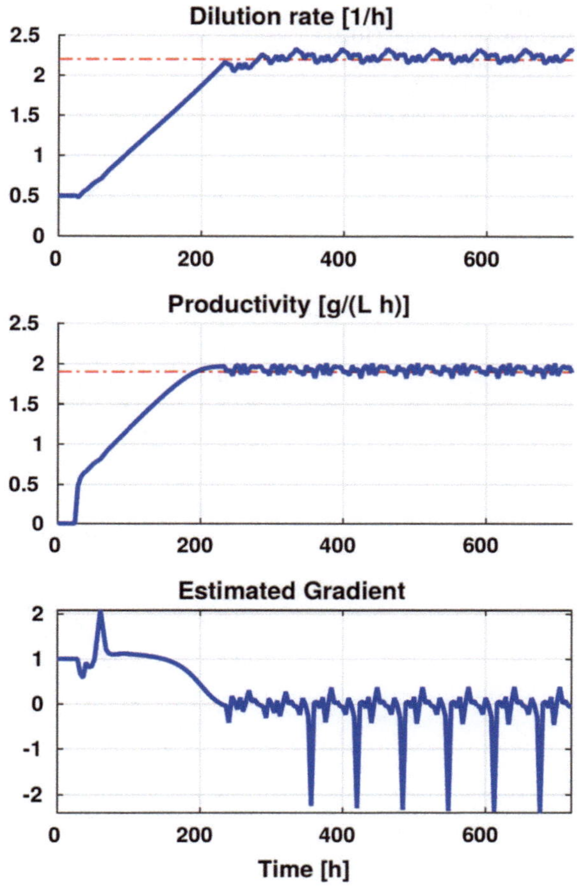

Fig. 10.14 Control input (10.13) and reactor productivity. For the upper and middle panels, the discontinuous line depicts the location of the optimal conditions. The lower panel shows the estimation of the gradient upon which is computed the control input

$5\text{x}10^{-4}$. When such a condition is met, the dilution rate keeps the last computed value. To let the differentiator converge, we allow for an initial period of $0.5[h]$ in which the dilution rate equals $0.5[1/h]$. In turn, the estimated gradient, σ, is computed via (10.19) and (10.16) with the following parameters values:

$$\{\eta, \beta, \gamma\} = \{5, 5, 10\}, \quad \{p, q\} = \{0.50, 0.50\}, \text{ and } \{k_1, k_2\} = \{2, 2\}.$$

Figures 10.15 and 10.16 show the closed loop dynamics and the input–output behavior, respectively.

By comparing Fig. 10.13 from Scenario 1 and Fig. 10.15 from Scenario 2, one may see that both controllers can reach optimum conditions. In Scenario 1, the convergence time is around $230[h]$ vis à vis the $1[h]$ required in Scenario 2. The latter one is around 2 orders of magnitude smaller as compared to the one in Scenario 1; however, such reduction comes at a cost. The online, continuous computation of

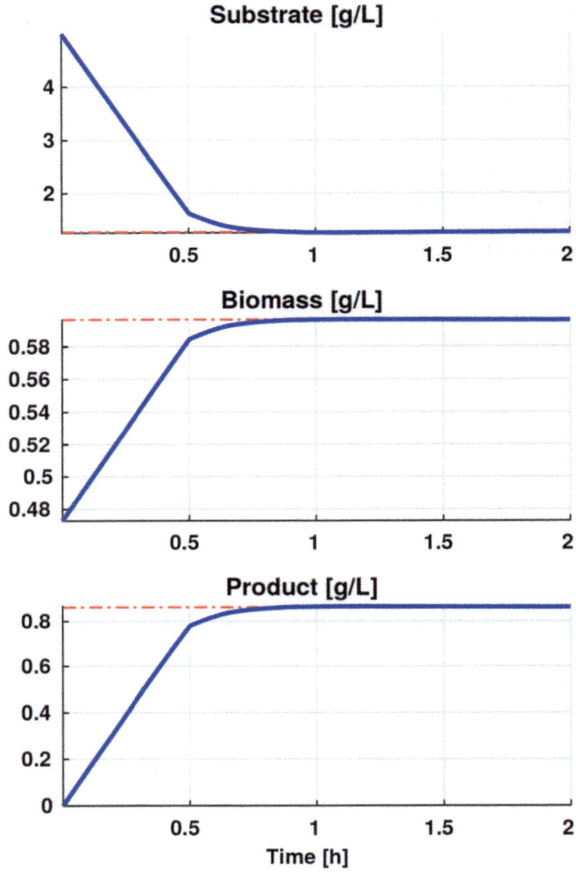

Fig. 10.15 Close loop performance of plant (10.21) with a continuous update of control (10.13). The organization of this figure follows that of Fig. 10.13

the control input requires larger computational power and, more importantly, a valve capable of adapting the dilution rate continuously.

10.5 Conclusions

In this chapter a fermentation model is presented that simulates the inhibition of alcohol production due to large concentrations of the substrate, biomass, and product. A parameter fitting procedure was developed for this model to reproduce data available in the literature [32]. With such parameters values we computed the steady-state and performed a sensitivity analysis in order to determine which parameters variation affect the location of the steady-state. Such analysis showed that the product dynamic is sensitive only to variations on the maximum substrate consumption rate (parameter k_{11}).

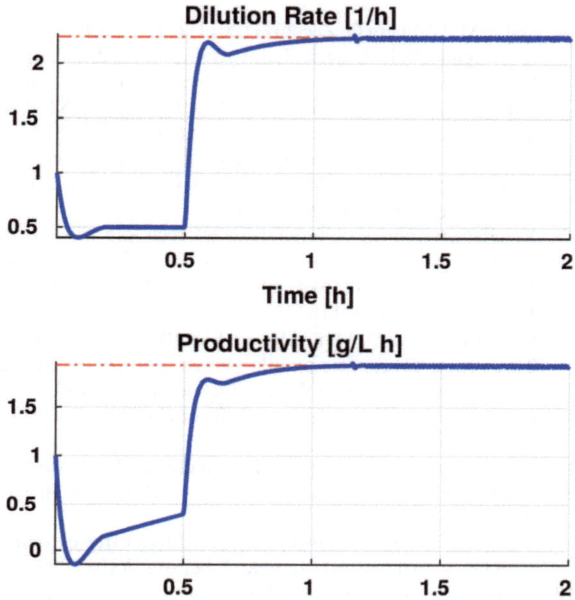

Fig. 10.16 Control input continuously computed with (10.13) and reactors productivity. The organization and legends of this figure follows that of Fig. 10.14

With the system control model presented in this chapter, we applied our new ESC to maximize different performance indices depending on the application considered. Furthermore, to improve the convergence of the ESC, we developed two simulation frameworks: in the first one, a common ESC strategy was applied, in which we wait until the reactor dynamics reaches its steady state to update the dilution rate. In the second strategy, the dilution rate of the reactor is updated continuously. In the latter approach, the convergence time is reduced by two orders of magnitude compared to the conventional ESC scheme, improving its applicability in real industrial applications.

Acknowledgements This chapter was supported by the National System of Researchers (SNII) program of CONAHCYT, Mexico.

References

1. Saini JK, Saini R, Tewari L (2015) Lignocellulosic agriculture wastes as biomass feedstocks for second-generation bioethanol production: concepts and recent developments. 3 Biotech 5(4):337–353
2. Joseph AM, Tulasi Y, Shrivastava D, Kiran B (2023) Techno-economic feasibility and exergy analysis of bioethanol production from waste. Energy Convers Manage X:100358
3. Içöz E, Tuğrul KM, Saral A, Içöz E (2009) Research on ethanol production and use from sugar beet in Turkey. Biomass Bioenergy 33(1):1–7

4. Galbe M, Wallberg O, Zacchi G (2011) 6.47—Techno-economic aspects of ethanol production from lignocellulosic agricultural crops and residues. In: Moo-Young M (ed) Comprehensive biotechnology, 2nd edn. Academic Press, Burlington, pp 615–628
5. Vohra M, Manwar J, Manmode R, Padgilwar S, Patil S (2014) Bioethanol production: feedstock and current technologies. J Environ Chem Eng 2(1):573–584
6. Prado-Rubio OA, Morales-Rodríguez R, Andrade-Santacoloma P, Hernández-Escoto H (2016) Process intensification in biotechnology applications. In: Process intensification in chemical engineering. Springer, pp 183–219
7. Regulatory Council of Tequila (2009) Tech Rep
8. López-López A, Dávila-Vázquez G, León-Becerril E, Villegas-García E, Gallardo-Valdez J (2010) Tequila vinasses: generation and full scale treatment processes. Rev Environ Sci Biotechnol 9(2):109–116
9. García-Depraect O, Muñoz R, van Lier JB, Rene ER, Diaz-Cruces VF, León-Becerril E (2020) Three-stage process for tequila vinasse valorization through sequential lactate, biohydrogen and methane production. Biores Technol 307:123160
10. Cinar A, Parulekar SJ, Undey C, Birol G (2003) Batch fermentation: modeling, monitoring, and control. CRC Press
11. Nosrati-Ghods N, Harrison ST, Isafiade AJ, Leng Tai S (2020) Mathematical modelling of bioethanol fermentation from glucose, xylose or their combination–a review. ChemBioEng Rev 7(3):68–88
12. Dewasme L, Vande Wouwer A (2020) Model-free extremum seeking control of bioprocesses: a review with a worked example. Processes 8(10):1209
13. Ariyur KB, Krstic M (2003) Real-time optimization by extremum-seeking control. Wiley
14. Dochain D, Perrier M, Guay M (2011) Extremum seeking control and its application to process and reaction systems: a survey. Math Comput Simul 82(3):369–380
15. Marcos N, Guay M, Dochain D, Zhang T (2004) Adaptive extremum-seeking control of a continuous stirred tank bioreactor with Haldane's kinetics. J Process Control 14(3):317–328
16. Pan Y, Kumar KD, Liu G (2012) Extremum seeking control with second-order sliding mode. SIAM J Control Optim 50(6):3292–3309
17. Angulo MT (2015) Nonlinear extremum seeking inspired on second order sliding modes. Automatica 57:51–55
18. Torres-Zúñiga I, Lopez-Caamal F, Hernandez-Escoto H, Alcaraz-Gonzalez V (2021) Extremum seeking control and gradient estimation based on the super-twisting algorithm. J Process Control 105:223–235
19. Levant A (1998) Robust exact differentiation via sliding mode technique. Automatica 34(3):379–384
20. Moreno JA (2012) Lyapunov approach for analysis and design of second order sliding mode algorithms. Sliding modes after the first decade of the 21st century: state of the art, pp 113–149
21. López-Caamal F, Moreno JA (2019) Generalised multivariable supertwisting algorithm. Int J Robust Nonlinear Control 29(3):634–660
22. Monod J (1949) The growth of bacterial cultures. Annu Rev Microbiol 3:371–394
23. Raposo S, Pardão JM, Diaz I, Lima-Costa ME (2009) Kinetic modelling of bioethanol production using agro-industrial by-products. Int J Energy Environ 3(8):1–8
24. Buehler EA, Mesbah A (2016) Kinetic study of acetone-butanol-ethanol fermentation in continuous culture. PLoS ONE 11(8):e0158243
25. Rorke DC, Gueguim Kana EB (2017) Kinetics of bioethanol production from waste sorghum leaves using saccharomyces cerevisiae by4743. Fermentation 3(2):19
26. Zentou H, Zainal Abidin Z, Yunus R, Awang Biak DR, Zouanti M, Hassani A (2019) Modelling of molasses fermentation for bioethanol production: a comparative investigation of Monod and Andrews models accuracy assessment. Biomolecules 9(8):308
27. Andrews JF (1968) A mathematical model for the continuous culture of microorganisms utilizing inhibitory substrates. Biotechnol Bioeng 10(6):707–723
28. Oliveira S, Oliveira R, Tacin M, Gattas E (2016) Kinetic modeling and optimization of a batch ethanol fermentation process. J Bioprocess Biotech 6(266):2

29. Chai A, Wong YS, Ong SA, Lutpi NA, Sam ST, Kee WC, Ng HH (2021) Haldane-Andrews substrate inhibition kinetics for pilot scale thermophilic anaerobic degradation of sugarcane vinasse. Biores Technol 336:125319
30. Galaction AI, Lupăşteanu AM, Caşcaval D (2010) Kinetic studies on alcoholic fermentation under substrate inhibition conditions using a bioreactor with stirred bed of immobilized yeast cells. Open Syst Biol J 3(1)
31. Levenspiel O (1980) The monod equation: a revisit and a generalization to product inhibition situations. Biotechnol Bioeng 22(8):1671–1687
32. Andrietta SR (1994) Modelagem, simulação e controle de fermentação alcoólica contínua em escala industrial. Ph.D. thesis, Universidade Estadual de Campinas
33. Sokol W, Howell J (1981) Kinetics of phenol oxidation by washed cells. Biotechnol Bioeng 23(9):2039–2049
34. Blanco M, Peinado AC, Mas J (2006) Monitoring alcoholic fermentation by joint use of soft and hard modelling methods. Anal Chim Acta 556(2):364–373
35. Bastin G, Dochain D (1990) On-line estimation and adaptive control of bioreactors. Elsevier
36. Zhang Q, Wu D, Lin Y, Wang X, Kong H, Tanaka S (2015) Substrate and product inhibition on yeast performance in ethanol fermentation. Energy Fuels 29(2):1019–1027
37. Abunde NF, Asiedu N, Addo A (2017) Dynamics of inhibition patterns during fermentation processes-zea mays and sorghum bicolor case study. Int J Ind Chem 8(1):91–99
38. Taherzadeh MJ, Karimi K (2011) Fermentation inhibitors in ethanol processes and different strategies to reduce their effects. In: Biofuels. Elsevier, pp 287–311
39. Lagarias JC, Reeds JA, Wright MH, Wright PE (1998) Convergence properties of the Nelder-Mead simplex method in low dimensions. SIAM J Optim 9(1):112–147
40. Powell MJ (1978) A fast algorithm for nonlinearly constrained optimization calculations. In: Numerical analysis. Springer, pp 144–157
41. Powell MJ (1978) The convergence of variable metric methods for nonlinearly constrained optimization calculations. In: Nonlinear programming 3. Elsevier, pp 27–63
42. Saltelli A, Tarantola S, Campolongo F, Ratto M (2004) Sensitivity analysis in practice: a guide to assessing scientific models. Wiley
43. López-Caamal F, Hernández-Escoto H, Torres I (2022) A gradient-based extremum seeking control of a substrate-and product-inhibited alcoholic fermentation process. IFAC-PapersOnLine 55(7):638–642
44. Torres I, López-Caamal F, Hernández-Escoto H, Vargas A (2020) Extremum seeking control based on the super-twisting algorithm. IFAC-PapersOnLine 53(2):1621–1626
45. Torres I, Vargas A, López-Caamal F, Hernández-Escoto H (2018) Comparison of two real-time optimization strategies to maximize the hydrogen production in a dark fermenter. IFAC-PapersOnLine 51(13):137–142
46. Shtessel Y, Edwards C, Fridman L, Levant A (2014) Sliding mode control and observation. Springer
47. Rice E, Bridgewater L, Association APH, Association AWW, Federation WE (2012) Standard methods for the examination of water and wastewater. American Public Health Association
48. Jett BD, Hatter KL, Huycke MM, Gilmore MS (1997) Simplified agar plate method for quantifying viable bacteria. Biotechniques 23(4):648–650
49. Miller GL (1959) Use of dinitrosalicylic acid reagent for determination of reducing sugar. Anal Chem 31(3):426–428

Chapter 11
Performance Evaluation of the Non-structured and Structured Kinetic Modelling for the ABE Process. From Batch to Continuous Fermentation

Hugo I. Velázquez-Sánchez, Alma R. Domínguez-Bocanegra, and Ricardo Aguilar-López

Abstract This work proposes the performance evaluation of three novel kinetic models of a biological process, one considering the traditional non-Structured Monod and Ludeking-Piret representations, also the named structured modelling approach which tries to include metabolic kinetic information, as possible and a final one utilising the so-called Non-Structured Phenomenological methodology that aim to become a middle-point between the both above mentioned, for the Acetone–Butanol–Ethanol (ABE) system as a study case. The obtained results under batch and continuous regimes with free cells and batch and continuous regimes with immobilized cells are considered. The above, for the determination of the reproducibility and predictability capabilities of the corresponding kinetic models via a statistical analysis based on the determination coefficient methodology (R^2), the modelling performance index (Π) and the ANOVA analysis for mean comparison and significance of the simulation results generated with the three model structures versus experimental data. These analyses indicate that the traditional non-structured modeling approach is just statistically significant for the ABE fermentation process operating with free cells, still the performance of the phenomenological kinetic model is more reliable for the predicting of the behavior of fermentations utilizing immobilized cells.

Keywords Biofuels · Butanol · Clostridium · Kinetic modeling · Comparison analysis · Selection criteria

H. I. Velázquez-Sánchez · A. R. Domínguez-Bocanegra · R. Aguilar-López (✉)
Biotechnology and Bioengineering Department, Centro de Investigación y de Estudios Avanzados del IPN, Av. Instituto Politécnico Nacional 2508 Col. San Pedro Zacatenco, 07360 Ciudad de México, México
e-mail: raguilar@cinvestav.mx

H. I. Velázquez-Sánchez
e-mail: hvelazquezs@ipn.mx

A. R. Domínguez-Bocanegra
e-mail: adomin@cinvestav.mx

V. Alcaraz Gonzalez et al. (eds.), *Wastewater Exploitation*, Springer Water,
https://doi.org/10.1007/978-3-031-57735-2_11

11.1 Introduction

Mathematical modeling and simulation techniques of biological systems are widely employed as a first approach in process design and operation analysis, through the evaluation of the behavior of the studied system under different process conditions and regimes. However, the constructing of an adequate model that can reproduce the evolution of biological entities under study is not always possible, due to the high complexity of the metabolic pathways and regulation mechanisms exhibited by them [1]. Nevertheless, the rising environmental and energetic issues in the industry demand the implementation of novel biological-based production strategies that can satisfy the increasing demand for quality products while using the least quantity of available resources.

Current developed strategies for the mathematical representation of biological systems can follow either a classical "black box" approach, where the intrinsic metabolic reactions involved in the transformation of substrates to the desired products are neglected, and the input–output relationships are generally based in mass balance approach and kinetic relationships composed of empirical equations; in "white box" approach where novel analysis techniques provide sufficient information of the corresponding metabolism and try to explain the cell dynamic response [2].

Even with the prior considerations regarding the advances in "white box" modeling and their potential, the use of such structures as a base for the engineering and implementation of novel bioprocesses is limited, as commonly to attain a great reproducibility of the intracellular behavior of the cells there is a need to either include a high number of ordinary differential equations or involve the use of complex structures such as artificial neural networks or logic relationships to achieve such goal, which in turn make the implementation of analysis, optimization and control process strategies another open challenge in current literature [3, 4].

On the other hand, traditional "black box" structures have the advantage of usually being low-order systems comprised of either algebraic or differential equations, which in turn lead to a lower number of free parameters and require little analytical effort to be solved and applied.

However, because of their relative simplicity, they are also poorly suited to represent the behaviour of the modelled system outside the constraints used for parametric identification of the equations. There are general problems with the linear dependence of the variables represented and do not explicitly implement enough knowledge about metabolism to adequately describe phenomena such as prolonged lag phases, cell death, the generation of secondary metabolites or the segregation of the cell population, and the switching of metabolic processes [2].

One biological system currently studied for its potential to solve environmental and energetic issues is the Acetone–Butanol–Ethanol (ABE) fermentation system by *Clostridium* bacteria, which are anaerobic gram-positive capable of generating butanol as biofuel [5]. This system has the particularity that the known metabolism of these microorganisms is mainly divided in: (1) acidogenesis where the bacteria uptakes the available carbon source and partially degrade it to organic acids such as butyric, lactic, and acetic ones and (2) solventogenic which reincorporates the produced acids to produce acetone, butanol and ethanol in a process closely related to the cell sporulation mechanism which is heavily affected by the pH activity within both the intracellular contents and the culture media [6]. Such complex behavior is the reason for the wide array of mathematical models for the ABE system reported in literature [7]. However, even this diversity shows that there is currently no consensus on which modelling strategy is best suited for engineering purposes.

Also, considering that the ABE fermentation system under a batch operational regime cannot generate more than 12 kg m^{-3} of butanol. There are several reports in the literature on the implementation of various modified strategies, including the use of in situ recovery strategies, genetically engineered strains, saccharification of agro-industrial residues for use as feedstock. They also include fed-batch fermentation, continuous fermentation in single or multiple stages, and the use of immobilized cells to improve production capacities. Nevertheless, the reported applications follow a heuristic approach and lack the support of a robust system analysis and synthesis procedure [8].

In this context, the so-called unstructured phenomenological kinetic modeling approach has attempted to find a middle ground between the complexity and predictive capabilities of kinetic models. Here, as much information about the metabolic pathways as possible was considered to represent the kinetic rates of the extracellular products. This approach has been successfully applied to several fermentation systems and is a solid perspective for evaluating their performance under more complex operating conditions outside the classical batch culture [9, 10].

The objective of this work is to compare the performance of three different modeling strategies for the ABE fermentation system. A traditional Monod level game model, a Ludeking-Piret model, a phenomenological unstructured model and a structured model were used. The experiments conducted for this purpose were performed with *Clostridium acetobutylicum* ATCC 824 considering the following process conditions:

(i) batch operation and continuous operation with free cells.
(ii) batch operation and continuous operation with immobilized cells using statistical analysis and simulation techniques to determine numerical criteria that can be used to establish the validity and potential of such process systems for bioprocess analysis, optimization, and control.

11.2 Methodology

11.2.1 Study Case

To compare both the descriptive and predictive capabilities of the different modeling strategies applied to the ABE fermentation process, the experimental data reported by [11] were used. The experiments were conducted in a stirred tank reactor with a nominal volume of one liter. The reaction volume contains 600 ml of P2 medium with glucose as the main carbon source. Inoculated cells were *Clostridium acetobutylicum* ATCC 824.

In the case study mentioned above, the reactor was operated in four different operating regimes: Batch and continuous with free cells and Batch and continuous with immobilized cells, respectively. Pulverized brick with particle diameter between 0.15 and 2.4 mm was used as immobilized material. In addition, the performance of the system was tested under different operating conditions, varying, for example, the dilution rate (D) and the initial and feed substrate concentrations (Sg_0 and Sg_{in}, respectively). This was to investigate the process/physical behavior of the reaction/reactor system and to calibrate the proposed kinetic models.

11.2.2 Model Development

The simplified metabolic pathway (Fig. 11.1) of *Clostridium acetobutylicum* reported by Haus et al. [12] was employed to implement the biochemical principles to posteriorly propose the reaction rates of the ABE process. Besides to the glycolysis metabolism shown in Fig. 11.1, another biochemical pathway is also included, it is called solventogenesis. This biochemical pathway occurs in conjunction with glycolysis reactions and is the trigger for sporulation phenomena that origin the uptake of organic acids into the cell, where they are transformed into butanol and acetone.

For the ethanol production, it is reported in the literature reports that its generation is constitutive, regardless of the metabolic state of the crop [13]. It is important to note that the strain *Clostridium acetobutylicum* ATCC 824 has not been genetically modified to inhibit its sporulation phenomena, so this effect was also considered within the model construction.

In characterizing the system using immobilized cells, a pseudo homogeneous approach was considered to describe the effects of diffusion phenomena on the system by applying the so-called efficiency factor (ηi).

The efficiency factor is defined as the quotient of the apparent reaction rate with diffusion limitations and the reaction rate without diffusion limitations [14], taking values into the closed interval [0 1].

To account for the fact that the biomass in the reactor remains immobilized, an additional parameter ψ is introduced to account for the immobilized mass fraction in the mass balance for the biomass concentration.

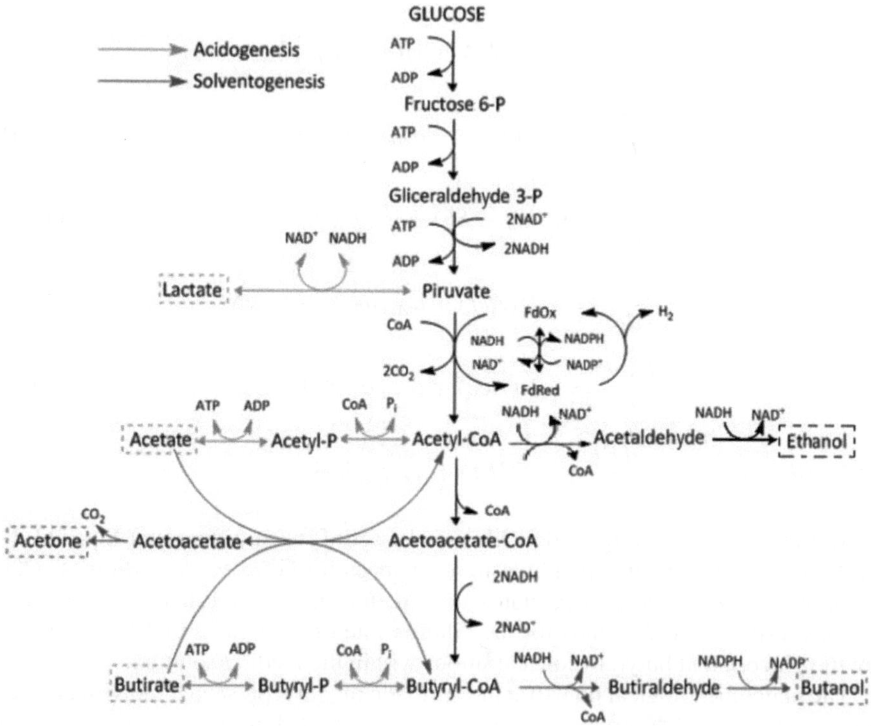

Fig. 11.1 Simplified metabolic pathway of *Clostridium acetobutylicum*

This parameter represents the mass fraction of biomass leaving the reactor due to the input/output mass flow under a continuous regime, $\psi = 1$ represents the case of an ideally mixed continuously operated bioreactor, $\psi = 0$ represents packed-bed flow dynamics [15].

To build the model, the classical Ludeking-Piret relationship was used for product formation and the Monod-Levenspiel growth rate was used to describe bacterial population dynamics within the fermentation. This accounts for the inhibitory effect of both substrate and product accumulation in the reaction vessel (11.1). This allowed us to investigate the effects of model complexity and structure on kinetic core parameters such as μmax and the inhibitory concentration of glucose and butanol.

The following equations then represent the traditional Ludeking-Piret model:

$$\mu_X = \eta_1\left(\mu_{maxX} * \left(\frac{Sg}{kSg + Sg}\right)\left(1 - \left(\frac{But}{kBut}\right)\right)\left(1 - \left(\frac{Sg}{ksi}\right)\right)\right) \tag{11.1}$$

$$\frac{dSg}{dt} = D(Sg_{in} - Sg) - \left(\frac{\mu_X \cdot X}{Y_{\frac{X}{Sg}}}\right) \tag{11.2}$$

$$\frac{dX}{dt} = -\psi DX + (\mu_X - k_{Spo})X \tag{11.3}$$

$$\frac{dBut}{dt} = -D(But) + \alpha_{But}\mu_X X + \beta_{But} X \tag{11.4}$$

$$\frac{dSb}{dt} = D(Sb_{in} - Sb) + \alpha_{Sb}\mu_X X + \beta_{Sb} X \tag{11.5}$$

$$\frac{dAct}{dt} = -D(Act) + \alpha_{Act}\mu_X X + \beta_{Act} X \tag{11.6}$$

$$\frac{dAce}{dt} = -D(Ace) + \alpha_{Ace}\mu_X X + \beta_{Ace} X \tag{11.7}$$

$$\frac{dEt}{dt} = -D(Et) + \alpha_{Et}\mu_X X + \beta_{Et} X \tag{11.8}$$

The kinetic growth rate model (11.1) and a set of seven differential Eqs. (11.2–11.8) describing the corresponding mass balances for Glucose (*Sg*), Biomass (*X*), Butanol (*But*), Butyrate (*Sb*), Acetate (*Act*), Acetone (*Ace*) and Ethanol (*Et*) concentrations respectively. *D* stands for the dilution rate of the reactor and Sg_{in} and Sb_{in} are the glucose and butyrate concentrations within the feeding solution.

The phenomenological non-structured kinetic model represented individual reaction rates for each of the produced molecules following a Michaelis–Menten approach (11.9–11.13) and a classical mass balance approach for each modelled variables (11.14–11.20). It must be noted that as the ABE pathway is segmented into the acidogenesis and solventogenesis stages both butyrate and acetate equations have a generation term and a consumption one to reflect such phenomenon.

$$v_{But} = \eta_2\left(v_{maxBut}\left(\frac{Sb}{kSb + Sb}\right)\right) \tag{11.9}$$

$$v_{Sb} = \eta_3\left(v_{maxSb}\left(\frac{Sg}{kSgSb + Sg}\right)\left(1 - \left(\frac{But}{kBut}\right)\right)\right) - \frac{\eta_2\left(v_{maxBut} * \left(\frac{Sb}{kSb+Sb}\right)\right)}{YButSb} \tag{11.10}$$

$$v_{Ace} = \eta_4\left(v_{maxAce}\left(\frac{Act}{kAA + Act}\right)\right) \tag{11.11}$$

$$v_{Act} = \eta_5\left(v_{maxAct}\left(\frac{Sg}{kSgAct + Sg}\right)\left(\frac{1}{1 + \left(\frac{Ace}{kiAce}\right)}\right)\right) - \frac{\eta_4\left(v_{maxAce} * \left(\frac{Act}{kAA+Act}\right)\right)}{YAceAct} \tag{11.12}$$

$$v_{Et} = \eta_6\left(v_{maxEt}\left(\frac{Sg}{kSgEt + Sg}\right)\right) \tag{11.13}$$

$$\frac{dSg}{dt} = D(Sg_{in} - Sg) - \left(\frac{\mu_X \cdot X}{Y_{\frac{X}{Sg}}}\right) \tag{11.14}$$

$$\frac{dX}{dt} = -\psi DX + (\mu_X - k_{Spo})X \tag{11.15}$$

$$\frac{dBut}{dt} = -D(But) + v_{But}X \tag{11.16}$$

$$\frac{dSb}{dt} = D(Sb_{in} - Sb) + v_{Sb}X \tag{11.17}$$

$$\frac{dAct}{dt} = -D(Act) + v_{Act}X \tag{11.18}$$

$$\frac{dAce}{dt} = -D(Ace) + v_{Ace}X \tag{11.19}$$

$$\frac{dEt}{dt} = -D(Et) + v_{Et}X \tag{11.20}$$

Lastly, for constructing the structured kinetic model, the approach made by Shinto et al. [16] represents the 19 enzymatic steps of the metabolic pathway (Fig. 11.1) with a Michaelis–Menten structure. In this scenario, it was also assumed that both reactions are carried out in all reversible steps by single enzymes and not by a single dual-function enzyme.

The result is a model comprised of 15 ordinary differential equations (ODEs) describing the behavior of both the extracellular (11.21–11.28) and intracellular (11.29–11.35) metabolites produced in the ABE fermentation.

$$\frac{dSg}{dt} = D(Sg_{in} - Sg) - \left(v_{max1}\left(\frac{Sg}{Km1 + Sg}\right)\right)X - \left(\frac{\mu_X}{Yxs}\right)X \tag{11.21}$$

$$\frac{dX}{dt} = -DX + (\mu_X - k_{Spo})X \tag{11.22}$$

$$\frac{dBut}{dt} = (D(-But)) + \left(v_{max19}\left(\frac{BCoA}{Km19 + BCoA}\right)\right)X \tag{11.23}$$

$$\begin{aligned}\frac{dSb}{dt} &= (D(Sb_{in} - Sb)) + \left(v_{max18}\left(\frac{BCoA}{Km18 + BCoA}\right)X\right) \\ &\quad - \left(v_{max17}\left(\frac{Sb}{Km17 + Sb}\right)\left(\frac{1}{1 + \left(\frac{But}{kBut}\right)}\right)\right)X\end{aligned} \tag{11.24}$$

$$\frac{dAct}{dt} = (D(-Act)) + \left(v_{max9}\left(\frac{ACoA}{Km9 + ACoA}\right)\right)X - \left(v_{max7}\left(\frac{Act}{Km7 + Act}\right)\right)X \tag{11.25}$$

$$\frac{dAce}{dt}(D(-Ace)) + \left(v_{max16}\left(\frac{AAto}{Km16 + AAto}\right)\right)X \tag{11.26}$$

$$\frac{dEt}{dt} = (D(-Et)) + \left(v_{max11}\left(\frac{ACoA}{Km11 + ACoA}\right)\right)X \tag{11.27}$$

$$\frac{dLac}{dt} = (D(-Lac)) + \left(v_{max5}\left(\frac{Pir}{Km5 + Pir}\right)\right)X - \left(v_{max4}\left(\frac{Lac}{Km4 + Lac}\right)\right)X \tag{11.28}$$

$$\frac{dF6P}{dt} = (D(-F6P)) + \left(v_{max1}\left(\frac{Sg}{Km1 + Sg}\right)\right)X - \left(v_{max2}\left(\frac{F6P}{Km2 + F6P}\right)\right)X \tag{11.29}$$

$$\frac{G3P}{dt} = (D(-G3P)) + \left(v_{max2}\left(\frac{F6P}{Km2 + F6P}\right)\right)X - \left(v_{max3}\left(\frac{G3P}{Km3 + G3P}\right)\right)X \tag{11.30}$$

$$\begin{aligned} \frac{dPir}{dt} &= (D(-Pir)) + \left(v_{max3}\left(\frac{G3P}{Km3 + G3P}\right)\right)X \\ &+ \left(v_{max4}\left(\frac{Lacc}{Km4 + Lacc}\right)\right)X - \left(v_{max5}\left(\frac{Pir}{Km5 + Pir}\right)\right)X \\ &- \left(v_{max6}\left(\frac{Pir}{Km6 + Pir}\right)\right)X \end{aligned} \tag{11.31}$$

$$\begin{aligned} \frac{dACoA}{dt} &= (D(-ACoA)) + \left(v_{max6}X\left(\frac{Pir}{Km6 + Pir}\right)\right) \\ &+ \left(v_{max7}X\left(\frac{Act}{Km7 + Act}\right)\right) - \left(v_{max9}X\left(\frac{ACoA}{Km9 + ACoA}\right)\right) \\ &- \left(v_{max10}\left(\frac{ACoA}{Km10 + ACoA}\right)\right)X - \left(v_{max11}\left(\frac{ACoA}{Km11 + ACoA}\right)\right)X \end{aligned} \tag{11.32}$$

$$\begin{aligned} \frac{dAACoA}{dt} &= (D(-AACoA)) + \left(v_{max10}\left(\frac{ACoA}{Km10 + ACoA}\right)\right)X \\ &- \left(v_{max8}\left(\frac{AACoA}{Km8 + AACoA}\right)\right)X \\ &- \left(v_{max14}\left(\frac{AACoA}{Km14 + AACoA}\right)\right)X \end{aligned} \tag{11.33}$$

$$\frac{dAAto}{dt} = (D(-AAto)) + \left(v_{max8}\left(\frac{AACoA}{Km8 + AACoA}\right)\right)X - \left(v_{max16}\left(\frac{AAto}{Km16 + AAto}\right)\right)X \tag{11.34}$$

$$\frac{dBCoA}{dt} = (D(-BCoA)) + \left(v_{max14}\left(\frac{AACoA}{Km14 + AACoA}\right)\right)X + \left(v_{max17}\left(\frac{Sb}{Km17 + Sb}\right)\right)X - \left(v_{max19}\left(\frac{BCoA}{Km19 + BCoA}\right)\right)X - \left(v_{max18}\left(\frac{BCoA}{Km18 + BCoA}\right)\right)X \tag{11.35}$$

Regarding the inclusion of the pseudo-homogeneous parameters to the structured model it should be mentioned that as the ODEs represent both the intracellular and extracellular products as taking part in the same "compartment" (i.e.: there is not an explicit discretization of the contents of the cell and the contents of the culture medium) then the assignment of efficiency factors to the intracellular rates seems inadequate, as the hypothesis is that the mass transfer limitation described for such parameters only affect the flux of molecules to and from the cell, not inside it; such consideration and also the fact that every extracellular compound is directly correlated to an intracellular metabolite in the selected equations limits its use to free cell systems only.

11.2.3 Model Validation and Performance Assessment

The Marquardt algorithm in the software Modelmaker® 3.0.3 is employed for the parametric identification of the proposed models. Numerical simulations were performed on a PC with an Intel® Xeon© E5460 processor and 8 GB RAM in MATLAB® 2020b software, the ode15s command of the ODE solver library is employed to solve differential equations.

The theoretical performance of the proposed models was evaluated by applying a standard determination coefficient approach to the three models compared with the experimental data obtained by Yen and Li [11] and Qureshi et al. [17], this, did not only include the global performance, but also the accuracy of the reported variables. Additionally, to further validate the proposed structures, a complementary analysis was performed considering the index reported by [18] and defined in Eq. 11.36. The Π index indicates that the model has a poor performance if its value is negative, If the value is zero, this indicates that the model can at least reproduce the mean value of the variable under study, if the value is greater than 0 but less than 0.5 the performance is considered as acceptable. If the value greater than 0.5 then the performance of the model is deemed as adequate.

$$\Pi = 1 - \frac{\sum_{i=1}^{N}[|Y - Y^*|]}{\sum_{i=1}^{N}[|Y^* - \bar{Y}|]} \in (-\infty,\ 1] \tag{11.36}$$

Finally, to analyze the predictive capabilities of the chosen structures both experimental and simulated butanol concentrations at equilibrium state in batch, stationary and continuous mode were investigated. The ANOVA essay was applied to determine punctual statistical differences at each operating condition. In addition, the goodness of fit of the models on the overall behavior of the reaction system was tested when manipulating the initial reactor substrate concentration (Sg_0) in batch mode or the dilution rate (D) and the added substrate concentration (Sg_{in}) in continuous mode.

11.3 Results

11.3.1 Free Cell Reactor Performance

Tables 11.1, 11.2, and 11.3 summarize the values obtained by parametric identification procedure for the traditional, phenomenological, and structured models, respectively. The validation of the applied models was made by numerical simulation considering the following initial conditions: $X_0 = 0.2$ g L^{-1}, $Sg_0 = 60$ g L^{-1}, $But_0 = 0.01$ g L^{-1}, $Sb_0 = 0.01$ g L^{-1}, $Ace_0 = 0.01$ g L^{-1}, $Act_0 = 0.01$ g L^{-1} and $Et_0 = 0.01$ g L^{-1} (Fig. 11.2). It should be noted that the parameters η_i and ψ are assigned a value of one when the structure is used to model the system with free cells. This is true in both batch and continuous modes. In addition, the performance of the intracellular variables described by the structured model is shown in Fig. 11.3. As it should be noted that all modeled metabolites show stable behavior.

Table 11.1 Parametric identification of the traditional Ludeking-Piret model considering experimental data of an ABE fermentation system operating with free cells reported by Yen and Li [11]

Parameter	Value	Units	Parameter	Value	Units
α_{Ace}	0.1934 ± 0.4552	g g^{-1}	β_{Sb}	−0.04 ± 0.0499	g h^{-1}
α_{Act}	0.4737 ± 0.3788	g g^{-1}	k_{But}	11.5657 ± 2.5350	g L^{-1}
α_{But}	0.8712 ± 0.7405	g g^{-1}	k_{Spo}	0.0834 ± 0.0646	g L^{-1}
α_{Et}	0.0185 ± 0.36	g g^{-1}	k_{Sg}	19.5091 ± 34.5349	g L^{-1}
α_{Sb}	0.4951 ± 0.3798	g L^{-1}	k_{si}	262.5 ± 63.197	g L^{-1}
β_{Ace}	0.1 ± 0.0620	g h^{-1}	μ_{maxX}	0.4759 ± 0.4133	h^{-1}
β_{Act}	-0.0354 ± 0.0493	g h^{-1}	Y_{XSg}	0.085 ± 0.0377	g g^{-1}
β_{But}	0.1364 ± 0.0725	g h^{-1}	ψ	1	–
β_{Et}	0.0386 ± 0.0497	g h^{-1}	η_1	1	–

Table 11.2 Parametric identification of the phenomenological model considering experimental data of an ABE fermentation system operating with free cells reported by Yen and Li [11]

Parameter	Value	Units	Parameter	Value	Units
k_{AA}	3.7924 ± 1.1382	$g\ L^{-1}$	v_{maxEt}	0.0368 ± 0.044	h^{-1}
k_{But}	9.8772 ± 1.1608	$g\ L^{-1}$	v_{maxSb}	0.7159 ± 0.0758	h^{-1}
k_{Spo}	0.1 ± 0.0590	h^{-1}	μ_{maxX}	0.6003 ± 0.0738	h^{-1}
k_{iAce}	16.4454 ± 6.4519	$g\ L^{-1}$	Y_{AceAct}	0.4182 ± 0.1699	$g\ g^{-1}$
k_{Sb}	3.5 ± 3.0195	$g\ L^{-1}$	Y_{ButSb}	0.4852 ± 0.0645	$g\ g^{-1}$
k_{Sg}	25 ± 4.9228	$g\ L^{-1}$	Y_{XSg}	0.1088 ± 0.0403	$g\ g^{-1}$
k_{SgAct}	0.5 ± 0.0684	$g\ L^{-1}$	ψ	1	–
k_{SgEt}	0.0726 ± 0.0018	$g\ L^{-1}$	η_1	1	–
k_{SgSb}	2.51 ± 0.9926	$g\ L^{-1}$	η_2	1	–
k_{si}	281.9999 ± 6.1425	$g\ L^{-1}$	η_3	1	–
v_{maxAce}	0.6361 ± 0.1581	h^{-1}	η_4	1	–
v_{maxAct}	0.3349 ± 0.1079	h^{-1}	η_5	1	–
v_{maxBut}	0.9949 ± 0.0684	h^{-1}	η_6	1	–

Table 11.3 Parametric identification of the Structured model considering experimental data of an ABE fermentation system operating with free cells reported by Yen and Li [11]

Parameter	Value	Units	Parameter	Value	Units
k_{But}	8.1600 ± 13.3063	$g\ L^{-1}$	k_{si}	251.7550 ± 62.3938	$g\ L^{-1}$
k_{Spo}	0.09256 ± 0.1402	h^{-1}	μ_{maxX}	0.5078 ± 0.2496	h^{-1}
k_{m1}	9.9610 ± 1.0583	$g\ L^{-1}$	v_{max1}	0.9929 ± 0.7247	h^{-1}
k_{m2}	0.0103 ± 0.0905	$g\ L^{-1}$	v_{max2}	0.9471 ± 0.8214	h^{-1}
k_{m3}	0.0151 ± 0.0702	$g\ L^{-1}$	v_{max3}	0.9732 ± 0.3080	h^{-1}
k_{m4}	0.0279 ± 0.0392	$g\ L^{-1}$	v_{max4}	0.0208 ± 0.0214	h^{-1}
k_{m5}	0.9889 ± 0.1094	$g\ L^{-1}$	v_{max5}	0.0101 ± 0.0096	h^{-1}
k_{m6}	0.0122 ± 0.0698	$g\ L^{-1}$	v_{max6}	0.9431 ± 0.4049	h^{-1}
k_{m7}	0.5324 ± 0.5232	$g\ L^{-1}$	v_{max7}	0.9692 ± 0.3387	h^{-1}
k_{m8}	0.0030 ± 0.0035	$g\ L^{-1}$	v_{max8}	0.2136 ± 0.1419	h^{-1}
k_{m9}	0.0102 ± 0.0333	$g\ L^{-1}$	v_{max9}	0.7668 ± 0.2779	h^{-1}
k_{m10}	0.0118 ± 0.0387	$g\ L^{-1}$	v_{max10}	0.8609 ± 0.3522	h^{-1}
k_{m11}	0.0119 ± 0.0388	$g\ L^{-1}$	v_{max11}	0.0769 ± 0.0313	h^{-1}
k_{m14}	0.0540 ± 0.0624	$g\ L^{-1}$	v_{max14}	0.8854 ± 0.9184	h^{-1}
k_{m16}	0.0109 ± 0.0125	$g\ L^{-1}$	v_{max16}	0.1651 ± 0.2201	h^{-1}
k_{m17}	0.1340 ± 0.1442	$g\ L^{-1}$	v_{max17}	0.0141 ± 0.0683	h^{-1}
k_{m18}	0.0102 ± 0.0232	$g\ L^{-1}$	v_{max18}	0.0675 ± 0.0170	h^{-1}
k_{m19}	0.0245 ± 0.0131	$g\ L^{-1}$	v_{max19}	0.9869 ± 0.5956	h^{-1}
k_{Sg}	24.7464 ± 3.0122	$g\ L^{-1}$	Y_{XSg}	0.1000 ± 0.2344	$g\ g^{-1}$

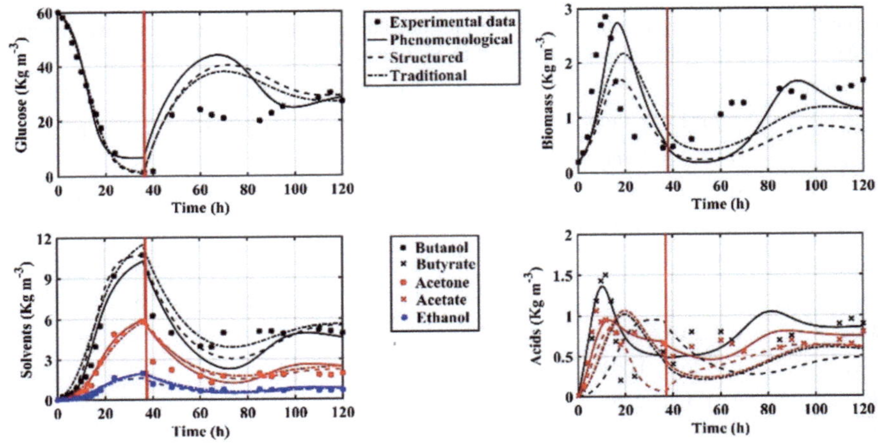

Fig. 11.2 Validation of the selected kinetic models versus *Clostridium acetobutylicum* ATCC 824 experimental data of a reactor operating with free cells reported by Yen and Li [11]. Horizontal line indicates the switch between batch and continuous regime (36 h)

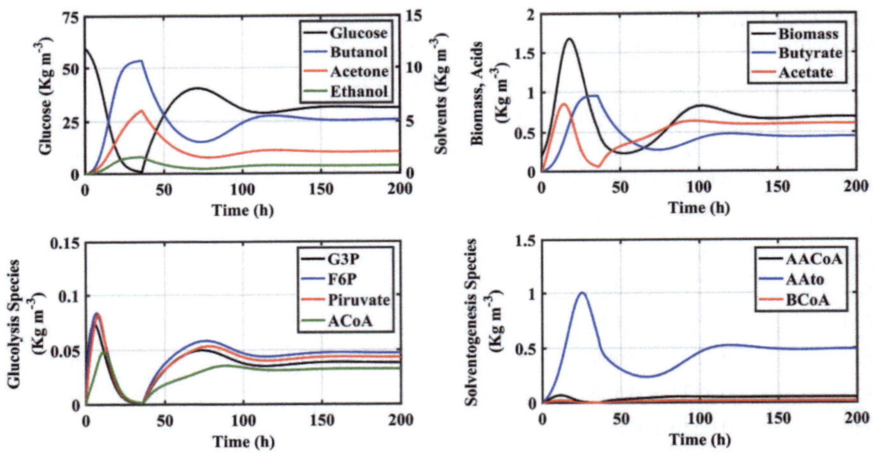

Fig. 11.3 Simulation results of the intracellular variables modelled using the structured kinetic model constructed in this work for *Clostridium acetobutylicum* ATCC 824. Horizontal line indicates the switch between batch and continuous regime (36 h)

The uncertainty present in some parameters can be attributed to the fact that underlying experimental data can only be measured with finite precision and only a subset of the state variables can be accessed experimentally "in-line", since samples must be removed from culture and processed offline. The inherent nonlinearity of biological systems also contributes [19].

For the obtained parameters note that the values obtained for μ_{maxX}, k_{But}, k_{si}, k_{Sg} and Y_{XSg} are all statistically equal based on the confidence intervals determined by the Marquardt algorithm. This is an expected result, because the growth rate kinetics are represented by the same structure in all the tested models. All the previously mentioned parameter values are within regions reported in the literature for Clostridium bacteria [20, 21].

To evaluate the three descriptive models, a goodness-of-fit calculation is performed using the standard coefficient of determination method and the Π-modelling performance index, the results of which are shown in Table 11.4. Notice that a negative value for R^2 indicates that the performance of the model fit is poor compared to the experimental data than when using a single horizontal line with a value equal to the mean of the data set, etc.

Such a condition would indicate that the selected model is inadequate to represent the system's behavior. Also, in principle, the three mathematical models achieve a global determination index $R^2 > 0.95$ and p-value < 0.001, which indicates that no significant difference exists between the predicted behaviour described by the proposed model and the experimental data [22].

An interesting outcome about the results of this analysis is the wide gaps between the modeling performance, particularly related to the approximation of both, biomass, and organic acids dynamics. In the case of biomass, the explanation is that the selected kinetic growth rate does not accurately reflect the exponential growth phase of the culture, as all three models indicate that the culture does not begin metabolic activity (growth) until 6–8 h later, even though the inoculum was added under optimal conditions to keep the lag phase as short as possible.

On other hand, the poor performance of both, the traditional and structured models regarding organic acid behavior the first has the disadvantage of representing such variables as a lineal dependency of cellular growth and quantity in the form of the α and β parameters from Ludeking-Piret, and while the values obtained make sense from the original model conception for metabolite generation. The above, because

Table 11.4 Statistical performance indexes of the evaluated kinetic models considering experimental data obtained under batch regime with free cells reported by Yen and Li [11]

Variable	Traditional		Phenomenological		Structured	
	R^2	Π	R^2	Π	R^2	Π
Glucose	0.9773	0.8723	0.9721	0.8568	0.9869	0.9030
Biomass	−0.0209	−0.1481	0.1277	0.1943	0.0793	−0.2110
Butanol	0.9632	0.8108	0.9729	0.8413	0.9163	0.7693
Butyrate	−0.2788	−0.2620	0.8117	0.5319	−1.1006	−0.2869
Acetate	−0.2886	0.1913	0.7735	0.5626	0.0892	0.2565
Acetone	0.9693	0.8304	0.9840	0.8859	0.9733	0.8461
Ethanol	0.9619	0.8344	0.9511	0.8179	0.8902	0.6329
Global	0.9732	0.8532	0.9630	0.8619	0.9807	0.8397

the positive α value and negative β one can be interpreted as a basic representation of the acidogenic and solventogenic phase of the culture respectively, is precisely the fact that β takes a negative value what makes the modeled curve to represent a lower acid concentration than the experimental one, even decaying into negative values at long simulation times (>300 h).

The structured model suffers from a low determination index for both acetate and butyrate reproduction since the metabolites explicitly mentioned in their mass balance equations are intracellular species that (1) have a high interconversion rate that makes them behave as pseudo-stationary variables and (2) the experimental determination of the concentration such molecules in real-time is not considered into the values reported for the parametric identification for being intracellular species.

With this exercise, it should be clear that guiding the selection of a kinetic model only considering a global determination index is ill-advised, especially if the goal of the model construction is using it as a base for the application of analysis, design, and intensification of bioprocesses. Such an issue led to the following analysis, where the performance of the three evaluated models was now determined by calculating their goodness of fit outside the operational region in which they were adjusted, with the hypothesis that if both the reactor type, operational regime, used strain and utilized culture medium does not change then the internal metabolism of the cell should not be affected. This hypothesis is vital in the determination of the predictive capabilities of any given model, as traditionally there is a wide array of works reported in literature where the authors make use of an iterative methodology for parametric identification and change all the parameter values to make their structures fit data under different process conditions, even if all the prior operational variables remain constant [23, 24].

Therefore for such exercise there were considered experimental data under batch regime at different initial glucose concentrations reported by Qureshi et al. [17], where the objective, from the goal mentioned in the prior paragraph, was to corroborate that the analyzed structures were capable of reproduce the growth inhibition effect reported in both the works of Qureshi et al. [17] and Lee et al. [25], as this step is crucial to determine if the models can be used at least to design and optimize conditions for batch culture with free cells. The analysed results are portrayed in Fig. 11.4, where the determination coefficient after the model's name represents their goodness of fit for the overall performance versus the change of operational conditions and the boxes over the bars help to compare the statistical significance between the experimental values (in black bars) and the predicted values of the three models, where overlapping box areas indicate statistical equivalence between them.

Considering the data obtained from Fig. 11.4 it is evident that the structured model is not adequate to represent the actual behavior of the butanol production over the experimental conditions tested, as even if it can reproduce the substrate inhibition effect over the fermentation the values it generates have a statistically significant difference for all the three initial glucose conditions. Such results can be taken as contradictory due to the hypothesis that a more complex model constructed under a more comprehensive knowledge of the metabolic pathway should lead to a better representation of the system. However, as mentioned previously the poor

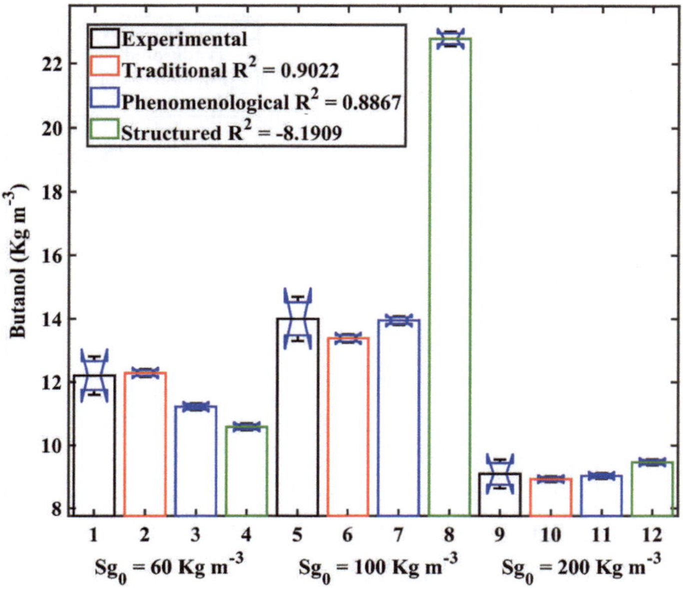

Fig. 11.4 Statistical comparison between experimental data from Qureshi et al. [17] and simulation values obtained by the selected kinetic models for *Clostridium acetobutylicum* ATCC 824 under batch operation with free cells at different initial glucose concentrations

performance can be attributed to the limited experimental information available for the parametric identification of the structured model.

A last numerical experiment to identify the performance limitations of the selected structures considering ABE fermentations operating with free cells was realized, the same analysis was made for the batch culture data but now considering a change into a continuous regime, whose results are summarized in Fig. 11.5. The experimental data available for the comparison comprised only a single operational set of variables the R^2 value cannot be determined. However, all three models achieve equivalent statistically significant values than the experimental ones. This shouldn't be taken as sufficient proof to guarantee that all the structures have an adequate performance for modelling fermentations operating in continuous regime, however this result provides a precedent for the incoming analysis for the modelling performance comparison against the system operating with immobilized cells.

11.3.2 Immobilized Cell Reactor Performance

For this study, the case was reconsidered with the hypothesis that even if the cells suffered from mass transfer limitation due the use of the immobilization matrix their intrinsic metabolism would not change and that the only parameters to be identified

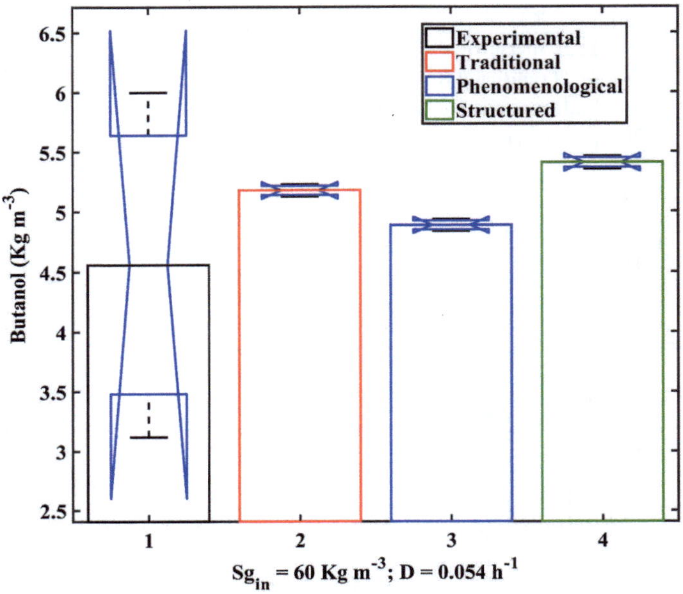

Fig. 11.5 Statistical comparison between experimental data from Yen and Li [11] and simulation values obtained by the selected kinetic models for *Clostridium acetobutylicum* ATCC 824 under continuous operation with free cells

under such scenario are the non-ideal flux and the efficiency factors denoted by ψ and η_i respectively. As an additional consideration, due reports in literature that attribute the increase of alcohol generation to a protection effect against product inhibition in both ABE and ethanol-producing systems, parameters k_{Spo} and k_{But} were also included in the set [11].

In this regard, Tables 11.5 and 11.6 indicate the results of the parametric identification of the immobilized-bound set of parameters mentioned in the prior paragraph for the traditional and phenomenological models respectively. The validation of the chosen models under this operational regime was made by numerical simulation considering the initial conditions reported by Yen and Li [11]: $X_0 = 0.5$ g L^{-1}, $Sg_0 = 60$ g L^{-1}, $But_0 = 0.01$ g L^{-1}, $Sb_0 = 0.01$ g L^{-1}, $Ace_0 = 0.01$ g L^{-1}, $Act_0 = 0.01$ g L^{-1} and $Et_0 = 0.01$ g L^{-1} (Fig. 11.6), where both structures achieve a global determination index $R^2 > 0.95$ and a p-value < 0.001.

Following the prior validation there was calculated of the individual determination coefficients for the seven modelled variables and for both evaluated models, whose results maintain the same behaviour observed during the calculation of goodness of fit for the free cell case: the traditional model has difficulties to reproducing accurately the dynamics of the production and consumption of butyrate and acetate due a negative β value present in those equations, which also cause a decrease in the predicted concentration of those acids below zero when the simulation time exceeds 300 h (Table 11.7).

Table 11.5 Parametric identification of the traditional model considering experimental data of an ABE fermentation system operating with immobilized cells reported by Yen and Li [11]

Parameter	Value	Units
k_{But}	21.1500 ± 3.2688	$g\,L^{-1}$
k_{Spo}	0.0718 ± 0.0086	h^{-1}
ψ	0.125	–
η_1	0.6268 ± 0.0211	–

Table 11.6 Parametric identification of the Phenomenological model considering experimental data of an ABE fermentation system operating with immobilized cells reported by Yen and Li [11]

Parameter	Value	Units
k_{But}	17.1115 ± 3.5613	$g\,L^{-1}$
k_{Spo}	0.0671 ± 0.0176	h^{-1}
ψ	0.35	–
η_1	0.6196 ± 0.0333	–
η_2	0.5551 ± 0.1206	–
η_3	0.7843 ± 0.0843	–
η_4	0.6140 ± 0.1692	–
η_5	0.995 ± 0.1658	–
η_6	0.7110 ± 0.2926	–

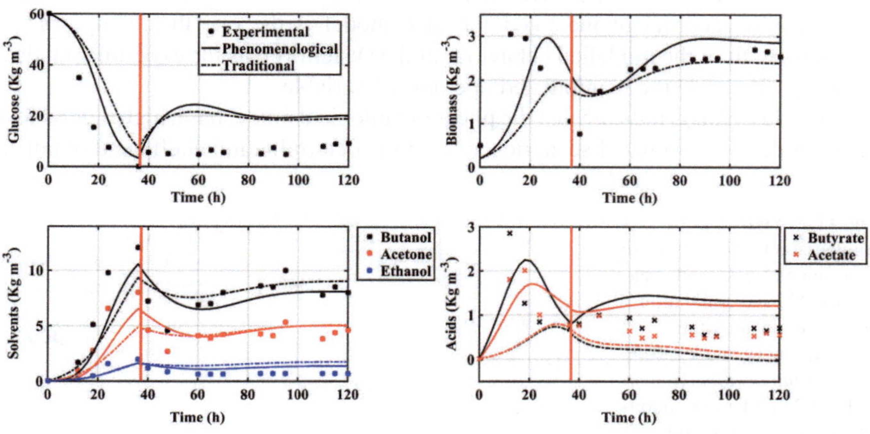

Fig. 11.6 Validation of the selected kinetic models versus *Clostridium acetobutylicum* ATCC 824 experimental data of a reactor operating with immobilized cells reported by Yen and Li [11]. Horizontal line indicates the switch between batch and continuous regime (36 h)

Table 11.7 Statistical performance indexes of the evaluated kinetic models considering experimental data obtained under batch regime with immobilized cells reported by Yen and Li [11]

Variable	Traditional		Phenomenological	
	R^2	Π	R^2	Π
Glucose	0.9984	0.9692	0.9953	0.9478
Biomass	0.8892	0.8070	0.9074	0.8343
Butanol	0.9580	0.8862	0.9928	0.9498
Butyrate	0.2511	0.1957	0.8274	0.7331
Acetate	0.4428	0.2761	0.8883	0.7859
Acetone	0.9495	0.8299	0.9505	0.8662
Ethanol	0.8580	1	0.9710	1
Global	0.9874	0.9035	0.9952	0.9418

Following the prior assumption that the selection of an adequate structure for use in the analysis and design of bioprocesses should not be made considering only the determination coefficients of the structures under the same conditions where they were adjusted made a statistical analysis comparing the results of numerical simulation obtained with both models versus experimental data against a continuously operated reactor with variations in a) Dilution rate (D) (Figs. 11.7 and 11.8b) Feeding glucose concentration in the inlet current (Sg_{in}) (Figs. 11.9 and 11.10). Again, the determination coefficient indicated for each model is the one that compares the goodness of fit of the modelled behaviour of the system versus the experimental data tendency alongside the manipulated operational variable.

These last results indicate that the phenomenological model has a higher goodness of fit for the two evaluated scenarios, even if the determination coefficient obtained

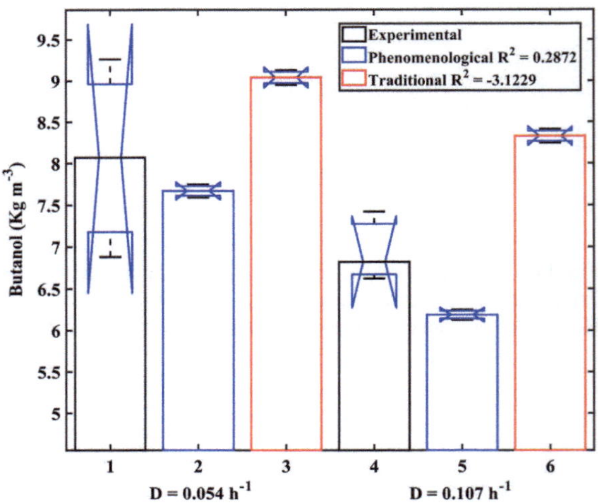

Fig. 11.7 Statistical comparison between butanol experimental data from Yen and Li [11] and simulation values obtained by the selected kinetic models for *Clostridium acetobutylicum* ATCC 824 under continuous operation with immobilized cells and different dilution rates (D). Overlapping box areas indicate statistical equivalence of the means of each set

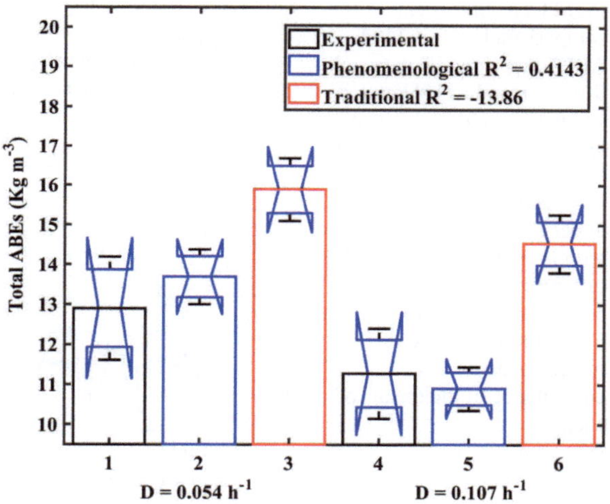

Fig. 11.8 Statistical comparison between total ABEs experimental data from Yen and Li [11] and simulation values obtained by the selected kinetic models for *Clostridium acetobutylicum* ATCC 824 under continuous operation with immobilized cells and different dilution rates (D). Overlapping box areas indicate statistical equivalence of the means of each set

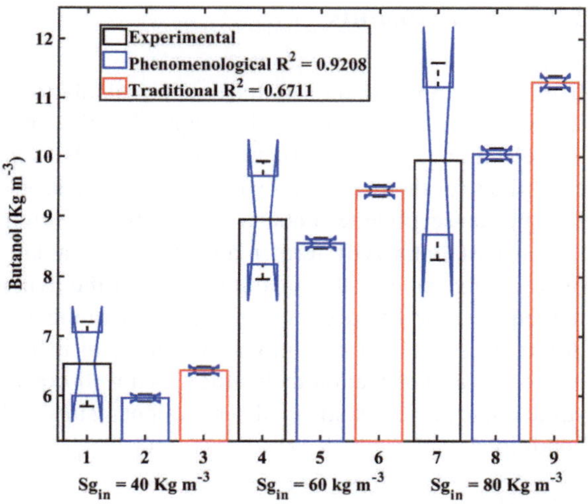

Fig. 11.9 Statistical comparison between butanol experimental data from Yen and Li [11] and simulation values obtained by the selected kinetic models for *Clostridium acetobutylicum* ATCC 824 under continuous operation with immobilized cells and different glucose feeding concentrations (Sg_{in}). Overlapping box areas indicate statistical equivalence of the means of each set

in Fig. 11.6 is lower than 0.95, as the negative value of the determination coefficient for the traditional model coupled with the fact that the ANOVA essay applied for the comparison between the individual treatments indicated that the simulation value obtained with the phenomenological model is statistically equal to the experimental one under operational conditions of $D = 0.054\ h^{-1}$ and $Sg_{in} = 60\ kg\ m^{-3}$ suggests that such modelling structure has a wider application potential regarding its use for predicting the ABE system performance under both batch and continuous regimes for fermentations operating with immobilized cells, while also maintaining a relatively simple structure and making use of little additional parameters to characterize such conditions.

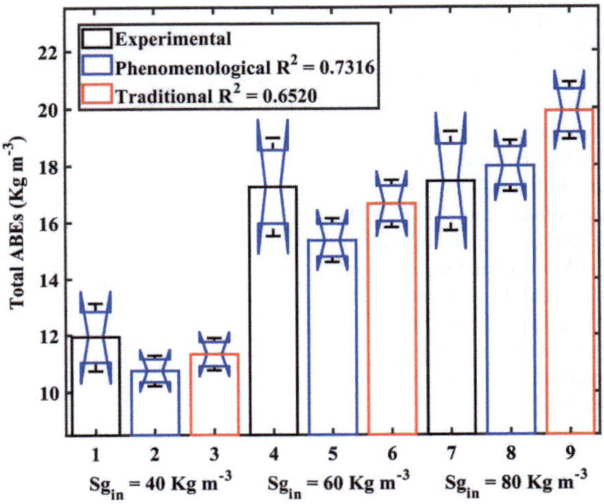

Fig. 11.10 Statistical comparison between total ABEs experimental data from Yen and Li [11] and simulation values obtained by the selected kinetic models for *Clostridium acetobutylicum* ATCC 824 under continuous operation with immobilized cells and different glucose feeding concentrations (Sg_{in}). Overlapping box areas indicate statistical equivalence of the means of each set

11.4 Conclusions

In this work, it was demonstrated that the calculation of the determination coefficient for the selection of kinetic models applied to diverse bioprocesses is not adequate for the selection of reliable structures for the analysis and optimization of operational conditions for them, especially if such determination coefficients are only relying on a single dataset obtained under a very confined and regime. Furthermore it was shown that, at least for the ABE fermentation process evaluated in this work, the development of structured models that take into account the dynamics of intracellular species is not required for the adequate representation of the overall culture performance under different operational conditions, which make them ill-suited for their application in process intensification techniques, as a well-constructed non-structured model is sufficient to achieve statistically equivalent results under a wider range of operational conditions.

Acknowledgements H.I. Velázquez-Sánchez is very grateful to Consejo Nacional de Humanidades, Ciencias y Tecnología for the financial support via a postgraduate scholarship and to CINVESTAV-IPN for supplying the research facilities to develop this work.

Nomenclature

D	Reactor's dilution rate, h^{-1}
k_{AA}	Acetate-acetone affinity constant, $g\ L^{-1}$
k_{But}	Butanol growth inhibition constant, $g\ L^{-1}$
k_{Spo}	Specific cell sporulation rate, h^{-1}
k_{Sg}	Glucose affinity constant, $g\ L^{-1}$

k_{si} Glucose growth inhibition constant, g L^{-1}
k_{iAce} Acetone inhibition constant, g L^{-1}
k_{Sb} Butanol-butyrate affinity constant, g L^{-1}
k_{SgAct} Glucose-acetate affinity constant, g L^{-1}
k_{SgEt} Glucose-ethanol affinity constant, g L^{-1}
k_{SgSb} Glucose-butyrate affinity constant, g L^{-1}
k_{mi} Affinity constant, g L^{-1}
Y_{AceAct} Acetone per acetate mass yield, g g^{-1}
Y_{ButSb} Butanol per butyrate mass yield, g g^{-1}
Y_{XSg} Biomass per glucose mass yield, g g^{-1}

Greek symbols

α_{Ace} Ludeking-Piret alfa for acetone production, g g^{-1}
α_{Act} Ludeking-Piret alfa for acetate production, g g^{-1}
α_{But} Ludeking-Piret alfa for butanol production, g g^{-1}
α_{Et} Ludeking-Piret alfa for ethanol production, g g^{-1}
α_{Sb} Ludeking-Piret alfa for butyrate production, g g^{-1}
β_{Ace} Ludeking-Piret beta for acetone production, g g^{-1} h^{-1}
β_{Act} Ludeking-Piret beta for acetate production, g g^{-1} h^{-1}
β_{But} Ludeking-Piret beta for butanol production, g g^{-1} h^{-1}
β_{Et} Ludeking-Piret beta for ethanol production, g g^{-1} h^{-1}
β_{Sb} Ludeking-Piret beta for butyrate production, g g^{-1} h^{-1}
μ_{maxX} Maximum specific cell growth rate, h^{-1}
Ψ Non-ideal flux constant, dimensionless
η_1 Efficiency factor, dimensionless
v_{maxAce} Maximum specific acetone production rate, h^{-1}
v_{maxAct} Maximum specific acetate production rate, h^{-1}
v_{maxBut} Maximum specific butanol production rate, h^{-1}
v_{maxEt} Maximum specific ethanol production rate, h^{-1}
v_{maxSb} Maximum specific butyrate production rate, h^{-1}
v_{maxi} Maximum specific *i-th* production rate, h^{-1}

References

1. Bastin G, Dochain D (1990) On-line estimation and adaptative control of bioreactors. Elsevier Sc. Publishers, Amsterdam
2. Almquist J, Cvijovic M, Hatzimanikatis V, Nielsen J, Jirstrand M (2014) Kinetic models in industrial biotechnology—improving cell factory performance. Metab Eng 24:38–60
3. Sánchez BJ, Nielsen J (2015) Genome scale models of yeast: towards standardized evaluation and consistent omic integration. Integr Biol 7:846–858

4. Sewsynker-Sukai Y, Faloye F, Kana EBG (2017) Artificial neural networks: an efficient tool for modelling and optimization of biofuel production (a mini review). Biotechnol Biotechnol Equip 31(2):221–235
5. Green EM (2011) Fermentative production of butanol–the industrial perspective. Curr Opin Biotechnol 22(3):337–343
6. Lütke-Eversloh T, Bahl H (2011) Metabolic engineering of *Clostridium acetobutylicum*: recent advances to improve butanol production. Curr Opin Biotechnol 22:634–647
7. Millat T, Winzer K (2017) Mathematical modelling of clostridial acetone-butanol-ethanol fermentation. Appl Microbiol Biotechnol 101:2251–2271
8. Khöler KAK, Rühl J, Blank LM, Schmid A (2015) Integration of biocatalyst and process engineering for sustainable and efficient n-butanol production. Eng Life Sci 15:4–19
9. Letisse F, Lindley ND, Roux G (2003) Development of a phenomenological modeling approach for prediction of growth and xanthan gum production using *Xanthomonas campestris*. Biotechnol Prog 19(3):822–827
10. Velázquez-Sánchez HI, Montes-Horcasitas MC, Aguilar-López R (2014) Development of a phenomenological kinetic model for butanol productiong using *Clostridium beijerinckii*. Revista Mexicana de Ingeniería Química 13(1):103–112
11. Yen HW, Li RJ (2011) The effects of dilution rate and glucose concentration on continuous acetone-butanol-ethanol fermentation by *Clostridium acetobutylicum* immobilized on bricks. J Chem Technol Biotechnol 86:1399–1404
12. Haus S, Jabbari S, Millat T, Janssen H, Fischer RJ, Bahl H, King JR, Wolkenhauer O (2011) A systems biology approach to investigate the effect of pH-induced gene regulation on solvent production by *Clostridium acetobutylicum* in continuous culture. BMC Syst Biol 5:5–10
13. Jones D, Wood D (1986) Acetone-butanol fermentation revisited. Microbiol Rev 50:484–524
14. Rout KR, Jakobsen HA (2012) Reactor performance optimization by the use of a novel combined pellet reflecting both catalyst and adsorbent properties. Fuel Process Technol 99:13–34
15. Méndez-Acosta HO, Campos-Delgado DU, Femat R, González-Alvarez V (2005) A robust feedforward/feedback control for an anaerobic digester. Comput Chem Eng 29:1613–1623
16. Shinto H, Tashiro Y, Yamashita M, Kobayashi G, Sekiguchi T, Hanai T, Kuriya Y, Okamoto M, Sonomoto K (2007) Kinetic modeling and sensitivity analysis of acetone-butanol-ethanol production. J Biotechnol 131:45–56
17. Qureshi N, Saha BC, Cotta MA (2007) Butanol production from wheat straw hydrolysate using *Clostridium beijerinckii*. Bioprocess Biosyst Eng 30:419–427
18. Köhne JM, Köhne S, Šimůnek J (2006) Multi-process herbicide transport in structured soil columns: experiments and model analysis. J Contam Hydrol 85:1–32
19. Vanlier J, Tiemann CA, Hilbers PAJ, van Riel NAW (2013) Parameter uncertainty in biochemical models described by ordinary differential equations. Math Biosci 246:305–314
20. Procentese A, Raganati F, Olivieri G, Russo ME, Salatino P, Marzocchella A (2014) Continuous xylose fermentation by *Clostridium acetobutylicum*—kinetics and energetics issues under acidogenesis conditions. Biores Technol 164:155–161
21. Raganati F, Procentese A, Olivieri G, Götz P, Salatino P, Marzocchella A (2015) Kinetic study of butanol production from various sugars by *Clostridium acetobutylicum* using a dynamic model. Biochem Eng J 99:156–166
22. Sellke T, Bayarri MJ, Berger JO (2001) Calibration of p-values for testing precise null hypotheses. Am Stat 55(1):62–71
23. López I, Borzacconi L (2010) Modelling of slaughterhouse solid waste anaerobic digestion: determination of parameters and continuous reactor simulation. Waste Manage 30:1813–1821
24. López I, Passeggi M, Borzacconi L (2015) Variable kinetic approach to modelling an industrial waste anaerobic digester. Biochem Eng J 96:7–13
25. Lee S, Cho MO, Park CH, Chung Y, Kim JH (2008) Continuous butanol production using suspended and immobilized *Clostridium beijerinckii* NCIMB 8052 with supplementary butyrate. Energy Fuels 13:3459–3464

Part VII
Microalgae

Chapter 12
Microalgae-Based Diesel: A Historical Perspective to Future Directions

Darissa Alves Dutra, Adriane Terezinha Schneider, Rosangela Rodrigues Dias, Leila Queiroz Zepka, and Eduardo Jacob-Lopes

Abstract Microalgae are one of the biggest promises for biodiesel, which is why leading researchers are looking to advance the science needed to provide a renewable alternative that significantly benefits the fuel sector. However, while some technological advances have been an important step, the still incipient technology still needs to reach the market. In light of difficult predictions of success, this chapter aims to provide a historical perspective of microalgae-based diesel for future directions. Initially, a retrospective of microalgae-based diesel is provided to better understand the benefits of using microalgae over conventional alternative sources. The chapter's core is the state of the art of microalgae-based diesel, as well as a critical discussion of the mosaic of solutions that can confirm the belief that it is possible to produce diesel from microalgae on a significant commercial scale. Future directions are presented and discussed in this chapter.

Keywords Algae · Biodiesel · Lipids · Sustainability · Technological nodes

D. A. Dutra · A. T. Schneider · R. R. Dias · L. Q. Zepka · E. Jacob-Lopes (✉)
Bioprocess Intensification Group, Federal University of Santa Maria, UFSM, Roraima Avenue 1000, Santa Maria, RS 97105-900, Brazil
e-mail: ejacoblopes@gmail.com

D. A. Dutra
e-mail: darissa.alves@acad.ufsm.br

A. T. Schneider
e-mail: adriane.schneider@acad.ufsm.br

R. R. Dias
e-mail: ro.rosangelard@gmail.com

L. Q. Zepka
e-mail: zepkaleila@yahoo.com.br

V. Alcaraz Gonzalez et al. (eds.), *Wastewater Exploitation*, Springer Water,
https://doi.org/10.1007/978-3-031-57735-2_12

12.1 Introduction

Economies worldwide face an insatiable demand for fuels due to their rapid development. In this way, the consumption of finite reserves of fossil fuels is still increasing, accompanied by massive CO_2 emissions into the atmosphere worldwide [1]. Although, in the global demand for replacing fossil fuels with a more sustainable energy source has increased in recent years. In this narrative, the use of microalgae as a source of third-generation biofuels attracted the attention of researchers in the present decade. This is due to the high lipid productivity of microalgae [2]. In reports, it has been found that algal biomass can contain 30 to 70% lipids [3].

Biodiesel based on microalgae, is a renewable fuel obtained from oils (lipids) through transesterification with short-chain alcohols (methanol or ethanol) [4] To have better performance in the transesterification process is necessary to use catalysts, which can be acidic or alkaline. The transesterification process of biodiesel based on microalgae also produces glycerin as a by-product, which can be used as a substrate for the cosmetic and food industries. Furthermore, what remains of the biomass from biodiesel production can be used for methane production by anaerobic digestion and bioethanol production by fermentation. With microalgae as raw material, three different energy sources can thus be obtained in the same cycle [5].

Seen as an attractive fuel due to its combustion producing less harmful emission of polluting gases when compared to traditional diesel production, its demand has been growing gradually. Biodiesel based on microalgae is considered a promising strategy for the eminent end of resources, the lack of energy availability, and the pollution of the environment [6]. This is in line with the intense search for new raw materials and process technology that enable economically competitive and environmentally sustainable production compared to mineral diesel [7].

In the status quo, first and second generation biofuels have significant environmental issues compared to microalgae biofuels—including biodiesel. The use of microalgae in the production of biofuels may be a response to the use of fossil fuels, which contribute to the emission of greenhouse gases. The gains from using biodiesel through microalgae lie in the non-competition for arable land for food crops [8] This third-generation biofuel, unlike the fourth generation, is at an advanced stage of technological development. Microalgae, among other advantages, have fast growth, and good biomass yield, and are adaptable to different cultivation conditions and environmental conditions. Furthermore, it is worth mentioning that from this potential raw material, bioethanol, biomethane, biohydrogen, and biobutanol can also be produced [9].

Currently, several studies are being carried out on a laboratory scale to verify the feasibility of using microalgae as raw material in the production of these biofuels [10–12]. The economic and environmental feasibility of producing biofuels specifically biodiesel, is one of the main bottlenecks to overcome to implement the process on large scale. Due to the technological nodes, the rise of structures and technologies is required, which may reduce production costs, thus making it economically attractive and environmentally benign [13].

In light of all this, in this chapter, we provide a historical perspective for future directions that may enable microalgae-based diesel production. We cover a retrospective of microalgae-based diesel, state of the art and solutions that can help microalgae-based biodiesel cross the valley of death between basic research and successful innovation.

12.2 A Retrospective on Microalgae-Based Diesel

Microalgae have a wide variety when it comes to industrial applications. The application of these microorganisms in the production of biofuels—more specifically biodiesel—has been widely reported in relevant studies [14] Historically, according to some researchers, it was in the 1970s, with the oil crisis, that studies about alternative fuel sources were stimulated. As a result, projects were launched to produce diesel on the basis of microalgae. These studies have enabled the identification and characterization of microalgae species with a more significant potential for extracting lipids for producing third-generation biodiesel [15].

Furthermore, going back to the 1960s, *Chlorella* culture, followed by *Arthospira* in the 1970s, kick-started microalgae for industrial applications. Protein was the main product directed to the industry, emphasizing food and prophylactic use. Fast forward a decade, in 1980, the commercial production of *Dunaliella salina* and *Haematococcus* became significant, with its direction towards the production of pigments, focusing on b-carotene and astaxanthin as food additives and animal feed. More recently, in early 1990s, the production of polyunsaturated fatty acids began, focusing on Docosahexaenoic acid (DHA) and Eicosapentaenoic acid (EPA) (Fig. 12.1) [16, 17].

According to the current state of the art, the above-mentioned products are still the most important in the global microalgae market [18]. Commercial facilities that produce microalgae are scattered around the world. In 2017, Taiwan, Japan, the USA, China, Brazil, Spain, Israel, Germany, and Myanmar were the forerunners in producing microalgae biomass and derived products. Currently, the United States, Asia, and Oceania dominate the market and Europe is expected to become one of the leaders, soon [19].

Regarding biodiesel, it can be classified into different categories known as the first, second, and third generation, depending on the raw materials used for its production. The first generation is limited to the use of edible plant materials. Second-generation ones are produced from non-food waste. On the other hand, third-generation biodiesel is being produced from microalgae, which accumulate lipids that, after extraction, are transesterified to obtain diesel based on microalgae. The raw material that most please biorefineries among the generations belongs to the second and third generations [20]. Looking to the future, investing in the third generation of biofuels is believing in a more promising alternative for obtaining renewable energy production [21].

Noteworthy, the production of first-generation biodiesel has limitations. The most common concern related to this generation is that as production capacities increase,

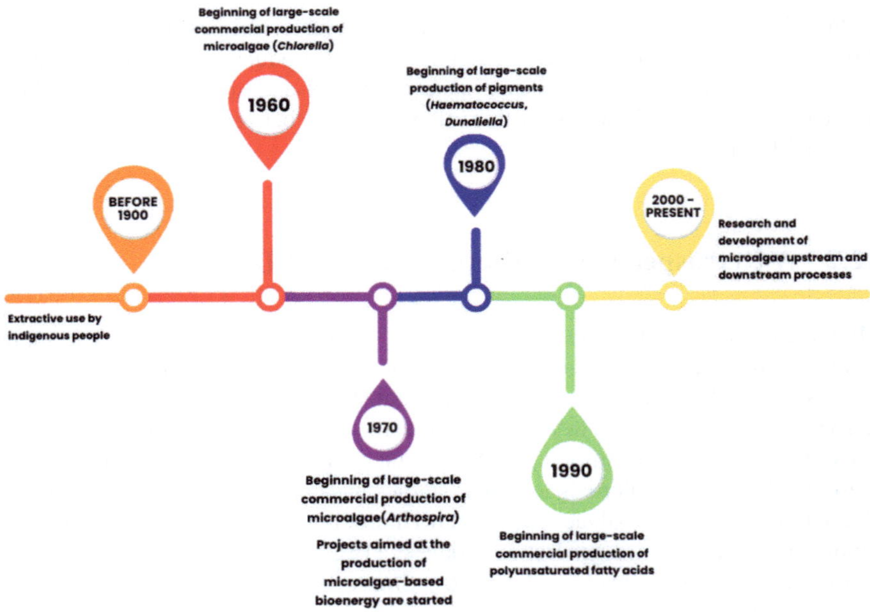

Fig. 12.1 The timeline of commercial microalgae development

so does competition with agriculture for arable land used for food production. As a result, enthusiasm for first-generation biodiesel has waned [22] The researchers then turned to second-generation biofuels. However, second-generation production involves costly processes, making this biodiesel generation inefficient [23]. In this sense, after repeated failures in the first and second generations, the researchers decided to focus their activities on third-generation production, using microalgae and, currently, it is considered a plausible option as a source of renewable energy, overcoming some of the weaknesses of the first generation and second generation [24].

Currently, the biggest bottleneck in the production of biodiesel from microalgae oil is the viability of the process. The high production cost, inefficient upstream and downstream processes, and the environmental sustainability of microalgae-based diesel production are part of the technological nodes for the implementation of the large-scale output and its consolidation in the global market [25]. Adopting more appropriate approaches, it may be possible to take a step forward to produce this renewable product sustainably [26]. The need to curb our massive addiction to fossil fuels and advance environmentally friendly practices can leverage research into microalgae biodiesel and make it the backbone of the transition from petroleum-based diesel to a source that meets energy requirements ecologically correct and sustainable [27].

12.3 State of the Art of Microalgae-Based Diesel: From Petri Dish to Commercial Product

Biofuels, such as biodiesel, can be produced from various raw materials, including biomass, which is a renewable energy source [28]. This initiative aims to reduce dependence on fossil fuels. Many advances have already been achieved in economically producing of biodiesel from microalgae. However, many bottlenecks still need to be overcome to cross the valley of death between research and commercial products [6, 29].

Currently, studies on the species and their real potential in the production of biofuels are more focused [30]. Microalgae diesel is mainly derived from oils present in biomass. Through its ability to take advantage of sunlight and capture carbon dioxide, lipid accumulation, according to the literature, can reach up to 80% of dry weight. On the other hand, when compared to vegetable cultures, these reach contents of 5–10% of oil. This result raises the great potential of microalgae for biodiesel production [26].

However, it is noteworthy that most lipid contents usually range between 10 and 30% of dry weight. But regardless, studies report that microalgae such as *Nannochloris* sp., and *Schizochytrium* sp. reached oil contents of 56 and 80% of dry weight, respectively. However, these have a slow growth rate. In contrast, the microalgae *Scenedesmus* sp. and *Chlorella* sp. have lower oil content but their growth rate is higher. In that regard, it would be necessary to balance the oil content produced and the growth rate [31]. In simultaneous, it is known that other species of microalgae also have auspicious characteristics to produce biodiesel, such as *Dunaliella, Amphora, Ettlia oleoabundans, Ankistrodesmus falcatus, Botryococcus braunii, Kirchneriella lunaris, Ankistrodesmus fusiformis*, and *Chlamydocapsa bacillus* [32].

Today, several research centers, universities, and companies join forces to leverage the production area of microalgae biofuels, as shown in Table 12.1. As an example, in mid-2017, research carried out by ExxonMobil and Synthetic Genomics to increase the oil production in microalgae, using strain modification technology, obtained promising results, reaching a production greater than twice as much as previously produced [33].

Fortunately, investments in the area seeking a future production of biodiesel based on microalgae on a large scale are significant. However, production costs, need to be lowered, not economically compensating for their use as an alternative biofuel [34]. Thus, there is a search for several alternatives to reduce the high production costs to consolidate its generation. In state of the art, the concept of biorefineries could favor the reduction of the expenses by producing products with high-added value concomitantly with the production of biodiesel [35, 36]. And there some patents currently assigned to microalgae biorefineries covering energy products (Table 12.2). There are also already some studies aimed at targeting suitable locations for establishing biofuel production facilities based on geographical parameters such as climate, availability of land and nutrients, and infrastructure. Table 12.3 presents some of these surveys.

Table 12.1 Main microalgae-based biofuels companies

Company	Location	References
Algae Biotecnologia	Brazil	https://agevolution.canalrural.com.br/startup/algae-biotecnologia/
Algae Floating Systems, Inc	United States	https://intengine.com/directory/profile/california/san-francisco/algae-floating-systems-inc/12290
Algaewheel	United States	https://algaewheel.com/
AlgaEnergy S.A	Spain	https://www.algaenergy.es/
Algaecake Tech. Corp	United States	http://www.oilgae.com/ref/cap/
Algenol	United States	https://www.algenol.com/
AlgalOilDiesel, LLP	United States	http://algaloildiesel.wikifoundry.com/
AlgoSource (Alpha Biotech)	France	https://algosource.com/
Aurora Algae, Inc	United States	https://ipira.berkeley.edu/aurora-algae-inc
Bio Fuel Systems	Spain	https://biopetroleo.com/
Cellana, LLC	United States	http://cellana.com/
Chevron Renewable Energy Group	United States and Europe	https://www.regi.com/
Dao Energy, LLC	China	http://www.opportunitycycle.com/
Eldorado Biofuels	United States	http://eldoradobiofuels.com/
ExxonMobil	United States	https://corporate.exxonmobil.com/Research-and-innovation/Advanced-biofuels
Green Star Products, Inc	United States	http://www.greenstarusa.com/
Inventure Chemical	United States	https://inventurechem.com/
Kent BioEnergy	United States	https://openei.org/wiki/Kent_BioEnergy
Manta Biofuel	United States	https://mantabiofuel.com/
Oilgae	India	http://www.oilgae.com/
Organic Fuel Technology	Denmark	https://www.organicfueltechnology.com/
Petroalgae	United States	http://www.petroalgae.com/
PetroSun	United States	https://petrosuninc.com/index.php
Sapphire Energy, Inc	United States	https://www.harrisgroup.com/projects/sapphire-energy
Seambiotic Ltd	Israel	http://www.seambiotic.com/
W2 Energy	Canada	https://w2energy.com/

Table 12.2 Patents deposited and granted on microalgae biorefineries

Number	Patent title
WO2010080377A1	Optimization of algal product production through uncoupling cell proliferation and algal product production
US7977076B2	Integrated processes and systems for production of biofuels using algae
US20120283493A1	Multiproduct biorefinery for synthesis of fuel components and chemicals from lignocellulosics via levulinate condensations
WO2013086302A1	Fractionation of proteins and lipids from microalgae
WO2015044721A1	Microalgae biorefinery for biofuel and valuable products production
US9719114B2	Tailored oils
CN109504527A	Microalgae based biomass method for refining and the system for implementing it
WO2013054170A2	A process for producing lipids suitable for biofuels
US20110275118A1	Method of producing fatty acids for biofuel, biodiesel, and other valuable chemicals
US20090211150A1	Method for producing biodiesel using high-cell-density cultivation of microalga *Chlorella protothecoides* in bioreactor
CN109735580B	Method for selectively catalyzing microalgae powder to coproduce docosahexaenoic acid and biodiesel by using lipase
US20220025414A1	Microalgal strain and its use for the production of lipids
CN113025665A	Method for producing biodiesel by using red tide algae and algicidal bacteria

Table 12.3 Identifying locations for large-scale microalgae cultivation for biofuel production

Goal	Factors	Study location	References
Assessment of the potential of propitious resources for microalgae biofuel production in the country and future contribution to decision-makers	CO_2, climate, nutrients, water, and land	India	Milbrandt and Jarvis [50]
Identify locations in geographic space with potential for large-scale production of microalgae for subsequent production of biofuel	Irradiation, climate, CO_2 and nutrient availability, soil workability, land use, labor, and infrastructure	Western Australia	Boruff et al. [51]
Performance evaluation of a microalgae-based biodiesel supply chain design model carried out in two-stage	Land use, roads, slope, saline water, wastewater, freshwater, CO_2, temperature, solar radiation	Iran	Mohseni et al. [52]
Evaluation of an economic structure of a microalgae-based biofuel supply chain carried out in three-stage	CO_2, freshwater, wastewater, saline water, nutrients, climate, and land	United States	Kang et al. [53]

Simultaneous with efforts to confirm the potential for biodiesel production from microalgae [37] report that combining effluent treatment with biodiesel production would also be a promising alternative for cost reduction. Its use would provide the necessary nutrients for the growth of algae, producing biomass for later use in the production of biodiesel, along with benefits for the environment, allied with the reduction of greenhouse gases (GHG), generation of carbon credits, and the treatment of effluents [38]. Benefiting more than one sector seems to bring light closer to the light at the end of the tunnel to achieve the consolidation of biodiesel production using microalgae [39].

12.4 Microalgae-Based Diesel: What Can We Approach to Make It Economically Viable and Eco-Friendly?

In the era of the effects of climate change, the transition from energy sources with high carbon dioxide (CO_2) emissions to low-emission sources are undoubtedly crucial. The transport sector has a large share in this ranking, contributing to a quarter of the world's CO_2 emissions [40]. Biofuels hold great promise for achieving emissions neutrality. In addition, it is worth mentioning that it is necessary to seek energy strategies that are mutually more economical and, environmentally correct. Therefore, microalgae can—who knows—consolidate these requirements when used for diesel production [6].

However, the premise that the production of microalgae and its products are more environmentally friendly than other raw materials need to be revised [41]. Regarding sustainability metrics and indicators, diesel production by microalgae still needs advances to reach a life cycle of low-emission pollutants. Of note, through the life cycle assessment (LCA) tool, it is possible to assess the environmental sustainability of microalgae-based processes and products according to different impact categories. Data from the literature reveal that to produce 1 kg of whole dry biomass of *C. vulgaris* using a tubular photobioreactor, impacts in the order of 1551.94 CTUe were obtained for the ecotoxicity category, 475.53 MJ for the energy resource category, 220.32 $kgCO_2eq$ for the global warming category and 5.05 kgO_3eq for the photochemical oxidant creation potential category [42].

These, advances in using clean energies in biomass production bring hope for achieving better environmental performance throughout the process, as shown by data [43]. When comparing energy consumption for biomass production using tubular photobioreactors and raceway ponds, consumption also is approximately 4.46 times lower in raceway ponds, thus resulting in a lower environmental impact [13].

Nazloo et al. [26] in turn, present values of 591 L of fresh water needed for the production of one liter of biodiesel based on the microalgae *C. Vulgaris*, while for the production of biodiesel based on soy, 33 times more fresh water is needed for the production of 1 L of biodiesel. [44] also describe some data on the impacts of the life cycle of biodiesel production using soy as raw material. The values for the terrestrial ecotoxicity, energy resource depletion, global warming, and photochemical oxidant creation potential were around 0.0003 CTUe, 0.00019 MJ, 0.035 $kgCO_2eq$, and 0.000076 kgO_3eq, respectively. Comparing with the data for microalgae biomass cited above, it is clear that there is a huge difference in the values.

It is clear that performance in terms of sustainability is still a hot spot blocking much of the advancement in the commercialization of microalgae biodiesel. However, the economic and technological challenge is a current barrier to achieving this feat [45]. Strategies of biochemical engineering, genetic engineering, and transcription factor engineering can be the solid point for solving the existing bottlenecks [46]. Currently, biochemical engineering is the most used, consisting of physiological stress techniques to increase oil [47]. In genetic engineering, we have a compilation of techniques for recombining the DNA of strains, thus improving the most appreciated and desired characteristics [48] And in turn, the goal of transition factor engineering is to change several components of a single metabolic pathway concomitantly [49] Among these three strategies, genetic engineering seems to be the one that will lay the groundwork to confirm the belief that it is possible to produce diesel from microalgae on a significant commercial scale. However, it is worth noting that, unfortunately, advances are still timid in this field.

In that regard, we must not lay hands resort to policy development that can generate greater technological impetus and market attraction in biodiesel. It is expected that in the not-so-distant future, new technologies will be able to increase the productivity of microalgae further, minimize impacts in the upstream and downstream stages, as well as increase the economic gains of production, also associated with a greater obtaining of co-products with high added value thus making them even more attractive as an alternative source of biodiesel.

12.5 Conclusion

The biggest challenge for diesel production based on microalgae is related to its economic and environmental sustainability. These challenges have been discussed by many researchers who seek to make real the expectations created in the biotechnological field of microalgae. However, it is noteworthy that after a few decades since the recognition of the main technological nodes, it is controversial that the processes remain plastered and far from the viability of producing biodiesel on a large scale. Despite the literature presenting a mosaic of solutions, the truth is that, in practice, they do not apply because they imply a myriad of new bottlenecks. In this sense, future directions point to a greater interface between the areas of knowledge since what is—essentially—lacking is a holistic approach created by transdisciplinary collaboration, that is, by professionals from different areas and sectors.

References

1. Wu W, Tan L, Chang H et al (2023) Advancements on process regulation for microalgae-based carbon neutrality and biodiesel production. Renew Sustain Energy Rev 171:112969. https://doi.org/10.1016/j.rser.2022.112969
2. Ananthi V, Raja R, Carvalho IS et al (2021) A realistic scenario on microalgae based biodiesel production: third generation biofuel. Fuel 284:118965. https://doi.org/10.1016/j.fuel.2020.118965
3. Muhammad G, Alam MdA, Xiong W et al (2020) Microalgae biomass production: an overview of dynamic operational methods. In: Microalgae biotechnology for food, health and high value products. Springer, Singapore, pp 415–432
4. Raheem A, Prinsen P, Vuppaladadiyam AK et al (2018) A review on sustainable microalgae based biofuel and bioenergy production: Recent developments. J Clean Prod 181:42–59. https://doi.org/10.1016/j.jclepro.2018.01.125
5. Katiyar R, Kumar A, Gurjar BR (2017) Microalgae based biofuel: challenges and opportunities, pp 157–175
6. Ali SS, Mastropetros SG, Schagerl M et al (2022) Recent advances in wastewater microalgae-based biofuels production: a state-of-the-art review. Energy Rep 8:13253–13280. https://doi.org/10.1016/j.egyr.2022.09.143
7. Skorupskaite V, Makareviciene V, Gumbyte M (2016) Opportunities for simultaneous oil extraction and transesterification during biodiesel fuel production from microalgae: a review. Fuel Process Technol 150:78–87. https://doi.org/10.1016/j.fuproc.2016.05.002
8. Cavalheiro LF, Misutsu MY, Rial RC et al (2020) Characterization of residues and evaluation of the physico chemical properties of soybean biodiesel and biodiesel: Diesel blends in different storage conditions. Renew Energy 151:454–462. https://doi.org/10.1016/j.renene.2019.11.039
9. Dias RR, Maroneze MM, de Oliveira ÁS et al (2021) Bioconversion of industrial wastes into biodiesel feedstocks, pp 109–120
10. Agrawal K, Bhatt A, Bhardwaj N et al (2020) Algal biomass: potential renewable feedstock for biofuels production—Part I, pp 203–237
11. Mudunkothge HP, Jayarathne DPRUAV, Shanthamareen M, Ketheesan B (2022) A preliminary study of cost and energy analysis of bio-fuel production from microalgae cultivated in parboiled rice mill wastewater. J Dry Zone Agric 8:16–29. https://doi.org/10.4038/jdza.v8i2.60
12. Chhandama MVL, Satyan KB, Changmai B et al (2021) Microalgae as a feedstock for the production of biodiesel: a review. Bioresour Technol Rep 15:100771. https://doi.org/10.1016/j.biteb.2021.100771
13. Dias RR, Deprá MC, Zepka LQ, Jacob-Lopes E (2022) Roadmap to net-zero carbon emissions in commercial microalgae-based products: environmental sustainability and carbon offset costs. J Appl Phycol 34:1255–1268. https://doi.org/10.1007/s10811-022-02725-y
14. Dickinson S, Mientus M, Frey D et al (2017) A review of biodiesel production from microalgae. Clean Technol Environ Policy 19:637–668. https://doi.org/10.1007/s10098-016-1309-6
15. Marques IM, Melo NR, Oliveira ACV, Moreira ÍTA (2020) Bioremediation of urban river wastewater using Chlorella vulgaris microalgae to generate biomass with potential for biodiesel production. Res Soc Dev 9(7) https://doi.org/10.33448/rsd-v9i7.4882
16. Jacob-Lopes E, Maroneze MM, Deprá MC et al (2019) Bioactive food compounds from microalgae: an innovative framework on industrial biorefineries. Curr Opin Food Sci 25:1–7. https://doi.org/10.1016/j.cofs.2018.12.003
17. Jacob-Lopes E, Zepka LQ (2023) Food and feed from microalgae: a historical perspective to future directions. In: Handbook of food and feed from microalgae. Elsevier, pp 3–7
18. Singh A, Rai A, Rai PK, Sharma NK (2023) Biodiversity and biogeography of microalgae with food and feed potential. In: Handbook of food and feed from microalgae. Elsevier, pp 9–21
19. Rumin J, de Oliveira G, Junior R, Bérard J-B, Picot L (2021) Improving microalgae research and marketing in the European Atlantic area: analysis of major gaps and barriers limiting sector development. Mar Drugs 19:319. https://doi.org/10.3390/md19060319

20. Nanda S, Rana R, Sarangi PK et al (2018) A broad introduction to first-, second-, and third-generation biofuels. In: Recent advancements in biofuels and bioenergy utilization. Springer, Singapore, pp 1–25
21. Deprá MC, dos Santos AM, Severo IA et al (2018) Microalgal biorefineries for bioenergy production: can we move from concept to industrial reality? Bioenergy Res 11:727–747. https://doi.org/10.1007/s12155-018-9934-z
22. Nanda S, Azargohar R, Dalai AK, Kozinski JA (2015) An assessment on the sustainability of lignocellulosic biomass for biorefining. Renew Sustain Energy Rev 50:925–941. https://doi.org/10.1016/j.rser.2015.05.058
23. Mat Aron NS, Khoo KS, Chew KW et al (2020) Sustainability of the four generations of biofuels—a review. Int J Energy Res 44:9266–9282. https://doi.org/10.1002/er.5557
24. Alam F, Mobin S, Chowdhury H (2015) Third generation biofuel from algae. Procedia Eng 105:763–768. https://doi.org/10.1016/j.proeng.2015.05.068
25. Severo IA, Siqueira SF, Deprá MC et al (2019) Biodiesel facilities: What can we address to make biorefineries commercially competitive? Renew Sustain Energy Rev 112:686–705. https://doi.org/10.1016/j.rser.2019.06.020
26. Nazloo EK, Moheimani NR, Ennaceri H (2022) Biodiesel production from wet microalgae: Progress and challenges. Algal Res 68:102902. https://doi.org/10.1016/j.algal.2022.102902
27. Galina D, Porto PS da S, Freitas RR (2018) Estudo das tecnologias para produção de biodiesel a partir de microalgas do gênero nannochloropsis. Res Soc Dev 7:e5712482. https://doi.org/10.33448/rsd-v7i12.482
28. Ruan R, Zhang Y, Chen P et al (2019) Biofuels: Introduction. In: Biofuels: alternative feedstocks and conversion processes for the production of liquid and gaseous biofuels. Elsevier, pp 3–43
29. Ayoola AA, Hymore FK, Omonhinmin CA et al (2019) Analysis of waste groundnut oil biodiesel production using response surface methodology and artificial neural network. Chem Data Collect 22:100238. https://doi.org/10.1016/j.cdc.2019.100238
30. Ganesan R, Manigandan S, Samuel MS et al (2020) A review on prospective production of biofuel from microalgae. Biotechnol Rep 27:e00509. https://doi.org/10.1016/j.btre.2020.e00509
31. Shokravi Z, Shokravi H, Chyuan OH et al (2020) Improving 'lipid productivity' in microalgae by bilateral enhancement of biomass and lipid contents: a review. Sustainability 12:9083. https://doi.org/10.3390/su12219083
32. Sharma S, Chauhan A, Dobbal S, Kumar R (2022) Biology of plants coping stresses: epigenetic modifications and genetic engineering. S Afr J Bot 144:270–283. https://doi.org/10.1016/j.sajb.2021.08.031
33. ExxonMobil and Synthetic Genomics algae biofuels program targets 10,000 barrels per day by 2025_ExxonMobil
34. Jeswani HK, Chilvers A, Azapagic A (2020) Environmental sustainability of biofuels: a review. Proc R Soc A Math Phys Eng Sci. https://doi.org/10.1098/rspa.2020.0351
35. Kandasamy S, Zhang B, He Z et al (2022) Microalgae as a multipotential role in commercial applications: current scenario and future perspectives. Fuel 308:122053. https://doi.org/10.1016/j.fuel.2021.122053
36. Cuevas-Castillo GA, Navarro-Pineda FS, Baz Rodríguez SA, Sacramento Rivero JC (2020) Advances on the processing of microalgal biomass for energy-driven biorefineries. Renew Sustain Energy Rev 125:109606. https://doi.org/10.1016/j.rser.2019.109606
37. Xin C, Addy MM, Zhao J et al (2016) Comprehensive techno-economic analysis of wastewater-based algal biofuel production: a case study. Bioresour Technol 211:584–593. https://doi.org/10.1016/j.biortech.2016.03.102
38. Scaglioni PT, Badiale-Furlong E (2023) Microalgae as a source of preservatives in food/feed chain. In: Handbook of food and feed from microalgae. Elsevier, pp 267–277
39. Liu Y, Cruz-Morales P, Zargar A et al (2021) Biofuels for a sustainable future. Cell 184:1636–1647. https://doi.org/10.1016/j.cell.2021.01.052
40. Cabrera-Jiménez R, Tulus V, Gavaldà J et al (2023) Microalgae biofuel for a heavy-duty transport sector within planetary boundaries. ACS Sustain Chem Eng 11:9359–9371. https://doi.org/10.1021/acssuschemeng.3c00750

41. Rodrigues Dias R, Schneider AT, Bittencourt Fagundes M (2023) Challenges and opportunities for microalgae biotechnology development. In: Microalgae-based systems. De Gruyter, pp 41–54
42. Deprá MC, Severo IA, dos Santos AM et al (2020) Environmental impacts on commercial microalgae-based products: sustainability metrics and indicators. Algal Res 51:102056. https://doi.org/10.1016/j.algal.2020.102056
43. Dias RR, Sartori RB, Deprá MC et al (2022) Biochemical engineering approaches to enhance the production of microalgae-based fuels. In: 3rd Generation Biofuels. Elsevier, pp 65–90
44. Hou J, Zhang P, Yuan X, Zheng Y (2011) Life cycle assessment of biodiesel from soybean, Jatropha and microalgae in China conditions. Renew Sustain Energy Rev 15:5081–5091. https://doi.org/10.1016/j.rser.2011.07.048
45. Sun J, Xiong X, Wang M et al (2019) Microalgae biodiesel production in China: a preliminary economic analysis. Renew Sustain Energy Rev 104:296–306. https://doi.org/10.1016/j.rser.2019.01.021
46. Dias RR, Deprá MC, Zepka LQ, Jacob-Lopes E (2023) Disruptive technologies to improve the performance of microalgae cultures. In: Handbook of food and feed from microalgae. Elsevier, pp 529–536
47. Queiroz MI, Maroneze MM, Manetti AG da S et al (2018) Enhanced single-cell oil production by cold shock in cyanobacterial cultures. Ciência Rural 48 https://doi.org/10.1590/0103-8478cr20180366
48. Kumar Sharma G, Ahmad Khan S, Kumar A, et al (2022) Algal biorefinery: a synergetic sustainable solution to wastewater treatment and biofuel production. In: Progress in microalgae research—a path for shaping sustainable futures. IntechOpen
49. Bajhaiya AK, Ziehe Moreira J, Pittman JK (2017) Transcriptional engineering of microalgae: prospects for high-value chemicals. Trends Biotechnol 35:95–99. https://doi.org/10.1016/j.tibtech.2016.06.001
50. Milbrandt A, Jarvis E (2010) Resource evaluation and site selection for microalgae production in India (No. NREL/TP-6A2-48380). National Renewable Energy Lab.(NREL), Golden, CO (United States)
51. Boruff BJ, Moheimani NR, Borowitzka MA (2015) Identifying locations for large-scale microalgae cultivation in Western Australia: a GIS approach. Appl Energy 149:379–391
52. Mohseni S, Pishvaee MS, Sahebi H (2016) Robust design and planning of microalgae biomass-to-biodiesel supply chain: a case study in Iran. Energy 111:736–755
53. Kang S, Heo S, Realff MJ, Lee JH (2020) Three-stage design of high-resolution microalgae-based biofuel supply chain using geographic information system. Appl Energy 265:114773

Chapter 13
Bioconversion of Industrial CO_2 into Synthetic Fuels

Alessandro A. Carmona-Martínez and Clara A. Jarauta-Córdoba

Abstract The energy-intensive industry (EIIs) is responsible for significant CO_2 emissions, which must be limited to avoid irreversible consequences. Implementing renewable energy (RE) technologies and avoiding CO_2 emissions by EIIs are necessary steps towards decarbonization. A potential solution for using renewable energy is to transform CO_2 into synthetic fuels and chemicals. However, CO_2 is a very stable molecule and it needs a lot of energy to change it. This means that the process is not sustainable unless it uses renewable energy sources. Therefore, it is important to find sustainable ways to transform CO_2 and to evaluate them based on scientific understanding and economic analysis. There are four main ways to use heat and catalysts to change CO_2 into fuels. They are Methane dry reforming, Reverse water gas shift reaction, Sabatier reaction process, and Direct methanol synthesis. These methods need H_2 or CH_4 from a renewable source. Another option is to use electricity or light to reduce CO_2 and make fuels. This chapter focuses on the Biological conversion of CO_2, a promising approach to produce synthetic fuels from industrial CO_2. Algae can efficiently capture CO_2 to produce biomass that can be used to produce biomethane. However, most of these technologies require green H_2 to produce synthetic fuels from captured industrial CO_2. In conclusion, decarbonization of EIIs is necessary to limit CO_2 emissions, and the implementation of renewable energy technologies and the use of renewable energy sources in CO_2 conversion processes are crucial. Biological conversion of CO_2 is a promising approach to produce synthetic fuels, but it requires green H_2 for its successful implementation.

Keywords Bioconversion · Bioelectrofuels · Carbon dioxide sequestration · Industrial CO_2

A. A. Carmona-Martínez (✉) · C. A. Jarauta-Córdoba
CIRCE—Technology Center, Parque Empresarial Dinamiza, Ave. Ranillas 3D, 1st Floor, 50018 Zaragoza, Spain
e-mail: acarmona@fcirce.es; cajarauta@fcirce.es

V. Alcaraz Gonzalez et al. (eds.), *Wastewater Exploitation*, Springer Water,
https://doi.org/10.1007/978-3-031-57735-2_13

13.1 Carbon Dioxide Sources and Conversion Processes

To achieve decarbonisation by 2050, the European Union (EU) has set high goals that depend on using renewable energy (RE) technologies instead of fossil-based energies. Another key factor for this decarbonisation path is to reduce the greenhouse gases (GHG) that come from Energy Intensive Industries (EIIs). These industries have a large share of the EU's GHG emissions historically [1].

The Chemical and Petrochemical sector was the biggest user of energy in the European Union in 2020, with 22% of the total energy consumption of about 2685 TWh. The Non-metallic minerals sector came second with 14%, followed by the Paper, pulp, and printing sector and the Food, beverages, and tobacco sector, both with 12%. These four industrial sectors used more than half of the energy that European industries consumed [2].

The world depends a lot on burning fossil fuels, which is a process that converts energy inefficiently and produces a lot of CO_2. This also causes economic and geopolitical problems because of the reliance on fossil energy. On the other hand, renewable energy sources are not enough to satisfy the increasing demand. Scientists agree that it is hard to know the exact effects of human-made climate change, but they also agree that it is very important to reduce CO_2 emissions.

An artificial carbon cycle that uses renewable energy to change CO_2 into synthetic fuels and chemicals is a potential solution to lower gas emissions. CO_2 is a source of carbon that is non-flammable and non-corrosive, and there are many resources available, such as gases from power plants that use coal or natural gas, biogas, and landfill gas (which have up to 50% CO_2), and also gases from some industrial processes like making ammonia or fermentation. CO_2 that comes from natural gas and oil wells can also be used; this CO_2 is usually released into the air.

Changing CO_2 into fuels and chemicals needs H_2, and this can be unsustainable because it needs a lot of energy. Also, renewable energy sources should be used for renewable CO_2 conversion processes, which means a high investment is needed. So, it is important to find sustainable ways to change CO_2 into fuels and chemicals by evaluating them carefully based on a scientific understanding of how CO_2 changes and also by looking at the economic aspects [3].

Industrial CO_2 comes from different sources, as shown in Fig. 13.1. These sources are carbon capture from different types of combustion: post-combustion, pre-combustion, and oxy-combustion. Additionally, various pathways for the conversion of CO_2 into fuels are shown. Thermocatalytic pathways have a high relevance since they are well established technologies that can be already applied. Syngas, a mixture mainly of H_2 and CO, can be made from different reactions involving methane (CH_4) and water or CO_2. One reaction is methane steam reforming (MSR), where CH_4 and water react together. Another reaction is methane dry reforming (MDR), where CH_4 and CO_2 react together. Syngas can also be made from CO_2 hydrogenation, using the reverse water–gas shift reaction (RWGS). Syngas is a useful chemical material that can be changed into fuels by Fischer–Tropsch synthesis (FTS). Biochemical pathways are also relevant due to their potential to convert industrial CO_2 into fuels.

An industrial example exists within the steel sector for converting CO_2 contained in flue gases into bioethanol, as described in a further section. Methane can be produced from CO_2 by hydrogenotrophic microorganisms fed with H2. It is worth noticing that the Sabatier reaction, known as CO_2 abiotic methanation, also produces synthetic CH4 or renewable natural gas through highly exothermic reactions.

Synthetic fuels can be made from CO_2 emissions that have a high concentration. Pre-combustion processes use industrial methods that are already used for making H_2 and chemical products. These processes change fuel materials like coal and natural gas into syngas (a mixture of hydrogen and carbon monoxide) by using different methods such as gasification, steam reforming, autothermal reforming, or partial oxidation. Subsequently, CO is converted into CO_2 through a water–gas shift reaction (WGS), generating additional H_2. The captured CO_2 is then removed through a carbon capture system. The fuel gas that has a lot of hydrogen can be used to make power and heat, for example, in boilers, gas turbines, and fuel cells. After the WGS,

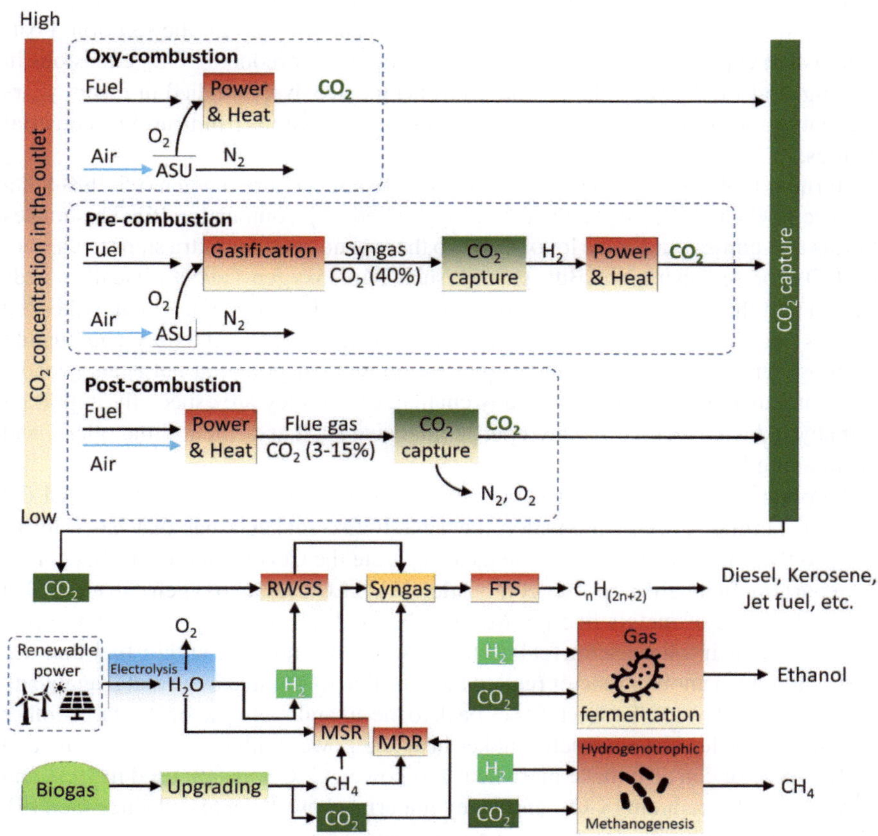

Fig. 13.1 Schematic diagram of the existing industrial CO_2 sources (above) and available pathways for CO_2 conversion into fuels. Redrawn and adapted from the original sources [3, 4]

the gas that comes out usually has a lot of CO_2, from 15 to 60% when it is dry, and the pressure is 2–7 MPa. As a result, physical solvents such as selexol and rectisol are commonly employed for pre-combustion capture, rather than chemical solvents, due to their effectiveness in this specific application [4].

Post-combustion capture, a widely studied method for carbon capture, involves the separation of CO_2 from flue gases after combustion in power plants, industrial sources, and other facilities. The CO_2 in the gases that come out after combustion is usually low in both pressure (from 0.03 to 0.2 bar) and concentration (from 3 to 20%). However, certain industrial processes, such as cement manufacturing and stainless-steel production, categorized as post-combustion capture, may have higher CO_2 concentrations than typical flue gases from power plants. For instance, the CO_2 concentration in flue gases from steel production blast furnaces can range from 20 to 27%, while aluminum production and cement processes can have CO_2 concentrations of 1–2 and 14–33%, respectively. Post-combustion capture can use chemical absorption-based solutions that have water and amines, such as a solution with 30% of monoethanolamine (MEA). These solutions are already used in the market [5]. These solutions capture CO_2 from flue gases by absorbing it into the solution. Post-combustion capture is considered a promising option for reducing CO_2 emissions in existing coal-fired power plants, and it has been extensively studied in recent years as a viable approach for mitigating greenhouse gas emissions from various industrial sources.

Burning fuel materials in an environment with almost pure oxygen (95–99%) or a mixture of O_2 and CO_2 is a new technology called oxy-combustion [6]. This makes the gas that comes out have a lot of CO_2, so there is no need for extra steps to capture the CO_2 for sequestration. But, getting such pure oxygen (>95%) usually needs using a unit that separates oxygen from the air by cooling it down, which makes the process more expensive. This shows the economic issues related to oxy-combustion as an option for CO_2 capture and sequestration that has potential but is also costly. Oxy-combustion processes have many challenges and they are especially important in many industry sectors that are energy-intensive such as cement, lime, glass, and steelmaking [7].

A combustion process that has potential is chemical looping combustion (CLC), which is similar to oxy-combustion, because it also makes a gas that has a lot of CO_2, so there is no need for extra steps to separate the CO_2 from the gas that comes from burning fuel before or after combustion [8]. CLC uses an oxygen carrier, which is usually made of metals like Fe, Mn, Cu, Ni, and Co. The oxygen carrier is first mixed with air in one reactor (called the air-reactor) and then with a fuel that has hydrocarbons in another reactor (called the fuel-reactor) to make the metal again and release CO_2 and water. The metal goes back to the air and fuel-reactors to start another combustion cycle. The air-reactor makes heat and power with high temperature, and the fuel-reactor can also make more heat and power. CLC was first used in 1951 and then in 1987 for reducing CO_2, and many materials have been tested since then [9].

CLC does not need an expensive unit to separate air, which makes it cheaper than oxy-combustion. But, the CLC process is more complicated and needs more efforts to reach the market. The choice of a good metal/metal-oxide pair as the oxygen carrier is very important for making the chemical looping process cost-effective.

Direct air carbon capture (DACC) has emerged as a promising technology for extracting or removing CO_2 directly from the atmosphere, initially proposed to mitigate climate change in 1999 [10]. DACC lies outside of the scope of the efforts by energy-intensive industries to mitigate emissions. DACC is a method that has generated interest because it can capture CO_2 from sources that are not fixed or move around, such as cars, boats, and planes. Some companies are working on making DACC systems at a big scale. But, the CO_2 in the air is very low (~400 ppm) and this makes it hard. To capture such a small amount of CO_2, air has to go through the capture unit. This needs heat for adsorption, which makes this method use a lot of energy and money compared to capturing CO_2 from sources that have more CO_2. There is an ongoing debate within the scientific community about whether DACC is an enabling factor or merely an expensive distraction in practical climate change mitigation efforts. Despite the challenges, DACC holds promise as a potential solution for capturing CO_2 from mobile and diffuse emission sources [11].

Other industrial processes that have a lot of CO_2 in their tail gases are H_2 production, ammonia synthesis, and methanol manufacture. These tail gases can be used directly or stored after they are cleaned. Like oxy-combustion and chemical looping combustion (CLC) processes, these industrial processes usually do not need extra units to capture CO_2.

Most of the efforts to lower CO_2 emissions are about capturing CO_2 from power stations and storing it. They use methods such as amine scrubbing, pressure swing adsorption, or membrane separation followed by cryogenic liquefaction to clean CO_2 for storage in places under the ground or for helping oil recovery. But, these methods are energy-intensive and require a high investment [12]. Also, it is not clear what will happen to CO_2 that is stored in underground places in the long term. A different way to lower CO_2 is to change it into fuels and chemicals that can be used instead of fossil materials in the chemical industry.

CO_2 is a carbon source that is non-flammable and abundant, but it also causes the greenhouse effect. It can come from different places, such as gases from power plants that use coal or natural gas, biogas and landfill gas, and also gases from some industrial processes like making ammonia or fermentation. Additionally, large quantities of CO_2 are vented into the atmosphere as flare gas from natural gas and oil production wells, presenting an opportunity for capture and utilization [13].

The conversion of captured CO_2 into synthetic fuels and chemicals has gained significant interest from research groups worldwide as a promising pathway for reducing CO_2 emissions. Many research efforts have focused on the photochemical and electrochemical reduction of CO_2 into products such as formic acid and methanol in aqueous environments, using water as a source of hydrogen for CO_2 reduction. While this approach seems attractive due to its utilization of water and CO_2 as starting materials, and potential use of solar or renewable electricity as an energy source, it has inherent limitations in solar energy utilization for photochemical processes and low efficiencies of electricity utilization for electrochemical processes, along with challenges related to CO_2 solubility in water and diffusion limitations [14].

An alternative approach is thermocatalytic conversion, which employs high temperatures and heterogeneous catalysts to achieve fast reaction rates, enabling large-scale production of fuels and chemicals from CO_2. This approach offers potential advantages in terms of process efficiency and scalability, making it an attractive avenue for CO_2 emissions reduction research.

As the world seeks sustainable solutions for mitigating climate change, the chemical transformation of CO_2 into valuable products through thermocatalytic conversion presents a promising direction for reducing CO_2 emissions and utilizing CO_2 as a valuable resource in the transition towards a low-carbon economy. Further research and development in this field are crucial for advancing the science and technology of CO_2 utilization and driving towards a more sustainable future.

13.2 Biological Conversion of CO_2

Interest has been generated in recent years by the idea of using CO_2 as a valuable raw material instead of a greenhouse gas. Most of the research and development in this area has been on chemical methods of converting CO_2, such as using heat, electricity, or light as catalysts. However, there is another way of turning CO_2 and water into useful organic substances, which is the natural process of photosynthesis. Photosynthetic organisms have been doing this for more than two billion years, and they can produce complex biological molecules from CO_2 [15]. Some of these molecules, such as bio-alcohols and bio-diesel, are especially important for their potential applications [16, 17].

The amount of CO_2 that is used in the chemical industry for making products like urea, methanol, pigments, and carbonates is over 100 million metric tons per year worldwide [18]. However, this is much less than the amount of CO_2 that is converted into biomass by photosynthetic organisms on land and in the ocean, which is over 100 billion metric tons per year [19]. A promising technology to produce biofuels directly from CO_2 is to use microorganisms that can fix CO_2 in their cells. This technology does not need sunlight and a bioreactor can be used to control the process. Unlike photosynthesis, which has a low efficiency of less than 1%, this technology can have a higher efficiency of CO_2 conversion. It's important to note that not only photosynthetic organisms such as plants, algae, and cyanobacteria can assimilate CO_2, but also other organisms that can do this.

Autotrophic bacteria have also gained momentum as research endeavors in this area have progressed. CO_2-utilizing microorganisms have been extensively studied and could be further developed for industrial-scale bioprocesses.

Table 13.1 shows the advantages and disadvantages of exploitation in biological carbon capture and utilization (CCU) [20]. However, it's important to note that these microorganisms have not naturally evolved for industrial-scale production of desired chemicals, and many of their inherent properties, such as growth characteristics, types of metabolites produced, thermo-stability, and tolerance to inhibitors, may not be optimized for this purpose. The application of genetic engineering has made it possible to improve their phenotypes and expand their repertoire of chemical syntheses.

The remarkable diversity of microorganisms capable of utilizing CO_2, coupled with the advancements in genetic modification techniques and protein engineering,

Table 13.1 Advantages and disadvantages of CO_2-utilizing microorganisms for industrial-scale chemical production

Organisms	Advantages	Disadvantages
Algae	+Widely available +Multiple ways of cultivation +Fast growth +High cell density +Good CO_2 uptake +High lipid content +High-value molecules producers +Genetically modifiable organisms	−Light dependent organisms −To grow they require water −Large amount of phosphorous required −Biodiesels produced from algae are considered of low quality
Cyanobacteria	+They grow with basic nutritional requirements +Cultivation is not expensive +Better growth rates when compared to algae and higher plants +Considerable amount of lipids +Broad range of derived fuels (H_2, ethanol, diesel, methane, etc.) +Genetically modifiable organisms	−Low productivity when not grown in their optimum temperature, pH and light intensity −Growth limitation by carbon and nitrogen (nitrate, ammonia, urea, etc.) −Agitation is required for growth which increases the operational cost
B-Proteobacteria	+Aerobic microorganisms (no requirements for anoxia) +Multiple carbon sources and utilization pathways +Carbon storage capacity as PHA +Genetically modifiable organisms	−Gas fermentation is still under development
Clostridia	+Diverse of carbon substrates (carbohydrates, CO_2/H_2 and CO) +Diverse pathways for production of useful metabolites (ethanol, acetate, acetone, lactate, butanol, etc.) +Certain tolerance to toxic metabolites +Genetically modifiable organisms	−Anerobic cultivation can increase the operational costs −Gas fermentation is still under development
Archaea	+Capable of methane production +Capable of accumulating PHA +Excellent sources of thermostable enzymes	−Anerobic cultivation can increase the operational costs −Gas fermentation is still under development −Difficult to emulate growth conditions −Non genetically modifiable organisms

Note The table has been adapted from its original source [20]. PHA stands for Polyhydroxyalkanoates which are considered a potential replacement for some petrochemical-based plastics [23]

has significantly broadened the spectrum of bio-based products that can be directly synthesized from CO_2. This biological CCU approach presents an appealing alternative to biomass-based biorefining methods. A few key observations can be drawn from this comparison: (i) certain chemicals such as salicylic acid and methane can be produced using either chemical or biological routes, (ii) biological routes complement chemical approaches and expand the range of chemicals that can be derived from CO_2, and (iii) complex molecules like proteins can only be synthesized through a biological route, as microorganisms can be considered as integrated chemical factories that harbor diverse biocatalysts [21, 22].

13.3 Utilization of Microorganisms for the Biological Fixation of CO_2

Autotrophic microorganisms can directly synthesize various compounds, such as ethanol, butanol, alkanes, and lipids. These microorganisms utilize sunlight through photosynthesis or chemical reactions through chemosynthesis to produce complex organic molecules. Among these microorganisms, microalgae stand out as promising candidates for direct biodiesel production due to their capacity to accumulate large amounts of lipids. However, the high production costs associated with algal-based biofuels have hindered their widespread commercialization [24].

The main goal of the research in this area is to find and select microalgae species that can produce a lot of lipids, grow fast, and use light efficiently. There is a possibility of improving the metabolism of microalgae in the future with the help of new technologies in genome sequencing and molecular biology. However, it is not easy to modify the metabolism of eukaryotic organisms [25].

In contrast, cyanobacteria, which are prokaryotic microorganisms, are comparatively easier to engineer genetically. Moreover, these photosynthetic organisms have the ability to directly produce fuels and chemicals without undergoing lipid synthesis pathways. Recent studies have successfully introduced synthetic pathways into cyanobacterial metabolism, producing various chemicals, including ethanol and butanol. Furthermore, genes associated with the production of alkanes in cyanobacteria have also been recently identified. Despite the relative simplicity of metabolic engineering in cyanobacteria compared to microalgae, it remains challenging compared to other prokaryotes (bacteria) that are already utilized on an industrial scale [27].

Besides plants, algae, and cyanobacteria, which are photosynthetic organisms, there are also other organisms that can use CO_2. Some microorganisms can use CO_2 and they have the potential to be used in industrial processes (Fig. 13.2). These microorganisms are proteobacteria, clostridia, and archaea, and they have different ways of fixing CO_2 and different advantages and disadvantages for using CO_2. Their growth characteristics, thermal stability, and tolerance to inhibitors are important factors to consider when using them in processes that use CO_2. Furthermore, most of

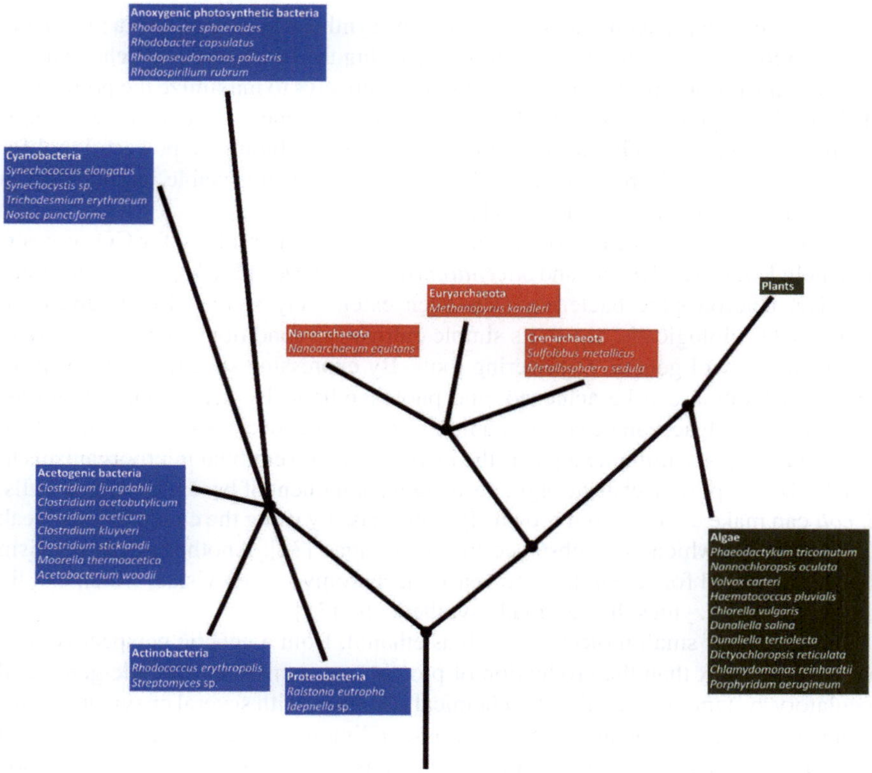

Fig. 13.2 Summary of microorganisms capable of CO_2 assimilation. Redrawn from its original source [3]

these microorganisms require optimization to produce desired chemicals, which can be achieved through genetic and metabolic engineering as well as synthetic biology tools.

13.4 Synthetic Biology Tools

Synthetic biology is a research field that involves many different disciplines such as biotechnology, systems biology, molecular biology, genetic engineering, metabolic engineering, biophysics, and computer engineering. It combines these disciplines to create new biological systems or modify existing ones. Its goal is to engineer artificially-modified microorganisms to create fully synthetic biological systems that can perform desired functions, such as the production of synthetic chemicals [28]. However, microorganisms often face limitations such as slow growth rates, demanding cultivation conditions, and low product yields, which hinder their direct application for synthetic fuel production. Synthetic biology offers a promising approach to overcome these constraints by altering the biochemical mechanisms of microorganisms or introducing new metabolic pathways to maximize the production of desired compounds. Recent advances in protein, genetic, and metabolic engineering, as well as system biology, have made synthetic biology a powerful tool for enhancing biological processes to make them economically viable, particularly in converting CO_2 into renewable fuels [29].

Notable examples of microorganisms used in synthetic biology for CO_2 conversion include *Escherichia coli* and *Saccharomyces cerevisiae* [30, 31]. *E. coli*, a Gram-negative, heterotrophic bacterium, has been extensively studied for its ability to convert CO_2 biologically due to its simple cultivation conditions, high growth rate, and availability of genetic engineering tools. By expressing specific enzymes in *E. coli*, CO_2 fixation can be achieved, and photosynthetically accumulated biomass, such as brown algae, can be converted into valuable chemicals, such as ethanol. One of the challenges in utilizing algae is the inability of conventional microorganisms to metabolize the polysaccharide alginate, a main component of brown algae cell walls. *E. coli* can make ethanol from brown algae biomass by using the enzymes that break down alginate, which is a substance in brown algae [32]. Another microorganism that is often used for research is the yeast Saccharomyces cerevisiae, which can fix CO_2 by using enzymes that are made synthetically [33].

Production of small molecules, such as ethanol, from a genetic perspective may be more complex than the production of proteins, as it involves multiple genes and regulatory systems and requires biochemical pathways with several enzymatic steps. Pathway engineering, which consists of assembling new or improved biochemical pathways using genes from one or more organisms, has been the focus of most efforts in synthetic biology to date. However, completely novel pathways are expected to emerge in the coming years, reflecting the dynamic nature of this rapidly evolving field [34].

In addition to their potential in biofuel production, engineered bacteria have been explored for applications in bioremediation, where they can degrade agricultural waste, pollutants, and excess herbicides. They have also been designed as biosensors or for antimicrobial purposes [35, 36]. Moreover, microorganisms are widely used in manufacturing various products, including alcohol, solvents, biofuels, food additives, dyes, and antibiotics. Efficient pathways that convert otherwise useless materials into valuable products, such as alcohols, have been developed through biotechnology, which can be considered one of the earliest ventures into synthetic biology [37, 38].

13.5 Microbial Electrofuels

A good way to make renewable chemicals is to use solar energy and CO_2 to make renewable fuels and platform chemicals. Photosynthetic microorganisms can do this by using light reactions, which change solar energy into biochemical energy, and dark reactions, which use that energy to change CO_2 into complex molecules, all in one biological cell [39]. The idea of microbial electrofuels is different from this by separating the light and dark reactions, and using artificial devices such as photovoltaic cells instead of biological light reactions. In this process, light energy is first changed into electrons by photovoltaic electricity, which makes the biological change of CO_2 possible by using engineered pathways [40].

This approach has a big advantage because it can use solar or wind energy, changed into electricity, to make dark reactions happen, without needing to store electricity in batteries. The dark reactions can happen in big bioreactors that do not have the problem of light transfer that photobioreactors have, which are usually small or medium-sized. Also, the efficiency of changing solar energy into fuels and chemicals can be improved by making microbial electrofuels. Photosynthesis is not very efficient, because it only changes a little part of sunlight into biomass [41]. On the other hand, artificial devices like photovoltaic cells can change solar energy into electricity with efficiencies from 10 to 40% [42]. This means that less land would be needed for making renewable chemicals.

Microbial electrofuels have another benefit, which is that they can use CO_2 with different kinds of microorganisms, not only photosynthetic ones. Lithoautotrophic microorganisms, which get their energy from minerals, can also be used. Lithoautotrophs have different features and pathways, which give more options for making renewable fuels and chemicals. Photosynthetic microalgae and cyanobacteria only use one pathway for fixing CO_2, which is the Calvin- Benson-Bassham cycle, but lithoautotrophic microorganisms use different pathways that can be more efficient [43–46]. For example, some bacteria and archaea use the Wood-Ljungdahl pathway, which uses hydrogen as an electron donor and CO_2 as an electron acceptor and as a material for making products. This reduces CO_2 to formic acid and formyl and methyl groups [47, 48].

The production of chemicals and fuels while simultaneously reducing CO_2 emissions from the atmosphere using renewable energy sources is crucial to achieving a sustainable society. However, CO_2, the most oxidized form of carbon, requires substantial energy inputs to convert it into valuable products [49]. One promising research avenue for CO_2 conversion is bioelectrochemistry, which enables the production of more complex chemical compounds than purely electrochemical methods. This technology relies on the ability of microorganisms to take up electrons from solid-state electrodes, utilize them in their metabolism to convert CO_2 and excrete a reduced chemical as an electron sink. This electricity-driven microbial conversion of CO_2 is known as microbial electrosynthesis [50]. Figure 13.3 illustrates the six main products formed in microbial electrosynthesis to date, along with their current industrial production methods (depicted in red). Acetate production has been the dominant product in MES studies, with a greater diversification of the product spectrum observed only in recent years. The main industrial processes used for producing these chemicals are primarily fossil fuel-based and often require high temperatures and/or high pressures.

There are two main ways of giving electrons to microbes: directly from an electrode or through a substance that can carry electrons. Some research has found that some microorganisms, such as *Geobacter metallireducens* and *Methanobacterium*

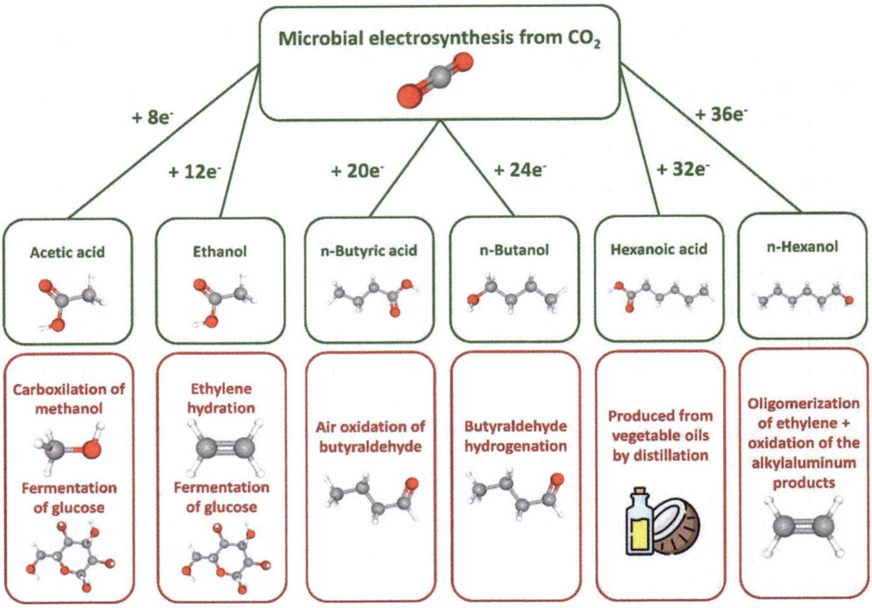

Fig. 13.3 Overview of the main products formed by microbial electrosynthesis from CO_2. Redrawn and adapted from its original source [51]. 3D structure compounds were obtained from the PubChem database [52]

palustre, can take electrons directly from electrodes, which makes them grow on a cathode as a microbial biofilm that can use electrons [53]. Other studies have also found many microorganisms, such as *Sporomusa* species, *Clostridium* species, and *Moorella thermoacetica*, that can take electrons directly from the electrode and make compounds like acetate from CO_2 [54]. These findings show the possibility of using the direct electron transfer way to make synthetic fuels from renewable electricity and CO_2.

Another way of giving electrons to the microbes in a water solution is by using substances that can carry electrons. These substances should be cheap, harmless to the environment, and easy to use, such as hydrogen (H_2) and formic acid (HCOOH). Hydrogen can be made by splitting water with light, which can also happen inside the bioreactor with the microorganisms, but it is hard to transfer it in water because it does not dissolve well [55]. Formic acid is a better option because it dissolves well in water and it is easy to handle as a liquid under normal conditions [56]. Also, many microorganisms can use formate-based pathways to make things. In general, finding substances that can carry electrons that are low-cost, non-toxic, and easy to handle is very important [57].

13.6 Industrialisation of CO_2 Bioconversion into Fuels

Examples of CO_2 bioconversion into fuels exist. One can be found within the steel sector carried out at ArcelorMittal facilities within the Steelanol project framework. It consists of a plant in Ghent (Belgium) aiming at producing 80 million liters of bioethanol which will be used in, for example, the transport sector (Fig. 13.4). This decarbonization initiative has received funding from various sources including, the European Union. The Steelanol plant is expected to avoid the emission of approximately 125 000 tons per year. The produced bioethanol is estimated to be blended with conventional gasoline and used as a low-carbon alternative fuel for the transport sector.

An example of innovation in biomethanation can be witnessed in the operations of Electrochaea. This company has demonstrated significant strides in mitigating risks associated with the biomethanation process through large-scale deployment for upgrading biogas waste streams into pipeline-quality renewable natural gas [27]. Biomethanation is a complex two-step process involving using hydrogenotrophic methanogenic microorganisms to convert renewable hydrogen and waste CO_2 into renewable methane, the primary component of natural gas. The approach involves harnessing biogenic CO_2 from various sources such as dairies, wastewater treatment plants, and landfills, to produce a drop-in direct replacement fuel that aligns with the emerging carbon market trend. However, in cases where industrial CO_2 utilization is pursued, a cleaning step similar to the one elucidated in the Steelanol project shown above may be necessary [59].

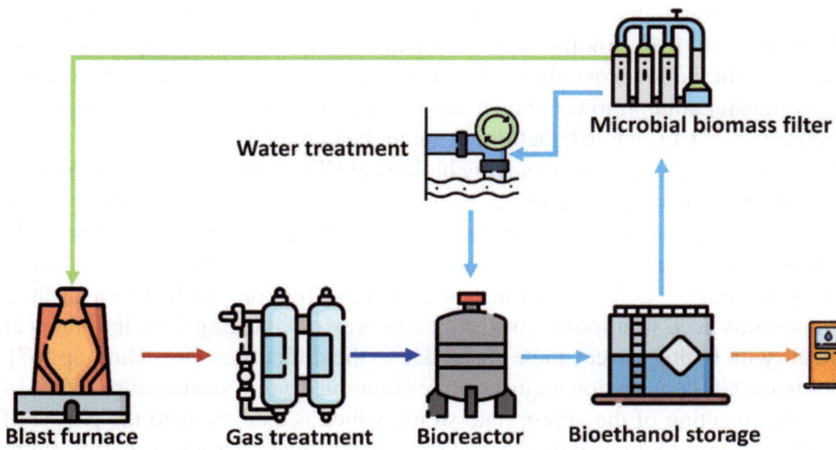

Fig. 13.4 Simplified flow sheet of Steelanol® process for the bioconversion of CO_2 within blast furnace flue gas into ethanol. Redrawn and adapted from its original source [58]

13.7 Issues to Be Solved by Research and Concluding Remarks

Using CO_2 as a raw material for making bio-based products is a promising opportunity, because there are many microorganisms that can use CO_2 and we can improve them with genetic engineering, metabolic engineering, and synthetic biology. This way, we can use CO_2 that is wasted by industrial processes and make it into something useful, instead of using fossil fuels that harm the environment. But, there are still some problems that need to solved before scaling-up the technology. Scientists, engineers, and governments need to work together and they need to have the right infrastructure and policies to support this.

Making biofuels from microalgae might be one promising method to use concentrated CO_2 streams biologically [60]. There has been progress in growing microalgae on a large scale, but most of the work has been on growing algae in open ponds with CO_2 from the air as a carbon source.

However, considering the substantial amount of CO_2 emitted by the industry, the utilization of concentrated CO_2 streams is highly relevant [61]. The use of photobioreactors, which offer a confined environment for microalgae cultivation, can overcome the land requirement limitations of open ponds, as well as issues of low biomass productivity and contamination. Photobioreactors need a higher investment to build and run, which can make the biofuels more expensive. But, photobioreactors can use CO_2 streams that are more concentrated, like the ones that come from industries. Challenges associated with photobioreactor design include optimizing light transfer efficiency, addressing concerns of oxygen poisoning and heat removal, and avoiding poor growth due to insufficient irradiation or photoinhibition from excessive light intensity [62]. Another approach that combines the advantages of photobioreactors with the low cost offered by high-rate algae ponds is the inclusion of a high-gas-transfer column/tower, which allows microalgae to be in contact with high concentrations of CO_2 while benefiting from high sunlight exposure [63].

There is not much progress in making biological CO_2-fixation processes based on lithoautotrophic microorganisms work on a large scale [64, 65]. Most of the research is in an early stage of development. In principle, lithoautotrophic microbes could be grown in big bioreactors with H_2 and CO_2 as the main feedstocks. Gases can be used to go through a culture with microbes that are free or attached to a surface. But, conditions need to improve [66]. Any new way of changing CO_2 into fuels or chemicals with biology needs to be tested due to the difficulties for scaling-up [67].

Biological CO_2 conversion requires some common design considerations. One of them is the selection of the appropriate strain, which is analogous to the choice of the catalyst for non-biological methods of CO_2 transformation, such as thermochemical, electrochemical, or photochemical processes [68]. Factors such as cell density, growth characteristics, CO_2 fixation rate, product type, and yield should be taken into account [69]. Another consideration is the design of the bioreactor, which can vary depending on the mode of operation (batch or continuous), the type of reactor (stirred tank, fixed bed, or photobioreactor), and the source of energy (light or electricity) [70, 71]. Finally, besides technical aspects, socio-economic factors such as research funding, government policies, and incentives for CO_2 utilization should also be considered [72].

Acknowledgements The project, RE4Industry, is supported by the Horizon 2020 research and innovation program of the European Union, with the grant number 952936. Fig. 13.4 has been designed using images from Flaticon.com.

References

1. Lamb WF, Wiedmann T, Pongratz J, Andrew R, Crippa M, Olivier JGJ, Wiedenhofer D, Mattioli G, Khourdajie AA, House J, Pachauri S, Figueroa M, Saheb Y, Slade R, Hubacek K, Sun L, Ribeiro SK, Khennasdela S, RueduCan S, Minx J (2021) A review of trends and drivers of greenhouse gas emissions by sector from 1990 to 2018. Environ Res Lett 16(7):73005. https://doi.org/10.1088/1748-9326/abee4e
2. Carmona-Martínez AA, Fresneda-Cruz A, Rueda A, Birgi O, Khawaja C, Janssen R, Davidis B, Reumerman P, Vis M, Karampinis E, Grammelis P, Jarauta-Córdoba C (2023) Renewable power and heat for the decarbonisation of energy-intensive industries. Processes 11(1). https://doi.org/10.3390/pr11010018
3. Simakov DSA (2017) Renewable synthetic fuels and chemicals from carbon dioxide: fundamentals, catalysis, design considerations and technological challenges. Springer
4. Wang X, Song C (2020) Carbon capture from flue gas and the atmosphere: a perspective. Front Energy Res 8.https://doi.org/10.3389/fenrg.2020.560849
5. Meng F, Meng Y, Ju T, Han S, Lin L, Jiang J (2022) Research progress of aqueous amine solution for CO_2 capture: a review. Renew Sustain Energy Rev 168:112902. https://doi.org/10.1016/j.rser.2022.112902
6. Nemitallah MA, Habib MA, Badr HM, Said SA, Jamal A, Ben-Mansour R, Mokheimer EMA, Mezghani K (2017) Oxy-fuel combustion technology: current status, applications, and trends. Int J Energy Res 41(12):1670–1708. https://doi.org/10.1002/er.3722
7. Pisciotta M, Pilorgé H, Feldmann J, Jacobson R, Davids J, Swett S, Sasso Z, Wilcox J (2022) Current state of industrial heating and opportunities for decarbonization. Progr Energy Combust Sci 91:100982. https://doi.org/10.1016/j.pecs.2021.100982
8. Abuelgasim S, Wang W, Abdalazeez A (2021) A brief review for chemical looping combustion as a promising CO_2 capture technology: fundamentals and progress. Sci Total Environ 764:142892. https://doi.org/10.1016/j.scitotenv.2020.142892
9. Lyngfelt A, Brink A, Langørgen Ø, Mattisson T, Rydén M, Linderholm C (2019) 11,000 h of chemical-looping combustion operation—where are we and where do we want to go? Int J Greenhouse Gas Control 88:38–56. https://doi.org/10.1016/j.ijggc.2019.05.023
10. Gambhir A, Tavoni M (2019) Direct air carbon capture and sequestration: how it works and how it could contribute to climate-change mitigation. One Earth 1(4):405–409. https://doi.org/10.1016/j.oneear.2019.11.006
11. Erans M, Sanz-Pérez ES, Hanak DP, Clulow Z, Reiner DM, Mutch GA (2022) Direct air capture: process technology, techno-economic and socio-political challenges. Energy Environ Sci 15(4):1360–1405
12. Kuramochi T, Ramírez A, Turkenburg W, Faaij A (2012) Comparative assessment of CO2 capture technologies for carbon-intensive industrial processes. Progr Energy Combust Sci 38(1):87–112. https://doi.org/10.1016/j.pecs.2011.05.001
13. Bains P, Psarras P, Wilcox J (2017) CO_2 capture from the industry sector. Progr Energy Combust Sci 63:146–172. https://doi.org/10.1016/j.pecs.2017.07.001

14. Yaashikaa PR, Senthil Kumar P, Varjani SJ, Saravanan A (2019) A review on photochemical, biochemical and electrochemical transformation of CO_2 into value-added products. J CO_2 Utilizat 33:131–147. https://doi.org/10.1016/j.jcou.2019.05.017
15. Farquhar GD, von Caemmerer S, Berry JA (2001) Models of photosynthesis. Plant Physiol 125(1):42–45. https://doi.org/10.1104/pp.125.1.42
16. Gebremariam SN, Marchetti JM (2018) Economics of biodiesel production: review. Energy Convers Manage 168:74–84. https://doi.org/10.1016/j.enconman.2018.05.002
17. Weber C, Farwick A, Benisch F, Brat D, Dietz H, Subtil T, Boles E (2010) Trends and challenges in the microbial production of lignocellulosic bioalcohol fuels. Appl Microbiol Biotechnol 87(4):1303–1315. https://doi.org/10.1007/s00253-010-2707-z
18. Nestler F, Krüger M, Full J, Hadrich MJ, White RJ, Schaadt A (2018) Methanol synthesis—industrial challenges within a changing raw material landscape. Chemie Ingenieur Technik 90(10):1409–1418. https://doi.org/10.1002/cite.201800026
19. Barber J (2009) Photosynthetic energy conversion: natural and artificial. Chem Soc Rev 38(1):185–196. https://doi.org/10.1039/B802262N
20. Jajesniak P, Ali H, Wong TS (2014) Carbon dioxide capture and utilization using biological systems: opportunities and challenges. J Bioprocess Biotech 4(155):2
21. Xie D (2022) Continuous biomanufacturing with microbes—upstream progresses and challenges. Curr Opin Biotechnol 78:102793. https://doi.org/10.1016/j.copbio.2022.102793
22. Zhang Y-HP, Sun J, Ma Y (2017) Biomanufacturing: history and perspective. J Ind Microbiol Biotechnol 44(4–5):773–784. https://doi.org/10.1007/s10295-016-1863-2
23. Reddy CSK, Ghai R, Rashmi C, Kalia VC (2003) Polyhydroxyalkanoates: an overview. Bioresource Technol 87(2):137–146. https://doi.org/10.1016/S0960-8524(02)00212-2
24. Jacob A, Ashok B, Alagumalai A, Chyuan OH, Le PTK (2021) Critical review on third generation micro algae biodiesel production and its feasibility as future bioenergy for IC engine applications. Energy Convers Manage 228:113655. https://doi.org/10.1016/j.enconman.2020.113655
25. Yew GY, Lee SY, Show PL, Tao Y, Law CL, Nguyen TTC, Chang JS (2019) Recent advances in algae biodiesel production: from upstream cultivation to downstream processing. Bioresour Technol Rep 7:100227. https://doi.org/10.1016/j.biteb.2019.100227
26. Arias DM, Ortíz-Sánchez E, Okoye PU, Rodríguez-Rangel H, Balbuena Ortega A, Longoria A, Domínguez-Espíndola R, Sebastian PJ (2021) A review on cyanobacteria cultivation for carbohydrate-based biofuels: cultivation aspects, polysaccharides accumulation strategies, and biofuels production scenarios. Sci Total Environ 794:148636. https://doi.org/10.1016/j.scitotenv.2021.148636
27. Ahrens T, Fontaine D, Hafenbradl D, Hoerl M, Pesic A, Silva KT (2021) Method to use industrial CO_2 containing gas for the production of a methane enriched gas composition. Google Patents
28. Zeng AP (2019) New bioproduction systems for chemicals and fuels: needs and new development. Biotechnol Adv 37(4):508–518. https://doi.org/10.1016/j.biotechadv.2019.01.003
29. Dong F, Simoska O, Gaffney E, Minteer SD (2022) Applying synthetic biology strategies to bioelectrochemical systems. Electrochem Sci Adv 2(6):e2100197. https://doi.org/10.1002/elsa.202100197
30. Liang L, Liu R, Freed EF, Eckert CA (2020) Synthetic biology and metabolic engineering employing escherichia coli for C2–C6 bioalcohol production. Front Bioeng Biotechnol 8. https://doi.org/10.3389/fbioe.2020.00710
31. Liu Z, Wang J, Nielsen J (2022) Yeast synthetic biology advances biofuel production. Curr Opin Microbiol 65:33–39. https://doi.org/10.1016/j.mib.2021.10.010
32. Lee OK, Lee EY (2016) Sustainable production of bioethanol from renewable brown algae biomass. Biomass Bioenergy 92:70–75. https://doi.org/10.1016/j.biombioe.2016.03.038
33. Bierbaumer S, Nattermann M, Schulz L, Zschoche R, Erb TJ, Winkler CK, Tinzl M, Glueck SM (2023) Enzymatic conversion of CO_2: from natural to artificial utilization. Chem Rev. https://doi.org/10.1021/acs.chemrev.2c00581

34. Sánchez-Muñoz S, Castro-Alonso MJ, Barbosa FG, Mier-Alba E, Balbino TR, Rubio-Ribeaux D, Hernández-De Lira IO, Santos JC, Aguilar CN, Da Silva SS (2022) Metabolic engineering of yeast, zymomonas mobilis, and clostridium thermocellum to increase yield of bioethanol. Bioethanol 97–139
35. Hicks M, Bachmann TT, Wang B (2020) Synthetic biology enables programmable cell-based biosensors. Chem Phys Chem 21(2):132–144. https://doi.org/10.1002/cphc.201900739
36. Rylott EL, Bruce NC (2020) How synthetic biology can help bioremediation. Curr Opin Chem Biol 58:86–95. https://doi.org/10.1016/j.cbpa.2020.07.004
37. Lv X, Wu Y, Gong M, Deng J, Gu Y, Liu Y, Li J, Du G, Ledesma-Amaro R, Liu L, Chen J (2021) Synthetic biology for future food: Research progress and future directions. Future Foods 3:100025. https://doi.org/10.1016/j.fufo.2021.100025
38. Prasad RK, Chatterjee S, Mazumder PB, Gupta SK, Sharma S, Vairale MG, Datta S, Dwivedi SK, Gupta DK (2019) Bioethanol production from waste lignocelluloses: a review on microbial degradation potential. Chemosphere 231:588–606. https://doi.org/10.1016/j.chemosphere.2019.05.142
39. Choi KR, Jiao S, Lee SY (2020) Metabolic engineering strategies toward production of biofuels. Curr Opin Chem Biol 59:1–14. https://doi.org/10.1016/j.cbpa.2020.02.009
40. Su-ungkavatin P, Tiruta-Barna L, Hamelin L (2023) Biofuels, electrofuels, electric or hydrogen?: a review of current and emerging sustainable aviation systems. Progr Energy Combust Sci 96:101073. https://doi.org/10.1016/j.pecs.2023.101073
41. Luan G, Zhang S, Lu X (2020) Engineering cyanobacteria chassis cells toward more efficient photosynthesis. Curr Opin Biotechnol 62:1–6. https://doi.org/10.1016/j.copbio.2019.07.004
42. Benda V, Černá L (2020) PV cells and modules—state of the art, limits and trends. Heliyon 6(12):e05666. https://doi.org/10.1016/j.heliyon.2020.e05666
43. Antonovsky N, Gleizer S, Milo R (2017) Engineering carbon fixation in E. coli: from heterologous RuBisCO expression to the Calvin–Benson–Bassham cycle. Curr Opinn Biotechnol 47:83–91. https://doi.org/10.1016/j.copbio.2017.06.006
44. Liao JC, Mi L, Pontrelli S, Luo S (2016) Fuelling the future: microbial engineering for the production of sustainable biofuels. Nat Rev Microbiol 14(5):288–304. https://doi.org/10.1038/nrmicro.2016.32
45. Liu X, Xie H, Roussou S, Lindblad P (2022) Current advances in engineering cyanobacteria and their applications for photosynthetic butanol production. Curr Opin Biotechnol 73:143–150. https://doi.org/10.1016/j.copbio.2021.07.014
46. Zeng J, Wang Z, Chen G (2021) Biological characteristics of energy conversion in carbon fixation by microalgae. Renew Sustain Energy Rev 152:111661. https://doi.org/10.1016/j.rser.2021.111661
47. Ragsdale SW (2008) Enzymology of the wood–ljungdahl pathway of acetogenesis. Ann New York Acad Sci 1125(1):129–136. https://doi.org/10.1196/annals.1419.015
48. Ragsdale SW, Pierce E (2008) Acetogenesis and the wood–ljungdahl pathway of CO_2 fixation. Biochimica et Biophysica Acta (BBA)—Prot Proteom 1784(12):1873–1898. https://doi.org/10.1016/j.bbapap.2008.08.012
49. Aresta M, Dibenedetto A, Angelini A (2013) The changing paradigm in CO_2 utilization. J CO_2 Utilizat 3–4:65–73. https://doi.org/10.1016/j.jcou.2013.08.001
50. Rabaey K, Rozendal RA (2010) Microbial electrosynthesis—revisiting the electrical route for microbial production. Nat Rev Microbiol 8(10):706–716. https://doi.org/10.1038/nrmicro2422
51. Jourdin L, Burdyny T (2021) Microbial electrosynthesis: where do we go from here? Trends Biotechnol 39(4):359–369
52. Compound summary (2023) National library of medicine, national center of biotechnology information
53. Moscoviz R, de Fouchécour F, Santa-Catalina G, Bernet N, Trably E (2017) Cooperative growth of Geobacter sulfurreducens and Clostridium pasteurianum with subsequent metabolic shift in glycerol fermentation. Sci Rep 7(1):44334. https://doi.org/10.1038/srep44334
54. Nevin KP, Woodard TL, Franks AE, Summers ZM, Lovley DR (2010) Microbial electrosynthesis: feeding microbes electricity to convert carbon dioxide and water to multicarbon extracellular organic compounds. MBio 1(2). https://doi.org/10.1128/mBio.00103-10

55. Fajrina N, Tahir M (2019) A critical review in strategies to improve photocatalytic water splitting towards hydrogen production. Int J Hydro Energy 44(2):540–577. https://doi.org/10.1016/j.ijhydene.2018.10.200
56. Bulushev DA, Ross JRH (2018) Towards sustainable production of formic acid. Chem Sus Chem 11(5):821–836. https://doi.org/10.1002/cssc.201702075
57. Mao W, Yuan Q, Qi H, Wang Z, Ma H, Chen T (2020) Recent progress in metabolic engineering of microbial formate assimilation. Appl Microbiol Biotechnol 104(16):6905–6917. https://doi.org/10.1007/s00253-020-10725-6
58. Steelanol recycles carbon into sustainable, advanced bio-ethanol (2023, April 29)
59. Lardon L (2017) Biocat–power to gas technology by biological methanation: integration to a resource treatment plant. Electrochaea GmbH, Planegg, Germany
60. Ganesan R, Manigandan S, Samuel MS, Shanmuganathan R, Brindhadevi K, Lan Chi NT, Duc PA, Pugazhendhi A (2020) A review on prospective production of biofuel from microalgae. Biotechnol Rep 27:e00509. https://doi.org/10.1016/j.btre.2020.e00509
61. Rissman J, Bataille C, Masanet E, Aden N, Morrow WR, Zhou N, Elliott N, Dell R, Heeren N, Huckestein B, Cresko J, Miller SA, Roy J, Fennell P, Cremmins B, Koch Blank T, Hone D, Williams ED, de la Rue du Can S, … Helseth J (2020) Technologies and policies to decarbonize global industry: review and assessment of mitigation drivers through 2070. Appl Energy 266:114848. https://doi.org/10.1016/j.apenergy.2020.114848
62. Sirohi R, Kumar Pandey A, Ranganathan P, Singh S, Udayan A, Kumar Awasthi M, Hoang AT, Chilakamarry CR, Kim SH, Sim SJ (2022) Design and applications of photobioreactors-a review. Bioresour Technol 349:126858. https://doi.org/10.1016/j.biortech.2022.126858
63. Marín D., Carmona-Martínez, A. A., Blanco, S., Lebrero, R., & Muñoz, R. (2021). Innovative operational strategies in photosynthetic biogas upgrading in an outdoors pilot scale algal-bacterial photobioreactor. Chemosphere 264.https://doi.org/10.1016/j.chemosphere.2020.128470
64. Kajla S, Kumari R, Nagi GK (2022) Microbial CO_2 fixation and biotechnology in reducing industrial CO_2 emissions. Arch Microbiol 204(2):149. https://doi.org/10.1007/s00203-021-02677-w
65. Liang B, Zhao Y, Yang J (2020) Recent advances in developing artificial autotrophic microorganism for reinforcing CO2 fixation. Front Microbiol 11. https://doi.org/10.3389/fmicb.2020.592631
66. Fackler N, Heijstra BD, Rasor BJ, Brown H, Martin J, Ni Z, Shebek KM, Rosin RR, Simpson SD, Tyo KE, Giannone RJ, Hettich RL, Tschaplinski TJ, Leang C, Brown SD, Jewett MC, Köpke M (2021) Stepping on the gas to a circular economy: accelerating development of carbon-negative chemical production from gas fermentation. Annu Rev Chem Biomol Eng 12(1):439–470. https://doi.org/10.1146/annurev-chembioeng-120120-021122
67. Stoll IK, Boukis N, Sauer J (2020) Syngas fermentation to alcohols: reactor technology and application perspective. Chemie Ingenieur Technik 92(1–2):125–136. https://doi.org/10.1002/cite.201900118
68. Gleizer S, Bar-On YM, Ben-Nissan R, Milo R (2020) Engineering microbes to produce fuel, commodities, and food from CO_2. Cell Rep Physic Sci 1(10):100223. https://doi.org/10.1016/j.xcrp.2020.100223
69. Köpke M, Simpson SD (2020) Pollution to products: recycling of 'above ground' carbon by gas fermentation. Curr Opin Biotechnol 65:180–189. https://doi.org/10.1016/j.copbio.2020.02.017
70. Morweiser M, Kruse O, Hankamer B, Posten C (2010) Developments and perspectives of photobioreactors for biofuel production. Appl Microbiol Biotechnol 87(4):1291–1301. https://doi.org/10.1007/s00253-010-2697-x
71. Prévoteau A, Carvajal-Arroyo JM, Ganigué R, Rabaey K (2020) Microbial electrosynthesis from CO_2: forever a promise? Curr Opin Biotechnol 62:48–57. https://doi.org/10.1016/j.copbio.2019.08.014
72. Ueckerdt F, Bauer C, Dirnaichner A, Everall J, Sacchi R, Luderer G (2021) Potential and risks of hydrogen-based e-fuels in climate change mitigation. Nat Clim Chang 11(5):384–393. https://doi.org/10.1038/s41558-021-01032-7

Part VIII
Future Trends

Chapter 14
Bioprocesses Coupling for Biohydrogen Production: Applications and Challenges

Jose Antonio Magdalena, María Fernanda Pérez-Bernal, María del Rosario Rodero, Eqwan Roslan, Alice Lanfranchi, Ali Dabestani-Rahmatabad, Margot Mahieux, Gabriel Capson-Tojo, and Eric Trably

Abstract The decarbonisation of industry based on the sustainable use of resources is one of the main objectives of our current society. To achieve this, rich-carbohydrate residual streams constitute a cost-effective feedstock from which hydrogen can be produced via dark fermentation (DF). In recent years, bench-scale testing has delivered encouraging results. Nonetheless, the low hydrogen productivity obtained still prevents the upscaling of this technology. A possible solution to overcome this technical barrier might be to couple DF with other available bioprocesses. The resulting coupling would enhance substrate exploitation and increase hydrogen productivity.

María Fernanda Pérez-Bernal, María del Rosario Rodero, Eqwan Roslan, Alice Lanfranchi, Ali Dabestani-Rahmatabad, Margot Mahieux and Gabriel Capson-Tojo authors contributed equally to this work

---

J. A. Magdalena (✉) · M. F. Pérez-Bernal · M. del Rosario Rodero · E. Roslan · A. Lanfranchi · A. Dabestani-Rahmatabad · M. Mahieux · G. Capson-Tojo · E. Trably (✉)
LBE, Univ Montpellier, INRAE, 102 Avenue des Étangs, 11100 Narbonne, France
e-mail: jose-antonio.magdalena-cadelo@inrae.fr

E. Trably
e-mail: eric.trably@inrae.fr

J. A. Magdalena
Vicerrectorado de Investigación y Transferencia de la Universidad Complutense de Madrid, 28040 Madrid, Spain

M. del Rosario Rodero
Institute of Sustainable Processes, University of Valladolid, 47011 Valladolid, Spain

E. Roslan
Department of Mechanical Engineering, College of Engineering, Universiti Tenaga Nasional, 43000 Kajang, Selangor, Malaysia

A. Lanfranchi
Dipartimento di Scienze Ambientali, Informatica e Statistica, Università Ca' Foscari Venezia, 30174 Mestre, Italy

M. Mahieux
ENGIE Lab CRIGEN, 4 Rue Joséphine Baker, 93240 Stains, France

V. Alcaraz Gonzalez et al. (eds.), *Wastewater Exploitation*, Springer Water,
https://doi.org/10.1007/978-3-031-57735-2_14

The biohydrogen produced could be used either as an energetic vector or as a platform molecule for added-value compound production. This chapter aims to comprehensively review the existing bioprocesses under investigation coupled with DF as a pivotal technology for biohydrogen production. More specifically, technologies such as microbial electrolysis cells, microalgae cultivation, biomethanation, photofermentation, and lactate production are evaluated. Aspects such as the optimal operational conditions that favour the coupling in each case and the hydrogen yields obtained, are reported. Furthermore, the advantages and disadvantages of the process couplings are also discussed. Finally, current challenges and future perspectives that each hydrogen production platform entails are pointed out to set the way forward in the coming years.

Keywords Bioeconomy · Biohydrogen · Bioproducts · Coupling processes · Dark fermentation

14.1 Dark Fermentation – Microbial Electrolysis Cells

Biological hydrogen production through the dark fermentation (DF) process is considered the most promising and viable method among other bioprocesses (*i.e.*, biophotolysis and photofermentation [1–3]. DF is a biological process where biomass can be anaerobically converted into hydrogen-rich biogas and a mixture of fermentative metabolites [4]. Nonetheless, the low hydrogen yield obtained due to a thermodynamic limit of 4 mol H_2 per mol of glucose is the main disadvantage of this technology [5–7]. The metabolites produced mainly comprise volatile fatty acids (VFAs) such as acetate and butyrate, propionate, and other acids such as lactate and ethanol. However, through DF, only 20–25% of the chemical oxygen demand (COD) of the initial organic substrate is converted into bio-H_2, while the remaining 75–80% is obtained in the form of the abovementioned fermentative metabolites [8]. For this reason, an integrated scheme treating the DF effluent with secondary processes is necessary to maximize COD recovery and ensure the economic viability of the process.

One of the most attractive options for the further use of VFAs is microbial electrochemical technologies (METs). [9]. These technologies are based on the ability of the so-called electroactive bacteria (EABs) to perform extracellular electron transfer (EET), which is a type of microbial respiration where electrons are transported through the cell wall to solid external electron donors or acceptors (*e.g.,* metals, electrodes) for energy metabolism [10, 11]. METs consist of a circuit between an anode and a cathode placed in one or two separate compartments, where redox reactions are bio-catalyzed in one or both electrodes. METs can be classified into two major categories according to the spontaneity of the reaction: (i) Microbial Fuel Cell (MFC), where the reactions take place spontaneously, and (ii) Microbial Electrolysis cell (MEC), where the reaction is not spontaneous, and energy input is required. The extra voltage is achieved by either setting the anode potential with a potentiostat and a reference electrode (three-electrode set-up) or by adding voltage with a Direct

Current (DC) power supply [12]. Both MFC and MEC technologies can be potentially used for treating DF effluents. While electrons provided by the oxidation of organic matter at the anode produce electricity in MFCs, hydrogen is produced in MECs at the cathode [3]. Overall, MECs show higher performance efficiencies [7]. Even though energy input is required for hydrogen formation at the cathode, it is minimal (0.2–0.8 V), especially when compared with traditional abiotic water electrolysis (1.23–1.8 V) [7, 13, 14]. Usually, MECs are designed as a two-chamber system (Fig. 14.1). In the anodic chamber, EABs defined as exoelectrogens [15] or anode-respiring microorganisms, develop a biofilm converting organic matter into protons and electrons, the latter cross the electric circuit from the anode to the cathode where protons are reduced into hydrogen. Other than avoiding short-circuiting between the electrodes, the separation between anodic and cathodic compartments, usually with an ion exchange membrane (IEM), keeps the purity of the hydrogen produced at the cathode.

Moreover, cathode colonization by electrotrophs is prevented, as well as hydrogen consumption by hydrogenotrophs such as homoacetogens or methanogens. The main drawbacks when using IEMs are internal resistance increases, substrate crossover from anode to cathode, biofouling, undesirable ion crossing, and pH splitting [16–18]. In this respect, pH is maybe the main disadvantage for bioanodes as they work more efficiently around neutrality [17]. Indeed, most known EABs are completely inhibited at pH below 6.

Two important parameters to evaluate MECs performances are the Coulomb efficiency (CE) and the current density (CD) [13] (see [19] for calculation). The CE represents to which extent the oxidized substrate is transformed into current (number of electrons delivered to the anode). At the same time, CD (A/m^2) indicates the

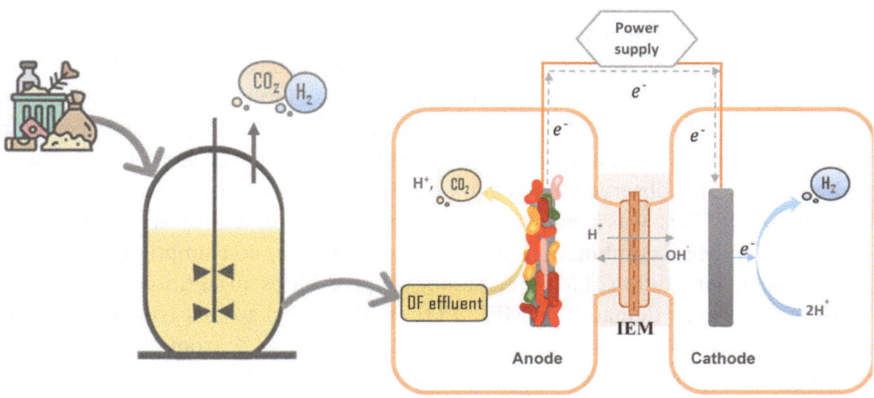

Fig. 14.1 Integrated DF and MEC process. During DF, glucose follows the overall reaction: $C_6H_{12}O_6 + 2H_2O \rightarrow 2CH_3COOH + 2CO_2 + 4H_2$. Acetate produced can be further oxidized at the anode: $CH_3COO^- + 2H_2O \rightarrow 2CO_2 + 7H^+ + 8e^-$, and electrons are delivered at the cathode to produce H_2: following: $2H_2O + 2e^- \rightarrow H_2 + 2OH^-$ With the integrated anode–cathode reactions: $CH_3COO^- + H^+ + 2H_2O \rightarrow 2CO_2 + 4H_2$ 4 mol H_2 are obtained per mol of acetate

number of electrons delivered per unit of time to the electrode, indicating how fast the substrate is oxidized. Cathodic hydrogen recovery is also routinely reported on MEC studies, indicating the ratio between the amount of hydrogen recovered and the theoretical amount based on the current measured. Ideally, this recovery should approach 100% when there is no hydrogen recycling, and the MEC is airtight enough to avoid losses. High CE (up to 90%) could be achieved when a true EAB community predominates in the system. Otherwise, substrate consumption diversion could occur in non-current generating reactions by non-EAB. On the other hand, CE above 100% could indicate MEC dysfunction (*e.g.*, hydrogen recycling).

A wide spectrum of substrates has been used as feedstock for MECs, from simple to complex industrial waste [20]. Even if complex substrates could be directly applied to MECs, performances are far behind those achieved with simpler ones due to the restricted substrate spectrum of EABs, which needs syntrophic partners to completely oxidize the most complex ones [7, 21]. For example, when it comes to VFAs, the model EAB *Geobacter sulfurreducens*, largely found in MECs, can only oxidize acetate. Other EABs, such as *Geobacter metallireducens* have a wider spectrum of substrates. Nevertheless, acetate remains the preferred organic acid as electron donor for MECs [20]. The need for a process like DF for obtaining a VFA-rich feedstock for MECs when treating complex substrates seems evident, making it a current research topic [3]. Some explicit DF + MEC coupling proposals are shown in Table 14.1. Certainly, DF well complements MEC as it efficiently breaks down large and complex organic compounds into low-molecular organic acids (*i.e.*, VFAs) that can be used by exoelectrogens [6]. This coupling greatly boosts hydrogen yields with a theoretical output of 12 mol H_2 per mol of glucose [7]. One of the first studies dealing with the coupling was carried out by Lu et al. in 2009 [14], reporting an overall hydrogen recovery of 96% of the maximum theoretical yield (0.125 gH_2/gCOD), with a buffered DF effluent.

The importance of acetate for the maintenance of an efficient anodic community has already been pointed out by Moreno et al. who worked with cheese whey in 2015 [22]. They diluted the DF effluent to reduce the effects of low pH on MEC. However, this resulted in a low CE and the need to add salts (K^+, Cl^-, $P0_4^{3-}$) and acetate to achieve an optimal acetate/lactate ratio as previously determined with a synthetic medium. They also observed high cathode methane production, probably due to H_2 reconsumption. This was confirmed by the occurrence of *CE* above 100%. Even though the DF-MEC combination can lead to lower energy consumption, a study has shown that the integration of MFC technology is possible as an additional technology to cover the energy demand for MEC [23]. With respect to this combination, the authors reported an overall hydrogen yield increase of 41% from cellulose. They also observed hydrogen reconsumption when working with a single chamber, as evidenced by an increasing CE of over 175% and zero hydrogen recovery at the end of the assays.

It has also been stated that the origin of the inoculum plays an important role in MEC performance [9]. However, the wastewater/nutrient influx is also an important issue, because its composition shapes the microbial structure by favouring or disadvantaging the electroactive community. This key feature was outrighted in a recent

Table 14.1. Examples of studies coupling DF+MEC for H2 production

DF type (substrate)	DF type (substrate)	Major DF metabolites/ MEC influent	MEC type/ operation	Anode/ cathode	Added/ applied voltage (V)	MEC conditions	CE (%)	H2cat recovery (%)	Total H2 recovery (%)	Ref
Molasses wastewater	Ethanol-type CSRT continuous ORL 22.8 kg COD/m3/d	EtOH, C2, C3, C4, C5, residual sugars	Single chamber Batch mode	Grafite brush/Carbon cloth with Pt	0.6 (DC power supply)	Buffered effluent pH 6.7–7.0 25 °C	87	83	69%	[14]
Cheese whey	Batch/35 °C	Lactate, C2, C4, C3	Membrane-free polycarbonate plates Continuous (10h, HRT)	Carbon felt /gas diffusion with Ni particles	1 (DC power supply)	Acetate and salts amended	80	N.M.	N.M.**	[22]
Cellulose	Continuous/ 60°C	C4, C2, C5, EtOH, C3	Batch membrane-free	Carbon brush/ Carbon cloth with Pt content	0.44 V (MFC supplying)	Buffered pH 7 25 °C	58–175	8.7–92	N.M.**	[23]

(continued)

Table 14.1. (continued)

DF type (substrate)	DF type (substrate)	Major DF metabolites/ MEC influent	MEC type/ operation	Anode/ cathode	Added/ applied voltage (V)	MEC conditions	CE (%)	H2cat recovery (%)	Total H2 recovery (%)	Ref
FJW, **VB2,** **CW,** **FPW,** **SW,** **PW**	Batch	**FJW:** C4, C2 **VB2:** C4, 1,3-PDO, C2, C3, Succ; C2; **CW:** EtOH, C4, C2, Succ; **FPW:** EtOH, C2, C4; **SW:** C2, C4, EtOH, C3, Succ; **PW:** C4, Succ, C2, EtOH, C3	Double chamber/ AEM	Carbon felt/ Pt-Ir mesh	0.44 V (anode potential vs SHE)	37 °C, pH adjusted (7)	76 75 75 80 38 33	101 65 62 53 61 53	115.02* 106.34* 59.84* 53.93* 28.33* 18.36*	[24]
Cassava starch (manioc)	Continuous/ UASB 55 °C 25.2 kg COD/ m3/d	C4, C2, C3, C5	Continuous up-low membrane-free	Graphite felt/Cu wire	0.6 V (DC power supply°	55 °C pH 6	N.M	N.M	33%*	[6]

CE = coulombic efficiency; H2cat= cathodic H2 recovery; *calculated with available data on the paper based on a molar volume (Vm=22.414) at standard temperature and pressure conditions; N.M. = not mentioned; ** not enough data to calculate; CW = cheese wey; FPW = fruit processing wastewater; SW = sugar production wastewater; FJW = industrial fruit juice production wastewater; VB2 = concentrated vinasse residue; PW = paper mill wastewater study where different effluent composition profiles from different substrates after undergoing DF, despite the same operational conditions [24]. These different profiles impacted MEC performances, with *CE* decreasing from 33 to 76%. It is important to mention that anodic enrichments were carried out under the same conditions, *i.e.*, the same type of inoculum, synthetic medium, and operational conditions.

MECs are usually operated under mesophilic temperature conditions and at neutral pH. Khongkliang et al. [6], demonstrated that MEC operation under thermophilic conditions is also possible. In their study, DF and MEC were fully integrated and operated in continuous mode (up-flow) under thermophilic conditions to treat a complex substrate (cassava starch processing wastewater). DF effluent was directly fed to MEC without pH amendment (pH 6). Interestingly, primary MEC enrichment was done at 55 °C and pH 6.5, which certainly favored the establishment of a thermophilic community and the acclimation to mildly acidic pH conditions. Concerning the microbial community composition found in the MEC, several specific representatives reported as thermophilic were observed with predomination of *Brevibacillus* sp., *Caloranaerobacter* sp., and *Geobacillus* sp. species that were very different from those "classically" found in MEC operated under mesophilic conditions.

Coupling DF and MEC instead of a single process to maximize hydrogen production is primarily advantageous. Mainly because this enables more efficient regulation of the individual processes. [18]. However, further efforts to improve overall hydrogen yields are required to scale up this two-process system, for example, by producing DF effluent with a profile composition favoring EABs. Moreover, studying the microbial community composition and the role of microbial interactions in electroactive biofilms are key aspects to better understand and improve MEC performances.

14.2 Dark Fermentation—Microalgae Cultivation

Microalgae cultivation coupled with DF is a promising technology to enhance substrate conversion to hydrogen and other high value-added compounds. Microalgae are unicellular eukaryotic microorganisms ubiquitously present in nature thanks to their metabolic versatility, exhibiting autotrophic, heterotrophic, and mixotrophic metabolisms. For simplicity, in this chapter, the term *microalgae* includes the prokaryotic cyanobacteria (green–blue algae) that share the same bioenergetic metabolism and biotechnological applications. Microalgae have gained attention because of their ability to convert carbon dioxide and organic compounds into high-added value molecules such as lipids, proteins, carbohydrates, and various secondary metabolites, among which are carotenoids (astaxanthin, β-Carotene), xanthophylls (lutein, zeaxanthin) and phycocyanin [25]. So far, the economic and environmental sustainability of large-scale microalgae farming has been hampered by the high energy requirements, especially in the harvesting and extraction phases, and the need for low-cost nutrient sources, especially nitrogen and phosphorus [26–29].

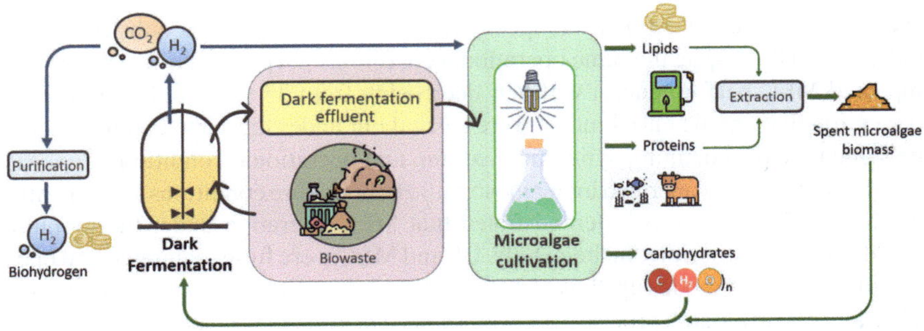

Fig. 14.2 Coupling of DF and microalgae cultivation: conceptual scheme outlining the main processes and outputs. DF generates biogas and an effluent which can be supplied as substrates for microalgae growth and storage of lipids, proteins, and carbohydrates. Carbohydrates can be eventually recirculated as DF feedstock

Coupling DF and microalgae cultivation (Fig. 14.2) can improve the sustainability of both processes in a biorefinery approach, which is envisaged for the transition to bioeconomy [30, 31]. DF effluents as cultivation media for microalgae provide VFAs as an inexpensive source of organic carbon, yielding higher biomass and added-value compounds concentrations and productivities concerning autotrophy [32]. This, in turn, can improve the efficiency of the harvesting and extraction steps. Moreover, DF effluents can contain enough N and P to sustain microalgae growth in ammonium and orthophosphate due to the mineralization occurring during DF [33]. As shown in Fig. 14.2, microalgae could also upgrade the biogas by fixing the carbon dioxide that it contains and providing a higher hydrogen content in the biogas (up to 85% v/v H2) [34]. This process has been extensively studied with methane-rich biogas produced by AD, obtaining a 54–99% v/v CO_2 removal and 65–97% v/v CH_4 recovery [35], while studies on the hydrogen-rich biogas generated via DF are moving their first steps, with a promising 85% v/v CO_2 removal and fixation rate of 95 mL CO_2/L/h [34].

Unlike the other processes coupled to DF, microalgae cultivation does not convert the remaining COD directly into hydrogen. The result is a biomass rich in valuable compounds, including up to 71.1% lipids, 63% proteins and 80% carbohydrates (DM basis) in a percentage that depends on the strain and culture conditions (Fig. 14.2) [36–39]. Carbohydrate-rich biomass might be recirculated as DF feedstock, thus enhancing the hydrogen yield of the whole process. For instance, an experimental yield of 0.93 mol H_2/mol reduced sugars was obtained from a recirculated hydrolyzed *C. vulgaris* [34, 40]. Also, the spent biomass after high-added value compounds extraction constitutes a suitable substrate for DF. For instance, the fermentation of *Dunaliella salina* lipid-extracted biomass resulted in a high biohydrogen yield of 192 mLH_2/gVS [41].

Multiple interacting factors affect the coupling of DF with microalgae cultivation, from both the abiotic (effluent composition, pH, C:N:P ratio, illumination conditions, feeding mode, and process configurations) and biotic (bacterial and microalgal strains and their interactions) environment [42]. In particular, for optimal coupling, DF should be directed towards the acetate hydrogen production pathway due to (i) the higher theoretical hydrogen yield and (ii) the high acetate assimilability by many microalgae species, which seems to boost lipids production [43, 44].

Conversely, butyrate uptake is a major bottleneck in coupling the two processes, and the underlying metabolic mechanisms, mainly studied for the model microalga *Chlamydomonas reinhardtii*, are only partially understood [44, 45]. A significant breakthrough has been reported by [46], who proposed a butyrate metabolic network for the non-photosynthetic microalga *Polytomella* sp. through a proteomics approach. After entering the cell through membrane-bound transport proteins, butyrate would be activated to butyryl-CoA before entering the β-oxidation pathway in the peroxisomes [46]. Unlike acetate, butyrate lowered the accumulation of storage products with simultaneous induction of fatty acids synthesis. These fatty acids probably served for peroxisomes reorganization and the production of enzymatic cofactors involved in butyrate assimilation. Analysis of the butyrate-related metabolic network of *Polytomella sp.* identified the issues to be tackled to understand the poor butyrate assimilation in green microalgae, potentially serving as a metabolic reference [46]. DF effluents are principally composed of a mixture of acetate and butyrate. In such cases, diauxic growth was observed, and butyrate consumption started only after acetate depletion [47–49]. Acetate was consumed after 1.5–3 days, sustaining a microalgal growth rate of 3.4–0.81 d^{-1} depending on the strain, while butyrate was consumed after 6–10 days and resulted in a lower growth rate of 1.28–0.28 d^{-1} [36, 47]. *Polytomella* sp. stood out as the most rapidly growing strain on acetate and butyrate, with a growth rate of 4.1 d^{-1} and 2.5 d^{-1}, respectively [36]. Whereas acetate concentrations as high as 30 g/L were used to support the growth of *C. sorokiniana* and *A. prototechoides* at pH 6.8 [49], butyrate was reported to inhibit their growth at concentrations as low as 0.1 and 0.5 g/L, respectively, at pH 6.5 [47]. The mechanism underlying butyrate inhibition has been recently clarified and is detailed elsewhere [50]. It is important to highlight that microalgae growth is strongly affected by the undissociated form of the organic acids (ROOH), which rises as the pH lowers (pKa of VFAs ~ 4.8–4.9). Therefore, ROOH concentration can be maintained under the inhibitory threshold by controlling the pH at alkaline values. The inhibitory threshold is species-specific, with maximum ROOH concentrations ranging from 71 to 207 mg/L for acetic acid and 13–25 mg/L for butyric acid for the five most commonly cultivated strains [50].

In microalgae cultivation on DF effluents, pH determines the chemical form of VFAs, the VFA's chemical species, and the total ammonium nitrogen content (TAN, *i.e.*, free ammonia and ammonium). Ammonium is the optimal nitrogen source for microalgae growth, while the other forms of nitrogen need to be reduced to ammonium ions [51]. Additionally, despite ammonium being the preferred form for microalgae utilization, high TAN levels can cause inhibition (50–260 mg TAN/L), varying remarkably depending on the microalgae strain and cultivation conditions [52]. Although some studies reported that the toxic effect of TAN is mainly attributable to ammonium [53], other researchers proposed ammonia as the major inhibitor of microalgal growth [54, 55]. Since ammonia concentration rises with the pH, this parameter should be monitored depending on the DF effluent composition concerning VFAs and TAN content.

The C:N:P ratio of the DF effluent depends on the substrates used in DF and strongly influences the microalgae growth and their macromolecular composition. Considering the Redfield ratio 106:16:6 as a reference for average algal biomass, nutrient-replete conditions support biomass growth. At the same time, nitrogen starvation seems to trigger storage product accumulation, namely lipids or carbohydrates, depending on the strain [48, 56, 57]. Therefore, optimizing a two-phase cultivation strategy with a nutrient-replete growth phase followed by a nitrogen starvation storage phase can improve the overall conversion of COD to storage products. Fed-batch cultivation mode can be applied in the first stage, thus achieving a high cell density that facilitates the harvesting and extraction [58, 59].

Illumination is a fundamental factor for microorganisms supporting an autotrophic metabolism. Mixotrophy can increase the titer of microalgae cultures by a yield equal to or higher than the sum of the yields obtained under autotrophy and heterotrophy [60, 61]. Under mixotrophy, autotrophic and heterotrophic metabolisms can boost each other. The organic substrate metabolization releases carbon dioxide, which is directly used in autotrophic metabolism. The oxygen produced during photosynthesis is available in turn for cell respiration. In continuous processes, respiratory oxygen consumption and phototrophic oxygen production can be counterbalanced by adjusting the rate at which the organic carbon source is provided to the microalgae culture [62]. Moreover, mixotrophy can alleviate butyrate inhibition thanks to the autotrophic generation of part of the biomass which can consume it [63]. Another advantage of mixotrophy is the enhancement of some cellular processes (*e.g.*, lipids and carbohydrates storage) and metabolite production associated with light (*e.g.*, astaxanthin, β-carotene) [43, 64–66].

Regarding biotic factors, microalgae-bacteria interactions play a fundamental role in the coupling, especially with a perspective of full-scale application, where effluent sterilization would be economically unsustainable. The process can be positively affected by synergistic interactions, such as the gas exchange between microalgae (O_2) and aerobic bacteria (CO_2), or negatively impacted by substrate competition, namely for acetate consumption. *C. sorokiniana* was shown to outcompete bacteria for acetate when heterotrophically grown on a real DF effluent containing acetate and butyrate [63]. When the aerobic bacterial strains become dominant in the originally anaerobic DF consortium, they can consume butyrate, but this ability depends on the microbial composition of the consortium [63, 67, 68]. When evaluating a microalgae-bacteria consortium, the main obstacle is to differentiate the VFAs uptake by microalgae and bacteria, respectively. Microalgae growth on labeled carbon (^{13}C/^{14}C) followed by flux cytometry and cell sorting could be a feasible approach to measure carbon incorporation [69]. Moreover, the carbon dioxide generated by VFAs degradation and not fixed by the biomass should be quantified. Finally, the selection of microalgae strains should focus on tipping the scales in microalgae's favor considering the consortium. This can be done by selecting or adapting strains able to consume butyrate or using microalgae strains that prey on bacteria such as *Ochromonas danica* [70]. A significant step forward in this direction has been made with the fast-butyrate consuming strain *Polytomella* sp., which yielded 0.65 g carbohydrates/g biomass [36]. However, lipid-accumulating microalgae with the same ability to consume butyrate remain to be found. This can be obtained by further exploring the biodiversity or by improving the already known strains (*i.e.*, through genetic modification or adaptive laboratory evolution (ALE)) [71].

To sum up, acetate and hydrogen should be properly targeted in the first step of DF. In contrast, a mixotrophic two-phase microalgae cultivation at controlled pH can favor a high biomass productivity and product storage in the second step. The main bottleneck is the metabolization of longer chain VFAs, which several strategies can overcome, eventually combined: (i) exploring the microalgal biodiversity to find butyrate-consuming strains producing the desired storage product, (ii) performing ALE on the strains already known and (iii) exploiting the synergistic interactions of microalgae-bacteria consortia.

14.3 Dark Fermentation—Anaerobic Digestion Based Processes

Anaerobic Digestion (AD) is a mature and well-established process that has been applied at an industrial scale for decades [72]. Nonetheless, the hydrolysis step still hinders the total exploitation of the substrate, especially when dealing with complex chemical structures. For this reason, hydrolysis enhancement, together with hydrogen and metabolite productions achieved in DF, has recently gained a lot of interest in the so-called two-stage AD concept [73, 74]. This configuration allows the energetic potential optimization of the organic matter employed as feedstock to increase the overall methane yield while producing hydrogen simultaneously in the first step. More recently, biomethanation (Fig. 14.3) has been considered as another strategy to benefit from hydrogen produced during DF by increasing the methane content in the biogas produced during AD thanks to hydrogen injections [75]. Coupling biomethanation with DF might thus allow the development of the next generation of two-stage AD processes, as depicted in Fig. 14.3 [76].

Two-stage DF-AD coupling

A wide range of organic residual streams (manure, straw, food waste, sewage sludge, among others) can be used as substrates in AD or DF [77–79]. However, due to their hydrogen or methane production potential (relying on factors such as their carbohydrate content), there might be an interest in deploying AD, DF, or both [80]. Coupling DF and AD can be an interesting way to improve waste management strategies. The main strength that justifies the coupling is that the effluent obtained from DF is the result of simultaneous partial degradation of organic matter and hydrogen production. As a result, high VFA (such as acetate, propionate, and butyrate), ethanol, and lactate concentrations in the effluent can be further valorized in the subsequent AD unit [74]. Some investigations have compared the Energy Recovery (ER) between

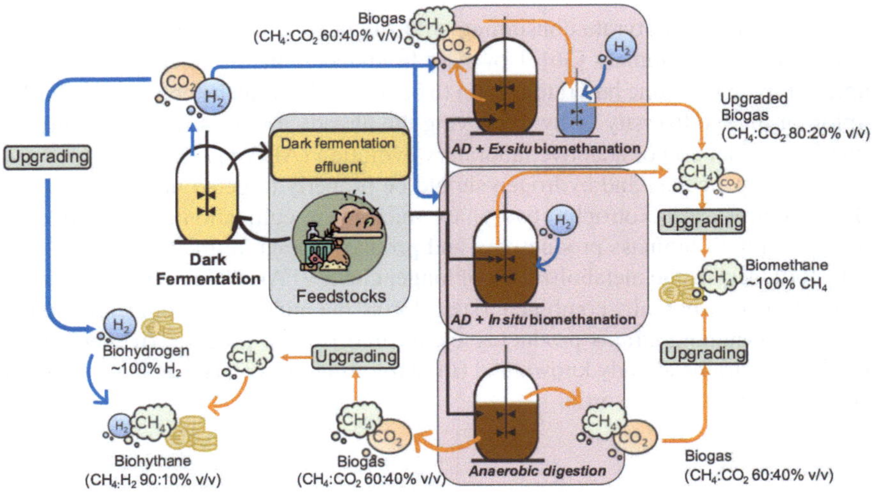

Fig. 14.3 Coupling of DF and AD based processes (two stage-AD and biomethanation); black arrows stand for organic matter flow whereas blue and orange arrows represent hydrogen and methane flow, respectively

single-stage AD and two-stage AD to determine the economic interest of the coupling [81]. Indeed, it was reported that the ER obtained from the coupling was 20–60% higher than in single-stage AD [79, 82–84]. Nonetheless, the ER was not equally distributed among the two stages. According to [79] and confirmed by [81], the highest ER was achieved when the hydrogen produced was a small fraction (*i.e.*, 5–10%vol). This result suggests that the optimal coupling configuration uses DF to improve substrate accessibility through the acidogenic environment in DF, further promoting the methanogenic step in AD. To achieve such improvement, accurate optimization of both stages is required. For this reason, operational parameters such as temperature (mesophilic or thermophilic) [85], pH (acidic in the first step vs. alkaline in the second) [79], ammonium and free ammonia concentrations (methane inhibition at NH_3 concentrations higher than 700 mg/L) [86, 87], initial substrate pretreatment (*e.g.*, thermal or chemical) [82, 84] and digestate recirculation [87, 88] have been shown to impact the yields, stability and efficiency of the coupling. In addition, applied Hydraulic Retention Time (HRT) and Organic Loading Rate (OLR) should be adjusted considering each process performance but also to ensure the coupling synergy [89–91]. As an illustration, Luo et al. [89] showed a 6.7% improvement in energy generation by applying an HRT ratio of 1:14 instead of 3:12 (days:days), respectively, to DF and AD.

The energy recovery of the two-stage AD can also be enhanced by improving the degree of degradation of organic matter. For that purpose, different reactor configurations for DF and AD can be used. Whereas conventional Continuous Stirred-Tank Reactors (CSTR) are mainly chosen to perform DF, several reactor configurations, such as Up-Flow Anaerobic Sludge blanket (UASB) or fixed bed reactors, studied at bench scale, were in favor of higher ER in AD. As shown by De Souza Almeida et al. [92] obtained an improvement of 47% of the ER when using an anaerobic fluidized bed reactor for co-digestion of cheese when and glycerol [92]. However, those configurations are not suitable for all feedstocks, which might limit their use at a larger scale [90].

Another key aspect that should be considered is the two types of microorganisms that should be promoted for each process: fermentative bacteria in DF and methanogenic archaea in AD [93]. It is well known that the growth rate of fermentative bacteria is much higher than the one observed for methanogenic archaea (*e.g.*, 0.125 vs. 1.5–7 days for fermentative bacteria and methanogenic archaea, respectively) [94]. As a result, fermentation kinetics are more rapid than methanogenesis, resulting in lower DF reactor volumes due to lower HRTs applied (hours or a few days in DF vs several weeks in AD) [5].

Finally, the coupling of DF and AD has also been applied to produce biohythane (*i.e.*, a mixture of hydrogen and methane containing 5 to 20% v/v of H_2) [95]. The added volume of H_2 in the gas grid will gradually increase in the coming years. Some recent estimations within the European Union suggest that this value can rise from 5–10% to 15–20% by 2030 [96]. Nevertheless, this theoretical value is never reached in natural gas and is subject to controversy regarding synthetic methane (CH_4 produced through methanation processes, either biologicals or chemicals). Biohythane production has several advantages over methane production, such as lower ignition temperature, a wide flammability range, and reduced NOx emissions [97, 98]. Moreover, the mass-specific heating value of biohythane (119.930 kJ/kg) is 2.5 times higher than the one of biomethane (50.020 kJ/kg) [97]. Furthermore, methane and hydrogen production from two-stage AD production through the coupling mentioned above allows anaerobic digestion to operate at higher OLR and solid removal efficiency, both in lower HRT [99, 100]. Subsequently, a technically relevant way to increase biohythane production will rely on addressing two-stage AD optimization.

Biomethanation

Biomethanation is a bioprocess in which hydrogen and carbon dioxide are converted into methane. From an operational point of view, biomethanation can be done either in situ [101] or ex situ [102]. During in situ biomethanation, hydrogen is injected in the same anaerobic digester where biogas is produced from organic substrates (Fig. 14.3). As for ex situ biomethanation, biogas is transferred to another bioreactor, and hydrogen is mixed with only the biogas allowing either pure culture or mixed culture of archaea to convert hydrogen and carbon dioxide into methane. Here, an external source of hydrogen is needed to perform biomethanation. A way to obtain this compound is from water electrolysis, where the excess electricity obtained from renewable resources is used to produce hydrogen, a concept referred to as Power-to-Gas (PtG) [35, 103]. PtG is the main coupling concept when referring to biomethanation [89]. Nonetheless, PtG projects associated with biomethanation still own the fewest installed power (in MW_{el} terms) compared to hydrogen and chemical methane formation [103]. Despite a rapid fall in the capital expenditure for electrolysis technology (*i.e.*, from 1300 €/kW_{el} in 2017 to 500 €/kW_{el} predicted by 2050 [103]), the electricity price and consumption, as well as the maintenance of those devices, still represent a major part of methane annual production cost with PtG [104, 105]. In addition, the combination of drastically different technologies (*i.e.*, water electrolysis and biomethanation) is a technical barrier at operational and societal levels [106, 107]. As a possible solution, DF could be used instead of water electrolysis as a bio-based technology to produce hydrogen. DF would contribute to better waste management and improve methane production (Table 14.1) [73]. Using DF, some associated costs derived from the upgrade and storage of the gas mixture generated in DF (H_2:CO_2, 50:50% v/v) should be considered [108, 109]. However, carbon dioxide presence might be useful to stabilize the H_2:CO_2 ratio during the biomethanation process, preventing carbon dioxide depletion from the gas phase and associated pH drop and acetate accumulation [110]. The compatibility between DF and biomethanation for feeding, maintenance, and gas production control is also crucial to envision the future use of this technology as it allows the industrial development of existing AD facilities [76]. The main challenges that should be faced in the coming years are related to hydrogen production and consumption optimization. In particular, specific objectives such as (i) which feedstocks should be employed considering their hydrogen and methane production potentials, (ii) to pursue the development of adapted equipment (responding to legislation about the use of hydrogen), and (iii) to overcome limitations resulting from the hydrogen low gas–liquid mass transfer, have to be faced to allow the development of this coupling at industrial scale. Indeed, the gas–liquid mass transfer rate remains the process bottleneck when hydrogen is converted to methane either by in situ or ex situ biomethanation [111]. The main reason lies in the physicochemical properties of hydrogen gas (solubility 1.6×10^{-4} g/100 g water, Henry constant 7.8×10^{-4} mol/kg/bar), which limit its methanogen consumption [112]. To overcome those boundaries, several strategies have been developed, such as different bioreactor configurations (membrane bioreactors [113, 114] and trickling bed bioreactors [115]) and optimization of operational parameters (mesophilic and thermophilic temperatures [116] and partial pressure of hydrogen [117]). Likewise, changes in the microbial community of mixed cultures are also influenced by hydrogen partial pressure in the system. Acetogenesis is carried out by syntrophic microorganisms, which are thermodynamically constrained by the H_2 partial pressure, which should remain under 10^{-4} atm to allow VFAs degradation and methanogenesis [117]. According to different authors, archaeal community adaptation to hydrogen inputs is required to avoid acetate accumulation and optimize methane production [110, 117, 118]. In the same way, during in situ biomethanation, continuous hydrogen injection into the anaerobic digester was reported to inhibit VFA degradation resulting in a pH decrease, which finally caused process failure. Therefore, coupling biomethanation with mixed cultures and DF might lead to biomethanation failure due to high VFA concentration in DF effluent without an adapted community. To avoid this accumulation, accurate choice of initial inoculum [117, 118], as well as pulsed hydrogen injection [119] and use of additives [120, 121], are strategies that are promising to promote community activity and adaptation during biomethanation processes. In addition, the feeding strategy of DF effluent to the biomethanation reactor could be adapted to avoid the increase of VFA concentration in the biomethanation reactor (*e.g.*, co-digestion with other substrates or slow stepwise feeding). Considering the Technology Readiness Level (TRL), the ex situ biomethanation is more advanced than in situ biomethanation. Whereas several industrial ex situ biomethanation units are currently operational (*e.g.*, DEMETHA project (mixed culture) [122] or Electrochaea company (pure culture) [123]), in situ processes are mainly performed at lab scale, with few trials at pilot scale [124, 125]. This delay in developing in situ biomethanation is due to the impact mentioned above of hydrogen on the AD process. On the contrary, with ex situ biomethanation, hydrogen injection does not inhibit the microbial community but at the expense of building a new reactor (Table 14.2).

Table 14.2 Opportunities and limits of coupling DF with AD and DF with biomethanation [35, 81, 97, 126–129]

Coupling	Opportunities	Limits
DF–AD	• Producing hydrogen and methane in separate processes but on the same plant • Improvement of the ER from residues • Avoiding methanogens inhibition • Producing biohythane with higher OLR and shorter HRT for AD	• Upgrading cost of gas produced with both processes • New constraints associated to hydrogen production and selling (storage, transports) or biohythane introduction in the gas network (restrictions from legislations) • Additional capital and operational expenditures (or CAPEX and OPEX) due to DF implementation and coupling control • Development is required to optimize DF (TRL 7) at industrial scale
DF–Biomethanation	• Increasing methane content in biogas • Decrease in carbon dioxide emissions • Improvement of the ER from residues • Avoiding methanogens inhibition (for ex situ biomethanation) • No investment for H_2 storage and distribution • Reduced upgrading cost for methane production	• Additional CAPEX and OPEX associated to the creation of DF reactor and biomethanation sub-reactor (for ex situ biomethanation) • Development is required to perform in situ biomethanation without process failure and optimize DF at industrial scale

14.4 Dark Fermentation—Photofermentation

Purple phototrophic bacteria (PPB) are diverse bacteria that can grow using various metabolic pathways. This versatility allows PPB to survive in various environments [130]. Figure 14.4 shows their most relevant metabolic features, highlighting those related to hydrogen production/consumption.

Their most unique characteristic feature is their capability to grow via anoxic photosynthesis. Under anaerobic conditions and in the presence of light (mostly infrared, with absorption peaks at 750–1100 nm), PPB can grow using light as an energy source and a wide range of electron donors. They can fix carbon dioxide when growing on inorganic electron donors (*i.e.*, photoautotrophy) or use organic carbon as a C source instead (*i.e.*, photoheterotrophy) [131]. In addition, in the absence of light and under anaerobic conditions, PPB can grow via fermentation in the presence of organic matter. If oxygen and organic matter are present, chemotrophic growth via respiration is the prevalent growth mode [132]. PPB can be classified into purple sulfur bacteria (PSB) and purple non-sulfur bacteria (PNSB). PSB grow mainly photoautotrophycally using reduced sulfur compounds as electron donors. At the same time, PNSB have a more versatile metabolism, using a wide range of organic and inorganic electron donors (*e.g.*, organic matter, hydrogen, hydrogen sulfide, reduced metals, etc.) [133]. PPB also have a diverse metabolism to regenerate reduced cofactors (*e.g.*, NADH or NADPH) [134]. If carbon dioxide is available, its fixation is the main mechanism for cofactor recycling [135]. In addition, in the presence of an excess of organic carbon, PPB can accumulate polyhydroxyalkanoates (PHAs) or produce hydrogen, both mechanisms serving as electron sinks [136].

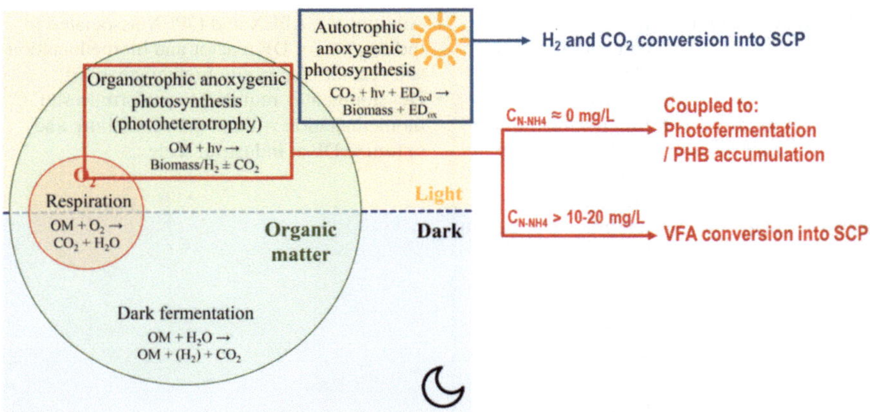

Fig. 14.4 Simplified representation of the main metabolic modes of PPB structured according to energy and carbon sources and electron acceptors. Relevant metabolic modes for hydrogen production/consumption are highlighted. Adapted from [131]. SCP stands for single-cell proteins, OM for organic matter, hv for light energy, ED for electron donor (reduced or oxidized), and $C_{N\text{-}NH4}$ for ammonium-N concentration

Regarding resource recovery, the outstanding ability of PPB to grow at high biomass yields (up to 1 g $COD_{biomass}$/g $COD_{removed}$) and to accumulate added-value products has attracted increased attention in recent years, particularly when growing PPB photoheterotrophically [137, 138]. Nevertheless, the most widely researched application of PPB involves hydrogen production. Hydrogen is synthetized by PPB, such as *Rhodobacter sp.*, *Rhodopseudomonas sp.*, or *Rhodospirillum sp.*, under anaerobic, illuminated, and ammonia-limited conditions (Fig. 14.4) [139]. During the so-called photofermentation, the nitrogenase enzyme can uptake electrons generated from the anaerobic oxidation of organic substrates, use protons as electron acceptors and light as an energy source, and generate molecular hydrogen [136]. The light energy collected by light-harvesting complexes is used to generate ATP via photophosphorylation, and high-energy electrons reduce ferredoxin through reverse electron flow. The reduced ferredoxin (electron carrier) and ATP are then used to produce hydrogen via proton reduction catalyzed by a nitrogenase [140]. This enzyme is also responsible for ammonia production from the reduction of molecular nitrogen. Therefore, molecular nitrogen decreases hydrogen production due to competition at the enzymatic reaction centres [141]. More importantly, hydrogen production via photofermentation must be performed at low ammonia concentrations (above 10–20 mg $N\text{-}NH_4^+$/L), as nitrogenase activity is inhibited due to product inhibition [131]. Therefore, efficient photofermentation is limited to low-N streams.

ATP generation from light makes photofermentation interesting compared to other processes because hydrogen production is not linked to catabolic processes. Therefore, simple organic compounds, including VFAs such as acetic acid and butyric acid, can be used as substrates for hydrogen production. Other organic substrates can also be consumed via photofermentation, including simple sugars (*e.g.*, glucose, sucrose) and alcohols. Despite the advantages, the low hydrogen production rates (maximum volumetric productivities of 3.6 L/Ld and average values of 2.2 L/Ld) hamper the cost-effective hydrogen production via photofermentation due to low biomass concentrations [131]. Direct use of complex substrates like food or agro-industrial waste requires a pretreatment, mainly hydrolysis, to enhance their biodegradability [142, 143] (see Fig. 14.5, process 1). The light requirement is another limitation of photofermentation, as it entails high operational and capital costs. All the challenges mentioned above limit the potential application of single-stage photofermentation.

Coupling DF with photofermentation might be a niche application of photofermentation. Thanks to the possibility of further consuming short-chain VFAs for hydrogen production, photofermentation can be used to overcome the main bottleneck of DF, which is characterised by lower hydrogen yields (0.11 g COD_{H2}/g CODfed on average). [74]. PPB can theoretically convert 1 mol of acetate into 4 mol of hydrogen, increasing the yield to 12 mol hydrogen/mol glucose [144]. However, this theoretical yield is hardly achieved in reality since the growth and maintenance of PPB require part of the electrons and carbon (and competition with PHA production always occurs to some extent) [145]. Thus, average hydrogen yields around 0.25 g COD_{H2}/g COD_{fed} are often reached in DF, followed by photofermentation [131]. Therefore, photofermentation can be used for the bioconversion of the VFAs

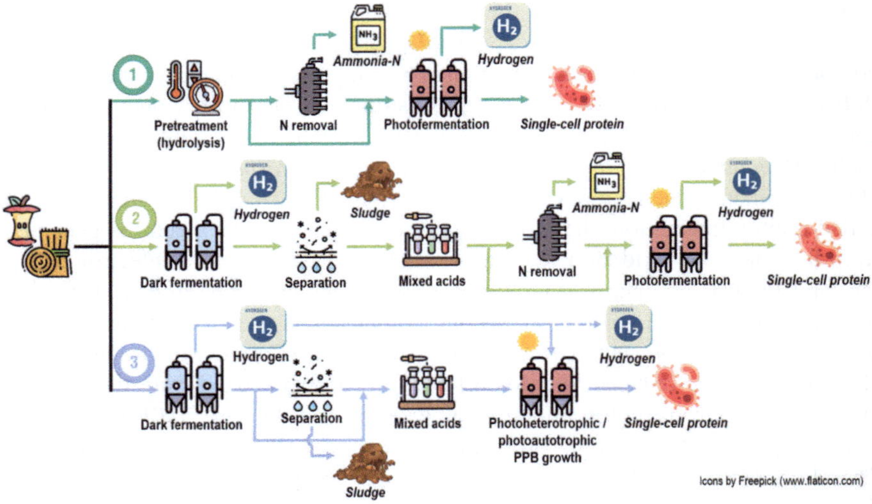

Fig. 14.5 Potential operational configurations for the bioconversion of organic waste into value-added products via (1) photofermentation, (2) sequential DF and photofermentation for hydrogen production, and (3) sequential DF and photofermentation for single-cell protein production

produced during DF into hydrogen, enhancing the overall yields without jeopardizing the overall rates (Fig. 14.5, process 2) [146]. As an additional benefit, the biomass obtained during photofermentation could be further valorized as an animal feed substitute due to the high protein content of PPB and its adequate amino acid profile [147].

Some technical barriers need to be overcome when considering the coupling DF-photofermentation. Before photofermentation, the separation of the sludge by filtration or centrifugation is required for effective light distribution. Moreover, dilution or a previous N removal step (*e.g.*, via membranes, adsorption, or stripping) is also necessary when using substrates with high N contents to avoid hydrogen production inhibition by ammonia [136, 145]. In addition, an important factor to consider in photofermentation is the energy consumption due to light supply. Artificial light for hydrogen production exhibits prohibitive costs [131]. Therefore, the economic feasibility of photofermentation after DF must be considered, and efforts should be carried out using natural light and optimal operational conditions to maximize production rates.

Optimal conditions for hydrogen production via photofermentation (Fig. 14.5, process 1) have intensively been studied (Fig. 14.6). pH values above 5.5 promote hydrogen production with an optimal range between 6.5 and 7.4 (Fig. 6a). Since low pH values lead to hydrogen production inhibition, the applicable OLRs must be limited due to the risk of reactor acidification. OLRs higher than ~ 2–6 g COD/Ld significantly decrease the hydrogen yields, although this drop depends on the photobioreactor configuration [131]. No pH control is required at appropriate loads since organic acid consumption increases the pH. Increasing in the light intensity up to 3,500 lx favors hydrogen production by photofermentation (Fig. 6b). It must be considered that light attenuation is particularly relevant in PPB-based processes, as near-infrared light (the main energy source for PPB) is more attenuated by water than light within the visible spectrum [148]. The increase in light intensity above 4000–4500 lx causes a decrease in the hydrogen yields due to photoinhibition [139].

Regarding temperature, high hydrogen yields are obtained even at low temperatures (<25 °C), whereas values over 40 °C result in decreasing hydrogen yields due to microbial inhibition (Fig. 6c). The operation at low temperatures is essential since no energy requirements for reactor heating might be needed for photofermentation. Organic matter concentrations in the substrates above 4–8 g COD/L have a negative impact on hydrogen yields (Fig. 6d). This factor, along with the inhibition due to ammonia-N, considerably limits the direct use of photofermentation to valorize DF effluents, restricting its application to streams with low organic and nitrogen contents. Dilution strategies could be applied, but they would increase the operational costs of the process, thus compromising the economic feasibility. The reduction state of the C

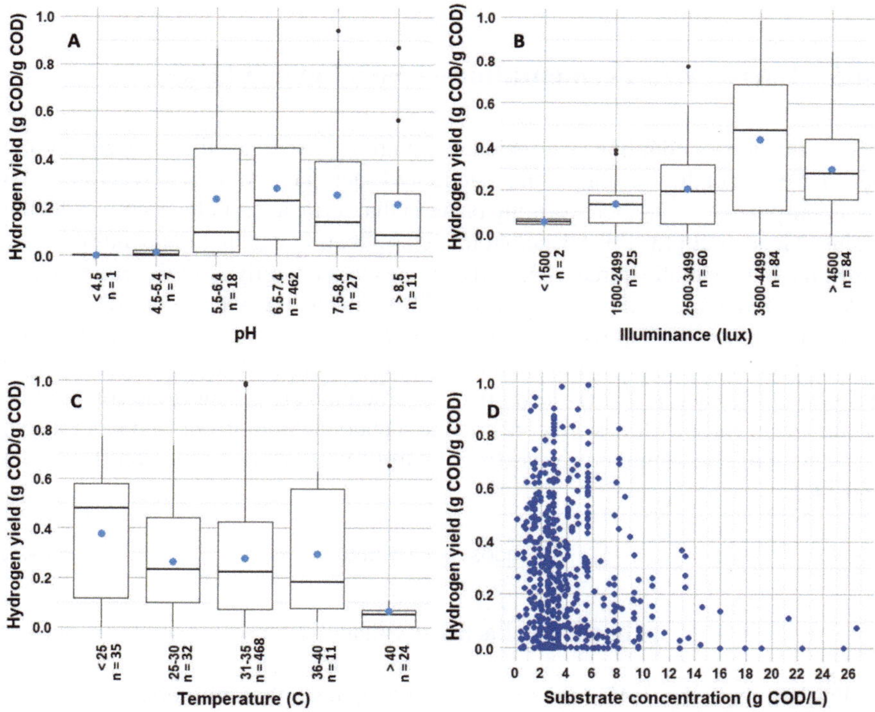

Fig. 14.6 Hydrogen yields produced by PPB in photofermentation processes at different (**a**) pH values, (**b**) illuminances, (**c**) temperatures, and (**d**) substrate concentrations. Light blue dots in boxplots represent mean values. COD stands for chemical oxygen demand, and "n" for the number of data points. Adapted from [131]

source impacts the carbon dioxide production/fixation by PPB, which might directly affect the hydrogen yields. This implies that the optimal conditions for individual DF and photofermentation processes might be different from those for the coupled process.

Another way to couple DF with PPB processes is the bioconversion of the DF gaseous effluents into single-cell protein. This approach has recently emerged as a promising solution for feed and food scarcity (Fig. 14.5, process 3). PPB can effectively use hydrogen as an electron donor and carbon dioxide as a carbon source for their growth (Fig. 14.4) [149]. Thanks to photophosphorylation, high yields of 1 g $COD_{biomass}/COD_{H2}$ have been achieved in mixed PPB cultures (own unpublished results). In addition, biomass productivities of 0.3–0.5 g VSS/Ld (own results), along with high protein and amino acid contents in the PPB biomass (50–60 and 40–50% on VS basis, respectively) have been reported, confirming the potential of this approach [147, 150]. However, autotrophic PPB growth entails lower biomass production rates (up to 0.5 g $COD_{biomass}$/Ld) than heterotrophic PPB growth (up to 6 g $COD_{biomass}$/Ld) [137]. As a result, more research is required to improve biological growth along with the gas–liquid mass transfer of hydrogen and carbon dioxide.

14.5 Lactic Acid Fermentation—Dark Fermentation

The motivation for coupling lactic acid fermentation (LAF) and DF is to overcome the negative effect of the accumulation of lactic acid bacteria (LAB) in DF reactors. LAB are gram-positive, non-spore forming bacteria that ferment carbohydrates producing mainly lactic acid [151]. LAB taxonomic classification has had many adjustments over time but generally agrees that LAB belongs to the family *Lactobacillaceae* and order *Lactobacillales* [151]. The presence of LAB in DF was widely considered detrimental to the process, but recently it was deemed inconclusive or poorly understood [152]. Three negative impacts of LAB in DF reactors are (i) substrate competition, (ii) bacteriocins release, and (iii) reactor over-acidification [152]. Substrate competition occurs when carbohydrates are converted to lactate via homolactic and heterolactic fermentations, steering the process from hydrogen to lactate production (Eqs. 14.1 and 14.2).

$$\text{Glucose} \rightarrow \text{Lactate} \tag{14.1}$$

$$\text{Glucose} \rightarrow \text{Lactate} + \text{Ethanol} + CO_2 \tag{14.2}$$

Furthermore, the LAB community outcompetes other microbial groups by releasing bacteriocins, specifically inhibiting hydrogen-producing bacteria (HPB), particularly *Clostridium* sp. [153]. Additionally, lactate production mediated by LAB can reduce the pH in DF reactors below the optimum range of 5.5–6.0 for hydrogen production. LAB might thrive at pH values as low as 3.5 [154]. Nonetheless, LAB were also reported to impact the HPB positively. A few of the positive relationships between LAB and HPB are (i) higher substrate hydrolysis, (ii) a contribution of LAB to oxygen depletion, (iii) a cross-feeding between LAB and HPB, and (iv) a direct contribution of lactic acid to hydrogen production [152]. Illustratively, a study evaluating starch as a substrate in DF concluded that *Bifidobacterium* assisted in breaking down starch into less complex molecules before being consumed by *Clostridium* for hydrogen production [155]. Facultative LAB *Lactobacillus* was suggested to consume oxygen producing lactate, thus providing an anaerobic environment for anaerobic HPB to produce hydrogen [156]. Cross-feeding of LAB and HPB was shown in multiple studies, where the lactate and acetate produced by LAB were subsequently consumed by HPB [157, 158]. The inability to convert lactate to hydrogen during DF was also associated with the common practice of heat-pretreatment of inoculum to deactivate methanogens and enrich HPB, which was found to also inhibit Lactate-Utilizing Hydrogen-Producing Bacteria (LU-HPB) such as *Megasphaera elsdenii* [159]. Circumventing this blind spot, lactate has successfully been converted to hydrogen by excluding heat pretreatment of inoculum for DF, with the suppression of hydrogenotrophic methanogens by incubation time [159]. Considering these positive findings, studies have been carried out to utilize lactate as one of the carbon sources for hydrogen production. There are multiple pathways reported for the conversion of lactate to hydrogen, summarized by [152], a few of which are as follows (Eqs. 14.3–14.5).

$$\text{Lactate} \rightarrow 0.5\,\text{Butyrate} + CO_2 + H_2 + 0.5\,H_2O \tag{14.3}$$

$$\text{Lactate} \rightarrow 0.5\,\text{Acetate} + 0.5\,\text{Ethanol} + CO_2 + H_2 \tag{14.4}$$

$$\text{Lactate} + H_2O \rightarrow \text{Acetate} + CO_2 + 2H_2 \tag{14.5}$$

A better understanding of the interrelated factors such as inoculum source, pretreatment or enrichment method, reactor configurations, and operational conditions is essential in achieving efficient lactate-driven DF (LD-DF) [152]. LAF-DF coupling was shown in different configurations according to (Fig. 14.7).

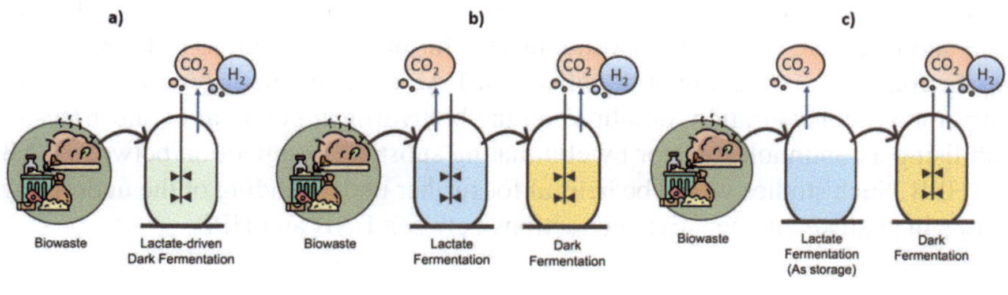

Fig. 14.7 Different process configurations assessed for the coupling of LAF-DF; **a** Lactate-driven DF, **b** two-step LAF + DF, and **c** LAF as storage method + DF

LD-DF in a single reactor (Fig. 14.7a) relies on a positive and balanced relationship between LAF and DF, at which lactate production and consumption rate do not cause instability. For instance, the importance of process pH to maintain a correct balance in a single reactor was highlighted [160]. The highest hydrogen yield was achieved at pH 7 (61.9 ± 0.2 NmL H_2/g VS) from fruit and vegetable waste, where simultaneous lactate production (below 10 g /L) and consumption were observed. The process was unbalanced at low pH values (uncontrolled, pH 5.5, pH 6.0, and pH 6.5), resulting in higher lactate accumulation (up to 17 g/L) and lower hydrogen production (41–59 NmL H_2/g VS).

In a two-stage LAF-DF (Fig. 14.7b), substrates are pre-fermented in a separate reactor to favor lactate production, and effluents are subsequently converted into hydrogen in a second DF reactor. Optimal operating conditions are essential in differentiating the two reactors, where lactate production is favored in the first reactor, and the second reactor is driven towards hydrogen production. In a recent study, tequila vinasse was pre-fermented at an HRT of 13.3 h and pH 5.5 to produce lactate-rich effluent (13.2 ± 1.7 g/L) [161]. This effluent was then fed to a CSTR (HRT 12 h and pH 5.8), which produced a maximum hydrogen yield of 109.8 ± 7.2 NmL H_2/g VS_{added}. Likewise, inoculum is important in providing suitable microbial communities for both carbohydrate and lactate conversions. Suitable inoculum can be obtained from the enrichment of various sources of wastewater or a specific mix of strains such as *Megasphaera elsdenii* with *Clostridium butyricum* as HPB together with *Lactobacillus delbrueckii*, *Lactococcus lactis*, *Leuconostoc mesenteroides*, *Enterococcus faecalis*, and *Enterococcus mundtii* as LAB [162, 163]. A key takeaway in two-stage LAF-DF is that the LAB in the first stage reactor did not negatively impact the performance of the second reactor for hydrogen production. It was inferred that the inhibiting effect of LAB on HPB was species-specific, but further research is required [161].

Finally, LAF has been used to preserve food and as part of the ensiling process to preserve crops for animal feed [164]. Recently LAF has been considered as a storage strategy to preserve the biomethane potential of organic substrate (Fig. 14.7c) [164]. However, information on utilizing LAF as a storage method before DF for hydrogen production is scarce. Storage is essential in allowing biorefineries to run continuously despite varying feedstock availability and is critical for easily biodegradable substrates such as food waste, where premature fermentation and organic carbon losses can occur during transportation [165–167]. With regard to the coupling of LAF and DF, there are many opportunities for further investigations of biohydrogen production, such as looking at the effects of LAF storage parameters (*e.g.*, storage temperature, concentration, duration) on the biohydrogen potential of substrates or stabilising a continuous reactor by eliminating substrate competition between LAB and HPB. Such studies would be helpful for further understanding of the underlying causes in positive and negative interactions between LAB and HPB.

14.6 Conclusions

The production and use of renewable hydrogen via coupling processes with DF technology is now regarded as an attractive biotechnological approach for the utilisation of residual streams. The metabolic profile and anaerobic microbiome obtained after DF are essential variables to optimize the coupling regardless of the second process. The change in operational conditions, separation step, and presence of unwanted microbial activity are some of the key challenges that deserve further specific investigation depending on the type of coupled process. Additionally, the grade of DF effluent purity needed is crucial to select a suitable separation technology to balance the economic cost. Overall, the potential benefits of coupling different biological processes with the DF studied in this chapter have been demonstrated, which may become a key biotechnological process in the future.

Acknowledgements Jose Antonio Magdalena would like to thank the Complutense University of Madrid for the financing of his contract at LBE-INRAE (France), with funds from the Ministry of Universities for the requalification of the Spanish University System for 2021–2023 (Modality 1. Margarita Salas), coming from the European Union-Next Generation EU funding. María del Rosario Rodero acknowledges the European Union-Next Generation EU for funding the Margarita Salas program for her research contract and the regional government of Castilla y León and the European FEDER Programme (CLU 2017-09, UIC 315 and CL-EI-2021-07) for their support. Eqwan Roslan would like to thank The Embassy of France in Malaysia, the AAIBE Chair of Renewable Energy, and Universiti Tenaga Nasional (UNITEN) for funding his stay at LBE-INRAE. Alice Lanfranchi would like to thank the Italian Ministry of Education and Merit for financing her PhD fellowship. Margot Mahieux would like to thank the ANRT (National Association for Research and Technology) and Engie for financing her Ph.D. fellowship (CIFRE N°2021/1463). Ali Dabestani-Rahmatabad expresses her deep thankfulness to ANR (National Agency of Research) in France to finance his Ph.D. fellowship. The icons depicted in the present chapter were extracted from www.flaticon.com (icons creators were Freepick, Eucalyp, Kiranshastry, and Vitaly Gorbachev).

References

1. Lenin Babu M, Venkata Subhash G, Sarma PN, Venkata Mohan S (2013) Bio-electrolytic conversion of acidogenic effluents to biohydrogen: an integration strategy for higher substrate conversion and product recovery. Bioresour Technol 133:322–331. https://doi.org/10.1016/J.BIORTECH.2013.01.029
2. Cao Y, Liu H, Liu W et al (2022) Debottlenecking the biological hydrogen production pathway of dark fermentation: insight into the impact of strain improvement. Microb Cell Fact 21:1–16. https://doi.org/10.1186/s12934-022-01893-3
3. Bakonyi P, Kumar G, Koók L et al (2018) Microbial electrohydrogenesis linked to dark fermentation as integrated application for enhanced biohydrogen production: a review on process characteristics, experiences and lessons. Bioresour Technol 251:381–389. https://doi.org/10.1016/J.BIORTECH.2017.12.064
4. Ramos-Suarez M, Zhang Y, Outram V (2021) Current perspectives on acidogenic fermentation to produce volatile fatty acids from waste. Rev Environ Sci Biotechnol 20:439–478

5. Tapia-Venegas E, Ramirez-Morales JE, Silva-Illanes F et al (2015) Biohydrogen production by dark fermentation: scaling-up and technologies integration for a sustainable system. Rev Environ Sci Biotechnol 14:761–785. https://doi.org/10.1007/s11157-015-9383-5
6. Khongkliang P, Kongjan P, Utarapichat B et al (2017) Continuous hydrogen production from cassava starch processing wastewater by two-stage thermophilic dark fermentation and microbial electrolysis. Int J Hydrogen Energy 42:27584–27592. https://doi.org/10.1016/j.ijhydene.2017.06.145
7. Koul Y, Devda V, Varjani S et al (2022) Microbial electrolysis: a promising approach for treatment and resource recovery from industrial wastewater. Bioengineered 13:8115–8134. https://doi.org/10.1080/21655979.2022.2051842
8. Sivagurunathan P, Kuppam C, Mudhoo A et al (2018) A comprehensive review on two-stage integrative schemes for the valorization of dark fermentative effluents. Crit Rev Biotechnol 38:868–882. https://doi.org/10.1080/07388551.2017.1416578
9. Ruiz V, Ilhan ZE, Kang DW et al (2014) The source of inoculum plays a defining role in the development of MEC microbial consortia fed with acetic and propionic acid mixtures. J Biotechnol 182–183:11–18. https://doi.org/10.1016/J.JBIOTEC.2014.04.016
10. Kato S (2015) Biotechnological aspects of microbial extracellular electron transfer. Microbes Environ Environ 30:133–139. https://doi.org/10.1264/jsme2.me15028
11. Kracke F, Lai B, Yu S, Krömer JO (2018) Balancing cellular redox metabolism in microbial electrosynthesis and electro fermentation—a chance for metabolic engineering. Metab Eng 45:109–120. https://doi.org/10.1016/j.ymben.2017.12.003
12. Nam JY, Tokash JC, Logan BE (2011) Comparison of microbial electrolysis cells operated with added voltage or by setting the anode potential. Int J Hydrogen Energy 36:10550–10556. https://doi.org/10.1016/j.ijhydene.2011.05.148
13. Flayac C, Trably E, Bernet N (2018) Microbial anodic consortia fed with fermentable substrates in microbial electrolysis cells: significance of microbial structures. Bioelectrochemistry 123:219–226. https://doi.org/10.1016/j.bioelechem.2018.05.009
14. Lu L, Ren N, Xing D, Logan BE (2009) Hydrogen production with effluent from an ethanol–H_2-coproducing fermentation reactor using a single-chamber microbial electrolysis cell. Biosens Bioelectron 24:3055–3060. https://doi.org/10.1016/J.BIOS.2009.03.024
15. de Fouchécour F, Larzillière V, Bouchez T, Moscoviz R (2022) Systematic and quantitative analysis of two decades of anodic wastewater treatment in bioelectrochemical reactors. Water Res 214.https://doi.org/10.1016/j.watres.2022.118142
16. Ramirez-Nava J, Martínez-Castrejón M, García-Mesino RL et al (2021) The implications of membranes used as separators in microbial fuel cells. Membranes (Basel) 11:1–27. https://doi.org/10.3390/membranes11100738
17. Rousseau R, Etcheverry L, Roubaud E, et al (2020) Microbial electrolysis cell (MEC): strengths, weaknesses and research needs from electrochemical engineering standpoint. Appl Energy 257.https://doi.org/10.1016/J.APENERGY.2019.113938
18. Liu H, Hu H, Chignell J, Fan Y (2010) Microbial electrolysis: novel technology for hydrogen production from biomass. Biofuels 1:129–142. https://doi.org/10.4155/bfs.09.9
19. Logan BE, Call D, Cheng S et al (2008) Microbial electrolysis cells for high yield hydrogen gas production from organic matter. Environ Sci Technol 42:8630–8640. https://doi.org/10.1021/es801553z
20. Satinover SJ, Rodriguez M, Campa MF et al (2020) Performance and community structure dynamics of microbial electrolysis cells operated on multiple complex feedstocks. Biotechnol Biofuels 13:1–21. https://doi.org/10.1186/s13068-020-01803-y
21. Obileke K, Nwokolo N, Makaka G, et al (2020) Anaerobic digestion: technology for biogas production as a source of renewable energy—a review. Energy Environ 32. https://doi.org/10.1177/0958305X20923117
22. Moreno R, Escapa A, Cara J et al (2015) A two-stage process for hydrogen production from cheese whey: integration of dark fermentation and biocatalyzed electrolysis. Int J Hydrogen Energy 40:168–175. https://doi.org/10.1016/j.ijhydene.2014.10.120

23. Wang A, Sun D, Cao G et al (2011) Integrated hydrogen production process from cellulose by combining dark fermentation, microbial fuel cells, and a microbial electrolysis cell. Bioresour Technol 102:4137–4143. https://doi.org/10.1016/J.BIORTECH.2010.10.137
24. Marone A, Ayala-Campos OR, Trably E et al (2017) Coupling dark fermentation and microbial electrolysis to enhance bio-hydrogen production from agro-industrial wastewaters and by-products in a bio-refinery framework. Int J Hydrogen Energy 42:1609–1621. https://doi.org/10.1016/j.ijhydene.2016.09.166
25. Park YH, Han S Il, Oh B, et al (2022) Microalgal secondary metabolite productions as a component of biorefinery: a review. Bioresour Technol 344:126206. https://doi.org/10.1016/j.biortech.2021.126206
26. Lardon L, Hélias A, Sialve B et al (2009) Life-cycle assessment of biodiesel production from microalgae. Environ Sci Technol 43:6475–6481. https://doi.org/10.1021/es900705j
27. da Cruz RVA, do Nascimento CAO (2012) Emergy analysis of oil production from microalgae. Biomass and Bioenergy 47:418–425.https://doi.org/10.1016/j.biombioe.2012.09.016
28. Ubando AT, Anderson S. Ng E, Chen WH, et al (2022) Life cycle assessment of microalgal biorefinery: a state-of-the-art review. Bioresour Technol 360:127615. https://doi.org/10.1016/j.biortech.2022.127615
29. Maiolo S, Cristiano S, Gonella F, Pastres R (2021) Ecological sustainability of aquafeed: an emergy assessment of novel or underexploited ingredients. J Clean Prod 294:126266. https://doi.org/10.1016/J.JCLEPRO.2021.126266
30. Okeke ES, Ejeromedoghene O, Okoye CO, et al (2022) Microalgae biorefinery: an integrated route for the sustainable production of high-value-added products. Energy Convers Manag X 16:100323. https://doi.org/10.1016/J.ECMX.2022.100323
31. Hussain F, Shah SZ, Ahmad H, et al (2021) Microalgae an ecofriendly and sustainable wastewater treatment option: biomass application in biofuel and bio-fertilizer production. A review. Renew Sustain Energy Rev 137:110603. https://doi.org/10.1016/J.RSER.2020.110603
32. Abreu AP, Morais RC, Teixeira JA, Nunes J (2022) A comparison between microalgal autotrophic growth and metabolite accumulation with heterotrophic, mixotrophic and photoheterotrophic cultivation modes. Renew Sustain Energy Rev 159:112247. https://doi.org/10.1016/J.RSER.2022.112247
33. Gonçalves AL, Pires JCM, Simões M (2017) A review on the use of microalgal consortia for wastewater treatment. Algal Res 24:403–415. https://doi.org/10.1016/J.ALGAL.2016.11.008
34. Liu CH, Chang CY, Liao Q et al (2013) Biohydrogen production by a novel integration of dark fermentation and mixotrophic microalgae cultivation. Int J Hydrogen Energy 38:15807–15814. https://doi.org/10.1016/J.IJHYDENE.2013.05.104
35. Angelidaki I, Treu L, Tsapekos P et al (2018) Biogas upgrading and utilization: current status and perspectives. Biotechnol Adv 36:452–466
36. Lacroux J, Jouannais P, Atteia A, et al (2022) Microalgae screening for heterotrophic and mixotrophic growth on butyrate. Algal Res 67:102843. https://doi.org/10.1016/j.algal.2022.102843
37. Cabanelas ITD, Marques SSI, de Souza CO et al (2015) Botryococcus, what to do with it? Effect of nutrient concentration on biorefinery potential. Algal Res 11:43–49. https://doi.org/10.1016/j.algal.2015.05.009
38. Tokuşoglu O and ÜMK (2003) Biomass nutrient profiles of three microalgae. J Food Sci 68:1144–1148
39. Šantek B, Felski M, Friehs K et al (2010) Production of paramylon, a β-1,3-glucan, by heterotrophic cultivation of Euglena gracilis on potato liquor. Eng Life Sci 10:165–170. https://doi.org/10.1002/elsc.200900077
40. Alibardi L, Cossu R (2016) Effects of carbohydrate, protein and lipid content of organic waste on hydrogen production and fermentation products. Waste Manag 47:69–77. https://doi.org/10.1016/j.wasman.2015.07.049
41. Chen S, Qu D, Xiao X, Miao X (2020) Biohydrogen production with lipid-extracted Dunaliella biomass and a new strain of hyper-thermophilic archaeon Thermococcus eurythermalis A501. Int J Hydrogen Energy 45:12721–12730. https://doi.org/10.1016/J.IJHYDENE.2020.03.010

42. Lacroux J, Llamas M, Dauptain K, et al (2023) Dark fermentation and microalgae cultivation coupled systems: outlook and challenges. Sci Total Environ 865:161136. https://doi.org/10.1016/j.scitotenv.2022.161136
43. Smith RT, Gilmour DJ (2018) The influence of exogenous organic carbon assimilation and photoperiod on the carbon and lipid metabolism of Chlamydomonas reinhardtii. Algal Res 31:122–137. https://doi.org/10.1016/J.ALGAL.2018.01.020
44. Li-Beisson Y, Thelen JJ, Fedosejevs E, Harwood JL (2019) The lipid biochemistry of eukaryotic algae. Prog Lipid Res 74:31–68. https://doi.org/10.1016/j.plipres.2019.01.003
45. Kato N, Nelson G, Lauersen KJ (2021) Subcellular localizations of catalase and exogenously added fatty acid in chlamydomonas reinhardtii. Cells https://doi.org/10.3390/cells10081940
46. Lacroux J, Atteia A, Brugière S, et al (2022) Proteomics unveil a central role for peroxisomes in butyrate assimilation of the heterotrophic Chlorophyte alga Polytomella sp. Front Microbiol 13.https://doi.org/10.3389/fmicb.2022.1029828
47. Turon V, Baroukh C, Trably E et al (2015) Use of fermentative metabolites for heterotrophic microalgae growth: yields and kinetics. Bioresour Technol 175:342–349. https://doi.org/10.1016/J.BIORTECH.2014.10.114
48. Lacroux J, Seira J, Trably E, et al (2021) Mixotrophic growth of Chlorella sorokiniana on acetate and butyrate: interplay between substrate, C:N ratio and pH. Front Microbiol 12.https://doi.org/10.3389/fmicb.2021.703614
49. Patel A, Krikigianni E, Rova U, et al (2022) Bioprocessing of volatile fatty acids by oleaginous freshwater microalgae and their potential for biofuel and protein production. Chem Eng J 438:135529. https://doi.org/10.1016/J.CEJ.2022.135529
50. Lacroux J, Trably E, Bernet N, et al (2020) Mixotrophic growth of microalgae on volatile fatty acids is determined by their undissociated form. Algal Res 47:101870. https://doi.org/10.1016/J.ALGAL.2020.101870
51. Cai T, Park SY, Li Y (2013) Nutrient recovery from wastewater streams by microalgae: status and prospects. Renew Sustain Energy Rev 19:360–369. https://doi.org/10.1016/J.RSER.2012.11.030
52. Xia A, Murphy JD (2016) Microalgal cultivation in treating liquid digestate from biogas systems. Trends Biotechnol 34:264–275
53. Zhao P, Wang Y, Lin Z et al (2019) The alleviative effect of exogenous phytohormones on the growth, physiology and gene expression of Tetraselmis cordiformis under high ammonia-nitrogen stress. Bioresour Technol 282:339–347. https://doi.org/10.1016/J.BIORTECH.2019.03.031
54. Jiang R, Qin L, Feng S, et al (2021) The joint effect of ammonium and pH on the growth of Chlorella vulgaris and ammonium removal in artificial liquid digestate. Bioresour Technol 325:124690. https://doi.org/10.1016/J.BIORTECH.2021.124690
55. Zhao XC, Tan XB, Bin YL et al (2019) Cultivation of Chlorella pyrenoidosa in anaerobic wastewater: the coupled effects of ammonium, temperature and pH conditions on lipids compositions. Bioresour Technol 284:90–97. https://doi.org/10.1016/J.BIORTECH.2019.03.117
56. Chen H-H, Jiang J-G (2017) Lipid accumulation mechanisms in auto-and heterotrophic microalgae. J Agric Food Chem 65:8099–8110. https://doi.org/10.1021/acs.jafc.7b03495
57. Li Z-Y, Zhang J, Shi J, et al (2022) MicroRNA expression profile analysis of Chlamydomonas reinhardtii during lipid accumulation process under nitrogen deprivation stresses. Bioengineering 9.https://doi.org/10.3390/bioengineering9010006
58. Zheng Y, Li T, Yu X et al (2013) High-density fed-batch culture of a thermotolerant microalga Chlorella sorokiniana for biofuel production. Appl Energy 108:281–287. https://doi.org/10.1016/J.APENERGY.2013.02.059
59. Chalima A, Boukouvalas C, Oikonomopoulou V, Topakas E (2022) Optimizing the production of docosahexaenoic fatty acid by Crypthecodinium cohnii and reduction in process cost by using a dark fermentation effluent. Chem Eng J Adv 11:100345. https://doi.org/10.1016/J.CEJA.2022.100345

60. Shen X-F, Qin Q-W, Yan S-K et al (2019) Biodiesel production from Chlorella vulgaris under nitrogen starvation in autotrophic, heterotrophic, and mixotrophic cultures. J Appl Phycol 31:1589–1596. https://doi.org/10.1007/s10811-019-01765-1
61. You X, Zhang Z, Guo L, et al (2021) Integrating acidogenic fermentation and microalgae cultivation of bacterial-algal coupling system for mariculture wastewater treatment. Bioresour Technol 320:124335. https://doi.org/10.1016/J.BIORTECH.2020.124335
62. Abiusi F, Wijffels RH, Janssen M (2020) Doubling of microalgae productivity by oxygen balanced mixotrophy. ACS Sustain Chem Eng 8:6065–6074. https://doi.org/10.1021/acssuschemeng.0c00990
63. Turon V, Trably E, Fayet A et al (2015) Raw dark fermentation effluent to support heterotrophic microalgae growth: microalgae successfully outcompete bacteria for acetate. Algal Res 12:119–125. https://doi.org/10.1016/J.ALGAL.2015.08.011
64. Cecchin M, Benfatto S, Griggio F, et al (2018) Molecular basis of autotrophic versus mixotrophic growth in Chlorella sorokiniana. Sci Rep 8.https://doi.org/10.1038/s41598-018-24979-8
65. Ip PF, Wong KH, Chen F (2004) Enhanced production of astaxanthin by the green microalga Chlorella zofingiensis in mixotrophic culture. Process Biochem 39:1761–1766. https://doi.org/10.1016/j.procbio.2003.08.003
66. Mokrosnop VM, Polishchuk AV, Zolotareva EK (2016) Accumulation of α-tocopherol and β-carotene in Euglena gracilis cells under autotrophic and mixotrophic culture conditions. Appl Biochem Microbiol 52:216–221. https://doi.org/10.1134/S0003683816020101
67. Chandra R, Arora S, Rohit MV, Mohan SV (2015) Lipid metabolism in response to individual short chain fatty acids during mixotrophic mode of microalgal cultivation: influence on biodiesel saturation and protein profile. Bioresour Technol 188:169–176. https://doi.org/10.1016/j.biortech.2015.01.088
68. Qi W, Mei S, Yuan Y et al (2018) Enhancing fermentation wastewater treatment by co-culture of microalgae with volatile fatty acid-and alcohol-degrading bacteria. Algal Res 31:31–39. https://doi.org/10.1016/J.ALGAL.2018.01.012
69. Hartmann M, Zubkov MV, Martin AP et al (2009) Assessing amino acid uptake by phototrophic nanoflagellates in nonaxenic cultures using flow cytometric sorting. FEMS Microbiol Lett 298:166–173. https://doi.org/10.1111/j.1574-6968.2009.01715.x
70. Wilken S, Schuurmans JM, Matthijs HCP (2014) Do mixotrophs grow as photoheterotrophs? Photophysiological acclimation of the chrysophyte *Ochromonas danica* after feeding. New Phytol 204:882–889. https://doi.org/10.1111/nph.12975
71. Zhang B, Wu J, Meng F (2021) Adaptive laboratory evolution of microalgae: a review of the regulation of growth, stress resistance, metabolic processes, and biodegradation of pollutants. Front Microbiol 12:1–8. https://doi.org/10.3389/fmicb.2021.737248
72. Aceves-Lara CA, Trably E, Bastidas-Oyenadel J-R et al (2008) Production de bioénergies à partir de déchets: Exemples du biométhane et du biohydrogène. J Soc Biol 202:177–189. https://doi.org/10.1051/jbio:2008020
73. Ueno Y, Tatara M, Fukui H et al (2007) Production of hydrogen and methane from organic solid wastes by phase-separation of anaerobic process. Bioresour Technol 98:1861–1865. https://doi.org/10.1016/j.biortech.2006.06.017
74. Moscoviz R, Trably E, Bernet N, Carrère H (2018) The environmental biorefinery: state-of-the-art on the production of hydrogen and value-added biomolecules in mixed-culture fermentation †. Green Chem 20.https://doi.org/10.1039/c8gc00572a
75. Luo G, Johansson S, Boe K et al (2012) Simultaneous hydrogen utilization and in situ biogas upgrading in an anaerobic reactor. Biotechnol Bioeng 109:1088–1094. https://doi.org/10.1002/bit.24360
76. (2022) Metha-HYn—Méthanation biologique In situ avec production d'hydrogène biologique. La Libr. ADEME
77. Nualsri C, Reungsang A, Plangklang P (2016) Biochemical hydrogen and methane potential of sugarcane syrup using a two-stage anaerobic fermentation process. Ind Crops Prod 82:88–99. https://doi.org/10.1016/j.indcrop.2015.12.002

78. Moletta RCN-665. 77 (2015) La méthanisation, 3e éd. Lavoisier-Médecine sciences, Paris
79. Liu D, Liu D, Zeng RJ, Angelidaki I (2006) Hydrogen and methane production from household solid waste in the two-stage fermentation process. Water Res 40:2230–2236. https://doi.org/10.1016/j.watres.2006.03.029
80. Eric T, Gwendoline C, Eric L, Christian L (2018) Production de biohydrogène Voie fermentaire sombre. Tech l'ingénieur Chim verte TIP142WEB. https://doi.org/10.51257/a-v2-bio3351
81. Schievano A, Tenca A, Lonati S et al (2014) Can two-stage instead of one-stage anaerobic digestion really increase energy recovery from biomass? Appl Energy 124:335–342. https://doi.org/10.1016/j.apenergy.2014.03.024
82. Pakarinen OM, Tähti HP, Rintala JA (2009) One-stage H_2 and CH_4 and two-stage H_2+ CH_4 production from grass silage and from solid and liquid fractions of NaOH pre-treated grass silage. Biomass Bioenerg 33:1419–1427. https://doi.org/10.1016/j.biombioe.2009.06.006
83. Rafieenia R, Girotto F, Peng W et al (2017) Effect of aerobic pre-treatment on hydrogen and methane production in a two-stage anaerobic digestion process using food waste with different compositions. Waste Manag 59:194–199. https://doi.org/10.1016/j.wasman.2016.10.028
84. Salem AH, Mietzel T, Brunstermann R, Widmann R (2018) Two-stage anaerobic fermentation process for bio-hydrogen and bio-methane production from pre-treated organic wastes. Bioresour Technol 265:399–406. https://doi.org/10.1016/j.biortech.2018.06.017
85. Kim M, Ahn Y-H, Speece RE (2002) Comparative process stability and efficiency of anaerobic digestion; mesophilic versus thermophilic. Water Res 36:4369–4385. https://doi.org/10.1016/S0043-1354(02)00147-1
86. Cavinato C, Giuliano A, Bolzonella D et al (2012) Bio-hythane production from food waste by dark fermentation coupled with anaerobic digestion process: a long-term pilot scale experience. Int J Hydrogen Energy 37:11549–11555. https://doi.org/10.1016/j.ijhydene.2012.03.065
87. Micolucci F, Gottardo M, Bolzonella D, Pavan P (2014) Automatic process control for stable bio-hythane production in two-phase thermophilic anaerobic digestion of food waste. Int J Hydrog Energy 39:17563. https://doi.org/10.1016/j.ijhydene.2014.08.136
88. Algapani DE, Qiao W, Ricci M et al (2019) Bio-hydrogen and bio-methane production from food waste in a two-stage anaerobic digestion process with digestate recirculation. Renew Energy 130:1108–1115. https://doi.org/10.1016/j.renene.2018.08.079
89. Luo G, Xie L, Zhou Q, Angelidaki I (2011) Enhancement of bioenergy production from organic wastes by two-stage anaerobic hydrogen and methane production process. Bioresour Technol 102:8700–8706. https://doi.org/10.1016/j.biortech.2011.02.012
90. Holl E, Steinbrenner J, Merkle W, et al (2022) Two-stage anaerobic digestion: state of technology and perspective roles in future energy systems. Bioresour Technol 360:127633. https://doi.org/10.1016/j.biortech.2022.127633
91. Zuo Z, Wu S, Zhang W, Dong R (2013) Effects of organic loading rate and effluent recirculation on the performance of two-stage anaerobic digestion of vegetable waste. Bioresour Technol 146:556–561. https://doi.org/10.1016/j.biortech.2013.07.128
92. Almeida P de S, de Menezes CA, Camargo FP, et al (2023) Biomethane recovery through co-digestion of cheese whey and glycerol in a two-stage anaerobic fluidized bed reactor: effect of temperature and organic loading rate on methanogenesis. J Environ Manage 330:117117. https://doi.org/10.1016/j.jenvman.2022.117117
93. Chen Y, Li L, Liu H, et al (2023) Regulating effects of Fe/C materials on thermophilic anaerobic digestion of kitchen waste: digestive performances and methanogenic metabolism pathways. Fuel 332:126140. https://doi.org/10.1016/j.fuel.2022.126140
94. Wu M, Fu Q, Huang J, et al (2021) Effect of sodium dodecylbenzene sulfonate on hydrogen production from dark fermentation of waste activated sludge. Sci Total Environ 799:149383. https://doi.org/10.1016/j.scitotenv.2021.149383
95. Paillet F, Barrau C, Escudié R, et al (2021) Robust operation through effluent recycling for hydrogen production from the organic fraction of municipal solid waste. Bioresour Technol 319:124196. https://doi.org/10.1016/j.biortech.2020.124196

96. Commission E, Centre JR, Kanellopoulos K, et al (2022) Blending hydrogen from electrolysis into the European gas grid. Publications Office of the European Union
97. Hans M, Kumar S (2019) Biohythane production in two-stage anaerobic digestion system. Int J Hydrogen Energy 44:17363–17380. https://doi.org/10.1016/j.ijhydene.2018.10.022
98. Burbano HJ, Amell AA, García JM (2008) Effects of hydrogen addition to methane on the flame structure and CO emissions in atmospheric burners. Int J Hydrogen Energy 33:3410–3415. https://doi.org/10.1016/j.ijhydene.2008.04.020
99. Basak B, Saha S, Chatterjee PK, et al (2020) Pretreatment of polysaccharidic wastes with cellulolytic *Aspergillus fumigatus* for enhanced production of biohythane in a dual-stage process. Bioresour Technol 299:122592. https://doi.org/10.1016/j.biortech.2019.122592
100. Ta DT, Lin C-Y, Ta TMN, Chu C-Y (2020) Biohythane production via single-stage anaerobic fermentation using entrapped hydrogenic and methanogenic bacteria. Bioresour Technol 300:122702. https://doi.org/10.1016/j.biortech.2019.122702
101. Fu S, Angelidaki I, Zhang Y (2021) In situ biogas upgrading by CO_2-to-CH_4 bioconversion. Trends Biotechnol 39:336–347. https://doi.org/10.1016/j.tibtech.2020.08.006
102. Kougias PG, Treu L, Benavente DP et al (2017) Ex-situ biogas upgrading and enhancement in different reactor systems. Bioresour Technol 225:429–437. https://doi.org/10.1016/j.biortech.2016.11.124
103. Thema M, Bauer F, Sterner M (2019) Power-to-gas: electrolysis and methanation status review. Renew Sustain Energy Rev 112:775–787. https://doi.org/10.1016/j.rser.2019.06.030
104. Ghafoori MS, Loubar K, Marin-Gallego M, Tazerout M (2022) Techno-economic and sensitivity analysis of biomethane production via landfill biogas upgrading and power-to-gas technology. Energy 239:122086. https://doi.org/10.1016/j.energy.2021.122086
105. Michailos S, Walker M, Moody A, et al (2021) A techno-economic assessment of implementing power-to-gas systems based on biomethanation in an operating waste water treatment plant. J Environ Chem Eng 9:104735. https://doi.org/10.1016/j.jece.2020.104735
106. Enevoldsen P, Sovacool BK (2016) Examining the social acceptance of wind energy: practical guidelines for onshore wind project development in France. Renew Sustain Energy Rev 53:178–184. https://doi.org/10.1016/j.rser.2015.08.041
107. Emodi NV, Lovell H, Levitt C, Franklin E (2021) A systematic literature review of societal acceptance and stakeholders' perception of hydrogen technologies. Int J Hydrogen Energy 46:30669–30697. https://doi.org/10.1016/j.ijhydene.2021.06.212
108. Lei L, Bai L, Lindbråthen A, et al (2020) Carbon membranes for CO_2 removal: status and perspectives from materials to processes. Chem Eng J 401:126084. https://doi.org/10.1016/j.cej.2020.126084
109. James BD, Houchins C, Huya-Kouadio JM, DeSantis DA (2016) Final report: hydrogen storage system cost analysis. Strategic Analysis Inc., Arlington, VA (United States)
110. Agneessens LM, Ottosen LDM, Andersen M et al (2018) Parameters affecting acetate concentrations during in-situ biological hydrogen methanation. Bioresour Technol 258:33–40. https://doi.org/10.1016/j.biortech.2018.02.102
111. Jensen MB, Ottosen LDM, Kofoed MVW (2021) H_2 gas-liquid mass transfer: a key element in biological power-to-gas methanation. Renew Sustain Energy Rev 147:111209. https://doi.org/10.1016/j.rser.2021.111209
112. Jensen MB, Kofoed MVW, Fischer K et al (2018) Venturi-type injection system as a potential H_2 mass transfer technology for full-scale in situ biomethanation. Appl Energy 222:840–846. https://doi.org/10.1016/j.apenergy.2018.04.034
113. Alfaro N, Fdz-Polanco M, Fdz-Polanco F, Díaz I (2019) H_2 addition through a submerged membrane for in-situ biogas upgrading in the anaerobic digestion of sewage sludge. Bioresour Technol 280:1–8. https://doi.org/10.1016/j.biortech.2019.01.135
114. Deschamps L, Imatoukene N, Lemaire J, et al (2021) In-situ biogas upgrading by biomethanation with an innovative membrane bioreactor combining sludge filtration and H_2 injection. Bioresour Technol 337:125444. https://doi.org/10.1016/j.biortech.2021.125444
115. Thapa A, Park J-G, Jun H-B (2022) Enhanced ex-situ biomethanation of hydrogen and carbon dioxide in a trickling filter bed reactor. Biochem Eng J 179:108311. https://doi.org/10.1016/j.bej.2021.108311

116. Jiang H, Wu F, Wang Y, et al (2021) Characteristics of in-situ hydrogen biomethanation at mesophilic and thermophilic temperatures. Bioresour Technol 337:125455. https://doi.org/10.1016/j.biortech.2021.125455
117. Braga Nan L, Trably E, Santa-Catalina G, et al (2020) Biomethanation processes: new insights on the effect of a high H_2 partial pressure on microbial communities. Biotechnol Biofuels 13:141.https://doi.org/10.1186/s13068-020-01776-y
118. Vechi NT, Agneessens LM, Feilberg A, et al (2021) In situ biomethanation: inoculum origin influences acetate consumption rate during hydrogen addition. Bioresour Technol Reports 14:100656. https://doi.org/10.1016/j.biteb.2021.100656
119. Agneessens LM, Ottosen LDM, Voigt NV et al (2017) In-situ biogas upgrading with pulse H_2 additions: the relevance of methanogen adaption and inorganic carbon level. Bioresour Technol 233:256–263. https://doi.org/10.1016/j.biortech.2017.02.016
120. Kutlar FE, Tunca B, Yilmazel YD (2022) Carbon-based conductive materials enhance biomethane recovery from organic wastes: a review of the impacts on anaerobic treatment. Chemosphere 290:133247. https://doi.org/10.1016/j.chemosphere.2021.133247
121. Tang J, Liu Z, Zhao M, et al (2022) Enhanced biogas biological upgrading from kitchen wastewater by in-situ hydrogen supply through nano zero-valent iron corrosion. J Environ Manage 310:114774. https://doi.org/10.1016/j.jenvman.2022.114774
122. DEMETHA—Programme de recherche et de démonstration de biométhanation sur des gaz renouvelables (2022) TBI, Toulouse Biotechnology Institute
123. Electrochaea GmbH—Power-to-Gas Energy Storage | (2022)
124. Tsapekos P, Treu L, Campanaro S, et al (2021) Pilot-scale biomethanation in a trickle bed reactor: process performance and microbiome functional reconstruction. Energy Convers Manag 244:114491. https://doi.org/10.1016/j.enconman.2021.114491
125. Tsapekos P, Alvarado-Morales M, Angelidaki I (2022) H_2 competition between homoacetogenic bacteria and methanogenic archaea during biomethanation from a combined experimental-modelling approach. J Environ Chem Eng 10:107281. https://doi.org/10.1016/j.jece.2022.107281
126. Kim HH, Saha S, Hwang J-H, et al (2022) Integrative biohydrogen- and biomethane-producing bioprocesses for comprehensive production of biohythane. Bioresour Technol 365:128145. https://doi.org/10.1016/j.biortech.2022.128145
127. Lawson N, Alvarado-Morales M, Tsapekos P, Angelidaki I (2021) Techno-economic assessment of biological biogas upgrading based on danish biogas plants. Energies 14:8252. https://doi.org/10.3390/en14248252
128. Rafrafi Y, Laguillaumie L, Dumas C (2021) Biological methanation of H_2 and CO_2 with mixed cultures: current advances, hurdles and challenges. Waste and Biomass Valorization 12:5259–5282. https://doi.org/10.1007/s12649-020-01283-z
129. Megret O, Hubert L, Calbry M, et al. (2015) RECORD, Production d'hydrogène à partir de déchets. Etat de l'art et potentiel d'émergence, vol 226. pp n°13–0239/1A
130. Sepúlveda-Muñoz CA, Ángeles R, de Godos I, Muñoz R (2020) Comparative evaluation of continuous piggery wastewater treatment in open and closed purple phototrophic bacteria-based photobioreactors. J Water Process Eng 38:101608. https://doi.org/10.1016/j.jwpe.2020.101608
131. Capson-Tojo G, Batstone DJ, Grassino M, et al (2020) Purple phototrophic bacteria for resource recovery: challenges and opportunities.https://doi.org/10.1016/j.biotechadv.2020.107567
132. Capson-Tojo G, Lin S, Batstone DJ, Hülsen T (2021) Purple phototrophic bacteria are outcompeted by aerobic heterotrophs in the presence of oxygen. Water Res 194:116941. https://doi.org/10.1016/j.watres.2021.116941
133. Madigan MT, Jung DO (2009) An overview of purple bacteria: systematics, physiology, and habitats. 1–15.https://doi.org/10.1007/978-1-4020-8815-5_1
134. Hunter CN, Daldal F, Thurnauer MC, Beatty JT (2009) The purple phototrophic bacteria
135. McKinlay JB, Harwood CS (2010) Carbon dioxide fixation as a central redox cofactor recycling mechanism in bacteria. Proc Natl Acad Sci 107:11669–11675

136. Ghimire A, Frunzo L, Pirozzi F et al (2015) A review on dark fermentative biohydrogen production from organic biomass: process parameters and use of by-products. Appl Energy 144:73–95. https://doi.org/10.1016/j.apenergy.2015.01.045
137. Hülsen T, Barnes AC, Batstone DJ, Capson-Tojo G (2022) Creating value from purple phototrophic bacteria via single-cell protein production. Curr Opin Biotechnol 76:102726. https://doi.org/10.1016/j.copbio.2022.102726
138. Hülsen T, Stegman S, Batstone DJ, Capson-Tojo G (2022) Naturally illuminated photobioreactors for resource recovery from piggery and chicken-processing wastewaters utilising purple phototrophic bacteria. Water Res 214:118194. https://doi.org/10.1016/j.watres.2022.118194
139. Jain A, Das E, Poosarla VG, Rajagopalan G (2022) Biohydrogen production technologies: current status, challenges, and future perspectives. Prod Technol Gaseous Solid Biofuels 115–168.https://doi.org/10.1002/9781119785842.ch5
140. Zhang Q, Liu H, Shui X, et al (2022) Research progress of additives in photobiological hydrogen production system to enhance biohydrogen. Bioresour Technol 362:127787. https://doi.org/10.1016/j.biortech.2022.127787
141. Eroglu E, Melis A (2011) Photobiological hydrogen production: recent advances and state of the art. Bioresour Technol 102:8403–8413. https://doi.org/10.1016/j.biortech.2011.03.026
142. Zhang Z, Yue J, Zhou X, et al (2014) Photo-fermentative bio-hydrogen production from agricultural residue enzymatic hydrolyzate and the enzyme reuse. BioResources 9:2299–2310. https://doi.org/10.15376/biores.9.2.2299-2310
143. Jiang D, Ge X, Zhang T et al (2016) Photo-fermentative hydrogen production from enzymatic hydrolysate of corn stalk pith with a photosynthetic consortium. Int J Hydrogen Energy 41:16778–16785. https://doi.org/10.1016/j.ijhydene.2016.07.129
144. Li S, Tabatabaei M, Li F, Ho SH (2022) A review of green biohydrogen production using anoxygenic photosynthetic bacteria for hydrogen economy: challenges and opportunities. Int J Hydrogen Energy.https://doi.org/10.1016/j.ijhydene.2022.11.014
145. Das SR, Basak N (2021) Molecular biohydrogen production by dark and photo fermentation from wastes containing starch: recent advancement and future perspective. Bioprocess Biosyst Eng 44:1–25. https://doi.org/10.1007/s00449-020-02422-5
146. Hay JXW, Wu TY, Juan JC, Md. Jahim J (2013) Biohydrogen production through photo fermentation or dark fermentation using waste as a substrate: overview, economics, and future prospects of hydrogen usage. Biofuels, Bioprod Biorefining 7:334–352. https://doi.org/10.1002/bbb.1403
147. Hülsen T, Züger C, Gan ZM, et al (2022) Outdoor demonstration-scale flat plate photobioreactor for resource recovery with purple phototrophic bacteria. Water Res 216:118327. https://doi.org/10.1016/j.watres.2022.118327
148. Capson-Tojo G, Batstone DJ, Grassino M, Hülsen T (2022) Light attenuation in enriched purple phototrophic bacteria cultures: implications for modelling and reactor design. Water Res 219:118572. https://doi.org/10.1016/j.watres.2022.118572
149. Spanoghe J, Ost KJ, Van Beeck W et al (2022) Purple bacteria screening for photoautohydrogenotrophic food production: are new H2-fed isolates faster and nutritionally better than photoheterotrophically obtained reference species? N Biotechnol 72:38–47. https://doi.org/10.1016/j.nbt.2022.08.005
150. Spanoghe J, Vermeir P, Vlaeminck SE (2021) Microbial food from light, carbon dioxide and hydrogen gas: kinetic, stoichiometric and nutritional potential of three purple bacteria. Bioresour Technol 337:125364. https://doi.org/10.1016/j.biortech.2021.125364
151. Gopal PK (2022) Bacteria, beneficial: probiotic lactic acid bacteria: an overview. Encycl dairy Sci 32–33. https://doi.org/10.1016/b978-0-12-818766-1.00018-0
152. García-Depraect O, Castro-Muñoz R, Muñoz R, et al (2021) A review on the factors influencing biohydrogen production from lactate: The key to unlocking enhanced dark fermentative processes. Bioresour Technol 324:124595. https://doi.org/10.1016/j.biortech.2020.124595
153. Castelló E, Nunes Ferraz-Junior AD, Andreani C, et al (2020) Stability problems in the hydrogen production by dark fermentation: Possible causes and solutions. Renew Sustain Energy Rev 119:109602. https://doi.org/10.1016/j.rser.2019.109602

154. Abdel-Rahman MA, Tashiro Y, Sonomoto K (2013) Recent advances in lactic acid production by microbial fermentation processes. Biotechnol Adv 31:877–902. https://doi.org/10.1016/j.biotechadv.2013.04.002
155. Cheng CH, Hung CH, Lee KS et al (2008) Microbial community structure of a starch-feeding fermentative hydrogen production reactor operated under different incubation conditions. Int J Hydrogen Energy 33:5242–5249. https://doi.org/10.1016/j.ijhydene.2008.05.017
156. Ohnishi A, Bando Y, Fujimoto N, Suzuki M (2010) Development of a simple bio-hydrogen production system through dark fermentation by using unique microflora. Int J Hydrogen Energy 35:8544–8553. https://doi.org/10.1016/j.ijhydene.2010.05.113
157. Detman A, Mielecki D, Chojnacka A et al (2019) Cell factories converting lactate and acetate to butyrate: *Clostridium butyricum* and microbial communities from dark fermentation bioreactors. Microb Cell Fact 18:1–12. https://doi.org/10.1186/s12934-019-1085-1
158. Detman A, Laubitz D, Chojnacka A et al (2021) Dynamics of dark fermentation microbial communities in the light of lactate and butyrate production. Microbiome 9:1–21. https://doi.org/10.1186/s40168-021-01105-x
159. Ohnishi A, Hasegawa Y, Abe S et al (2012) Hydrogen fermentation using lactate as the sole carbon source: solution for 'blind spots' in biofuel production. RSC Adv 2:8332–8340. https://doi.org/10.1039/C2RA20590D
160. Martínez-Mendoza LJ, Lebrero R, Muñoz R, García-Depraect O (2022) Influence of key operational parameters on biohydrogen production from fruit and vegetable waste via lactate-driven dark fermentation. Bioresour Technol 364:128070. https://doi.org/10.1016/j.biortech.2022.128070
161. García-Depraect O, Muñoz R, Rodríguez E et al (2021) Microbial ecology of a lactate-driven dark fermentation process producing hydrogen under carbohydrate-limiting conditions. Int J Hydrogen Energy 46:11284–11296. https://doi.org/10.1016/j.ijhydene.2020.08.209
162. García-Depraect O, León-Becerril E (2018) Fermentative biohydrogen production from tequila vinasse via the lactate-acetate pathway: operational performance, kinetic analysis and microbial ecology. Fuel 234:151–160. https://doi.org/10.1016/J.FUEL.2018.06.126
163. Ohnishi A, Hasegawa Y, Fujimoto N, Suzuki M (2022) Biohydrogen production by mixed culture of *Megasphaera elsdenii* with lactic acid bacteria as Lactate-driven dark fermentation. Bioresour Technol 343:126076. https://doi.org/10.1016/j.biortech.2021.126076
164. Villa R, Ortega Rodriguez L, Fenech C, Anika OC (2020) Ensiling for anaerobic digestion: a review of key considerations to maximise methane yields. Renew Sustain Energy Rev 134:110401. https://doi.org/10.1016/j.rser.2020.110401
165. Thompson VS, Volk TA, Wendt LM (2021) Editorial: storage of biomass feedstocks: risks and opportunities. Front Bioeng Biotechnol 9:657342
166. Parthiba Karthikeyan O, Trably E, Mehariya S et al (2018) Pretreatment of food waste for methane and hydrogen recovery: a review. Bioresour Technol 249:1025–1039. https://doi.org/10.1016/j.biortech.2017.09.105
167. Noblecourt A, Christophe G, Larroche C, Fontanille P (2018) Hydrogen production by dark fermentation from pre-fermented depackaging food wastes. Bioresour Technol 247:864–870. https://doi.org/10.1016/j.biortech.2017.09.199

Chapter 15
Harvesting Biofuels with Microbial Electrochemical Technologies (METs): State of the Art and Future Challenges

Clara Marandola, Lorenzo Cristiani, Marco Zeppilli, Marianna Villano, Mauro Majone, Elio Fantini, Loretta Daddiego, Loredana Lopez, Roberto Ciccoli, Antonella Signorini, Silvia Rosa, and Antonella Marone

Abstract Microorganisms and microbial technologies have accompanied mankind throughout its history. The growing demand for energy and increasing waste production require combined eco-friendly smart solutions. Microbial technologies have been involved in the waste treatment process since the beginning of industrialization. However, their natural capacity to recover nutrients and energy has not been fully exploited in recent decades. Indeed, energy recovery by biogas production and nutrient recovery from wastewater treatment plants and anaerobic digestors have considerably increased their capacity. An emerging biotechnology field for environmental applications is represented by microbial electrochemical technologies, which are based on using so-called electroactive microorganisms, which can exchange electrons with conductive materials under particular environmental conditions. If the electrodic material is adopted as the final acceptor of electrons produced by oxidizing metabolism, the bio-electrochemical interphase is named bioanode. On the other hand, if microorganisms use the electrode as a source of reducing power, it is called a biocathode. Microbial electrochemical technologies are promising for their ability to harvest energy from waste organic materials reducing the energy costs of producing high-value products such as biofuels and chemicals. Despite the possibility of directly converting the organic waste into electricity in microbial fuel cells

C. Marandola · L. Cristiani · M. Zeppilli · M. Villano · M. Majone
Department of Chemistry, Sapienza University of Rome, P.le Aldo Moro 5, 00185 Rome, Italy

E. Fantini · L. Daddiego · L. Lopez · R. Ciccoli · A. Signorini · S. Rosa · A. Marone (✉)
Biotechnological Processes for Energy and Industry Laboratory (PBE), Department of Energy Technologies and Renewables, ENEA, Casaccia R.C., 00123 Rome, Italy
e-mail: antonella.marone@enea.it

Biotechnological Processes for Energy and Industry Laboratory (PBE), Department of Energy Technologies and Renewables, ENEA, Trisaia R.C., 75026 Rotondella, MT, Italy

Present Address:
L. Cristiani
Department of Bio Pilot Plant, Leibniz Institute for Natural Product Research and Infection Biology—Hans-Knöll Institute, Adolf-Reichwein-Str. 23, 07745 Jena, Germany

V. Alcaraz Gonzalez et al. (eds.), *Wastewater Exploitation*, Springer Water,
https://doi.org/10.1007/978-3-031-57735-2_15

(MFCs), in a microbial electrolysis cell (MEC) several products can be obtained by applying an external electric voltage to the cell. MECs can produce hydrogen in the cathodic compartment of the cell while treating wastewater on the anodic side; moreover, by choosing optimal operating conditions, additional processes like biogas upgrading into biomethane can be performed simultaneously. Even if many environmental applications have been successfully investigated in METs, those technologies are going through the upscaling process. To avoid the technological valley of death, the systems must continue improving. This chapter, reports state of the art and the future challenges of the most recent advances in biofuel generation through MECs. An overview of the microbiology of METs and the metagenomic approaches to characterize their microbial communities are included. Moreover, particular attention will be addressed to highlight the most recent scale-up attempt of MECs devoted to hydrogen and/or methane generation.

Keywords Hydrogen · Methane · Microbial communities · Microbial electrolysis cells · Scale-up

15.1 Introduction

Modern society generates large amounts of waste that that have a significant impact on the environment and human health [1]. At the same time, today's world is facing an energy crisis involving all fossil fuels. Therefore, finding solutions to move towards renewable energy sources and a green and circular economy become increasingly important. An innovative technology that couples waste management and energy production is represented by microbial electrochemical technologies (METs), which exploit the ability of electroactive bacteria (EAB) to interact with a solid-state electrode [2]. In the field of bio-energy generation, microbial electrolysis cells (MECs) represent a sustainable and eco-friendly tool for biological hydrogen (Bio-H_2) production [3]. Green H_2 represents the most promising fuel of the future [4] because it is not linked with GHGs emissions. Nowadays, the lack of feasible storage and no efficient ways of producing it in a renewable way are the main limitations of the H_2 –producing technologies [5]. H_2 is mainly produced by steam reforming of natural gas, which produces CO_2 as a reaction by-product. On the contrary green hydrogen (H_2), i.e., H_2 produced by water electrolysis, results in an energy-intensive process that requires low-cost electrical energy to be competitive [6]. Bio-H_2, also included in the green H_2 definition, can be produced through biological routes involving bacterial and algal metabolism [3, 7]. In the last decades, an innovative bioelectrochemical process named electro-hydrogenesis has been identified as a promising strategy for green H_2 production. Biological electro-hydrogenesis involves the utilization of electroactive microorganisms (EAMs) capable of converting organic matter into CO_2, protons, and electrons in the anodic compartment of a microbial electrolysis cell, whose cathode can be utilized for H_2 production by applying an external voltage. Indeed, electrons are conveyed to the anode while protons flow to the cathode where, in the absence of oxygen, H^+ merges in H_2 according to the following equation: $2H^+ + 2e^- \rightarrow H_2$ [7]. This process is called hydrogen evolution reaction (HER) [8]. The oxidation of organic matter at the anode compartment of a MEC partially sustains the energy demand of the process [9], lowering the theoretical voltage required for microbial electrolysis concerning conventional electrolysis (e.g., in the case of acetate oxidation at the anode, − 0.187 vs. + 1.23 V) with a consequent increase in terms of energy efficiency. Moreover, bio-H_2 produced in MECs can be utilized for biogas upgrading. Indeed, hydrogenotrophic methanogens, players of the last step of anaerobic digestion (AD), are able, in the presence of H_2, to convert carbon dioxide (CO_2) into methane (CH_4) and water (H_2O) according to the following reaction: $4H_2 + CO_2 \rightarrow CH_4 + 2H_2O$ [10]. In MECs, EAMs reduce CO_2 into CH_4 [11] by using the cathode as an electron donor and/or the electrochemically produced hydrogen. This technology perfectly fits with the UE guidelines for GHGs emission reduction and renewable energy production [12]. Indeed, in 2021 AD plants in Europe accounted for around 20.000 units in operation, but just 1.023 plants present upgrading technology to convert biogas (50–70% CH_4; 30–50% CO_2) [13] into bio-CH_4 (Bio-CH_4 = CH_4 > 90%; CO_2 < 8%) [14] mostly because external upgrading technologies such as cryogenic treatment, chemical absorption, membrane separation, pressure swing adsorption, and water scrubbing are energy-cost and expensive [15]. In this context, MECs represent a cost-effective and environmental-friendly biological alternative to physical–chemical techniques. However, H_2 supply to methanogens represents the limiting step of the process because of the low H_2 solubility in the liquid phase [16]. Nowadays, the possibility of coupling AD and MET (MET-AD) is gaining more and more attention. The digestate deriving from AD can be directed to the anodic chamber of a MEC where the oxidation of organic matter, applying an external power supply, occurs [17]. It partially sustains the energy demand for CH_4 production, supplying the reducing power "in situ" to the methanogenic biofilm growing on the electrode's surface in the cathodic chamber. The result is biogas enriched with CH_4. The first papers on H_2 production in MECs were published in 2005 by a research group from Penn State University [18]and another in the Netherlands at Wageningen University [19]. In 2008 Logan et al., reviewed the first-time materials, architectures, performance, and energy efficiencies of MEC systems that showed promise as a method for renewable and sustainable energy production and wastewater treatment [20]. Although the high MEC versatility has been demonstrated in several environmental applications (bioremediation, bio-electrosynthesis bio-H_2, and bio-CH_4 production), few successful scaled-up processes have been reported in the literature [21]. Consistent scientific literature is available on possible MEC configurations for biofuel production. In recent years, different substrate compositions and different electrode materials (anode and cathode) have been investigated for microbe-electrode interaction, leading to important contributions to optimise and increase the performance of these systems. Indeed, many technical challenges need to be faced by scientists before MECs scale-up from the laboratory to practical application [22]. In this chapter, we will give an overview of all the different biotic and abiotic parameters that affect MECs performances and a state of the art of this technology.

15.2 Theoretical and Practical Considerations for Harvesting Biofuels with METs

Microbial electrochemical technologies (METs) can remarkably transform low-value feedstocks, like wastewaters and waste gases into valuable biofuels like H_2 and CH_4. The properties of these feedstocks can significantly vary based on their origin or the season. While there are general guidelines to optimize system performance and adjust bioreactor parameters, it is crucial to study each case individually and be prepared to respond to external disturbances. Selecting the appropriate environmental conditions is contingent upon the specific application and objectives of the MET, ensuring optimal outcomes.

15.2.1 Environmental Variables

When carrying out a biological process on site (real environment at industrial scale), there are several environmental variables that usually cannot be replicated in the laboratory, such as daily variations in process parameters or mass transfer restrictions are relevant on a real scale. The most variable parameter is probably the temperature, which significantly changes during the year, the season, and even during the day. Microorganisms have a range of temperatures in which they grow optimally; for example, *Geobacter sulfurreducens* and *Shewanella oneidensis* have the optimal growth temperature at around 30–35 °C even though it is possible for them to also grow at higher (such as 42 °C) or lower temperatures (4 °C minimum). The growth temperature may also depend on various factors such as the availability of nutrients, pH, and other environmental stressful conditions [23, 24]. The temperature control can be ensured by heating or cooling the reactor. However, this type of operation is associated with high energy requirements and is therefore not always favourable. A second parameter is the pH. The pH range for optimal growth may vary depending on the specific *genera* of bacteria, environmental conditions, and other stresses. Therefore, it is essential to determine the specific pH requirements for each microbial genus and species of interest. For example, *Geobacter* and *Shewanella* grow optimally in a pH range between 6.5 and 7.5, this range is the same for *Methanosarcina* and *Methanococcus* species. *M. labreanum*, however, has a higher pH range, between pH 7.8 and 8.0 [25, 26]. Moreover, methanogens are highly sensitive to changes in pH, and even small fluctuations can impact their growth and CH_4 production. Therefore, it is crucial to maintain a stable pH range within the optimal value to ensure maximum microbial growth and productivity [25]. To obtain different environments for different types of microorganisms, it is possible to divide the bioreactor into two (or more) chambers with an ion exchange membrane (AEM anion, CEM cation, or bipolar BEM). The pH split phenomena inside METs leads to an acidification of the anolyte and an alkalinisation of the catholyte. The anolyte pH can be regulated by changing the hydraulic retention time (HRT), and the catholyte pH can be decreased by flushing CO_2 inside the cathode chamber. This last operation is also necessary to provide CO_2 as a carbon source for the autotrophic microorganisms and to ensure the anaerobic condition. The anaerobic condition inside METs is necessary to avoid low coulombic efficiency (which represents the fraction of the removed substrate diverted into electric current, CE) or microbial inhibition due to oxygen as an inhibitor or electron acceptor [26].

Moreover, inhibitors such as antibiotics, heavy metals, and toxic compounds can affect microbial activity and electrochemical reactions. Those inhibitors are often present in wastewater due to human activity. Therefore, it is essential to analyse the feedstocks. The substrates' concentration and their availability inside the feeding solution are important parameters. Usually, wastewaters have a soluble COD (Chemical Oxygen Demand) concentration of around 0.5 gCOD/L, but the COD concentration does not indicate which nutrient type is present [27] and in which ratio. Microbial growth is impacted by macronutrients, such as carbon, nitrogen, and phosphorous, and micronutrients, such as iron, zinc, and manganese. The limited availability of nutrients can reduce metabolic reaction rates and act as a stressor for EAMs, which have proven to possess resistance and resilience [28].

15.2.2 Electrons Sources and Feedstocks Origins

Some microorganisms are able to obtain their energy from organic compounds such as sugars, amino acids and fatty acids. Part of this group of microbes is represented by the EAMs used to generate an electric current inside a MET by using the anode as the final electron acceptor. These microorganisms usually use an oxidized molecule or atom as the final electron acceptor, as *Geobacter* can reduce Fe^{3+} into Fe^{2+} or as *Shewanella* can use fumarate ($C_4H_4O_4$) as an electron acceptor, transforming it into succinate ($C_4H_6O_4$) [23]. *Shewanella* is a facultative anaerobe, capable of also using O_2 as an electron acceptor to reduce it into H_2O [29]. The reducing power extracted from organic feedstocks can be directed to the cathode using a potentiostat. Inside the cathodic chamber, in which the products of interest are produced, different *genera* of microorganisms can reduce CO_2 into CH_4 (like *M. maripaludis* or *M. formicicum*), acetate (CH_3COOH) (like *A. woodii* or *A. carbinolicum*), or ethanol (CH_3CH_2OH) (like *C. autoethanogenum* or *C. ljungdahlii*) [30–32]. As a source of electrons, those systems can harvest energy from substrates present in wastewater. Therefore, digestate, fermented biomass, or domestic wastewater coming from a digestor, a fermenter, and a sewer, respectively, can provide a nutrient solution for the anodic biofilm [33]. Theoretically, each molecule can participate in a redox reaction, but to harvest energy from reduced compounds, microorganisms must metabolize those nutrients oxidizing them. Therefore, a maximum of electrons is extractable from a determined redox reaction. The most common carbon and electron source reported in literature is acetate which can be oxidized into CO_2, extracting 8 electrons [34]. Following the same reasoning, glucose ($C_6H_{12}O_6$) can be oxidized into CO_2 extracting 24 electrons. However, wastewaters are complex mixtures containing many different chemical compounds from which microorganisms can harvest energy with different metabolic pathways and energetic gains [35].

To summarize every possible oxidizable molecule present in wastewater as a unique parameter, BOD (Biological Oxygen Demand) or COD are used [36]. Those parameters represent the oxygen necessary to oxidize biologically or chemically, respectively, the compounds inside wastewater. Therefore, to simplify the calculation for the theoretical number of electrons harvestable from wastewater, it is possible to use the oxygen redox reaction:

$$O_2 + 4H^+ + 4e^- \rightarrow 2H_2O$$

This means that 32 g of COD ($\cong$ 32 g of glucose) present in wastewater can generate 4 mol of electrons, which have 386 kC of charge, and if oxidized in 1 h (3600 s), they can generate 107 A (C/s). METs have demonstrated to be very flexible and resilient regarding the feeding solution. The main problem while feeding the anodic biofilm with wastewater consists of the relatively low conductivity of those solutions, around 500 μS/cm [37]. Higher conductivities would lead to lower resistances inside the BES, reducing the energetic demand of the process. In the cathodic chamber, waste gases like biogas from a fermenter or a digestor or gases coming from urea, steel, energy, or cement production plants can be used as carbon and electron acceptors [38, 39].

Moreover, those streams are usually warmer than the environmental temperature. Therefore, they can heat the METs, which usually work better at temperatures between 30 and 55 °C. In addition to the example reported in paragraph 2.1, *M. marburgensis* and *M. wolfeii* have optimal growth at temperatures around 55 °C, and this implies that to obtain higher performance, the METs aimed to produce bio-CH_4 must be heated somehow [40, 41]. Analysing the content of the inlet gasses is fundamental because they can contain toxic compounds such as H_2S or inhibiting substances like O_2 [42]. The provenience of the flue gas will determine the content. For example, for biogas, the type of organic waste fed into the digestor can significantly change the composition of the produced biogas [43]. The METs are resilient and can work with a variety of effluents, different wastewaters and flue gas. However, it is important to control the quality of these feedstocks and consider purification steps if needed [44].

15.2.3 Redox Reaction

As previously mentioned, METs, involving the exchange of electrons, strictly depend on the outcome of redox reactions. The metabolic pathways have as their primary goal the synthesis of adenosine triphosphate (ATP), which is often referred to as the "energy currency" of cells because it provides the necessary energy for various cellular processes and activities [45]. The standard free energy released by the hydrolysis of ATP to ADP and Pi under physiological conditions is approximately -30.5 kJ/mol. Various molecules, such as electron carriers like NADH and $FADH_2$, are involved in cellular redox reactions, with measurable midpoint potentials that play a significant role in energy metabolism [46]. The electrons are transported through the electron transport chain (ETC), a crucial cellular metabolism process that occurs in prokaryotes' plasma membrane [47]. Microorganisms harvest energy from reduced substrates gaining electrons with high potential, transporting them through the ETC in which the electrons lose potential, and the energy released is used to pump protons across the inner mitochondrial membrane. Later the electrochemical gradient formed is used to produce ATP, and the electrons are donated to a final electron acceptor [48]. In the case of the anodic electroactive biofilm, the electrode represents the final electron acceptor, while in the cathodic chamber it may be an oxidised molecule such as CO_2 or SO_4^{2-} [49]. Electrochemical potential reaction (E) is easily calculated through the Nernst equation, which takes into account the temperature, the pH (if the protons H^+ are involved in the reaction), the concentration of products, and reagents and the pressure (in case the reaction involves gases):

$$aA + bB \leftrightarrow cC + dD \qquad E = E^0 - \frac{RT}{nF} ln \frac{[C]^c[D]^d}{[A]^a[B]^b}$$

In the equation, "E" stands for the cell potential associated with the electrochemical reaction, "E^0" for the standard potential, "R" for the universal gas constant, "n" for the number of electrons involved in the redox reaction, "F" for the Faraday constant and "T" for the temperature. With this equation, the potential (E) of a certain redox reaction can be calculated theoretically. As explained before, the redox potential is directly linked to the energy (Gibbs free energy "ΔG") necessary or releasable by a reaction, following the relationship:

$$\Delta G = -nFE$$

Microorganisms gain energy by conducting redox metabolic reactions, and those equations can theoretically calculate the amount of energy. Obviously, under non-ideal condition, the 'systems' do not have an efficiency of 100% and some of the substrates are not only used for metabolism but also to synthesize complex molecules (anabolism) [50]. For those reasons, the energy necessary to sustain the reactions inside METs will be higher than the theoretical energetic consumption. Moreover, electrochemical systems have potential losses due to various reasons like Ohmic losses, which are strictly correlated with the internal resistance of the system to the charge movement (resistance of wires, conductivity of the media, and all the electric connections). These activation losses are due to the electron transferring reactions or concentration losses dependent on the reagent supply [49].

When a MET contains a membrane, the system also exhibits membrane resistance due to the diffusion of charges through the membrane, as well as resistances due to the different conductivity and pH of the liquid phases in the MET chambers [51]. As a result, METs operate at a higher potential than theoretically predicted.

For example, *G. sulfurreducens, G. metallireducens, G. daltonii, G. bemidjensis, G. chapellei, G. pelophilus*, and *S. oneidensis* (which are the most common EAB) have a formal potential between − 0.25 V and − 0.07 V versus the Standard hydrogen electrode (SHE) [51–55]. However, the most utilized anodic potentials are between 0 and + 0.4 V versus SHE to ensure the success of the reaction. Moreover, it was demonstrated that a non-controlled bioanode tends to work at a potential around + 0.5 V versus SHE. In the case of METs with an abiotic anode, the most common oxidation reactions used are the oxidation of water (E^0 + 1.2 V vs. SHE) or the oxidation of a sacrificial electrode (frequently made of iron or graphite) [56, 57]. The higher anodic potential leads to higher energetic consumption.

Moreover, the oxidation reaction has to be carefully chosen, considering the reactions' products and how they can affect the microorganisms (for example, the oxidation of water produces oxygen which can inhibit the microbial activity) [58]. To win the overpotentials, the cathodic potential will also be higher than predicted. Therefore, in systems with a two electrodes configuration (in which the potential difference ΔV is controlled) METs operate with higher potential differences than the one predictable:

$$\Delta V = E_{cat} - E_{an}$$

The two-electrode configuration does not permit control of the anodic or the cathodic potential, meaning that they are free to change. If the cathodic potential decreases by 200 mV (for example, from – 1.3 to – 1.5 V), the anodic potential must decrease by the exact 200 mV. A reference electrode permits to work in a three-electrode configuration, controlling the cathodic or anodic potential. Some METs require the control of the cathodic potential to maintain a certain reducing potential and to control the cathodic reaction [59]. Some METs exploit the ability of some microorganisms to interact with the electrodes through a redox mediator, this theoretically permits to work at the redox potential of the mediator and to work with the entire volume of the reactor and not only with the electrode surface. Some microorganisms can synthesize mediators, and others can only use them [60]. The most common mediator is the H_2, with a redox potential at pH 7 of around - 0.44 V versus SHE. Two mediator groups are the flavines (among which FAD: flavin adenine dinucleotide) and the phenazines which are largely utilized by EAMs as electron donors or acceptors [23, 61]. Due to their ambivalent midpoint potentials, the possibility of being either the electron donor or acceptor, leads to their utilization in reduction or oxidation reactions.

15.3 Microbiology of Electrode-Associated Biomass and Planktonic Microbial Communities in METs

Bio-electrochemical processes depend on particular features of various microorganisms, namely "electrogenic" or "electroactive" bacteria (EABs), as biocatalysts, which can exchange electrons with insoluble minerals or metals (e.g., electrodes). In the case of a MEC, where microbiological oxidation of the organic matter occurs at the anode, EABs are responsible for electron transfer from the substrates to the anode on which they develop, forming a biofilm, namely "electroactive biofilm".

If the cathode is biotic, EABs grown at the cathode are responsible for electron transfer from the cathode to the protons or the carbon dioxide to reduce H_2 and CH_4, respectively.

The mechanisms EABs use for electron transfer between microorganisms and electrodes are different: (1) direct contact due to the presence of redox-active proteins; (2) indirectly thought electronic shuttles (flavins, riboflavins, or polysaccharides); (3) direct transfer to long-range via conductive pili also called nanowires [62].

The most common and studied EABs are *Geobacter sulfurreducens* and *Shewanella oneidensis*, which are also associated to their preference for specific substrates, acetate, and lactate respectively [63]. In particular, the genus *Geobacter* (*δ-Proteobacteria*) is ubiquitous in METs, and it can develop nanowires, allowing direct-interspecies electron transfer. *Geobacter* typically dominates the microbial communities in METs when acetate serves solely as the electron donor [64, 65] but is also often the most abundant microorganism in the anodic community in METs fed with complex substrate [63, 66, 67]. However, the microbe-electrode exchange arises from an evolutionary process developed over millions of years, which resulted in a high diversity of EABs. As the number of studies increases, new EABs are identified in the *Proteobacteria, Firmicutes, Cyanobacteria, Deferribacteres*, and *Actinobacteria* phyla. At the same time, microorganisms belonging to other domains of life, such as *Archaea* (*Euryarchaeota* phylum) and *Eukarya* (fungal species in the *Ascomycota* phylum), have been identified in suspended cells and/or biofilms in METs [68].

It is now recognised that EABs are likely functioning as consortia in natural environments. The symbiosis of the other microorganisms with EABs can be attributed to the fact that EABs can utilize only few and simple substrates to biodegrade and produce electrons. Thus, the presence of other microorganisms allows the utilisation of a broader range of organic materials for biodegradation and improving bio-electrochemical cells' efficiency. Indeed, when fed with complex substrates (e.g., wastewater), microbial ecosystems in METs are very complex. In the anode of a MEC, fed with complex substrates, the microbial communities are divided into biofilm and planktonic microorganisms and, in addition to EABs, include hydrolytic and fermentative bacteria that convert complex organic molecules into simpler molecules such as volatile fatty acids (VFAs) which EABs can easily use.

The composition of anodic microbial communities is very diverse according to the composition of the substrate compounds to be oxidized, the origin of the inoculum, the type of MET, the electrode material, and the operating conditions. Bacteria belonging to the phyla *Proteobacteria* and *Firmicutes,* well-known fermenters and syntrophic bacterial partners, often dominate anode biofilms fed with complex substrates [65, 69]. *Bacteroidetes* or *Actinobacteria* are also found in the biofilm but represent a low fraction of total *Bacteria* [70, 71]. All these populations have positive interactions that promote efficient substrate conversion to electrons and, thus, allow for reduction reactions at the cathode. However, it should be highlighted that the CE decreases in MEC fed with complex substrates compared to simple substrates, such as VFA, due to the loss of electrons in the different reactions.

Degradation pathways of fermentation for end products are specific to each substrate, with a common final point, acetate, and H_2; both products enable a rapid transfer through EABs [66]. The main concurrent organisms are hydrogenotrophic methanogens, able to use H_2 efficiently, causing a decrease of the CE [72, 73]:

$$4H_2 + HCO_3^- + H^+ \rightarrow CH_4 + 3H_2O \qquad \Delta G^{\circ\prime} = -135.6\ \text{kJ/mol}$$

Moreover, in competition with EABs metabolism, the acetate can be converted to CH_4 by acetoclastic methanogens:

$$CH_3COO^- + H_2O \rightarrow CH_4 + HCO_3^- \qquad \Delta G^{\circ\prime} = -31.0\ \text{kJ/mol}$$

However, hydrogenotrophic methanogens are more often detected in anodic microbial communities than acetoclastic methanogens, probably because exoelectrogens could outcompete acetoclastic methanogens.

In addition to the above-mentioned hydrogenotrophic methanogens, the H_2 can also be used by homoacetogenic bacteria to form acetate:

$$4H_2 + 2HCO_3^- + H^+ \rightarrow CH_3COO^- + 4H_2O \qquad \Delta G^{\circ\prime} = -104.5\ \text{kJ/mol}$$

Homoacetogenesis also competes with hydrogenotrophic methanogenesis, but homoacetogens appear to establish positive syntrophy with exoelectrogens [74]. Indeed, good performances were obtained in a MEC fed with digestate thanks to the establishment of efficient syntrophic interactions between fermenters, homoacetogens, and exoelectrogens [75]. However, homoacetogenic pathways have only been demonstrated when methanogenesis is inhibited since, due to thermodynamic and kinetic advantages, hydrogentrophic methanogens have been shown to outcompete homoacetogens [76]. Therefore, if H_2 is the desired output, if a membrane does not physically separate the anode and cathode, H_2 scavenging occurs [76–80], decreasing H_2 efficiency.

However, when CH_4 is the target fuel, methanogens are responsible for CH_4 production in the cathode of MECs, where they can directly convert current into CH_4 via electromethanogenesis [68, 81, 82]. Indeed, microorganisms in the domain of *Archaea* have been found of great interest in METs systems, such as *Methanobacterium* sp. and *Methanococcus maripaludis,* which have been reported to be able to perform electromethanogenesis in MECs [82, 83]. In contrast, *Methanosarcina barkeri* has been recently found to have the ability to use electrons from the cathode via Direct Interspecies Electron Transfer (DIET) [84].

Moreover, depending on the composition of the substrate used to feed the MEC, alternative electron acceptors to the anode can be present, such as nitrate or sulfate, responsible for a decrease in the CE. Potential oxygen leakages could also favor heterotrophic bacteria and decrease the CE [85]. Thus, the anodic microbial community may also include microorganisms whose activity can be detrimental to the process, causing electrons to be diverted to alternative final electron acceptors through competing metabolic pathways, lowering the CE.

To improve stability, performance, and CE, it is necessary to achieve a functional mixed community whose stability depends on the exchange of metabolic products between microorganisms. These synergistic interactions are important to carry out processes such as H_2 and CH_4 production and organic matter oxidation.

Therefore, a comprehensive understanding of these communities, including their biodiversity, genetics, and metabolism, and identifying specific genes, proteins, and metabolites involved in biodegradation processes can contribute to the optimization and further improvement of METs (see Sect. 15.4).

15.4 Sequencing-Based Approaches to Characterize Microbial Communities in METs

The characterization of EABs and other microorganisms involved in improving the efficiency of bio-electrochemical cells is crucial to establish the relationships between the enriched microbial community and operating parameters such as inoculum, substrate [86], electrode material [87], electrode potential [88] and operating mode [65]. This section describes up-to-date sequencing-based approaches utilized for microbial consortia identification and characterization that can be applied to METs.

15.4.1 Nucleic Acid-Based Techniques

Nucleic acid-based techniques are the most used approaches to characterize microbial consortia. Several commonly used techniques have been employed, providing a range of tools to study microbial communities associated with MECs, including Denaturing Gradient Gel Electrophoresis (DGGE), Temperature Gradient Gel Electrophoresis (TGGE), Terminal Restriction Fragment Length Polymorphism (T-RFLP), Single-Strand Conformation Polymorphism (SSCP), Fluorescence In Situ Hybridization (FISH), Marker Hybridization, Quantitative Polymerase Chain Reaction (qPCR) and high-throughput sequencing. Over the last decades, the introduction of sequencing technologies has led to a drastic change in how microbial communities are explored. Since most environmental microbes remain uncultured [89], nucleic acid sequencing analysis has become the most used approach for studying complex microbial communities.

High throughput Next-Generation Sequencing (NGS) technologies and, successively, Third-Generation Sequencing (TGS) have undoubtedly led to a real revolution of microbial community analysis. Among NGS technologies, characterized by the production of short sequencing reads, noteworthy techniques are Pyrosequencing (Roche) and Reversible termination sequencing (Illumina). The latter has become the dominant NGS platform for microbial community characterization due to its ability to generate massive amounts of sequencing data in a single run.

Currently, the most widely used TGS technologies for microbial communities sequencing are single-molecule real-time sequencing (SMRT) and nanopore sequencing (NS), developed by Pacific Biosciences (PacBio) and Oxford Nanopore Technologies (ONT), respectively. Compared to NGS, these technologies are characterized by real-time long read sequencing that overcomes limitations associated with short read lengths, enabling the assembly of complex genomic regions, the detection of gene isoforms, and the sequencing of full-length or near full-length transcripts, providing a more comprehensive insight into genomic and transcriptomic characterizations of microbial communities.

While each technology is characterized by specific chemistry and sequencing device, in both NGS and TGS approaches, nucleic acid isolation, sequencing library preparation, and bioinformatics analysis follow common steps to ensure accurate representation of the original community composition, genetic diversity, and functional potential of the microbiome in METs samples.

15.4.2 Sequencing-Based Approaches

Microbial communities can be characterized by two different metagenomic approaches: targeted (metataxonomics or metabarcoding) and shotgun. Metataxonomics is based on the amplification and sequencing of conserved marker genes (amplicon sequencing), allowing the taxonomic profiling of a microbiome. Shotgun metagenomics consists of a comprehensive capture of the genomes present in a sample, providing insights into the taxonomic composition and the microbial community's functional potential (genes and functional pathways). In addition to taxonomic and functional profiling derived from metagenomic data, metatranscriptomics provides valuable information on gene expression patterns, helping to understand the relationships and metabolic network within the community.

15.4.2.1 Metataxonomics

Among the marker genes used to identify bacteria and archaea, the *16S* rRNA gene is the most used in metataxonomics. It consists of ~ 1500 bp, including nine hypervariable segments (V1–V9), each flanked by highly conserved regions useful for PCR amplification. Since NGS technologies produce short reads, only brief segments of the *16S* rRNA gene can be targeted. In contrast, SMRT and NS long-read sequencing can target the full-length *16S* rRNA gene, allowing a more realistic representation of the taxa composition in a sample [90].

In addition to the widely used *16S* rRNA, other bacterial marker genes are commonly used in metataxonomic studies to provide a more comprehensive understanding of microbial communities. These include the *rpo*B gene, which encodes the β-subunit of the RNA polymerase enzyme [91], the *rec*A gene, which encodes the recombinase [92], and the *gyr*B gene, which encodes the B-subunit of DNA gyrase [93]. These additional markers provide insight into specific taxonomic groups or functional traits within the bacterial domain, as they exhibit higher variability than the *16S* rRNA gene, allowing for finer taxonomic resolution, particularly at the species and strain level. Although most studies in METs characterization are based on *16S* rRNA gene sequencing, a recent paper observed that the *mcr*A gene improves taxonomic resolution and provides accurate real-time information on methanogens activity and community diversity [94].

Even though targeted *16S* rRNA gene sequencing approaches only characterize the taxonomic profile of a microbiome, it is a cost-effective option to exhaustively capture biodiversity across many samples, providing valuable insights into microbial community composition. Metataxonomy based on the *16S* rRNA gene is a widely employed method for investigating the microbial communities inhabiting anodes and/or cathodes in METs, which include MFCs, MECs, bioelectrochemical dechlorination systems, and Microbial Electrosynthesis systems (MESs). Within these METs, studies spanning references [82, 95–110] have unveiled considerable diversity in the microbial composition. This diversity is influenced by various factors, including biological origin, environmental conditions, and electrochemical parameters. These parameters encompass factors such as the source of the inoculum, the composition of the electrolyte and substrate, and the potential of the electrodes. Even within the same category of METs, notable variations in microbiome composition have been observed.

From metataxonomics data, predictive functional profiling of microbial communities can be generated utilizing software such as PICRUSt, Tax4Fun2, and MicFunPred [111, 112]. This approach has been recently used to analyze functional genes involved in bioelectricity production in MFCs [113] and valuable bioproducts (volatile fatty acids and alcohols) synthesis in acidogenic processes [114].

15.4.2.2 Shotgun Metagenomics

Shotgun metagenomics is a powerful approach to studying microbial communities' composition and functional potential, providing a detailed overview of their genetic diversity and ecological dynamics.

It is based on sequencing of the entire DNA content of the community (randomly fragmented DNA), followed by bioinformatic analyses to identify and classify microbial taxa and to predict genes and pathways. Shotgun metagenomics requires higher sequencing depth and computational resources compared to targeted sequencing approaches; in addition, data analysis and interpretation are more challenging due to the complexity of the data generated. Unlike the extensive use of *16S* rRNA sequencing, the application of shotgun sequencing for microbial community analysis has remained relatively limited. Such applications include research focused on microbial communities, as those growing on the biocathode in a MET treating Trichloroethylene (TCE) and TCE/Cr(VI)-contaminated water [115], or those able to switch between catalyzing anodic and cathodic reactions depending on the potential applied to the electrode [100] or MESs-associated metagenomes [116–119].

Shotgun sequencing data can also be processed to reconstruct complete and draft MET-associated genomes by genome binning [120, 121].

Ishii et al. [122] as metagenomics provides an overview of the potential functional capabilities of the microbial community, metatranscriptomics offers insights into the metabolic and functional activities at a given time and condition, provides a more detailed understanding of the microorganisms' interaction within the METs and their influence on the ongoing electrochemical processes. For example, a combined meta-omics approach has been used to characterize a microbial community that catalyzes energy-producing and energy-storing electrode reactions [100]. Metagenomic, genome binning, and stimulus-induced metatranscriptomic approaches have been applied to describe stimuli-induced metabolic switches in a complex microbial biofilm producing electrical current via extracellular electron transfer (EET) to a solid electrode surface [122].

15.4.2.3 Metatranscriptomics

Metatranscriptomics is a valuable tool based on the sequencing and analysis of mRNA transcripts produced by actively expressed genes within a microbial community at a given time and condition [123]. This can help to understand molecular interactions and predict the metabolic network within the community, including pathways involved in electron production or pollutant removal, and elucidate the metabolic role of uncultured members. This information is crucial for understanding the dynamics of electrochemical biofilms and their potential applications in various fields.

In recent years, metatranscriptomics has been employed in METs to elucidate the gene expression profiles associated with methanogenic performance in single and dual chamber MES under low-temperature conditions [124] to identify metabolic pathways related to wastewater treatment in METs [120, 125], to validate a pathway for coupling extracellular electron transfer to carbon fixation [121], to reveal the functional genes of the denitrifying MET obtained by polarity reversal [126], and to evaluate the effect of Fe^{2+} and Ni^{2+}on acetate production by CO_2 eduction in MES [127]. Overall, these studies demonstrate the potential of metatranscriptomics to understand microbial communities' functional capabilities and responses in METs, allowing a comprehensive understanding of metabolic activities, adaptation mechanisms, and interactions within a biofilm. Ongoing advances in metatranscriptomic techniques and data analysis offer exciting prospects for improving the electrochemical performance of METs and exploring increasingly impactful applications in wastewater treatment, energy production, and bioremediation.

15.5 Reactors and Electrode Configurations

In 2008, for the first time, Logan et al. reviewed the materials, architectures, performance, and energy efficiencies of MECs [20]. Since that moment, many reactor designs have been explored [128] to optimize first H_2, then CH_4 production (bioelectrochemical CO_2 conversion into CH_4 is strongly dependent on H_2 evolution) [129]. The biggest issue in MECs regards H_2 solubility in the liquid phase (0–1.5 mg/L at $T = 25$ °C and $P_{H2} = 1$ bar [130]), which makes H_2 a limiting substrate for methanogens [131]. Reactor design also affects the internal resistance and current density, thus H_2 yield [3]. Here, we review possible configuration options, electrodic materials, and membrane used to overcome this problem and optimize the process.

15.5.1 MEC Configuration

Geometry is an essential factor that influences reactor performance [132]. Reactors geometry can be planar or tubular. The planar configuration is usually used in laboratory-scale research and development studies. In contrast, the tubular configuration is the preferred option for pilot and industrial scale due to its higher surface area-to-volume ratio compared to planar geometry [133]. This ratio corresponds to the projected electrode surface and the working volume. The higher this ratio is, the more electrodes occupy the working volume, thus by the electroactive biofilm. Keeping a high surface-to-volume ratio is one of the most significant issues for MET scale-up [134]. Despite the geometry, the type of reactors is another important factor. Reactors can be batch, fed-batch, or continuous reactors [135]. Batch fermentation is the most often used at industrial scale due to its high efficiency and good control of operational parameters [136]. One advantage of batch reactors is their versatility. A single vessel can carry out a sequence of different operations and reactions. The reactor is initially loaded with substrates, and there is no addition or withdrawal until the end of the reaction [137]. When working with microorganisms, as in MECs, the biggest disadvantage of batch reactors is nutrient limitation, resulting in lower reactor performance [138]. In fed-batch reactors, reactions start in a batch mode with a bulk carbon source. Once this bulk is depleted, feeding starts to supplement the system with carbon sources for product formation and, eventually, biomass growth and maintenance. A substrate is resupplied once the second bulk is depleted [139]. The volume of substrate fed depends on different factors, such as the desired hydraulic retention time (HRT) and organic loading rate [140]. Continuous flow reactors are the best at the industrial scale because they allow for the continuous production of products or chemicals of interest, crossing the problem of inhibition due to product accumulation, resulting in a larger amount of desired product for mass unit [141]. In this case, the reactor is supported by an inlet and outlet for substrate supply and product recovery. The major problem linked to continuous reactors is their management and the control of the conditions (temperature, pH, etc.) [142]. Another feature of MET reactors is the presence of a membrane. The presence of a membrane determines the chamber's number of reactors. The most common reactors are single or double-chamber reactors [143]. In the last case, a membrane separates the anodic and cathodic chambers with consequences on reactor performance [144]. Further membranes result in additional chambers (modular reactors) [145]. Single chamber reactors usually consist of a single vessel hosting both the anode and the cathode [130]. This configuration reduces the capital cost of the process. Still, at the same time, it implies less H_2 availability due to the possible presence of hydrogenotrophic methanogens that compete with electrochemically active bacteria for both substrate (such as acetate) and product (such as H_2) [146]. To bypass this problem, the largest part of studies reported in the literature were carried out in two-chamber reactors separated by a membrane. The first two-chamber reactors reported in literature consisted of H-type cells, basically two glass bottles separated by a membrane. The presence of the membrane led to the phenomena of pH split, according to which acidification occurs at the anodic chamber and alkalization at the cathodic one [147]. This phenomenon affects the biofilm stability. Indeed, oxidation at the anode leads to the acidification of the chamber limiting the activity of the electroactive biofilms [148, 149]. At the same time, the alkalization of the cathodic chamber also decreases the microorganisms' activity, promoting a partial or complete inhibition of the biofilm. Indeed, strong alkaline pH is incompatible with methanogen growth [150]. pH split phenomenon can be exploited for biogas upgrading; indeed, if the cathode of a MEC is fed with CO_2 to mime biogas gained by anaerobic digestion (AD), pH split allows the removal of 9 mol of CO_2 for each mole of CH_4 produced [17, 151].

15.5.2 Electrodic Materials

The choice of the best electrodic materials in MET is essential because it influences reactor performance [152]. Indeed, electrodes are the surface on which EAMs attach, developing a biofilm [153] and working as an electron acceptor or electron donor [154]. Here, we review the most common anodic and cathodic materials utilized in MECs.

15.5.2.1 Anodic Electrodic Materials

The anode is established as a very significant part of the MECs, both functionally and structurally, due to its role. The materials mostly used for anodic electrodes are of carbon-based origin: carbon cloth, carbon fibers, carbon brush, graphite-fiber brush, graphite granules, and graphite plates [155, 156]. Highly pure graphite is chemically more stable than carbon (i.e., carbon cloth or carbon felt); despite the higher cost of high-grade graphite, it is widely used as a bioanode [157]. These materials were selected due to their conductivity, biocompatibility, morphological versatility, sufficiently low over-potentials, and low costs [20]. At the same time, they present some limitations due to low-moderate electrical conductivity and mechanical strength [158]. Among them, electrodes with a planar geometry are more suitable for basic research on microbe–electrode interactions. However, three-dimensional electrodes, porous electrodes, and foaming electrodes are more favorable because they offer a high specific surface area for the formation of electrocatalytically active biofilms [159, 160]. In some cases, a chemical–thermal pretreatment is required to increase the performance of some electrodic materials that show strong resistance (i.e., carbon felt, carbon fiber brush, etc.) [161, 162]. It has been demonstrated that the pretreatment also influenced the microbial community, enriching specific genera and reducing bacterial diversity in the anolyte [163]. Even metallic and composite materials, or their modification, were successful as anodic electrodes in the literature, showing higher current density and better microbial adhesion on the electrode surface [157, 161, 162].

15.5.2.2 Cathodic Electrodic Materials

Cathodic electrodic material is fundamental for hydrogen evolution reaction (HER) [164]. The key features of cathodic electrodes are high HER rate, non-corrosive, non-toxic, highly conductive, and low cost [165]. Cathodic materials include both carbon-based and metallic electrodes. However, cathodic H_2 production on plain carbon materials is often associated with a high over-potential [20], slowing the H_2 production rate in MECs. To bypass this issue, a catalyst is needed. Therefore, metal-based catalysts are usually utilized to catalyze HER [166]. The most common and efficient metal-based cathodic electrode is platinum (Pt). Indeed, it was shown that platinum as a cathode material is excellent in HER, and it also acts as a catalyst [167]. However, platinum is a precious metal with high cost and strong environmental impact due to mining [168]. In the last decade, a lot of efforts have been made by scientists to find out other no-precious metals able to act as a catalyst. Different studies reported nickel (Ni), stainless steel (SS), and hybrid materials (mixes of metals, carbon, organics, and other materials, such as alloys) as good candidates to replace Pt electrode [169, 170]. For instance, Selembo, et al. [171], demonstrated that Ni powders with different particle sizes at an external voltage of 0.6 V, Ni 210 achieved the highest H_2 production rate of 1.3 m^3/m^3d, close to the value of 1.6 m^3/m^3d of a Pt cathode.Even alloys gained interest as a catalyst in the last few years. For example, Call et al., showed that stainless steel brush cathodes had an H_2 production rate (1.7 m^3/m^3d) and a cathodic efficiency (84%) close to those with Pt cathodes in single-chamber MECs [172]. Hybrid materials also proved to be suitable candidates to replace Pt cathode catalysts. For instance, in carbon–metal hybrids, carbon can protect the metals from corrosion [173], resulting in continuous H_2 production [174].

Another possibility is using a biocathode that imply microorganisms as a catalyst. The first description of the development of a microbial biocathode for H_2 production was carried out in 2007 by Rozendal [175]. They obtained a current density of about − 1.2 A/ m^2 at a potential of − 0.7 V versus SHE, which was 3.6 times higher than that of a control electrode (− 0.3 A/m^2), then the proof of principle of an MEC in which microorganisms catalyzing both the anodic and cathodic reaction arrived in 2010, when Jeremiasse, et al. [176], demonstrated that at a cathode potential of – 0.7 V, the biocathode in the MECs had a higher current density (1.9 and 3.3 A/m^2) than a control cathode (0.3 A/m^2, graphite felt without biofilm) in an electrochemical half-cell. This indicates that H_2 production is catalyzed by electrochemically active microorganisms [176]. The most common electrodic materials for the biocathode are graphite felt, graphite granules, carbon-based materials, and stainless steel mesh [177]. EAMs interact with them through van der Waals interactions, steric and electrostatic interactions leading to electroactive biofilm formation on the electrode surface [153]. Biocathodes can potentially be used to produce valuable chemicals. Indeed, microorganisms harboring hydrogenases may play a key role in biocathode performance since H_2 generated on the electrode surface can act as an electron donor for CO_2 reduction [178]. In the context of biomethanation, the biocathode plays a pivotal role in the surface-driven electron uptake by CO_2-reducing bacterial species used in the integrated MET–AD cathode to increase the CH_4 content in biogas [179]. Microorganisms also interact with the electrode surface through physical appendices such as flagella or conductive pili [180]. Therefore, electrodes with modified surfaces (e.g., surface roughness, nanoparticles-doped, foaming, and extra-porous materials) provide additional surface for microbe-electrode interaction. Different electrode materials (plain and modified) used in MET-AD systems, along with their advantages and disadvantages, are largely discussed in the literature. For example, graphite rods/pillars/plates present low cost, high conductivity, and less corrosion but are rigid and have a low surface area. Hence, the interaction with microorganisms is weak. Carbon brush and carbon felt are highly flexible, they have a large surface area, but they need pretreatment. Porous metal electrodes (Cu/Ni or foaming SS) have higher surface area; they are semi-flexible and present a high conductivity but can be corrosive and poisonous. The last, the best, carbon fabric/felt with conductive nanoparticles or functional groups present the highest surface area, excellent electron transfer, and enhanced microbe-electrode interaction for the biomethanation of CO_2 to CH_4. However, such modifications can lead to the increased cost of electrode fabrication [179].

15.5.3 Membranes

Ion exchange membranes (IEM) are non-porous membranes that allow the selective transport of ions to ensure electroneutrality maintenance between the anode and cathode compartment of an electrochemical device. IEMs are characterized by hydrophobic polymeric materials with fixed charges on the surface that promote selective ion flux. IEMs are usually classified as cation exchange membranes (CEM), anion exchange membranes (AEM), and bipolar exchange membranes (BPM), depending on the ionic species that can be selectively transported. Membrane accounts for 35–40% of the cost in MECs [181]. Therefore, membrane-less MECs are cost-effective, easy to handle, and free of pH concerns [3]. Still, the absence of the membrane allows the diffusion of H_2 from the cathode to the anode (back diffusion), limiting the efficiency of the process [182]. Indeed, at the anode, hydrogenotrophic methanogens can utilize H_2 as a substrate to produce CH_4. To avoid it and to maintain H_2 purity, membranes are utilized. Membranes are divided considering their ion selectivity; the most common in MET are cation exchange membranes (CEMs), anion exchange membranes (AEMs), and composite membranes [183]. As already discussed, the presence of the membrane led to the phenomenon of pH split [184]. Among CEMs, PEMs are the most utilized. PEMs work as a barrier that allows the protons to pass through to ensure a redox reaction[65]. AEMs work as a barrier that allows the anions to pass through. Zeppilli et al. [17] demonstrated that PEMs allow higher COD oxidation and CH_4 production. Instead, AEMs allow a higher CO_2 removal due to HCO_3^- transport from the cathode to the anode [17]. The most utilized membranes in MECs are Nafion and Fumasep [3]. Nafion is the most widely used proton exchange membrane (PEM). It has a high H^+ conductivity due to the sulfonic groups attached to the inert structure of polytetrafluoroethylene [185]. AEMs, instead, perform better due to a lower internal resistance due to the lower transport resistance of ions through the membrane's constituent materials [186]. Indeed, Sleutels et al. [187] demonstrated that the better performance was mainly caused by the much lower internal resistance of the AEM configuration (192 $m\Omega/m^2$) compared to the CEM configuration (435 $m\Omega/m^2$) [188].

15.6 Biofuel Production

H_2 and CH_4 are the most promising fuels of the future. The EU Renewable Energy Directive states that future biofuel systems should have at least 60% GHG savings compared with fossil fuel replacement [189]. Bio H_2 produced by MEC is a promising alternative to produce H_2 in the future, as well as biogas produced by the AD, since they have a double positive effect coupling the wastewater treatment with bio-energy production [190]. Moreover, biogas can be biologically upgraded, but as previously discussed, different factors limit biogas upgrading into bio-CH_4, i.e., H_2 bio-availability. The key point making these processes effective from environmental and economic points of view is that H_2 derives from renewable energy sources. In this context, the excess of electricity deriving from wind or solar power can be used to power a MEC for H_2 production or can be exploited to hydrolyze water and produce H_2 that could be directly injected into the anaerobic digester [13].

15.6.1 Data Elaboration and Performance Assessment

To understand anodic and cathodic performance in MECs, it is important to describe some parameters. The most important are described hereafter.

The COD removal efficiency is the capacity of the anode to oxidize the organic matter, and it is described by the following equation:

$$COD_{removalefficiency} = \frac{F_{in} * COD_{in} - F_{out} * COD_{out}}{F_{in} * COD_{in}} \times 100$$

where the COD_{in} and COD_{out} represent the anodic influent and effluent COD concentration (gCOD/L), whereas F_{in} and F_{out} are the influent and effluent flow rates in the anodic chamber (L/d).

Considering the water oxidation reaction, the COD converted into electric current is expressed as electron equivalents. Then, the meq_{COD} is calculated by using a theoretical conversion factor of 4 meq/32 mgO_2.

The percentage of oxidized COD directly converted into current is represented by the CE (%). It is calculated as the ratio between the cumulative electric charge transferred at the anode (meq_i) and the cumulative equivalents released by the COD oxidation (meq_{COD}):

$$\%CE = \frac{meq_i}{meq_{COD}} \times 100$$

The cumulative electric charge (meq_i) is calculated by integrating the current (A) over time (s) and dividing it by the Faraday's constant (F = 96,485 C/eq).

The H_2/CH_4 production rates at the cathode are calculated in terms of electron equivalents considering the theoretical conversion factor of 2 meq/$mmol_{H2}$ and 8 meq/$mmol_{CH4}$ according to the following equations:

$$2e^- + 2H^+ \rightarrow H_2 \qquad r\,H_{2(mmol)} \times 2 = r\,H_{2(meq)}$$

$$CO_2 + 8e^- + 8H^+ \rightarrow CH_4 + 2H_2O \qquad rCH_{4(mmol)} \times 8 = rCH_{4(meq)}$$

In literature, values related to production rates are typically expressed as m^3/m^3 d or L/m^2 d (Tables 15.1 and 15.2). In the first case, the H_2/CH_4 production rate is normalized for the volume of the cathodic chamber (m^3); in the second one, the gas production rate is normalized for the cathodic electrode surface (m^2).

The Cathode Capture Efficiency (CCE %) is the fraction of generated current converted into H_2 ($\%CCE_{H2}$) and CH_4 ($\%CCE_{CH4}$). It is calculated by the ratio between the cumulative equivalents of produced H_2 or CH_4 ($meq_{H2/CH4}$) and the cumulative equivalents of current (i.e., the charge):

$$\%CCE = \frac{meq_{H2/CH4}}{meq_i} \times 100$$

Table 15.1 MECs performance for H_2 production

References	Current density	Cod Rem.Eff%	Coulombic Eff.%	Cathodic Config	H_2 Prod. Rate	CCE %	Energy efficiency%
[194]	163 A/m^3	88	47.2	Abiotic	1.04 m^3/m^3 d	57,2	89
[191]	312 ± 9 A/m^3	94.5 ± 2	75 ± 4	Biotic–SC[a]	4.18 ± 1 m^3/m^3 d	89.3 ± 4	94.1 ± 3.3
[188]	11.59 ± 0.72 mA	26.7 ± 0.4	117.7 ± 13.8	Biotic–SC	0.38 ± 0.09 m^3/m^3 d	$49.5 \pm 11,3$	26.1 ± 1.3
[195]	11.23 A/m^2	49.56	59.18	Abiotic	1.594 mmol/Ld	18.88	126.76
[192]	26.2 ± 2 A/m^3	50	62 ± 6	Biotic–SC	0.24 $m^3/$ m^3d	84 ± 4	–
[196]	27 ± 1.4 A/m^2	72 ± 6	99 ± 5	Biotic–SC	1.29 ± 0.13 $m^3/$ m^3d	–	53 ± 2
[197]	9.96 ± 0.65 A/m^3	–	89.1 ± 5.87	Abiotic	0.42 ± 0.04 m^3/m^3 d	78.9 ± 2.49	77.37 ± 1.5
[198]	7.27 A/m^2	89.3 ± 1.9	91.7 ± 3.6	Biotic–SC	2.29 ± 0.11 m^3/m^3 d	94.7 ± 4.5	64.5 ± 2.4
[174]	9.21 A/m^2	–	83 ± 7	Abiotic	0.63 ± 0.11 m^3/m^3 d	83 ± 9	71 ± 6
[199]	~ 75 A/m^3	–	87 ± 10.2	Abiotic	0.22 ± 0.02 m^3/m^3 d	39 ± 2.7	40 ± 4
[200]	811.33 ± 20 mA/ m^2	77.5 ± 1.0	18.35 ± 0.15	Abiotic	1.42 ± 0.02 mmol/ L d	49.1 ± 0.8	–
[201]	0.51 A	25.4	44	Abiotic	4.1 L/d	33	29.7
[105]	71.4 A/m^3	–	55	Abiotic	46 L_{H2} /L/d	85	–

(continued)

Table 15.1 (continued)

References	Current density	Cod Rem.Eff%	Coulombic Eff.%	Cathodic Config	H_2 Prod. Rate	CCE %	Energy efficiency%
[202]	0.012 A	44.92	78.89	Abiotic	0.38 m^3 /m^3 d	20.05	
[203]	336.3 ± 16.0 to 7.46 ± 1.76 A/m^2	35 to 63	32 ± 22 to 81 ± 5	Abiotic	219 ± 140 to 1478 ± 268 (mLH_2/ $gCOD_{consumed}$)	52.0 ± 0.0 101.4 ± 0.9	− 336.3 ± 16.0 to 725.0 ± 7.8

[a]Single chamber reactor

Table 15.2 MECs performance for CH_4 production

References	Current density	Cod Rem.Eff %	Coulombic Eff.%	Cathodic config	Ch_4 production	CCE %	Energy efficiency%
[208]	6.8 ± 0.3 A/m^2	–	–	Biotic	13.61 L m^2/L	83.4	–
[211]	4.9 mA	67	–	Biotic	87 ± 25 umol/L	55	
[209]	79 ± 2 mA	75	86 ± 8	Biotic	48 ± 2 meq/d	68 ± 4	99
[212]	0.74 A/m^2	75.3	13.5	Biotic–SC[a]	79 ± 1 mL	–	96.6
[213]	30 mA	79.9 ± 0.8	65 ± 1	Biotic–SC	59 ± 6 mL/L	–	–
[214]	24 mA	26.12	–	Biotic–SC	178 mL/Ld	–	–
[2]	241 ± 25 mA	41 ± 1	121 ± 1	Biotic	11 ± 1 mmol/L	41 ± 1	
[215]	145 ± 22 mA	125 ± 7	72	Biotic–2side	11 ± 1 mmol/d	69 ± 4	47 ± 2
[216]	–	91	–	Biotic–SC	283 ml/Ld	17.9	95.2
[217]	43.88 A/m^3	125.54	97.46	Biotic	39 ml/Ld	–	–
[218]	177 ± 17 mA	80	17 ± 0.1	Biotic	10 ± 1 mmol/d	51 ± 4	–
[210]	35 A/m^2	66	–	Biotic	4.3 L/L/d	67	–
[219]	0.13 A/m^2	–	–	Biotic	140 ± 13 $mmol/m^2d$	77	–

[a]Single chamber reactor

The energy efficiency ($\%\eta E$) is calculated considering the energy theoretically recoverable from the produced H_2 or CH_4 ($W_{H2/CH4}$) and the energy consumption of the system (W_{in}). It is described by the following equation:

$$\%\eta \mathrm{E} = \frac{W_{\mathrm{H2/CH4}}}{W_{in}} \times 100$$

$$W_{\mathrm{H2/CH4}} = r_{\mathrm{H2/CH4}(mmol)} \times \Delta G_{\mathrm{H2/CH4}} \qquad W_{in} = \Delta V \times \mathrm{i}$$

where $\Delta G_{H2/CH4}$ is the Gibbs free energy for the combustion of H_2 (kJ/mol) and CH_4 (kJ/mol), respectively. ΔV is the average potential difference established between the cathodic and anodic electrodes, and i represents the average current flowing in the circuit expressed as Coulombs and calculated by integrating the current over time.

15.6.2 Bio-H_2 Production

In order to maximize bio-H_2 production at the cathode, the optimization of some parameters in MECs is required. Table 15.1 reports some cardinal parameters that define MECs performance for H_2 production in recent literature papers. For H_2 production, the cathodic chamber configuration can be biotic (inoculated with microorganisms) or abiotic. For example, Kadier et al. [191], developed a biotic single-chamber reactor equipped with a Ni mesh cathode. They compared the performance of the Ni mesh cathode with that of platinum/carbon cloth, obtaining similar results for H_2 production rate (4.18 ± 1 m^3H_2/m^3d versus 4.25 ± 1 m^3H_2/m^3d, respectively) and volumetric current density (312 ± 9 A/m^3 versus 314 ± 5 A/m^3, respectively). Interestingly, the CE observed in Ni mesh cathode ($75 \pm 4\%$) was slightly higher than that of Pt/CC cathode ($72.7 \pm 1\%$). These results were higher than the results obtained from Zhen et al. [188] or the ones obtained from Chaurasia et al. [192], who utilized the same reactor configuration, but H_2 production rate was only 0.38 ± 0.09 m^3/m^3 d and 0.24 ± 0.005 m^3/m^3 d. Even regarding COD removal, Kadier et al. [191] reached a value of $94.5 \pm 2\%$, while Zhen et al. and Chaurasia et al. [192], just $26.7 \pm 0.4\%$ and the 50%, respectively. The difference in these studies is mainly due to the current density, which is much higher in Kadier et al. [191], who utilized as inoculum a pure culture of *G. sulfurreducens*, a well-known EAM that produces high power densities at moderate temperatures [193]. It is worthwhile to underline that Charasia et al. [192] also utilized a modified Ni-cathodic electrode. Still, using mixed microbial culture from sugar industry wastewater affected the reactor performance. Indeed, looking at Table 15.1, COD removal efficiency is a non-homogenous parameter among the studies selected. This is basically due to the inoculum type and the applied organic loading rate. Ni–P coated Ni foam cathodic electrode was also tested by Li et al. [132], who obtained results similar to those of Kadier et al. [191]. Indeed, the H_2 production rate accounted for 2.29 ± 0.11 m^3/m^3d with a corresponding CCE of $94.7 \pm 4.5\%$. Even the COD removal efficiency is comparable with the one obtained by Kadier et al. [191], 89.3 ± 1.9 and 94.5 ± 2%, respectively. It is interesting to underline that Li et al. [132] utilized the effluent of a MFC as inoculum derived from a wastewater treatment plant. The utilization of that effluent assured the presence of already selected EAMs able to utilize the electrode as an electron donor/electron acceptor. For example, Yang et al. [194] obtained good results, especially in terms of current density (163 A/ m^3) and COD removal efficiency (88%). Even in this case, the anodic chamber was enriched with exoelectrogenic bacteria from MFCs. This could explain the high current density. This study reached an H_2 production rate of 1.04 m^3/m^3d using a modified electrode with polyaniline (PANI)/multi-walled carbon nanotube (MWCNT) cathode catalysts. The results were compared with a Pt/C cathode, exhibiting a comparable performance with the Pt/C cathode regarding H_2 production rate and electrochemical tests.

15.6.3 Bio-CH_4 Production

When METs are used for the bioconversion of CO_2 into CH_4, the most debated issue linked to H_2 is its low solubility in the liquid phase, which makes H_2 availability a limiting factor [204]. For this reason, the integrated system AD-MECs is gaining more attention [11, 205]. Indeed, raw biogas from AD can be addressed to the cathode where hydrogenotrophic methanogens can directly use the H_2 produced in situ (by using at least partially the energy deriving from the oxidation of organic matter that occurs at the anode), bypassing the problems of H_2 transport, injection and solubilization [206]. As discussed in Sect. 5, geometry is an essential factor that affects MECs performance; indeed, the presence of the membrane can define the output in the case of CH_4 production [154]. Indeed, the possible retro-diffusion of H_2 in the anodic chamber due to low membrane selectivity limits the substrate availability for hydrogenotrophic methanogens [207] and leads to the pH split [184], as previously discussed. For instance, as reported in Table 15.2, the higher CCE% is reached in two/multiple chamber reactors. The Cathode Capture Efficiency (CCE %), the fraction of generated current converted into H_2 and CH_4, is a good indicator of MECs performance. Tian et al. [208] demonstrated that by applying a potential of − 1 V versus Ag/AgCl, the current density increased from − 0.5 A /m^2 in the abiotic reactor up to − 6.8 A/ m^2 in the biotic one, indicating the catalytic role of microorganisms in the bioconversion of H_2 and CO_2 into CH_4. This study reached a CCE of 83.4% with a CH_4 production rate of 13.61 L/m^2 d. Homogenous results in terms of CCE are presented by Zeppilli et al. [209], Cristiani et al. [2], and Liu et al. [210], where it is set between 50 and 70%. In Zeppilli et al. [209] the authors proposed an innovative two-side cathode configuration with two cathodic chambers connected in parallel by a titanium wire and separated from the anodic compartment by an anion and a cation exchange membrane (AEM and CEM). The two-side configuration has been proposed as a post-treatment unit capable of performing the biogas upgrading through the CO_2 reduction and removal into the cathodic chambers and the ammonium recovery from a liquid waste stream due to its migration for the electroneutrality maintenance of the cell.

Compared with a previous study [220] investigating a three-chamber MEC equipped with an accumulation chamber, the two-side chamber MEC resulted in higher performance regarding current density, COD removal efficiency, CH_4 production rate, and CCE. Regarding CH_4 production rate and CCE%, similar results were reached by Cristiani et al. [2]. This study focused on the role of the organic loading rate and the electrodes' potential control strategy on the performance of a micro-pilot tubular microbial electrolysis cell for biogas upgrading. The study was carried out with the anode potential controlled at + 0.2 V versus SHE, and three different organic loading rates (i.e., 2.55, 3.82, 5.11 gCOD/Ld) were explored. The best performance was at 2.55 gCOD/Ld (data reported in Table 15.2), and once the optimal OLR was selected, three different cathodic potentials (− 1.3, − 1.8, − 2.3 V vs. SHE) were investigated (switching from the potentiostatic control of anode to the cathode) with the aim of CO_2 abatement. The lowest energy consumption for CO_2 removal was obtained during the − 1.3 V versus SHE condition [2]. These are just some examples of the large applicability of MEC; indeed, this technology represents a flexible, eco-friendly, and effective way not only for biofuel production but also to tackle the current environmental challenge that we face.

15.7 Scale Up Experience and Perspectives

The design and scale-up of the electrochemical reactor play an important role in developing industrial electrochemical processes. Therefore, studying the controlling factors on scale-up makes the system's operation more efficient and economical in the commercialization stage. The pilot scale plant is a scale model of an industrial plant. The pilot scale operation must be conducted, simulating the operating conditions and performance of the full-scale plant. Developing a full-scale process is usually a pioneering approach that can follow several approaches. The similarity theory is based on the idea that specific parameters need to be constant during the technology transfer from one system to another, for example a pilot plant. To use the similarity theory for the scale-up of MET, which has not been shown before, new ratios or dimensionless numbers must be defined, which are kept constant during scale-up. The main challenge is combining a volume-depending process (gas transfer) and a surface-depending process (electron transfer). Indeed, it is necessary to find further scale-up criteria to keep constant to obtain improved results. These criteria should especially include electrode resistances, which seemed to be a major issue in this case. Additionally, the pilot scale reactor constructed as described in the literature showed shortcomings not covered by similarity theory, especially the choice of material that needs improvement. Only a few authors applied the similarity theory for scaling up a bioelectrochemical reactor to confirm the possible use of scale-up criteria to develop pilot-scale reactor with similar geometrical properties and abiotic parameters. Power consumption is a strong candidate for use as a criterion for (bio)reactor design and process scale-up. Often referred to as volumetric power consumption (*P*/*V*), it is defined as the amount of energy required to generate movement of a fluid within a vessel in a given period of time. Along with the power drawn by the fluid relevant to the outcome of a given process, further power is required to account for energy losses, mostly due to friction and power consumption by motors and gearboxes. Although often relevant in terms of overall power consumption, this excess power is not considered for the design or scale-up of the process. The volumetric power consumption is representative of the turbulence degree and media circulation in vessels and influences heat and mass transfer, mixing, and circulation times. Despite constant volumetric power input was applied successfully as the scale-up parameter for the early industrial penicillin fermentations (1 hp/gallon, equivalent to 1.8 kW/m^3), and in fermentations with low energy inputs, in the presence of polarized bio-reactors, i.e. like in the case of bioelectrochemical processes for bio-H_2 and bio-CH_4 production, the volumetric power consumption includes the electrical power provided to electrodes polarization giving important information for the scale-up procedure. One of the pilot-scale MEC configurations for H_2 [221] production was based on stack configuration in which several anode–cathode modules are connected and submerged in a tank. The use of six MEC cells contained in a polypropylene tank having a total volume equal to 100 L and functioning as a single MEC cell. Each of the six cells consists of two anodes in carbon felt placed on the two sides of the cathode compartment, which is constituted by a 10 mm thick plastic sheet containing the cathode constituted by stainless steel. Between the cathode and the two anodes, there is a non-ion selective membrane, which allows the transfer of cathodic and anodic liquid. Furthermore, adhering to the surface of the anode, there is a stainless-steel mesh that acts as an electrical collector. The anode is fed by domestic wastewater, to which acetate and propionate are added with concentrations of 28.6 and 12.0 mg/L, respectively. The pilot plant was installed in Howdon, Great Britain, in the municipal wastewater treatment plant. The site treats approximately 250,000 m^3 of domestic wastewater from the City of New Castle. The plant was kept in operation for 12 months, during which the ability to produce H_2 even in the coldest months, at temperatures between 1 and 5 °C, was evaluated. Another example of a scale-up process by stack configuration has been proposed using a rectangular reactor with a 1000 L volume and an operating volume of 910 L [222]. The stack configuration contained 24 modules placed perpendicular to the flow, each comprises six anodes and six cathodes, for 144 electrode pairs. The anodes are graphite, while the cathodes are stainless-steel (304). The pilot scale MEC addressed to H_2 production was powered by wastewater from a distillery in Oakland, California, which processes 7000 tons of grapes annually and produces 5.1×10^4 m^3 of wastewater annually. During the start-up period, synthetic substrates like acetate and fumarate were added to the feed, with concentrations of 8 and 1.4 g/L, to stimulate the growth of *Geobacteriaceae* responsible for the oxidation of organic material. Another scale up attempt with a different anode–cathode configuration was adopted in a 10L reactor, with carbon and titanium anode and carbon cathode coated with molybdenum phosphide as a catalyst. Two configurations were installed, the 3C and 5C configurations. The first was arranged as if it consisted of three MECs, while in the second, the electrodes were arranged as if there were five MECs. In the 3C configuration, there are three cathodes, each surrounded by a set of two anodes. The first anodic set is located at a distance of 0.9 cm from the cathode. In this case, the separation between anode and cathode is carried out thanks to a cloth. The last set is 1.2 cm from the cathode, and a separator between the electrodes is not used. An intermediate separation method separates the core anode set from the cathode. An anode is separated from the cathode by cloth at 0.9 cm. However, an anode is kept at 1.2 cm without of a separator. In the 3C and 5C configurations, the MECs are connected in parallel, and a voltage is applied between the two electrodes using direct current (DC, Yescom). The feed consists of domestic wastewater, to which acetate with a concentration of 50 mM has been added for the 3C configuration, while for the 5C configuration, acetate, and glucose with a concentration of 50 mM and 2 g/L, respectively. The anode was inoculated with a bacterial culture transferred from another functioning MEC. One of the most important examples of bioelectromethanogenesis was conducted by Enzmann and co-authors, which also showed one of the most significant tentative to apply similarity theory for processes scale-up in the field of BES. In the published work [223], authors adopted a rational scale-up based on the similarity theory of a 1-L lab-scale bioelectrochemical reactor to a 50L pilot-scale. The pilot scale reactor was characterized abiotically before using it as BES to allow comparison to the small-scale reactor. As an exemplary process, the pilot plant was used for bioelectromethanogenesis with a pure culture of electroactive methanogens, *Methanococcus maripaludis*. The 50 L working volume pilot-plant was constructed using polypropylene pipe of 0.31 m inner diameter and 0.77 m height (REHAU Unlimited polymer solutions, Rehau, Germany). 20 membrane windows (3 cm * 23 cm each) were arranged staggered in two rows.

As for the comparison of different pilot-scale MEC for H_2 production, the study conducted by Singh et al. [224], which provides for the application of a similar potential difference for the two configurations equal to 1.0 V, the quantity of H_2 produced is greater for the 5C configuration than for the 3C configuration, equal to 5.9 and 3.9 L/Ld, respectively. The greater quantity of H_2 produced with the 5C configuration is due to the presence of two additional cathodes to concerning the 3C configuration, which provides a greater surface area available for the proton reduction reaction to the addition of chloroform, which inhibits the growth of homoacetogenic bacteria, responsible for the conversion of H_2 into acetate, with a consequent reduction in yield. However, if chloroform acted by inhibiting the growth of homoacetogens, the H_2 production would be equal to 10.57 L/Ld. The current density achieved is also higher for the 5C configuration and equal to 34 A/m^2, compared to 31 A/m^2 for the 3C configuration. Despite this, higher energy efficiency (94%) and CCE at the cathode (67%) are achieved with the latter. Therefore, the 3C configuration turns out to be more advantageous from the point of view of the energy spent compared to the energy recovered (energy efficiency) and from the point of view of the electric current generated converted into H_2 (CCE at the cathode). In both configurations, the CE at the anode is greater than 100%. This could be due to homoacetogenic microorganisms converting H_2 into acetate, which is oxidized again at the anode by exoelectrogenic bacteria. The study conducted by Heidreich et al. [221] provides for a variable H_2 production over the 12 months the plant has been in operation. In fact, during the year, there was a deterioration in the plant's performance. In the acclimatization phase, the applied voltage is increased from 0.6 to 0.9 V until it reaches the final value at which H_2 production occurs, i.e., 1.1 V. Hydrogen production decreased from 0.8 L/Ld in the first six months of the year to 0.4 L/Ld in the last six months, with an average of H_2 produced equal to 0.007 L/Ld. However, the purity of the product obtained is high; in fact, H_2 constitutes about 98–99% of the recovered gas. The high purity is due to the low activity of methanogens, whose microbial growth is inhibited at low temperatures. The poor performance of the pilot-scale plant is confirmed by the measured current value, by the removal of the substrate, and by the CE at the anode. Indeed, during the year, the current generated by the oxidation of the substrate decreases from 35 to 24 mA. Substrate removal (COD) is also reduced at 33%; possibly due to issues related mainly to implant placement and pumping. While the CCE at the cathode, i.e., the current converted into H_2, equals 41.2%. This low value could be attributed to H_2 losses caused by the deterioration of the plastic components of the reactor. Therefore, the possibility of operating at low temperatures was confirmed in the following work. At the same time, several problems emerged, which will be studied to develop a process that can be implemented on an industrial scale. Regarding the work conducted by Cusick et al. [222] numerous problems emerged in 2011, especially regarding the duration of the trial start-up period. In fact, despite the enrichment of the feed with acetate and fumarate, no more than 0.4 A/m^2 of current is produced during the first 20 days. During the start of the process, the pH was maintained at values greater than six, while the temperature must be maintained between 20 and 30 °C using a thermostat. Despite this, the implemented measures do not contribute to increasing the current density. An increase in current density was only observed when diluted vinegar (acetic acid) was added to the wastewater, further stimulating the growth of bacteria belonging to *Geobacter* species. Following the addition of diluted vinegar, an increase in the current density is observed, reaching a maximum of 7.4 A/m^3 (i.e., 0.41 A/m^2). Therefore, it is demonstrated that only by adding a further carbon source is it possible to observe a satisfactory growth of exoelectrogenic bacteria and obtain a higher current density, even if much lower than the studies previously analyzed on a pilot-scale and laboratory scale. Lower current density values could be due to the materials used in the pilot scale. In fact, with the stainless-steel cathode used in the pilot-scale, the measured current intensity is reduced compared to that measured in the lab-scale in which the cathode is made up of platinum. Using pilot-scale stainless-steel results in a 44% loss of current density. However, despite the low measured current density, the substrate removal (COD) is higher than the value obtained in the study conducted by Heidrich et al. [221] and equal to 62%, while the CE at the anode and the cathode have a value of 11 and 59%, respectively. In addition to the parameters mentioned earlier, in Table 15.3, the production of H_2 can be compared with the other works. In particular, with the work of Cusick et al. [222] the average quantity of H_2 produced at 0.9 V is equal to 0.19 L/Ld. Although the H_2 production increases up to the maximum value of 0.21 L/Ld at the current density of 5 A/m^3, once the value of 7.4 A/m^3 is reached, the production decreases, as the H_2 is converted from the methanogens into CH_4 via the hydrogenotrophic route. Consequently, the purity of the H_2 produced is reduced.

Table 15.3 MECs performance for H_2 and CH_4 production AT PILOT SCALE

References	Current density	Cod Eem.Eff %	Coulombic Eff.%	Ch_4/h_2 production	CCE %	Energy efficiency%
[224]	31 A/m^2	–	116	3.9 L $_{H2}$/Ld	67	94
	34 A/m^2	–	127	5.9 L $_{H2}$/Ld	56	82
[221]	30 mA	33	–	0.007 L $_{H2}$/Ld	41.2	–
[222]	0.41 A/ m^2	62	11	0.19 L_{CH4}/Ld	59	–
[33]	0.29 A/ m^2	63,5	27.7	0.003 L_{H2}/Ld	< 10	–
[223]	85 mA/ m^2	–	–	10.24 $mmol_{CH4}/m^2d$	113	27
[225]	0.05 A/ m^2	84 ± 5	7–9	0.02 L_{H2}/Ld	81	–
[226]	0.3 A/m^2	37	23	0.028 $m^3{}_{H2}$ / m^3d	49	76

In reality, it has been observed that the experimental equipment and the developed process are more suitable for the production of biogas rather than the production of H_2; in fact, about 86% of the gas obtained is CH_4, unlike the anaerobic digestion process of sedimentation sludge, with which a biogas yield of 75% is obtained. Among the three pilot-scale studies analyzed, the most promising, from the point of view of the quantity of H_2 produced and the energy supplied compared to the energy obtained, is the study by Singh et al. [224]. Instead, the experimental equipment and operating conditions employed in the works of Heidrich et al. [221] and Cusick et al. [222] provide significantly lower H_2 production values. This evidence is easy to interpret, given the smaller reactor volume of Singh et al. [224].

15.8 Conclusive Remarks

Nowadays, biofuel recovery from waste streams (such as wastewater) is a key research topic to develop new technologies to face the energy crisis and depletion of non-renewable natural resources. Microbial electrochemical technologies (MET) could be defined as new generation conversion technologies since they provide a promising futuristic approach to valorize organic wastes and CO_2 into green energy. In recent years, many efforts have been made to improve these systems, including reactor designs, cost-effective materials for electrodes and membranes, and anodic and cathodic reaction optimization, leading to promising results for H_2 and CH_4 production at the lab-scale. However, many technical challenges need to be faced by scientists before METs scale-up from the laboratory to practical application, such as removing bottlenecks due to the consumption of H_2 produced by competing pathways or facing the low H_2 solubility in the liquid phase when METs are used for the bioconversion of CO_2 into CH_4. Besides the reactor configuration, METs rely on the abilities of electroactive microorganisms to catalyze oxidation and/or reduction reactions. When METs are fed with real complex substrates, microbial communities involved are very complex, including EABs and not-EABs, planktonic and biofilm microorganisms. A comprehensive understanding of these communities is required to control the metabolic pathways to increase current generation, prevent competitive pathways, and improve stability, performance, and coulombic efficiencies, enhancing H_2 and CH_4 production. Nowadays, the availability of sequencing-based techniques, and advanced bioinformatics tools offer exciting prospects for improving the electrochemical performance of METs.

Moreover, up to now, attempts to scale-up these technologies are limited. In general, lower current density and H_2 recovery performances are reported at the pilot-scale compared to lab-scale results. This is probably due to the materials used in the pilot-scale, the reactor design of pilots and the higher ohmic losses that will reduce the overall efficiency.

Further advancements will be required to improve METs efficiency and reduce the capital cost. In particular, large-scale studies in which renewable energy sources are used to power METs are needed.

Acknowledgements This research was funded by the European Union–NextGeneration EU from the Italian Ministry of Environment and Energy Security POR H2 AdP MMES/ENEA with involvement of CNR and RSE, PNRR—Mission 2, Component 2, Investment 3.5 "Ricerca e sviluppo sull'idrogeno" CUP: I83C22001170006.

Finanziato dall'Unione europea NextGenerationEU

References

1. Ahring BK (2003) Perspectives for anaerobic digestion
2. Cristiani L, Zeppilli M, Villano M, Majone M (2021) Role of the organic loading rate and the electrodes' potential control strategy on the performance of a micro pilot tubular microbial electrolysis cell for biogas upgrading. Chem. Eng. J 426 (2021). https://doi.org/10.1016/j.cej.2021.131909
3. Gautam R, Nayak JK, Ress NV, Steinberger-Wilckens R, Ghosh UK (2023) Bio-hydrogen production through microbial electrolysis cell: structural components and influencing factors. Chem Eng J 455 (2023). https://doi.org/10.1016/j.cej.2022.140535.
4. Jain IP (2009) Hydrogen the fuel for 21st century. Int J Hydrogen Energy 34:7368–7378. https://doi.org/10.1016/J.IJHYDENE.2009.05.093
5. Sołowski G (2018) Biohydrogen production-sources and methods: a review. Int J Bioproc and Biotech Rev Article Sołowski G. IJBBT-101. Citation: Sołowski G 101. https://doi.org/10.20911/IJBBT-101

6. Midilli A, Kucuk H, Topal ME, Akbulut U, Dincer I (2021) A comprehensive review on hydrogen production from coal gasification: Challenges and Opportunities. Int J Hydrogen Energy 46:25385–25412. https://doi.org/10.1016/J.IJHYDENE.2021.05.088
7. Yaashikaa PR, Keerthana Devi M, Senthil Kumar P (2022) Biohydrogen production: an outlook on methods, constraints, economic analysis and future prospect. Int J Hydrogen Energy 47:41488–41506. https://doi.org/10.1016/J.IJHYDENE.2022.07.082
8. Lasia A (2019) Mechanism and kinetics of the hydrogen evolution reaction. Int J Hydrogen Energy 44:19484–19518. https://doi.org/10.1016/J.IJHYDENE.2019.05.183
9. Zeppilli M, Cristiani L, Dell'Armi E, Majone M (2020) Bioelectromethanogenesis reaction in a tubular microbial electrolysis cell (MEC) for biogas upgrading, Renew. Energy 158:23–31. https://doi.org/10.1016/J.RENENE.2020.05.122
10. Villano M, Aulenta F, Ciucci C, Ferri T, Giuliano A, Majone M (2010) Bioelectrochemical reduction of CO_2 to CH_4 via direct and indirect extracellular electron transfer by a hydrogenophilic methanogenic culture. Bioresour Technol 101:3085–3090. https://doi.org/10.1016/j.biortech.2009.12.077
11. Batlle-Vilanova P, Rovira-Alsina L, Puig S, Balaguer MD, Icaran P, Monsalvo VM, Rogalla F, Colprim J (2019) Biogas upgrading, CO_2 valorisation and economic revaluation of bioelectrochemical systems through anodic chlorine production in the framework of wastewater treatment plants. Sci Total Environ 690:352–360. https://doi.org/10.1016/J.SCITOTENV.2019.06.361
12. Schmid C, Horschig T, Pfeiffer A, Szarka N, Thrän D (2019) Biogas upgrading: a review of national biomethane strategies and support policies in selected countries. Energies (Basel) 12. https://doi.org/10.3390/en12193803
13. Angelidaki I, Treu L, Tsapekos P, Luo G, Campanaro S, Wenzel H, Kougias PG (2018) Biogas upgrading and utilization: current status and perspectives. Biotechnol Adv 36:452–466. https://doi.org/10.1016/j.biotechadv.2018.01.011
14. European Biogas Association EBA (2023) About biogas and biomethane, About Biogas and Biomethane. https://www.europeanbiogas.eu/about-biogas-and-biomethane/ (Accessed 15 May 2023)
15. Ardolino F, Cardamone GF, Parrillo F, Arena U (2021) Biogas-to-biomethane upgrading: a comparative review and assessment in a life cycle perspective. Renew Sustainable Energy Rev 139. https://doi.org/10.1016/j.rser.2020.110588
16. Díaz I, Fdz-Polanco F, Mutsvene B, Fdz-Polanco M (2020) Effect of operating pressure on direct biomethane production from carbon dioxide and exogenous hydrogen in the anaerobic digestion of sewage sludge. Appl Energy 280 (2020). https://doi.org/10.1016/j.apenergy.2020.115915
17. Zeppilli M, Lai A, Villano M, Majone M (2016) Anion vs cation exchange membrane strongly affect mechanisms and yield of CO_2 fixation in a microbial electrolysis cell. Chem Eng J 304:10–19. https://doi.org/10.1016/J.CEJ.2016.06.020
18. Liu H, Grot S, Logan BE (2005) Electrochemically assisted microbial production of hydrogen from acetate. Environ Sci Technol 39:4317–4320. https://doi.org/10.1021/es050244p
19. Rozendal R, Buisman C (2005) Process for producing hydrogen, WO2005005981
20. Logan BE, Call D, Cheng S, Hamelers HVM, Sleutels THJA, Jeremiasse AW, Rozendal RA (2008) Microbial electrolysis cells for high yield hydrogen gas production from organic matter. Environ Sci Technol 42:8630–8640. https://doi.org/10.1021/es801553z
21. Dattatraya Saratale G, Rajesh Banu J, Nastro RA, Kadier A, Ashokkumar V, Lay CH, Jung JH, Seung Shin H, Ganesh Saratale R, Chandrasekhar K (2022) Bioelectrochemical systems in aid of sustainable biorefineries for the production of value-added products and resource recovery from wastewater: a critical review and future perspectives. Bioresour Technol 359. https://doi.org/10.1016/j.biortech.2022.127435
22. Jadhav DA, Park SG, Pandit S, Yang E, Ali Abdelkareem M, Jang JK, Chae KJ (2022) Scalability of microbial electrochemical technologies: applications and challenges. Bioresour Technol 345. https://doi.org/10.1016/j.biortech.2021.126498

23. Yu YY, Zhang Y, Peng L (2022) Investigating the interaction between Shewanella oneidensis and phenazine 1-carboxylic acid in the microbial electrochemical processes. Sci Total Environ 838:156501. https://doi.org/10.1016/j.scitotenv.2022.156501
24. Marsili E, Sun J, Bond DR (2010) Voltammetry and growth physiology of Geobacter sulfurreducens biofilms as a function of growth stage and imposed electrode potential. Electroanalysis 22:865–874. https://doi.org/10.1002/elan.200800007
25. Menezes AA, Cumbers J, Hogan JA, Arkin AP (2015) Towards synthetic biological approaches to resource utilization on space missions. J R Soc Interface 12 (2015). https://doi.org/10.1098/rsif.2014.0715
26. Rimboud M, Desmond-Le Quemener E, Erable B, Bouchez T, Bergel A (2015) Multi-system Nernst-Michaelis-Menten model applied to bioanodes formed from sewage sludge. Bioresour Technol 195:162–169. https://doi.org/10.1016/j.biortech.2015.05.069
27. Mancini G, Luciano A, Bolzonella D, Fatone F, Viotti P, Fino D (2021) A water-waste-energy nexus approach to bridge the sustainability gap in landfill-based waste management regions. Renew Sustain Energy Rev 137:110441. https://doi.org/10.1016/j.rser.2020.110441
28. Guo F, Liu Y, Liu H (2021) Hibernations of electroactive bacteria provide insights into the flexible and robust BOD detection using microbial fuel cell-based biosensors. Sci Total Environ 753:142244. https://doi.org/10.1016/j.scitotenv.2020.142244
29. Ross DE, Flynn JM, Baron DB, Gralnick JA, Bond DR (2011) Towards electrosynthesis in Shewanella: Energetics of reversing the Mtr pathway for reductive metabolism, PLoS One. 6. https://doi.org/10.1371/journal.pone.0016649
30. Batlle-Vilanova P, Ganigué R, Ramió-Pujol S, Bañeras L, Jiménez G, Hidalgo M, Balaguer MD, Colprim J, Puig S (2017) Microbial electrosynthesis of butyrate from carbon dioxide: Production and extraction. Bioelectrochemistry 117:57–64. https://doi.org/10.1016/j.bioelechem.2017.06.004
31. Choi KS, Kondaveeti S, Min B (2017) Bioelectrochemical methane (CH_4) production in anaerobic digestion at different supplemental voltages. Bioresour Technol 245:826–832. https://doi.org/10.1016/j.biortech.2017.09.057
32. Luo H, Qi J, Zhou M, Liu G, Lu Y, Zhang R, Zeng C (2020) Enhanced electron transfer on microbial electrosynthesis biocathode by polypyrrole-coated acetogens. Bioresour Technol 309:123322. https://doi.org/10.1016/j.biortech.2020.123322
33. Cotterill SE, Dolfing J, Jones C, Curtis TP, Heidrich ES (2017) Low temperature domestic wastewater treatment in a microbial electrolysis Cell with 1 m_2 anodes: towards system scale-up. Fuel Cells 17:584–592. https://doi.org/10.1002/fuce.201700034
34. Snider RM, Strycharz-Glaven SM, Tsoi SD, Erickson JS, Tender LM (2012) Long-range electron transport in Geobacter sulfurreducens biofilms is redox gradient-driven. Proc Natl Acad Sci U S A 109:15467–15472. https://doi.org/10.1073/pnas.1209829109
35. Schröder U (2007) Anodic electron transfer mechanisms in microbial fuel cells and their energy efficiency. Phys Chem Chem Phys 9:2619–2629. https://doi.org/10.1039/b703627m
36. Callegari A, Cecconet D, Molognoni D, Capodaglio AG (2018) Sustainable processing of dairy wastewater: long-term pilot application of a bio-electrochemical system. J Clean Prod 189:563–569. https://doi.org/10.1016/j.jclepro.2018.04.129
37. Rousseau R, Etcheverry L, Roubaud E, Basséguy R, Délia ML, Bergel A (2020) Microbial electrolysis cell (MEC): strengths, weaknesses and research needs from electrochemical engineering standpoint. Appl Energy 257:113938. https://doi.org/10.1016/j.apenergy.2019.113938
38. Sarker S, Lamb JJ, Hjelme DR, Lien KM (2018) Overview of recent progress towards in-situ biogas upgradation techniques. Fuel 226:686–697. https://doi.org/10.1016/j.fuel.2018.04.021
39. Martens JA, Bogaerts A, De Kimpe N, Jacobs PA, Marin GB, Rabaey K, Saeys M, Verhelst S (2017) The chemical route to a carbon dioxide neutral world. Chemsuschem 10:1039–1055. https://doi.org/10.1002/cssc.201601051
40. Gao T, Zhang H, Xu X, Teng J (2021) Integrating microbial electrolysis cell based on electrochemical carbon dioxide reduction into anaerobic osmosis membrane reactor for biogas upgrading. Water Res 190:116679. https://doi.org/10.1016/j.watres.2020.116679

41. Zhang Y, Jiang Q, Gong L, Liu H, Cui M, Zhang J (2020) In-situ mineral CO_2 sequestration in a methane producing microbial electrolysis cell treating sludge hydrolysate. J Hazard Mater 394:122519. https://doi.org/10.1016/j.jhazmat.2020.122519
42. Hansen KH, Angelidaki I, Ahring BK (1998) Anaerobic digestion of swine manure: inhibition by ammonia 32:5–12
43. Gandiglio M, Lanzini A, Soto A, Leone P, Santarelli M (2017) Enhancing the energy efficiency of wastewater treatment plants through co-digestion and fuel cell systems, front. Environ Sci 5:1–21. https://doi.org/10.3389/fenvs.2017.00070
44. Ryckebosch E, Drouillon M, Vervaeren H (2011) Techniques for transformation of biogas to biomethane. Biomass Bioenergy 35:1633–1645. https://doi.org/10.1016/j.biombioe.2011.02.033
45. He AY, Yin CY, Xu H, Kong XP, Xue JW, Zhu J, Jiang M, Wu H (2016) Enhanced butanol production in a microbial electrolysis cell by Clostridium beijerinckii IB4. Bioprocess Biosyst Eng 39:245–254. https://doi.org/10.1007/s00449-015-1508-2
46. Korth B, Harnisch F (2019) Spotlight on the energy harvest of electroactive microorganisms: the impact of the applied anode potential. Front Microbiol 10:1–9. https://doi.org/10.3389/fmicb.2019.01352
47. ter Heijne A, Pereira MA, Pereira J, Sleutels T (2021) Electron storage in electroactive biofilms. Trends Biotechnol 39:34–42. https://doi.org/10.1016/j.tibtech.2020.06.006
48. Rabaey K, Bioelectrochemical systems: from extracellular electron transfer to biotechnological application. https://doi.org/10.2166/9781780401621
49. Noori MT, Vu MT, Ali RB, Min B (2020) Recent advances in cathode materials and configurations for upgrading methane in bioelectrochemical systems integrated with anaerobic digestion. Chem Eng J 392:123689. https://doi.org/10.1016/j.cej.2019.123689
50. Madigan MT, Bender KS, Buckley DH, Sattley WM, Stahl DA (2016) Brock biology of microorganisms. Nat Acad Sci. https://doi.org/10.1073/pnas.1522149113
51. Kato S (2017) Influence of anode potentials on current generation and extracellular electron transfer paths of geobacter species. Int J Mol Sci 18. https://doi.org/10.3390/ijms18010108
52. Fricke K, Harnisch F, Schröder U (2008) On the use of cyclic voltammetry for the study of anodic electron transfer in microbial fuel cells, energy. Environ Sci 1:144–147. https://doi.org/10.1039/b802363h
53. Wagner RC, Call DF, Logan BE (2010) Optimal set anode potentials vary in bioelectrochemical systems. Environ Sci Technol 44:6036–6041. https://doi.org/10.1021/es101013e
54. Peng L, Zhang Y (2017) Cytochrome OmcZ is essential for the current generation by geobacter sulfurreducens under low electrode potential. Electrochim Acta 228:447–452. https://doi.org/10.1016/j.electacta.2017.01.091
55. Scarabotti F, Rago L, Bühler K, Harnisch F (2021) The electrode potential determines the yield coefficients of early-stage Geobacter sulfurreducens biofilm anodes. Bioelectrochemistry 140:107752. https://doi.org/10.1016/j.bioelechem.2021.107752
56. Fu Q, Xiao S, Li Z, Li Y, Kobayashi H, Li J, Yang Y, Liao Q, Zhu X, He X, Ye D, Zhang L, Zhong M (2018) Hybrid solar-to-methane conversion system with a Faradaic efficiency of up to 96%. Nano Energy 53:232–239. https://doi.org/10.1016/j.nanoen.2018.08.051
57. Holladay JD, Hu J, King DL, Wang Y (2009) An overview of hydrogen production technologies. Catal Today 139:244–260. https://doi.org/10.1016/j.cattod.2008.08.039
58. Gil-Carrera L, Escapa A, Carracedo B, Morán A, Gómez X (2013) Performance of a semi-pilot tubular microbial electrolysis cell (MEC) under several hydraulic retention times and applied voltages. Bioresour Technol 146:63–69. https://doi.org/10.1016/j.biortech.2013.07.020
59. Srikanth S, Venkateswar Reddy M, Venkata Mohan S (2012) Microaerophilic microenvironment at biocathode enhances electrogenesis with simultaneous synthesis of polyhydroxyalkanoates (PHA) in bioelectrochemical system (BES). Bioresour Technol 125:291–299. https://doi.org/10.1016/j.biortech.2012.08.060
60. Yang HY, Wang YX, He CS, Qin Y, Li WQ, Li WH, Mu Y (2020) Redox mediator-modified biocathode enables highly efficient microbial electro-synthesis of methane from carbon dioxide. Appl Energy 274:115292. https://doi.org/10.1016/j.apenergy.2020.115292

61. Franco A, Elbahnasy M, Rosenbaum MA (2022) Screening of natural phenazine producers for electroactivity in bioelectrochemical systems. Microb Biotechnol. https://doi.org/10.1111/1751-7915.14199
62. Lovley DR (2012) Electromicrobiology. Annu Rev Microbiol 66:391–409. https://doi.org/10.1146/annurev-micro-092611-150104
63. Yan YQ, Wang X (2019) Ecological responses to substrates in electroactive biofilm: a review. Sci China Technol Sci. 62:1657–1669. https://doi.org/10.1007/s11431-018-9410-6
64. Lovley DR, Coates JD, BluntHarris EL, Phillips EJP, Woodward JC (1996) Humic substances as electron acceptors for microbial respiration. Nature. https://doi.org/10.1038/382445a0
65. Kiely PD, Cusick R, Call DF, Selembo PA, Regan JM, Logan BE (2011) Anode microbial communities produced by changing from microbial fuel cell to microbial electrolysis cell operation using two different wastewaters. Bioresour Technol 102:388–394. https://doi.org/10.1016/J.BIORTECH.2010.05.019
66. Flayac C, Trably E, Bernet N (2018) Microbial anodic consortia fed with fermentable substrates in microbial electrolysis cells: significance of microbial structures. Bioelectrochemistry 123. https://doi.org/10.1016/j.bioelechem.2018.05.009
67. Jung S, Regan JM (2007) Comparison of anode bacterial communities and performance in microbial fuel cells with different electron donors. Appl Microbiol Biotechnol 77:393–402. https://doi.org/10.1007/s00253-007-1162-y
68. Castellano-Hinojosa A, González-Martínez A, Pozo C, González-López J (2022) Diversity of electroactive and non-electroactive microorganisms and their potential relationships in microbial electrochemical systems: a review. J Water Process Eng 50. https://doi.org/10.1016/j.jwpe.2022.103199
69. Freguia S, Teh EH, Boon N, Leung KM, Keller J, Rabaey K (2010) Microbial fuel cells operating on mixed fatty acids. Bioresour Technol 101:1233–1238. https://doi.org/10.1016/j.biortech.2009.09.054
70. Torres CI, Krajmalnik-Brown R, Parameswaran P, Marcus AK, Wanger G, Gorby YA, Rittmann BE (2009) Selecting anode-respiring bacteria based on anode potential: Phylogenetic, electrochemical, and microscopic characterization Environ Sci Technol 43:9519–9524. https://doi.org/10.1021/es902165y
71. Ruiz V, Ilhan ZE, Kang D-W, Krajmalnik-Brown R, Buitrón G (2014) The source of inoculum plays a defining role in the development of MEC microbial consortia fed with acetic and propionic acid mixtures. J Biotechnol 182–183:11–18. https://doi.org/10.1016/j.jbiotec.2014.04.016
72. Parameswaran P, Torres CI, Lee H-S, Krajmalnik-Brown R, Rittmann BE (2009) Syntrophic interactions among anode respiring bacteria (ARB) and non-ARB in a biofilm anode: electron balances. Biotechnol Bioeng 103:513–523. https://doi.org/10.1002/bit.22267
73. Lee HS, Torres CI, Parameswaran P, Rittmann BE (2009) Fate of H_2 in an upflow single-chamber microbial electrolysis cell using a metal-catalyst-free cathode. Environ Sci Technol 43:7971–7976. https://doi.org/10.1021/es900204j
74. Parameswaran P, Torres CI, Lee H-S, Rittmann BE, Krajmalnik-Brown R (2011) Hydrogen consumption in microbial electrochemical systems (MXCs): the role of homo-acetogenic bacteria. Bioresour Technol 102:263–271. https://doi.org/10.1016/j.biortech.2010.03.133
75. Gao Y, Ryu H, Santo Domingo JW, Lee HS (2014) Syntrophic interactions between H_2-scavenging and anode-respiring bacteria can improve current density in microbial electrochemical cells. Bioresour Technol 153:245–253. https://doi.org/10.1016/j.biortech.2013.11.077
76. Parameswaran P, Zhang H, Torres CI, Rittmann BE, Krajmalnik-Brown R (2010) Microbial community structure in a biofilm anode fed with a fermentable substrate: the significance of hydrogen scavengers. Biotechnol Bioeng 105:69–78. https://doi.org/10.1002/bit.22508
77. Ruiz Y, Baeza JA, Guisasola A (2013) Revealing the proliferation of hydrogen scavengers in a single-chamber microbial electrolysis cell using electron balances. Int J Hydrogen Energy 38:15917–15927. https://doi.org/10.1016/j.ijhydene.2013.10.034

78. Lu L, Ren N, Zhao X, Wang H, Wu D, Xing D (2011) Hydrogen production, methanogen inhibition and microbial community structures in psychrophilic single-chamber microbial electrolysis cells, energy. Environ Sci 4:1329–1336. https://doi.org/10.1039/c0ee00588f
79. Lu L, Xing D, Ren N (2012) Pyrosequencing reveals highly diverse microbial communities in microbial electrolysis cells involved in enhanced H_2 production from waste activated sludge. Water Res 46:2425–2434. https://doi.org/10.1016/j.watres.2012.02.005
80. Lee H-S, Rittmann BE (2010) Characterization of energy losses in an upflow single-chamber microbial electrolysis cell. Int J Hydrogen Energy 35:920–927. https://doi.org/10.1016/j.ijhydene.2009.11.040
81. Zhou H, Xing D, Xu M, Su Y, Ma J, Angelidaki I, Zhang Y (2021) Optimization of a newly developed electromethanogenesis for the highest record of methane production. J Hazard Mater 407:124363. https://doi.org/10.1016/j.jhazmat.2020.124363
82. Cheng S, Xing D, Call DF, Logan BE (2009) Direct biological conversion of electrical current into methane by electromethanogenesis. Environ Sci Technol 43:3953–3958. https://doi.org/10.1021/ES803531G
83. Zheng S, Liu F, Wang B, Zhang Y, Lovley DR (2020) Methanobacterium capable of direct interspecies electron transfer. Environ Sci Technol 54:15347–15354. https://doi.org/10.1021/ACS.EST.0C05525/SUPPL_FILE/ES0C05525_SI_001.PDF
84. Yee MO, Snoeyenbos-West OL, Thamdrup B, Ottosen LDM, Rotaru AE (2019) Extracellular electron uptake by two Methanosarcina species. Front Energy Res 7:1–10. https://doi.org/10.3389/fenrg.2019.00029
85. Guisasola A, Baeza JA, Marone A, Trably É, Bernet N (2020) Opportunities for hydrogen production from urban/industrial wastewater in Bioelectrochemical systems. Microbial Electrochem Technol 225–243. https://doi.org/10.1201/9780429487118-15
86. Kiely PD, Rader G, Regan JM, Logan BE (2011) Long-term cathode performance and the microbial communities that develop in microbial fuel cells fed different fermentation endproducts. Bioresour Technol 102:361–366. https://doi.org/10.1016/J.BIORTECH.2010.05.017
87. Larrosa-Guerrero A, Scott K, Katuri KP, Godinez C, Head IM, Curtis T (2010) Open circuit versus closed circuit enrichment of anodic biofilms in MFC: effect on performance and anodic communities. Appl Microbiol Biotechnol 87:1699–1713. https://doi.org/10.1007/S00253-010-2624-1/TABLES/6
88. Jung S, Regan JM (nd) Comparison of anode bacterial communities and performance in microbial fuel cells with different electron donors. Applied Microbial and Cell Physiology (n.d.). https://doi.org/10.1007/s00253-007-1162-y
89. Liu S, Moon CD, Zheng N, Huws S, Zhao S, Wang J (2022) Opportunities and challenges of using metagenomic data to bring uncultured microbes into cultivation. Microbiome 10:1–14. https://doi.org/10.1186/S40168-022-01272-5.
90. Johnson JS, Spakowicz DJ, Hong BY, Petersen LM, Demkowicz P, Chen L, Leopold SR, Hanson BM, Agresta HO, Gerstein M, Sodergren E, Weinstock GM (2019) Evaluation of 16S rRNA gene sequencing for species and strain-level microbiome analysis. Nat Commun 10. https://doi.org/10.1038/s41467-019-13036-1
91. Ogier JC, Pagès S, Galan M, Barret M, Gaudriault S (2019) RpoB, a promising marker for analyzing the diversity of bacterial communities by amplicon sequencing, BMC Microbiol 19. https://doi.org/10.1186/s12866-019-1546-z
92. Eisen JA (nd) The RecA protein as a model molecule for molecular systematic studies of bacteria: comparison of trees of RecAs and 16S rRNAs from the Same Species
93. Peeters K, Willems A (2011) The gyrB gene is a useful phylogenetic marker for exploring the diversity of Flavobacterium strains isolated from terrestrial and aquatic habitats in Antarctica. FEMS Microbiol Lett 321:130–140. https://doi.org/10.1111/j.1574-6968.2011.02326.x
94. Cai W, Liu W, Zhang Z, Feng K, Ren G, Pu C, Sun H, Li J, Deng Y, Wang A (2018) McrA sequencing reveals the role of basophilic methanogens in a cathodic methanogenic community. Water Res 136:192–199. https://doi.org/10.1016/j.watres.2018.02.062

95. Izadi P, Fontmorin JM, Godain A, Yu EH, Head IM (2020) Parameters influencing the development of highly conductive and efficient biofilm during microbial electrosynthesis: the importance of applied potential and inorganic carbon source NPJ Biofilms Microbiomes 6 (2020). https://doi.org/10.1038/s41522-020-00151-x
96. Godain A, Haddour N, Fongarland P, Vogel TM (2022) Bacterial competition for the anode colonization under different external resistances in microbial fuel cells. Catalysts 12. https://doi.org/10.3390/catal12020176
97. Rago L, Cristiani P, Villa F, Zecchin S, Colombo A, Cavalca L, Schievano A (2017) Influences of dissolved oxygen concentration on biocathodic microbial communities in microbial fuel cells. Bioelectrochemistry 116:39–51. https://doi.org/10.1016/j.bioelechem.2017.04.001
98. Pepè Sciarria T, Arioli S, Gargari G, Mora D, Adani F (2019) Monitoring microbial communities' dynamics during the start-up of microbial fuel cells by high-throughput screening techniques. Biotech Rep 21. https://doi.org/10.1016/j.btre.2019.e00310
99. Mei X, Xing D, Yang Y, Liu Q, Zhou H, Guo C, Ren N (2017) Adaptation of microbial community of the anode biofilm in microbial fuel cells to temperature. Bioelectrochemistry 117:29–33. https://doi.org/10.1016/j.bioelechem.2017.04.005
100. Mickol RL, Eddie BJ, Malanoski AP, Yates MD, Tender LM, Glaven SM (2021) Metagenomic and metatranscriptomic characterization of a microbial community that catalyzes both energy-generating and energy-storing electrode reactions. Appl Environ Microbiol 87. https://doi.org/10.1128/AEM.01676-21.
101. Sharma M, Nandy A, Taylor N, Venkatesan SV, Ozhukil Kollath V, Karan K, Thangadurai V, Tsesmetzis N, Gieg LM (2020) Bioelectrochemical remediation of phenanthrene in a microbial fuel cell using an anaerobic consortium enriched from a hydrocarbon-contaminated site. J Hazard Mater 389 (2020). https://doi.org/10.1016/j.jhazmat.2019.121845
102. Wang C, Dong J, Hu W, Li Y (2021) Enhanced simultaneous removal of nitrate and perchlorate from groundwater by bioelectrochemical systems (BESs) with cathodic potential regulation. Biochem Eng J 173. https://doi.org/10.1016/j.bej.2021.108068
103. Zakaria BS, Ranjan Dhar B (2021) An intermittent power supply scheme to minimize electrical energy input in a microbial electrolysis cell assisted anaerobic digester. Bioresour Technol 319. https://doi.org/10.1016/j.biortech.2020.124109
104. Lusk BG, Colin A, Parameswaran P, Rittmann BE, Torres CI (2018) Simultaneous fermentation of cellulose and current production with an enriched mixed culture of thermophilic bacteria in a microbial electrolysis cell, Microb. Biotechnol 11:63–73. https://doi.org/10.1111/1751-7915.12733
105. Badia-Fabregat M, Rago L, Baeza JA, Guisasola A (2019) Hydrogen production from crude glycerol in an alkaline microbial electrolysis cell. Int J Hydrogen Energy 44:17204–17213. https://doi.org/10.1016/j.ijhydene.2019.03.193
106. Hari AR, Venkidusamy K, Katuri KP, Bagchi S, Saikaly PE (2017) Temporal microbial community dynamics in microbial electrolysis cells—influence of acetate and propionate concentration. Front Microb 8. https://doi.org/10.3389/fmicb.2017.01371
107. Di Franca ML, Matturro B, Crognale S, Zeppilli M, Dell'Armi E, Majone M, Petrangeli Papini M, Rossetti S (2022) Microbiome composition and dynamics of a reductive/oxidative bioelectrochemical system for perchloroethylene removal: effect of the feeding composition. Front Microbiol 13. https://doi.org/10.3389/fmicb.2022.951911
108. Lin XQ, Li ZL, Liang B, Zhai HL, Cai WW, Nan J, Wang AJ (2019) Accelerated microbial reductive dechlorination of 2, 4, 6-trichlorophenol by weak electrical stimulation. Water Res 162:236–245. https://doi.org/10.1016/j.watres.2019.06.068
109. Mills S, Dessì P, Pant D, Farràs P, Sloan WT, Collins G, Ijaz UZ (2022) A meta-analysis of acetogenic and methanogenic microbiomes in microbial electrosynthesis, NPJ Biofilms Microbiomes 8. https://doi.org/10.1038/s41522-022-00337-5
110. Marshall CW, Ross DE, Fichot EB, Norman RS, May HD (2013) Long-term operation of microbial electrosynthesis systems improves acetate production by autotrophic microbiomes. Environ Sci Technol 47:6023–6029. https://doi.org/10.1021/es400341b

111. Langille MGI, Zaneveld J, Caporaso JG, McDonald D, Knights D, Reyes JA, Clemente JC, Burkepile DE, Vega Thurber RL, Knight R, Beiko RG, Huttenhower C (2013) Predictive functional profiling of microbial communities using 16S rRNA marker gene sequences. Nat Biotech 31(9):814–821. https://doi.org/10.1038/nbt.2676
112. Mongad DS, Chavan NS, Narwade NP, Dixit K, Shouche YS, Dhotre DP (2021) MicFunPred: a conserved approach to predict functional profiles from 16S rRNA gene sequence data. Genomics 113:3635–3643. https://doi.org/10.1016/j.ygeno.2021.08.016
113. Xin X, Hong J, Liu Y (2019) Insights into microbial community profiles associated with electric energy production in microbial fuel cells fed with food waste hydrolysate. Sci Total Environ 670:50–58. https://doi.org/10.1016/j.scitotenv.2019.03.213
114. Yin Q, Wu G, Lens PNL (2022) Characterization of the core microbial community governing acidogenic processes for the production of valuable bioproducts. NPJ Clean Water 5. https://doi.org/10.1038/s41545-022-00180-3.
115. Matturro B, Zeppilli M, Lai A, Majone M, Rossetti S (2021) Metagenomic analysis reveals microbial interactions at the biocathode of a bioelectrochemical system capable of simultaneous trichloroethylene and Cr(VI) reduction. Front Microbiol 12. https://doi.org/10.3389/fmicb.2021.747670
116. Alqahtani MF, Bajracharya S, Katuri KP, Ali M, Ragab A, Michoud G, Daffonchio D, Saikaly PE (2019) Enrichment of *Marinobacter* sp. and Halophilic Homoacetogens at the biocathode of microbial electrosynthesis system inoculated with red sea brine pool. Front Microbiol 10. https://doi.org/10.3389/fmicb.2019.02563
117. Marshall CW, Ross DE, Handley KM, Weisenhorn PB, Edirisinghe JN, Henry CS, Gilbert JA, May HD, Norman RS (2017) Metabolic reconstruction and modeling microbial electrosynthesis. Sci Rep 7. https://doi.org/10.1038/s41598-017-08877-z
118. Cao X, Zhang C, Zhang S, Sakamaki T, Wang H, Li XN (2023) Simultaneous removal of sediment and water contaminants in a microbial electrochemical system with embedded active electrode by in-situ utilization of electrons. J Hazard Mater 443. https://doi.org/10.1016/j.jhazmat.2022.130172
119. Ragab A, Shaw DR, Katuri KP, Saikaly PE (2020) Effects of set cathode potentials on microbial electrosynthesis system performance and biocathode methanogen function at a metatranscriptional level. Sci Rep 10 (2020). https://doi.org/10.1038/s41598-020-76229-5
120. Pei Y, Tao C, Ling Z, Yu Z, Ji J, Khan A, Mamtimin T, Liu P, X. Li, Exploring novel Cr(VI) remediation genes for Cr(VI)-contaminated industrial wastewater treatment by comparative metatranscriptomics and metagenomics. Sci Total Environ 742. https://doi.org/10.1016/j.scitotenv.2020.140435
121. Eddie BJ, Wang Z, Hervey WJ, Leary DH, Malanoski AP, Tender LM, Lin B, Strycharz-Glaven SM (2017) Metatranscriptomics Supports the Mechanism for Biocathode Electroautotrophy by " Candidatus Tenderia electrophaga" , MSystems 2. https://doi.org/10.1128/msystems.00002-17
122. Ishii S, Suzuki S, Tenney A, Norden-Krichmar TM, Nealson KH, Bretschger O (2015) Microbial metabolic networks in a complex electrogenic biofilm recovered from a stimulus-induced metatranscriptomics approach. Sci Rep 5. https://doi.org/10.1038/srep14840
123. Stewart FJ, Sharma AK, Bryant JA, Eppley JM, DeLong EF (2011) Community transcriptomics reveals universal patterns of protein sequence conservation in natural microbial communities. Genome Biol 12:1–24. https://doi.org/10.1186/GB-2011-12-3-R26/FIGURES/14
124. Wang H, Zeng S, Pan X, Liu L, Chen Y, Tang J, Luo F (2022) Bioelectrochemically assisting anaerobic digestion enhanced methane production under low-temperature, renew. Energy 194:1071–1083. https://doi.org/10.1016/j.renene.2022.05.118
125. Brunner S, Klessing T, Dötsch A, Sturm-Richter K, Gescher J (2019) Efficient bioelectrochemical conversion of industrial wastewater by specific strain isolation and community adaptation. Front Bioeng Biotechnol 7. https://doi.org/10.3389/fbioe.2019.00023
126. Zhao T, Xie B, Yi Y, Zang Y, Liu H (2023) Two polarity reversal modes lead to different nitrate reduction pathways in bioelectrochemical systems. Sci Total Environ 856. https://doi.org/10.1016/j.scitotenv.2022.159185

127. Zhang J, Liu H, Zhang Y, Fu B, Zhang C, Cui M-H, Wu P, Guan Z-W (2023) Metatranscriptomic insights into the microbial electrosynthesis of acetate by Fe^{2+}/Ni^{2+} addition, (2023). https://doi.org/10.21203/rs.3.rs-2499559/v1
128. Ayol A, Peixoto L, Keskin T, Abubackar HN (2021) Reactor designs and configurations for biological and bioelectrochemical c1 gas conversion: a review. Int J Environ Res Public Health 18 https://doi.org/10.3390/ijerph182111683
129. Tanguay-Rioux F, Nwanebu E, Thadani M, Tartakovsky B (2023) On-line current control for continuous conversion of CO^2 to CH^4 in a microbial electrosynthesis cell. Biochem Eng J 197:108965. https://doi.org/10.1016/J.BEJ.2023.108965
130. Kadier A, Simayi Y, Abdeshahian P, Azman NF, Chandrasekhar K, Kalil MS (2016) A comprehensive review of microbial electrolysis cells (MEC) reactor designs and configurations for sustainable hydrogen gas production. Alex Eng J 55:427–443. https://doi.org/10.1016/J.AEJ.2015.10.008
131. Ishak MAM, Ani AY, Syed Ismail SNA, Ali MLM, Ahmad R (2022) Conversion of biomass to biofuels, value-chain of biofuels: fundamentals, technology, and standardization, pp 49–67. https://doi.org/10.1016/B978-0-12-824388-6.00005-1
132. Sen Li J, Zhu LT, Luo ZH (2022) Effect of geometric configuration on hydrodynamics, heat transfer and RTD in a pilot-scale biomass pyrolysis vapor-phase upgrading reactor. Chem Eng J 428:131048. https://doi.org/10.1016/J.CEJ.2021.131048
133. Gallucci F, Fernandez E, Corengia P, van Sint Annaland M (2013) Recent advances on membranes and membrane reactors for hydrogen production. Chem Eng Sci 92:40–66 (2013). https://doi.org/10.1016/j.ces.2013.01.008
134. de Fouchécour F, Larzillière V, Bouchez T, Moscoviz R (2022) Systematic and quantitative analysis of two decades of anodic wastewater treatment in bioelectrochemical reactors. Water Res 214:118142. https://doi.org/10.1016/J.WATRES.2022.118142
135. Li SY, Srivastava R, Suib SL, Li Y, Parnas RS (2011) Performance of batch, fed-batch, and continuous A-B–E fermentation with pH-control. Bioresour Technol 102:4241–4250. https://doi.org/10.1016/J.BIORTECH.2010.12.078
136. Cardona CA, Sánchez ÓJ (2007) Fuel ethanol production: process design trends and integration opportunities. Bioresour Technol 98:2415–2457. https://doi.org/10.1016/J.BIORTECH.2007.01.002
137. Liu S (2017) Batch reactor. In: Bioprocess engineering, Elsevier, 139–178. https://doi.org/10.1016/b978-0-444-63783-3.00004-6
138. Singh V, Ormeci B, Radadiya P, Ranjan Dhar B, Sangal A, Hussain A (2023) Leach bed reactors for production of short-chain fatty acids: A review of critical operating parameters, current limitations and challenges, and prospects. Chem Eng J 456:141044. https://doi.org/10.1016/J.CEJ.2022.141044
139. Mears L, Stocks SM, Sin G, Gernaey KV (2017) A review of control strategies for manipulating the feed rate in fed-batch fermentation processes. J Biotechnol 245:34–46. https://doi.org/10.1016/J.JBIOTEC.2017.01.008
140. Parajuli A, Khadka A, Sapkota L, Ghimire A (2022) Effect of hydraulic retention time and organic-loading rate on two-staged, semi-continuous mesophilic anaerobic digestion of food waste during start-up. Fermentation 8. https://doi.org/10.3390/fermentation8110620
141. Porta R, Benaglia M, Puglisi A (2016) Flow chemistry: recent developments in the synthesis of pharmaceutical products. Org Process Res Dev 20:2–25. https://doi.org/10.1021/acs.oprd.5b00325
142. Costandy JG, Edgar TF, Baldea M (2019) Switching from batch to continuous reactors is a trajectory optimization problem. Ind Eng Chem Res 58:13718–13736. https://doi.org/10.1021/acs.iecr.9b01126
143. Liu Z, Xue X, Cai W, Cui K, Patil SA, Guo K (2023) Recent progress on microbial electrosynthesis reactor designs and strategies to enhance the reactor performance. Biochem Eng J 190. https://doi.org/10.1016/j.bej.2022.108745
144. Ramirez-Nava J, Martínez-Castrejón M, García-Mesino RL, López-Díaz JA, Talavera-Mendoza O, Sarmiento-Villagrana A, Rojano F, Hernández-Flores G (2021) The implications

of membranes used as separators in microbial fuel cells. Membranes (Basel) 11. https://doi.org/10.3390/membranes11100738
145. IAEA (2018) Advances in small modular reactor technology developments
146. Zhang Y, Angelidaki I (2014) Microbial electrolysis cells turning to be versatile technology: recent advances and future challenges. Water Res 56:11–25. https://doi.org/10.1016/j.watres.2014.02.031
147. Zeppilli M, Paiano P, Torres C, Pant D (2021) A critical evaluation of the pH split and associated effects in bioelectrochemical processes. Chem Eng J 422:130155. https://doi.org/10.1016/J.CEJ.2021.130155
148. Torres CI, Marcus AK, Rittmann BE (2008) Proton transport inside the biofilm limits electrical current generation by anode-respiring bacteria. Biotechnol Bioeng 100:872–881. https://doi.org/10.1002/bit.21821
149. Erable B, Féron D, Bergel A (2012) Microbial catalysis of the oxygen reduction reaction for microbial fuel cells: a review. Chemsuschem 5:975–987
150. Li X, Lu Y, Luo H, Liu G, Torres CI, Zhang R (2021) Effect of pH on bacterial distributions within cathodic biofilm of the microbial fuel cell with maltodextrin as the substrate. Chemosphere 265:129088. https://doi.org/10.1016/J.CHEMOSPHERE.2020.129088
151. Kumaravel V, Bartlett J, Pillai SC (2020) Photoelectrochemical conversion of carbon dioxide (CO_2) into fuels and value-added products. ACS Energy Lett 486–519. https://doi.org/10.1021/acsenergylett.9b02585
152. Logan BE (2012) Essential data and techniques for conducting microbial fuel cell and other types of bioelectrochemical system experiments. Chemsuschem 5:988–994. https://doi.org/10.1002/cssc.201100604
153. Kurbanalieva S, Arlyapov V, Kharkova A, Perchikov R, Kamanina O, Melnikov P, Popova N, Machulin A, Tarasov S, Saverina E, Vereshchagin A, Reshetilov A (2022) Electroactive biofilms of activated sludge microorganisms on a nanostructured surface as the basis for a highly sensitive biochemical oxygen demand biosensor. Sensors 22. https://doi.org/10.3390/s22166049
154. Borja-Maldonado F, López Zavala MÁ (2022) Contribution of configurations, electrode and membrane materials, electron transfer mechanisms, and cost of components on the current and future development of microbial fuel cells. Heliyon 8. https://doi.org/10.1016/j.heliyon.2022.e09849
155. Idris MO, Guerrero-Barajas C, Kim HC, Yaqoob AA, Ibrahim MNM (2023) Scalability of biomass-derived graphene derivative materials as viable anode electrode for a commercialized microbial fuel cell: a systematic review. Chin J Chem Eng 55:277–292. https://doi.org/10.1016/J.CJCHE.2022.05.009
156. Kadier A, Simayi Y, Kalil MS, Abdeshahian P, Hamid AA (2014) A review of the substrates used in microbial electrolysis cells (MECs) for producing sustainable and clean hydrogen gas, Renew. Energy 71:466–472. https://doi.org/10.1016/j.renene.2014.05.052
157. Rozenfeld S, Hirsch LO, Gandu B, Farber R, Schechter A, Cahan R (2019) Improvement of microbial electrolysis cell activity by using anode based on combined plasma-pretreated carbon cloth and stainless steel. Energies (Basel) 12 (2019). https://doi.org/10.3390/en12101968
158. Guo K, Prévoteau A, Patil SA, Rabaey K (2015) Engineering electrodes for microbial electrocatalysis. Curr Opin Biotechnol 33:149–156. https://doi.org/10.1016/J.COPBIO.2015.02.014
159. Rabiee H, Ge L, Hu S, Wang H, Yuan Z (2022) Microtubular electrodes: an emerging electrode configuration for electrocatalysis, bioelectrochemical and water treatment applications. Chem Eng J 450:138476. https://doi.org/10.1016/J.CEJ.2022.138476
160. Escapa A, Mateos R, Martínez EJ, Blanes J (2016) Microbial electrolysis cells: an emerging technology for wastewater treatment and energy recovery from laboratory to pilot plant and beyond. Renew Sustainable Energy Rev 55:942–956. https://doi.org/10.1016/J.RSER.2015.11.029

161. Wang X, Cheng S, Feng Y, Merrill MD, Saito T, Logan BE (2009) Use of carbon mesh anodes and the effect of different pretreatment methods on power production in microbial fuel cells. Environ Sci Technol 43:6870–6874. https://doi.org/10.1021/es900997w
162. Fonseca EU, Yang W, Wang X, Rossi R, Logan BE (2021) Comparison of different chemical treatments of brush and flat carbon electrodes to improve performance of microbial fuel cells. Bioresour Technol 342:125932. https://doi.org/10.1016/J.BIORTECH.2021.125932
163. Spiess S, Kucera J, Seelajaroen H, Sasiain A, Thallner S, Kremser K, Novak D, Guebitz GM, Haberbauer M (2021) Impact of carbon felt electrode pretreatment on anodic biofilm composition in microbial electrolysis cells. Biosensors (Basel) 11. https://doi.org/10.3390/bios11060170
164. Zhao G, Rui K, Dou SX, Sun W (2018) Heterostructures for electrochemical hydrogen evolution reaction: a review. Adv Funct Mater 28. https://doi.org/10.1002/adfm.201803291
165. Lekshmi GS, Bazaka K, Ramakrishna S, Kumaravel V (2022) Microbial electrosynthesis: carbonaceous electrode materials for CO_2 conversion. Mater Horiz 10:292–312. https://doi.org/10.1039/d2mh01178f
166. Karthikeyan R, Cheng KY, Selvam A, Bose A, Wong JWC (2017) Bioelectrohydrogenesis and inhibition of methanogenic activity in microbial electrolysis cells—a review. Biotechnol Adv 35:758–771. https://doi.org/10.1016/J.BIOTECHADV.2017.07.004
167. Ribot-Llobet E, Nam JY, Tokash JC, Guisasola A, Logan BE (2013) Assessment of four different cathode materials at different initial pHs using unbuffered catholytes in microbial electrolysis cells. Int J Hydrogen Energy 38:2951–2956. https://doi.org/10.1016/J.IJHYDENE.2012.12.037
168. Glaister BJ, Mudd GM (2010) The environmental costs of platinum–PGM mining and sustainability: Is the glass half-full or half-empty? Miner Eng 23:438–450. https://doi.org/10.1016/J.MINENG.2009.12.007
169. Yuan H, He Z (2017) Platinum group metal–free catalysts for hydrogen evolution reaction in microbial electrolysis cells. Chem Rec 17:641–652. https://doi.org/10.1002/tcr.201700007
170. Tang J, Bian Y, Jin S, Sun D, Ren ZJ (2022) Cathode material development in the past decade for H_2 production from microbial electrolysis cells. ACS Environ Au 2:20–29. https://doi.org/10.1021/acsenvironau.1c00021
171. Selembo PA, Merrill MD, Logan BE (2010) Hydrogen production with nickel powder cathode catalysts in microbial electrolysis cells. Int J Hydrogen Energy 35:428–437. https://doi.org/10.1016/J.IJHYDENE.2009.11.014
172. Call DF, Merrill MD, Logan BE (2009) High surface area stainless steel brushes as cathodes in microbial electrolysis cells. Environ Sci Technol 43:2179–2183. https://doi.org/10.1021/es803074x
173. Dwivedi D, Lepková K, Becker T (2017) Carbon steel corrosion: a review of key surface properties and characterization methods. RSC Adv 7:4580–4610. https://doi.org/10.1039/C6RA25094G
174. Dai H, Yang H, Liu X, Jian X, Liang Z (2016) Electrochemical evaluation of nano-$Mg(OH)_2$/graphene as a catalyst for hydrogen evolution in microbial electrolysis cell. Fuel 174:251–256. https://doi.org/10.1016/J.FUEL.2016.02.013
175. Rozendal RA (2007) Hydrogen production through biocatalyzed electrolysis
176. Jeremiasse AW, Hamelers HVM, Buisman CJN (2010) Microbial electrolysis cell with a microbial biocathode. Bioelectrochemistry 78:39–43. https://doi.org/10.1016/J.BIOELECHEM.2009.05.005
177. Jafary T, Daud WRW, Ghasemi M, Kim BH, Md Jahim J, Ismail M, Lim SS (2015) Biocathode in microbial electrolysis cell; present status and future prospects. Renew Sustainable Energy Rev 47:23–33. https://doi.org/10.1016/J.RSER.2015.03.003
178. Saheb-Alam S, Persson F, Wilén BM, Hermansson M, Modin O (2019) A variety of hydrogenotrophic enrichment cultures catalyse cathodic reactions. Sci Rep 9. https://doi.org/10.1038/s41598-018-38006-3
179. Noori MT, Vu MT, Ali RB, Min B (2020) Recent advances in cathode materials and configurations for upgrading methane in bioelectrochemical systems integrated with anaerobic digestion. Chem Eng J 392. https://doi.org/10.1016/j.cej.2019.123689

180. Zheng S, Li M, Liu Y, Liu F (2021) Desulfovibrio feeding Methanobacterium with electrons in conductive methanogenic aggregates from coastal zones. Water Res 202:117490. https://doi.org/10.1016/J.WATRES.2021.117490
181. Rozendal RA, Hamelers HVM, Rabaey K, Keller J, Buisman CJN (2008) Towards practical implementation of bioelectrochemical wastewater treatment. Trends Biotechnol 26:450–459. https://doi.org/10.1016/j.tibtech.2008.04.008
182. Chouhan A, Bahar B, Prasad AK (2020) Effect of back-diffusion on the performance of an electrochemical hydrogen compressor. Int J Hydrogen Energy 45:10991–10999. https://doi.org/10.1016/J.IJHYDENE.2020.02.048
183. Luo T, Abdu S, Wessling M (2018) Selectivity of ion exchange membranes: a review. J Memb Sci 555:429–454. https://doi.org/10.1016/J.MEMSCI.2018.03.051
184. Zeppilli M, Paiano P, Torres C, Pant D (2021) A critical evaluation of the pH split and associated effects in bioelectrochemical processes. Chem Eng J 422 (2021). https://doi.org/10.1016/j.cej.2021.130155
185. Ivanov I, Ahn YT, Poirson T, Hickner MA, Logan BE (2017) Comparison of cathode catalyst binders for the hydrogen evolution reaction in microbial electrolysis cells. Int J Hydrogen Energy 42:15739–15744. https://doi.org/10.1016/J.IJHYDENE.2017.05.089
186. Chacón-Carrera RA, López-Ortiz A, Collins-Martínez V, Meléndez-Zaragoza MJ, Salinas-Gutiérrez J, Espinoza-Hicks JC, Ramos-Sánchez VH (2019) Assessment of two ionic exchange membranes in a bioelectrochemical system for wastewater treatment and hydrogen production. Int J Hydrogen Energy 44:12339–12345. https://doi.org/10.1016/J.IJHYDENE.2018.10.153
187. Sleutels THJA, Hamelers HVM, Rozendal RA, Buisman CJN (2009) Ion transport resistance in microbial electrolysis cells with anion and cation exchange membranes. Int J Hydrogen Energy 34:3612–3620. https://doi.org/10.1016/J.IJHYDENE.2009.03.004
188. Zhen G, Kobayashi T, Lu X, Kumar G, Hu Y, Bakonyi P, Rózsenberszki T, Koók L, Nemestóthy N, Bélafi-Bakó K, Xu K (2016) Recovery of biohydrogen in a single-chamber microbial electrohydrogenesis cell using liquid fraction of pressed municipal solid waste (LPW) as substrate. Int J Hydrogen Energy 41:17896–17906. https://doi.org/10.1016/J.IJHYDENE.2016.07.112
189. Xia A, Cheng J, Murphy JD (2016) Innovation in biological production and upgrading of methane and hydrogen for use as gaseous transport biofuel. Biotechnol Adv 34:451–472. https://doi.org/10.1016/J.BIOTECHADV.2015.12.009
190. Poblete IBS, de Q.F. Araújo O, de Medeiros JL (2022) Sewage-water treatment with bioenergy production and carbon capture and storage. Chemosphere 286:131763. https://doi.org/10.1016/J.CHEMOSPHERE.2021.131763
191. Kadier A, Simayi Y, Chandrasekhar K, Ismail M, Kalil MS (2015) Hydrogen gas production with an electroformed Ni mesh cathode catalysts in a single-chamber microbial electrolysis cell (MEC). Int J Hydrogen Energy 40:14095–14103. https://doi.org/10.1016/J.IJHYDENE.2015.08.095
192. Chaurasia AK, Mondal P (2022) Enhancing biohydrogen production from sugar industry wastewater using Ni, Ni–Co and Ni–Co–P electrodeposits as cathodes in microbial electrolysis cells. Chemosphere 286:131728. https://doi.org/10.1016/J.CHEMOSPHERE.2021.131728
193. Logan BE, Rossi R, Ragab A, Saikaly PE (2019) Electroactive microorganisms in bioelectrochemical systems. Nat Rev Microbiol 17:307–319. https://doi.org/10.1038/s41579-019-0173-x
194. Yang Q, Jiang Y, Xu Y, Qiu Y, Chen Y, Zhu S, Shen S (2015) Hydrogen production with polyaniline/multi-walled carbon nanotube cathode catalysts in microbial electrolysis cells. J Chem Technol Biotechnol 90:1263–1269. https://doi.org/10.1002/jctb.4425
195. Jayabalan T, Matheswaran M, Naina Mohammed S (2019) Biohydrogen production from sugar industry effluents using nickel based electrode materials in microbial electrolysis cell. Int J Hydrogen Energy 44:17381–17388. https://doi.org/10.1016/J.IJHYDENE.2018.09.219
196. Cui W, Lu Y, Zeng C, Yao J, Liu G, Luo H, Zhang R (2021) Hydrogen production in single-chamber microbial electrolysis cell under high applied voltages. Sci Total Environ 780:146597. https://doi.org/10.1016/J.SCITOTENV.2021.146597

197. Hong-Yan D, Hui-Min Y, Xian L, Xiu-Li S, Zhen-Hai L (2021) Preparation and electrochemical evaluation of MoS_2/graphene quantum dots as a catalyst for hydrogen evolution in microbial electrolysis cell. J Electrochem 27:429–438. http://electrochem.xmu.edu.cn/EN/Y2021/V27/I4/429 (Accessed 7 May 2023)
198. Li F, Liu W, Sun Y, Ding W, Cheng S (2017) Enhancing hydrogen production with Ni–P coated nickel foam as cathode catalyst in single chamber microbial electrolysis cells. Int J Hydrogen Energy 42:3641–3646. https://doi.org/10.1016/J.IJHYDENE.2016.10.163
199. Liang D, Zhang L, He W, Li C, Liu J, Liu S, Lee HS, Feng Y (2020) Efficient hydrogen recovery with CoP-NF as cathode in microbial electrolysis cells. Appl Energy 264:114700. https://doi.org/10.1016/J.APENERGY.2020.114700
200. Spurgeon SNJ, Matheswaran M, Satyavolu J (2020) Simultaneous biohydrogen production with distillery wastewater treatment using modified microbial electrolysis cell. Int J Hydrogen Energy 45:18266–18274. https://doi.org/10.1016/J.IJHYDENE.2019.06.134
201. Chen J, Xu W, Wu X, Lu JEN, Wang T, Zuo H (2019) System development and environmental performance analysis of a pilot scale microbial electrolysis cell for hydrogen production using urban wastewater. Energy Convers Manag 193:52–63. https://doi.org/10.1016/J.ENCONMAN.2019.04.060
202. Hu K, Xu L, Chen W, Qiu Jia S, Wang W, Han F (2018) Degradation of organics extracted from dewatered sludge by alkaline pretreatment in microbial electrolysis cell. Environ Sci Pollut Res 25:8715–8724. https://doi.org/10.1007/s11356-018-1213-1
203. Marone A, Ayala-Campos OR, Trably E, Carmona-Martínez AA, Moscoviz R, Latrille E, Steyer JP, Alcaraz-Gonzalez V, Bernet N (2017) Coupling dark fermentation and microbial electrolysis to enhance bio-hydrogen production from agro-industrial wastewaters and by-products in a bio-refinery framework. Int J Hydrogen Energy 42:1609–1621. https://doi.org/10.1016/j.ijhydene.2016.09.166
204. Tucci M, Colantoni S, Cruz Viggi C, Aulenta F (2023) Improving the kinetics of H_2-fueled biological methanation with quinone-based redox mediators. Catalysts 13:859. https://doi.org/10.3390/catal13050859
205. Wang W, Chang JS, Lee DJ (2022) Integrating anaerobic digestion with bioelectrochemical system for performance enhancement: a mini review. Bioresour Technol 345:126519. https://doi.org/10.1016/J.BIORTECH.2021.126519
206. Osman AI, Mehta N, Elgarahy AM, Hefny M, Al-Hinai A, Al-Muhtaseb AH, Rooney DW (2022) Hydrogen production, storage, utilisation and environmental impacts: a review. Environ Chem Lett 20:153–188. https://doi.org/10.1007/s10311-021-01322-8
207. El-Qelish M, Hassan GK, Leaper S, Dessì P, Abdel-Karim A (2022) Membrane-based technologies for biohydrogen production: a review. J Environ Manage 316:115239. https://doi.org/10.1016/J.JENVMAN.2022.115239
208. Tian Y, Wu J, Liang D, Li J, Liu G, Lin N, Li D, Feng Y (2022) Insights into the electron transfer behaviors of a biocathode regulated by cathode potentials in microbial electrosynthesis cells for biogas upgrading. Environ Sci Technol. https://doi.org/10.1021/acs.est.2c09871
209. Zeppilli M, Paiano P, Villano M, Majone M (2019) Anodic vs cathodic potentiostatic control of a methane producing microbial electrolysis cell aimed at biogas upgrading. Biochem Eng J 152:107393. https://doi.org/10.1016/J.BEJ.2019.107393
210. Liu D, Roca-Puigros M, Geppert F, Caizán-Juanarena L, Na Ayudthaya SP, Buisman C, Heijne A (2018) Granular carbon-based electrodes as cathodes in methane-producing bioelectrochemical systems. Front Bioeng Biotechnol 9. https://doi.org/10.3389/fbioe.2018.00078
211. Seelajaroen H, Spiess S, Haberbauer M, Hassel MM, Aljabour A, Thallner S, Guebitz GM, Sariciftci NS (2020) Enhanced methane producing microbial electrolysis cells for wastewater treatment using poly(neutral red) and chitosan modified electrodes, Sustain. Energy Fuels 4:4238–4248. https://doi.org/10.1039/d0se00770f
212. Arvin A, Hosseini M, Amin MM, Najafpour Darzi G, Ghasemi Y (2019) Efficient methane production from petrochemical wastewater in a single membrane-less microbial electrolysis cell: The effect of the operational parameters in batch and continuous mode on bioenergy recovery. J Environ Health Sci Eng 17:305–317. https://doi.org/10.1007/s40201-019-00349-y

213. Dou Z, Dykstra CM, Pavlostathis SG (2018) Bioelectrochemically assisted anaerobic digestion system for biogas upgrading and enhanced methane production. Sci Total Environ 633:1012–1021. https://doi.org/10.1016/j.scitotenv.2018.03.255
214. Hou H, Li Z, Liu B, Liang S, Xiao K, Zhu Q, Hu S, Yang J, Hu J (2020) Biogas and phosphorus recovery from waste activated sludge with protocatechuic acid enhanced Fenton pretreatment, anaerobic digestion and microbial electrolysis cell. Sci Total Environ 704. https://doi.org/10.1016/j.scitotenv.2019.135274.
215. Zeppilli M, Simoni M, Paiano P, Majone M (2019) Two-side cathode microbial electrolysis cell for nutrients recovery and biogas upgrading. Chem Eng J 370:466–476. https://doi.org/10.1016/j.cej.2019.03.119
216. Krishnan S, Md Din MF, Taib SM, Nasrullah M, Sakinah M, Wahid ZA, Kamyab H, Chelliapan S, Rezania S, Singh L (2019) Accelerated two-stage bioprocess for hydrogen and methane production from palm oil mill effluent using continuous stirred tank reactor and microbial electrolysis cell. J Clean Prod 229:84–93. https://doi.org/10.1016/J.JCLEPRO.2019.04.365
217. Mansoorian HJ, Mahvi A, Nabizadeh R, Alimohammadi M, Nazmara S, Yaghmaeian K (2020) Evaluating the performance of coupled MFC-MEC with graphite felt/MWCNTs polyscale electrode in landfill leachate treatment, and bioelectricity and biogas production. https://doi.org/10.1007/s40201-020-00528-2/Published
218. Cristiani L, Leobello L, Zeppilli M, Villano M (2023) Role of C/N ratio in a pilot scale microbial electrolysis cell (MEC) for biomethane production and biogas upgrading, renew. Energy 210:355–363. https://doi.org/10.1016/J.RENENE.2023.04.049
219. Alqahtani MF, Katuri KP, Bajracharya S, Yu Y, Lai Z, Saikaly PE (2018) Porous hollow fiber nickel electrodes for effective supply and reduction of carbon dioxide to methane through microbial electrosynthesis. Adv Funct Mater 28. https://doi.org/10.1002/adfm.201804860
220. Zeppilli M, Mattia A, Villano M, Majone M (2017) Three-chamber bioelectrochemical system for biogas upgrading and nutrient recovery. Fuel Cells 17:593–600. https://doi.org/10.1002/fuce.201700048
221. Heidrich ES, Dolfing J, Scott K, Edwards SR, Jones C, Curtis TP (2013) Production of hydrogen from domestic wastewater in a pilot-scale microbial electrolysis cell. Appl Microbiol Biotechnol 97:6979–6989. https://doi.org/10.1007/s00253-012-4456-7
222. Cusick RD, Bryan B, Parker DS, Merrill MD, Mehanna M, Kiely PD, Liu G, Logan BE (2011) Performance of a pilot-scale continuous flow microbial electrolysis cell fed winery wastewater. Appl Microbiol Biotechnol 89:2053–2063. https://doi.org/10.1007/s00253-011-3130-9
223. Enzmann F, Holtmann D (2019) Rational Scale-Up of a methane producing bioelectrochemical reactor to 50 L pilot scale. Chem Eng Sci 207:1148–1158. https://doi.org/10.1016/J.CES.2019.07.051
224. Singh L, Miller AG, Wang L, Liu H (2021) Scaling-up up-flow microbial electrolysis cells with a compact electrode configuration for continuous hydrogen production. Bioresour Technol 331:125030. https://doi.org/10.1016/J.BIORTECH.2021.125030
225. San-Martín MI, Sotres A, Alonso RM, Díaz-Marcos J, Morán A, Escapa A (2019) Assessing anodic microbial populations and membrane ageing in a pilot microbial electrolysis cell. Int J Hydrogen Energy 44:17304–17315. https://doi.org/10.1016/J.IJHYDENE.2019.01.287
226. Baeza JA, Martínez-Miró À, Guerrero J, Ruiz Y, Guisasola A (2017) Bioelectrochemical hydrogen production from urban wastewater on a pilot scale. J Power Sources 356:500–509. https://doi.org/10.1016/J.JPOWSOUR.2017.02.087

Chapter 16
Evolution of the Biorefinery Concept and Tools for Its Evaluation Toward a Circular Bioeconomy

Idania Valdez-Vazquez, Leonor Patricia Güereca, Carlos E. Molina-Guerrero, Alejandro Padilla-Rivera, and Héctor A. Ruiz

Abstract In its simplest definition, biorefinery was coined as a processing facility that receives biomass for its transformation into products and energy. This work summarizes the evolution of biorefinery processing in the past twenty years from three perspectives. From the feedstock point of view, biorefineries first processed dedicated crops, but now they include a great variety of non-conventional biomasses. From the process point of view, pioneer schemes considered well-established technologies such as enzymatic saccharifications, yeast fermentation, and purifications. Currently, schemes incorporate novel intensified strategies for converting biomass with higher yields at lower costs. From the product point of view, biorefineries are well known to be coproduction facilities, from commodities such as cooking oil, sugar, or ethanol to high-value-added chemical specialties. Climate change has recently stressed the importance of assessing the sustainability of the biorefining processes.

I. Valdez-Vazquez (✉)
Juriquilla Academic Unit, Ingeneering Institute Universidad Nacional Autónoma de México, Blvd. Juriquilla 3001, 76230 Querétaro, Mexico
e-mail: ivaldezv@iingen.unam.mx

L. P. Güereca
Engeneering Institute, Universidad Nacional Autónoma de México, Coyoacán Ciudad de México, Av. Universidad 3000, Ciudad Universitaria, 04510 Coyoacán, Mexico
e-mail: LGuerecaH@iingen.unam.mx

C. E. Molina-Guerrero
Departamento de Ingenierías Química, Electrónica y Biomédica, División de Ciencias E Ingenierías, Campus León, Universidad de Guanajuato, Lomas del Bosque 103, Col. Lomas del Campestre, 37150 León, Mexico
e-mail: ce.molina@ugto.mx

A. Padilla-Rivera
School of Architecture, Planning and Landscape, University of Calgary, Calgary, AB, Canada
e-mail: padillariv@ucalgary.ca

H. A. Ruiz
Biorefinery Group, Food Research Department, School of Chemistry, Autonomous University of Coahuila, 25280 Saltillo, Coahuila, Mexico
e-mail: hector_ruiz_leza@uadec.edu.mx

V. Alcaraz Gonzalez et al. (eds.), *Wastewater Exploitation*, Springer Water,
https://doi.org/10.1007/978-3-031-57735-2_16

Developing sustainability evaluation methods (technical, environmental, and social) has received great attention in the past fifteen years to understand the real impacts of these biorefining processes. Finally, biorefineries represent a real opportunity for countries from all regions and levels of economic development to adopt the Circular Economy in the coming decades.

Keywords Biofuels · Life cycle assessment · Lignocellulosic biomass · Mass flow · Social life cycle assessment · Techno-economic evaluation · Value-added products

16.1 Revisiting the Biorefinery Concept

The concept of biorefineries has been in existence for more than 35 years. The first publications mentioning this philosophy were published by Rexen and Munk [1]. This report, "The agricultural refinery concept," mentioned the need for integrated and development in agricultural production processes for food, feed, and non-food industrial products such as chemicals and fiber. Koukios [2] reported that "biomass refining" was defined as an integrated process that separates the major components of biomass and converts them into biofuels, food, feed, and chemicals through physical, chemical, or biochemical platforms. According to Munk [3], Rexen [4] stated that during the period from 1985 to 1987, an economic feasibility study sponsored by the Danish Government was carried out in an "Agricultural refinery" in Denmark. The plant would provide grain processing, starch extraction, and an oil extraction facility diversification of the agricultural refinery. In addition, Gylling [5] reported the whole crop biorefinery project. That project was aimed to find new uses and markets for crops for energy production. Papatheofanous et al. [6] proposed "biomass refining" or "biorefining" as a process for the fractionation and production of high-value fractions from biomass for industrial utilization, specifically from agricultural materials. Audsley and Sells [7] mentioned that the "bio-refinery" could be introduced where the crop can be separated into different components, like an oil refinery, in which the oil is separated into components for different uses. Biorefining was developed as a "Green Chemistry" (no use of organic solvents) technology from oilseed crops. The objective was to convert the oilseed into lipids, oils, protein products, carbohydrates, and low molecular weight compounds (Bagger et al. [8]). Carlson [9] and Kiel [10], cited by Andersen and Kiel [11], reported that the green crop facilities are the first stage of "biorefinery" for multipurpose uses as phytochemicals and fermentation products. Andersen and Kiel [11] concluded that green biorefinery is an important alternative in producing organic acids (as lactic acid), and amino acids parallel with pellet production. Starke et al. [12] defined the "Green Biorefinery" as a system for the total and energetic valorization from biomass, especially from grass (alfalfa); they also commented on the importance of fractionation of biomass into water soluble compounds (carbohydrate and protein). Lactic acid was one of the products of greatest interest in green biorefineries. Danner et al. [13] reported the extraction yield of lactic acid between 31 and 96 g per Kg of silages. Furthermore, they proposed the "Green Biorefinery," applying the biotechnological, chemical, or physical fractionation and valorization of biomass to produce compounds of industrial and commercial interest as organic acids, amino acids, proteins, enzymes, biodegradable plastics, alcohols, fertilizers, and biofuels (biogas, solid fuel pellets, liquids biofuels, and electricity).

On the other hand, the production of high-added value compounds and bioenergy (biofuels) in parallel is the main idea of the biorefinery concept. In this context, Mielenz [14] published a review of technologies for bioethanol production from biomass and concluded that during the next decade "2010" bioethanol technology was likely to be commercialized in terms of biorefinery concepts around the world due to the need in the transportation sector.

Moreover, the biorefineries from biomass have been classified depending on the raw materials, processes, technological maturity, and the final product, among others. Kamm and Kamm [15] classified biorefineries as follows: Phase I biorefineries (dry-milling ethanol plant, using grain as feedstock) for ethanol production, feed-coproducts, and carbon dioxide; Phase II biorefineries (wet-milling technology) for starch production, high fructose, corn syrup, ethanol, corn oil, and corn gluten feed and meal; Phase III biorefineries (the use of different types of raw materials and process in the production of compounds for the industrial market). They mentioned that the flexibility of feedstock is a factor for adaptability in producing feed, food, and industrial commodities. In addition, they comment that the "bioeconomy" will be used within this concept of biorefinery. Fernando et al. [16] showed the differences between "Whole-Crop Biorefinery" (this type processes use wheat, rye, triticale, and corn as feedstock), "Green Biorefinery" (processes all fractions and components of a vegetal material, from grass, green plants or green crops, using a wet fraction to produce fiber rich press cake and nutrients rich green juice); "Lignocellulose Feedstock (LCF) Biorefinery" (this biorefinery primarily revolves around the use of hemicellulose-five sugar, cellulose-glucose and lignin-polyphenols compounds); "Integrated Biorefinery" (this concept is based on the conversion technology "thermochemical or biochemical) to produce electricity from the combination of existing technology platforms. In summary, Kamm et al. [17] mentioned that biorefinery technologies are based on using biomass and integrating processes. Additionally, they presented the pillar model for a future sustainable biobased economy in producing bioenergy, biofuels, and bioproducts.

According to Task 42: Biorefinery in a circular economy of the International Energy Agency Bioenergy: "*Biorefining is the sustainable processing of biomass into a spectrum of marketable products and energy*". The biorefineries can be classified in (1) *feedstocks* [first generation-1G (from corn, sugar cane), second generation-2G (agroindustrial residues—lignocellulosic biomass, organic residues), and third generation-3G (macroalgae and microalgae); (2) *processes* [Thermochemical: combustion, gasification, pyrolysis; Biochemical: fermentation, anaerobic digestion, enzymatic process; Chemical process: catalytic, pulping, esterification, hydrogenation, water electrolysis, etc.; Mechanical/physical process: extraction, fiber separation, mechanical fractionation, pretreatment, etc.] (3) *Platforms* [C5,

C6 sugars, oils, biogas, syngas, electricity, and heat, etc.] and (4) *Products* [biodiesel, bioethanol, biomethane, synthetic biofuels, electricity and heat; food, feed, fertilizer, biomaterials, chemicals and building blocks, polymers and resins] [18].

On the other hand, first generation biorefineries have been consolidated worldwide, mainly in the USA and Brazil, to produce ethanol and compounds of industrial interest. For example, bioethanol production at an industrial level has grown in the last 20 years, and the Renewable Fuels Association (RFA) reported the global ethanol production of 28 billion gallons in 2022 [19]. The USA is the main producer with 55% (15.4), followed by Brazil with 26% (7.420) and the European Union with 5% (1.330) of billion gallons, respectively. In the USA, ethanol production is based on the following raw materials: 94% (corn starch), 4% (cellulosic biomass/starch), 2% (corn, sorghum, and wheat), < 1% (waste sugars/starch) through two technologies, (1) dry mill with 92.6% and wet mill 7.4% in the use of operation for ethanol production, distillers grains, and corn distillers oil. In 2022, 199 first-generation commercial-scale biorefineries were in operation. According to the United States Department of Agriculture in the USA (USDA), there are 269 sugarcane ethanol plants, six flex plants of corn and sugar cane, and five corn-ethanol plants in Brazil [20].

16.2 Feedstocks

One of the main advantages of biorefinery facilities is that they receive any type of biomass available regionally. In this sense, the larger producers of cereals, such as China, the United States, India, and Russia, can exploit biorefineries that receive straws, producers of vegetable oils, such as Indonesia, Malaysia, Thailand, and Colombia can exploit oil cakes, producers of sugarcane such as Brazil, India, and China can exploit sugarcane bagasse, and so on. Similarly, some types of biomasses are common among nations, such as the organic fraction of municipal solid wastes (OFMSW); in contrast, other biomasses, such as microalgae, can be produced as demanded. The feedstock flexibility of a biorefinery also has a bottleneck; the biomass composition changes depending on several combined factors. For example, the composition of biomasses that derive from agro-industrial processes depends on the type of process applied, the composition of straws depends on the agrometeorological conditions that prevail regionally, and the composition of the OFMSW depends on the geographic region, the number of inhabitants, seasonality, and their social condition [21, 22]. In biochemical-based biorefineries, biomasses are characterized based on their macromolecule content as carbohydrates, proteins, and lipids (Fig. 16.1). Lignocellulosic biomasses include hardwoods, softwoods, straws, and grasses, all of which have been the workhorses in the design of biorefineries during the last few decades. These kinds of biomasses share cellulose, hemicellulose, extractives, and lignin as the main components (Fig. 1a), and their contents differ depending on the species, tissue, and growth stage [23]. Hemicellulose varies in its molecular mass and composition and distribution of the substituents of 5 carbon sugars (xylose, rhamnose, and arabinose), 6 carbon sugars (glucose, mannose, and galactose), and uronic acids (4-O-methylglucuronic, D-glucuronic, and D-galactouronic acids) with some degree of acetylation [24]. Lignin is an aromatic polymer derived from the phenyl propanolic alcohols p-coumaryl, coniferyl, and sinapyl alcohols. A comparison among straws and grasses showed that the content of extractives largely varied, 35% depending on the source. The contents of xylan and cellulose varied by 29 and 27%, respectively, while the content of lignin barely varied by 5% [23]. In the case of the OFMSW, the carbohydrate fraction comprises the major component, including starch > free sugars > raw fiber (cellulose and xylan). Proteins, followed by oils, are the second and third largest components of the OFMSW (Fig. 1b). These fractions present the largest variations depending on the origin of the OFMSW; the contents of lignin, raw fiber, and free sugars have the largest variation between 51 and 57%, oils and proteins vary between 31 and 38%, and the starch content has the lowest variation of 15% [21]. Oil cakes represent a non-conventional feedstock in biorefineries and derive from the processing of edible and non-edible vegetable oils such as soybeans, rapeseed/mustard, cottonseed, groundnuts, sunflower seed, palm kernels, and copra/coconut [25]. The major components of oil cakes are proteins, carbohydrates, and oils, which vary depending on the pre-processing and mode of oil extraction (Fig. 1c). The composition of the protein fraction varies among oil cakes since some of them are deficient in certain amino acids; for instance, cottonseed and groundnut cakes are deficient in methionine and tryptophan, sunflower cakes are deficient in alanine and tryptophan, palm kernel cakes are deficient in alanine and serine, and copra cake are deficient in tryptophan, alanine, and serine. The composition of the carbohydrate fraction also varies depending on the type of oil cake; for example, soybean cakes contain sucrose and raffinose, rapeseed cakes contain pectins and palm kernel cakes contain galactomannans. The fiber content in oil cakes shows the largest variation of 65% because the hull weight largely differs among seeds and also due to differences in the dehulling process efficiency. The protein content in oil cakes varies by 32% among sources, while the contents of oil and carbohydrates only vary by 12 and 2%, respectively. Some non-edible oil cakes contain anti-nutrient toxic compounds that decrease their nutritional value for livestock, such as poultry, non-ruminants, and aquaculture. These anti-nutrients include lectins, saponins, and phytates in soybean cakes, tannins, erucic acid, sinapine, phytates in rapeseed cakes, and gossypol in cottonseed cakes. Seaweed biomass (marine macroalgae) comprises red algae (Rhodophyceae), brown algae (Phaeophyceae), and green algae (Chlorophyceae), which possess unique compositions that make them an interesting feedstock in biorefineries [26]. The carbohydrate fraction is the major component in seaweed species (Fig. 1d): green algae contain starch, cellulose, and ulvan; red algae contain carrageenan, agar, and cellulose; and brown algae contain laminarin, mannitol, alginate, fucoidan, and cellulose [27]. The carbohydrate contents vary from 64 to 77% in red algae, 45–66% in brown algae, and 37–63% in green algae. Then, generally speaking, the ash fraction is the second most abundant component in seaweed biomass, with the highest content of 32% in brown algae. The protein fraction averages 15% with a variation of 27% among algae, and oils constitute the less abundant fraction with an average of 1%. So far, the descriptions of these four representative biomasses indicate that their composition may differ despite being the same type of biomass. Consequently, researchers have tested different processing technologies under a wide range of operational conditions for each kind of biomass.

Each of the fractions previously described is a candidate for being processed and converted into bioenergy or bioproducts through mutually exclusive processes. Each available option implies a different mass flow with different percentages of material losses. In the selected biorefining option depicted in Fig. 1a, the feedstock receives a pretreatment that solubilizes the extractive and xylan fractions, which are then methanized. The cellulose fraction experiences losses during three consecutive processes that totalized ca. 40% of total losses before being converted into ethanol

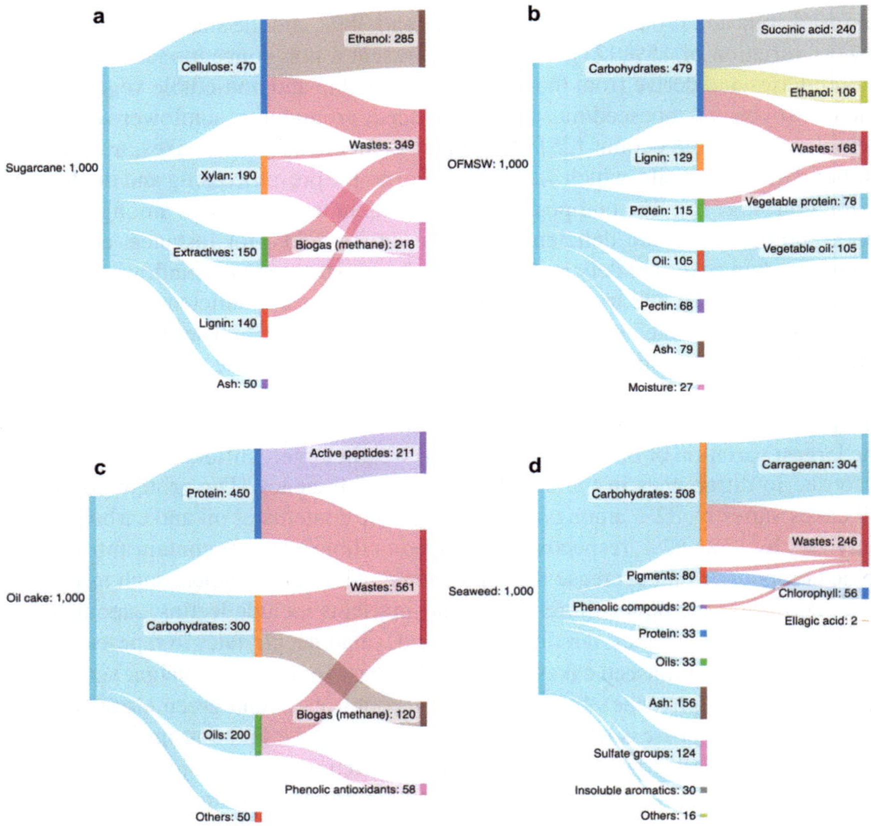

Fig. 16.1 Sankey diagrams provide an overview of the mass flow during the biorefining of sugarcane (**a**), the organic fraction of the municipal solid waste (OFMSW, (**b**), oil cake (**c**), and seaweed (**d**). The Sankey diagrams show the quantity of feedstock converted into different products and material losses in the form of waste. The biomasses are divided into their main components: carbohydrates, proteins, oils, and others. Flows are expressed in 1000 tons. Sankey diagrams were elaborated based on data from [28–34]

(Fig. 1a, [28]). For this biorefining option that uses sugarcane, the analysis of mass flows indicates that almost 40% of feedstock is lost during the processing. The lignin fraction still needs to be processed and valorized, and depending on the chosen biorefining option, the lignin fraction experiences adverse effects that reduce its quality.

The carbohydrate fraction of the OFMSW is suitable to be converted into ethanol, hydrogen, biogas (methane), or other bioproducts such as short-chain fatty acids. In the biorefining option chosen in Fig. 1b, the carbohydrate fraction is converted as follows: 20% into ethanol (corresponding to free sugars), 50% into succinic acid (corresponding to starch and glucan), and 30% of feedstock is lost [29, 30]. According to the source, 68% of protein and 100% of oils are recovered [29]. Pectins also represent an opportunity for valorizing. In this option that uses the OFMSW, the analysis of mass flows indicates that ca. 25% of feedstock is lost during its biorefining. Unlike the biomasses above, the non-conventional feedstocks of oil cakes primarily consist of proteins (Fig. 1c). Therefore, the primary focus has been biorefining the protein fraction to recover antioxidants and bioactive peptides, which have promising applications as food additives, food enrichment agents, nutraceuticals, and food packaging [31]. The carbohydrate fraction is suitable for producing ethanol, biogas, or dietary fiber in the case of edible oil cakes [25, 31]. The biorefining option chosen in Fig. 1c recovered active peptides with 50% of material loss [31]. Then, the feedstock has been poorly converted into biogas, with 40% of material loss [25]. Also, the oil fraction is suitable for recovering antioxidants such as phenolic compounds [32]. Finally, the non-conventional feedstock of seaweed presents a great diversity of biorefining options depending on the type of algae available regionally. The biorefining option chosen in Fig. 1d processed the algae *Kappaphycus alvarezii* to extract carrageenan, pigments, and ellagic acid with 25% material loss [33, 34].

By comparing the mass flows across various biorefining options, it is possible to observe that some biomass components need to be included in the processing. For example, the biorefining options available for lignocelluloses prioritize the valorization of the carbohydrate fraction, relegating the lignin ones. These early approaches alter the lignin quality and limit its chemical upgrading into value-added chemicals. Therefore, novel approaches now prioritize the depolymerization of lignin—the lignin-first approach—to obtain phenolic units and monolignol that served as raw materials to produce aromatic chemicals, biobased fuels, polymers, and drugs [35]. In some cases, the feedstock composition allows greater flexibility in selecting different biorefining options, as is the OFMSW case. The chosen options will depend on the sustainability performance and the regional context that dictates the need to be covered. Non-conventional feedstocks, such as oil cakes, encounter the challenge of enhancing their process efficiency to minimize material loss. Finally, other feedstocks, such as macroalgae, possess unique compositions that dictate the choice of biorefining options, which in many cases consists mainly of extraction processes more than conversion processes. In the literature, alternative biorefining options for each feedstock are yet to be evaluated to determine the most sustainable approach.

16.3 Extraction of High-Value Biomolecules Using Non-conventional Biomass

In recent years, innovative biorefinery systems have been used to produce a range of value-added products such as fuels, platform chemicals, polymers, materials and pharmaceuticals. Biomass generally undergoes chemical pre-treatment prior to enzymatic hydrolysis to produce free sugars. After sugar fermentation, the biomolecules of interest (high-value products) are produced [36, 37]. However, different types of biomolecules can also be extracted from the biomass before the lignocellulose transformation occurs. Different reports indicate that apple and grape pomace, broccoli agro-waste, oats, agave waste, and micro and macroalgae, which are abundant in different parts of the world, can be used for this purpose [38–42]. These biomasses have advantages over conventional crops, such as wheat straw since they contain molecules of interest to the food and pharmaceutical industries and as chemical building blocks. These biomolecules include pectin, fructans, polyphenols, flavonoids, amino acids, chlorophylls, carotenoids, proteins, pigments, and lipids [43, 44].

Fructans and pectin are of interest to the food industry. Fructans, derived from agave waste, consist of re-polyfructose molecules produced by the action of fructosyltransferases on sucrose [45]. The fructans are widely used for their prebiotic effect. For this reason, they are considered functional foods [38]. Pectin, a fiber found in the primary cell wall and intracellular layer of plant cells, is obtained from apple pomace [46]. This compound can be used as an absorbent, emulsifier, stabilizer, bulk-forming agent, and during the production of tablets, pellets, microparticles, and beads as wells [47].

Also, polyphenols are classified as phenolic acids, flavonoids, stilbenes, and lignans [48]. These compounds are beneficial to human health, as demonstrated by their probiotic, antioxidant, anticancer, anti-inflammatory, neuroprotective, antimicrobial and antiviral properties [49–52]. Some of these molecules, which can be obtained from agro-waste (i.e., broccoli agro-waste), have been used as chemical building blocks (CBB). The National Laboratory of Renewable Energy (NLRE) identified thirty biomolecules as the most important for the industry. The first list was published in 2004 [53], and ten years later, some chemicals with relatively small growth markets were removed, and others with potential use in the industry were added. The most important molecules add a total of twelve, and they are succinic acid, lactic acid, itaconic acid, 3-hydroxypropionic acid, isoprene, farnesene, glycerol, sorbitol, xylitol, levulinic acid, glucaric acid, hydroxymethyl furfural, and furfural [37, 54].

Sanford et al. [55] reported the production of these molecules at a commercial scale. For example, succinic acid has been produced after starch and corn syrup fermentation using low-pH recombinant yeast, Cargill recombinant yeast, and recombinant *Escherichia coli*. Data indicate that Cassano Spinola (Italy), Lake Providence (USA), and Sarnia (Canada) have achieved productions of 10,000, 13,600, and 30,000 ton/year, respectively [55]. Lactic acid is obtained using fermentation and polymerization technologies. During this process, two chiral lactic acid isomers are obtained after glucose fermentation. Later, these products are combined with other lactic isomers and polymerized into polylactic acid. NatureWorks, LLC (Blair, USA) reported a production capacity of 140,000 ton/year [55]. Itaconic acid is a monomer obtained through fermentation using *Aspergillus terreus* [56]. These molecules have been used to fabricate adhesives, detergents, plastics, elastomers, and fiberglass. Recently, they have been applied in ophthalmic, dental, and drug delivery fields [57]. 3-Hydroxypropionic acid is a three-carbon acid building block used to produce 1–3 propanodiol and acrylic acid. Isoprene can be obtained via fermentation using the recombinant *Saccharomyces cerevisiae* [58]. Isoprene is naturally found in poplars, oaks, eucalyptus, and leguminous plants [59]. For example, this molecule is widely used in the industry to produce rubber, tires, coatings, elastomer adhesives, and as an additive in the air yet fuels [59]. Farnesene ($C_{15}H_{24}$) is mainly found in sweet orange, rose, and orange essential oils. This sesquiterpene isoprenoid has been extensively used in the manufacturing of diesel, medicine, and cosmetics [58]. Farnesene has been successfully produced by the company Amyris through the fermentation of sucrose using recombinant yeasts. This company has reported a production capacity of 8000 ton/year [55]. Glycerol can be obtained as a byproduct during biodiesel production, and it serves as a precursor of succinic and propionic acids [60]. In addition, it can be fermented in the presence of an engineered *E. coli* to produce bioethanol. Moreover, *Klebsiella oxytoca, Klebsiella pneumoniae, Citrobacter freundii, Enterobacter agglomerans*, and *Clostridium butyricum* can convert glycerol into 1,3 propanediol and butanediol. Sorbitol is a natural sugar alcohol (polyol-$C_6H_{14}O_6$) with an estimated production of 500,000 ton/year. It is mainly obtained after catalytic hydrogenation of glucose [61], and is primarily used as a humectant, plasticizer, sweetening agent, and as tablet and capsule diluent. It is widely used in pharmacy, cosmetics, and the food industry [62]. Xylitol is a water-soluble sugar alcohol used as a gum sweetener since it does not induce caries like other sugars. Naturally, xylitol is contained in plums, raspberries, spinach, and carrots [63]. In vegetables, xylitol is present in cauliflower, spinach, carrot, onion, white mushroom, eggplant, lettuce, and pumpkin [64]. Another molecule, levulinic acid, is a byproduct generated during the pretreatment of lignocellulosic biomass using dilute acid [60]. This compound can be produced by the acidic hydrolysis of glucose when dilute sulfuric acid is used as a catalyst [54]. The major producer of levulinic acid is Biofine Inc (USA) [55]. The calculated world production is 3000 ton/year [60]. Solis-Sánchez et al. [54] reported that this molecule maintains the acidic balance of pharmaceuticals, agrochemicals, and solvents. In addition, it is useful when manufacturing personal care products, food additives, resins, coatings, fuel additives, and biofuels [54]. Glucaric acid can be produced by oxidation of glucose using nitric acid. The company Rivertop Renewables (USA) [55] has reported a production of 4500 ton/year. Additionally, some fruits and vegetables contain glucaric acid in small quantities. This molecule can potentially be used as a metal sequestering agent, retarding agent, and for corrosion inhibition in metals. Furthermore, glucaric acid is the starting material for producing other chemicals, particularly polyhydroxypolyamides [65].

The biomolecules mentioned above have been produced via fermentation and chemical transformation or identified in small amounts (as micronutrients) in different crops. However, the large amount of biomass generated by agricultural activities may represent an important source of building blocks for several industries. These biomolecules can be obtained using conventional extraction methods. (1) Solvent extraction uses a solvent to extract the molecule of interest. The solvent is then separated from the solution using several techniques, including evaporation, distillation, and filtration. (2) Steam distillation extracts volatile components, such as essential oils from plants. Herein, the mixture is heated, the oils are carried within the steam, and afterward, they are condensed and collected. (3) Soxhlet extraction is a technique that separates insoluble compounds in solvents. Other non-conventional extraction methods include ultrasound, microwaves, and supercritical fluids [38, 39]. Ultrasound uses high-frequency sound waves, and has been used in fructan and polyphenol extraction [66]. Microwaves are electromagnetic radiation with wavelengths between one millimeter and one meter. Microwaves have been useful in fructan and polyphenols extraction [67]. Supercritical extraction is a process that uses a supercritical fluid (carbon dioxide) and has been used in polyphenols extraction [39]. Some authors have reported the efficiency of these methodologies when they are used alone or in combination. According to their data, the most efficient and environmentally friendly process corresponds to the combination of ultrasound and microwaves [38].

In this context, a conventional biorefinery configuration represents an attractive and viable alternative for extracting biomolecules from residual biomass. Figure 16.2 shows the proposed block diagram to produce and extract biomolecules using a parallel route for fermentation, chemical transformation, and extraction. This sequence is similar to that of conventional processes. It should be noted that only a fraction of the biomass is sent to the extraction stage as it has been demonstrated that this part of the process is unprofitable at large scales (100, 500, 1000 ton/day) [68]. Thus, the remaining biomass is sent to a conventional biorefinery. Therefore, a balance between the amount of biomass sent to the extraction process and the biomass sent to the fermentation process should be considered.

To the best of our knowledge, the concept of biorefineries for the extraction of biomolecules and the simultaneous conversion of lignocellulosic biomass rich in biomolecules has not been studied in depth. In new plant configurations and more profitable processes with conventional and non-conventional extraction methods will be designed in the coming years. Given that plant residues contain many compounds, biomolecule purification may be challenging. Thus, purification is another area of interest for researchers and industry.

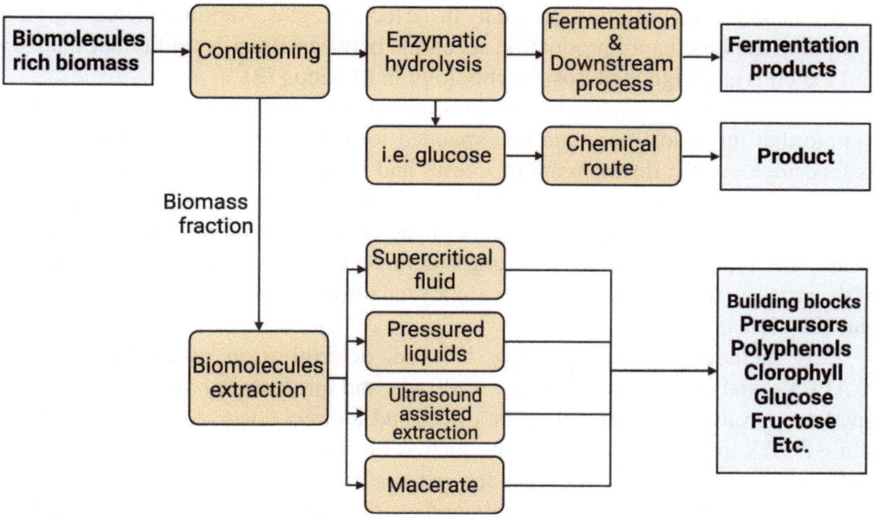

Fig. 16.2 Process block diagram for a biorefinery route using biomolecule-rich biomass

16.4 Tools of Evaluation

16.4.1 Economic Tools

Biorefineries are still emerging technologies at the early stages of maturity that must be evaluated to maximize their economic benefits. Techno-economic analysis (TEA) enables a thorough examination of the thermodynamics of the process, the kinetics of the reactions, and the equipment design [69]. When TEA is accompanied by process simulation, it plays a pivotal role when introducing new biorefinery designs [70, 71]. The first motivator for the development of biorefineries was to replace the oil-based industry with a biomass-based industry. In this technological transition, it has been desirable that new technologies achieve at least the same profitability as the existing industries, which is not necessarily reasonable as the existing industries have their path to maturity. In this tone, the USA Department of Energy (DOE) Advanced Manufacturing Office (AMO) introduces a cost benchmarking approach within which the costs of a new biorefinery are compared to those of existing commercial technologies. This assessment aims to evaluate the competitiveness of the new biorefinery. An interesting example of this approach that included an environmental analysis was published by Sanchez et al. [72], who compared a novel lignocellulosic biorefinery against commercial-scale technologies for hydrogen production. The authors also analyzed different alternatives of pretreatment and biogas production for self-consumption or sale. The authors demonstrated that the fossil fuel-based technologies for hydrogen production have lower total production costs than the novel biorefining schemes. However, this low cost does not compensate in monetary terms for the air and water pollution caused by these technologies. Another interesting TEA compared the bioethanol production scenarios between Brazil's traditional industry and the emerging lignocellulosic technology in Mexico [73].

While TEA often emphasises the production stage (technology), it is important to note that the assessment can be expanded to include other critical factors such as feedstock supply, downstream processes and final distribution [74]. In literature, there is an increasing number of TEA reports mainly focused on a single product where the authors analyzed the effects of the plant capacity, potential biorefining routes, and types of feedstocks, among others, on the profitability or unprofitability of the process [69]. Fewer reports have analyzed the coproduction of energy, biofuels, and chemicals.

TEA includes mass and energy balance, determination of capital expenses (CAPEX) and operative expenses (OPEX), and profitability analysis. CAPEX involves purchased equipment costs, direct and indirect costs, and start-up capital, while OPEX includes operative costs, such as raw materials, utility, royalties, labor, and administration costs [74]. Once the process flowsheet is defined for each biorefining option to be evaluated, the mass and energy balances are commonly resolved by using mainly the specialized software, Aspen Plus® (AspenTech Inc., Bedford, MA, USA) and SuperPro Designer (Intelligen, Inc., Scotch Plains, NJ, USA). These simulators are used for modeling and evaluating various chemical, biochemical, and related processes integrated into a biorefinery. The simulators generate comprehensive mass and energy balances, demand profiles for labor, raw materials, utilities, etc., and equipment sizing. The software estimates equipment costs, capital investment, and operating costs. Once CAPEX and OPEX are calculated for each biorefining option, the engineers evaluate the economic performance of each project using, for example, the Net Present Value (NPV) method [75]. The common outputs of TEA reports in the literature are the total production cost per unit (i.e., $/kg), NPV, capital and operating costs, and the payback period. The projects can also be evaluated using the Sustainability Analysis Method (SAM), which considers, economic and environmental domains [76]. The SAM can provide indicators that permit comparison between plant schemes to determine the most sustainable option [73]. Additionally, sensitivity analyses are performed manually or automatically to identify the variables (such as feedstocks, product yields, and operational conditions [T, P, solid loadings, among others]) with the greatest impact on the total production cost, energy consumption, CO_2 emissions, among others [71].

16.4.2 Environmental Tools

Biorefinery systems are strategic mechanisms for implementing the circular economy because they represent the efficient use of natural resources and the reduction of waste, air emissions, and water discharges [77]. However, the performance of the biorefineries still presents environmental concerns, such as greenhouse gas emissions, impacts on air quality, effects on soils, availability and quality of water resources, deforestation, and loss of biodiversity. Related to this, when biomass from crops is used, there is the risk of sacrificing natural areas to managed monocultures, polluting waterways with agricultural pollutants, threatening food supplies or farm lifestyles through competition for land, and increasing net carbon emissions into the atmosphere as a consequence of increased deforestation or demand for energy and fuels for manufacturing technologies [78, 79]. Therefore, it is necessary to evaluate the environmental impacts of biorefineries from a holistic, objective, and systematic approach that allows identifying environmental hotspots and proposes technological improvement throughout the complete production chain [79, 80].

Life Cycle Assessment (LCA) is a methodological platform with scientific support that has proven its effectiveness in the evaluation of bioenergy and bioproducts since it is a holistic, quantitative, and systematic methodology that determines all the environmental loads associated with the production of goods and services, since the obtaining raw materials to the End-of-Life stage, analyzing all the vectors involved: water, air, and soil [81]. LCA definitions and methodological requirements are presented in two standards from the International Organization for Standardization: ISO 14040 and ISO 14044 [82, 83].

A vast number of LCA studies have been conducted to evaluate the environmental impacts of biorefineries [84, 85], bioethanol [86–91], biogas [92–96], and biodiesel [97–100]. The LCA results are valuable in all cases and support decision-making and technological improvements. However, LCA studies require a large amount of quality data that are typically unavailable for the preliminary assessment at the early stages of process design. To tackle this problem, Karka et al. [101] mention that streamlined LCA studies have been reported considering proxy data and qualitative models [102]. These methods typically refer to specific application domains, such as water treatment [103], power generation [104], or chemical process synthesis [105, 106].

Other methodological tools have been developed to assess the environmental impact of processes and systems in a holistic way, such as:

(1) The Waste Reduction Algorithm (WAR) [107, 108] evaluates processes in terms of their potential environmental impacts (PEI), which is defined as the effect that a chemical would have on the environment if it were simply emitted into the environment. This methodology aims to minimize the PEI for a process instead of minimizing the waste (pollutants) generated by a process.
(2) The ecoefficiency assessment tool was developed by BASF, whose purpose is to harmonize the economy with ecology. This tool involves a comprehensive study of alternative solutions including a total cost determination and the calculation of ecological impact over the entire life cycle [109] (https://www.basf.com/).
(3) The methodological framework of eco-efficiency based on fuzzy logic and Life Cycle Assessment developed by the Engineering Institute of the Universidad Nacional Autónoma de México and the Tecnológico de Monterrey to be applied by Mexican small and medium enterprises [110]. It consists of four stages: (1) perform LCA and basic cost analysis, (2) normalize the environmental and economic impact categories results, (3) integrate economic and environmental

impact categories using a fuzzy treatment, and (4) obtain the fuzzy eco-efficiency index. The result is a preference hierarchy that indicates an order of scenarios according to their degree of eco-efficiency.

In parallel, other methodological frameworks have considered sustainability assessment, such as:

(1) The proposed by Azapagic and Perdan [111] is a general framework with a comprehensive set of indicators that cover the three aspects of sustainability, considering the most important stages in the life cycle and their impacts.
(2) Later, in 2012, the United States Environmental Protection Agency (US EPA) developed the Gauging Reaction Effectiveness for the Environmental Sustainability of Chemistries with a Multi-Objective Process Evaluator (GREENSCOPE), which is a sustainability assessment tool used to evaluate and assist in the design of chemical processes. The goal is to minimize resource use, prevent or reduce releases, and increase the economic feasibility of a chemical process [112, 113].
(3) The Prospective Sustainability assessment of Technologies project (PROSUITE) was set up by the European Commission as part of the 7th Framework program [114]. It resulted in the creation of a new methodology to assess the sustainability of technology, considering five sustainability pillars: human health, social well-being, natural environment, exhaustible resources, and prosperity. PROSUITE was developed by PreConsultants (https://pre-sustainability.com).

All the frameworks above are based on the Life Cycle Assessment approach, which has been recognized as quite comprehensive; however, it still needs additional research and development to increase robustness and reliability further.

The issues that persist as the most discussed in the LCA research community are the allocation procedures, uncertainty, data variability, regionalization, the heterogeneity of the modeling approaches, and the need to adapt LCA to circularity and interpretation.

In the case of LCA of biorefineries, the issue of interpretation is very relevant since important limitations have been identified in the LCA studies reported, which agree with Agostini et al. [115]. These limitations refer to the fact that in many of the studies, the results are discussed superficially, without analyzing the influence of the assumptions, allocation methods used, data quality and uncertainty, and the regional context. Additionally, comparisons are usually made with other studies based on different functions and functional units. Accordingly, the LCA of biorefineries constitutes a research line that must be developed, considering the principles of the circular economy.

16.4.3 Social Tools

This section discusses various social tools developed to assess the biorefinery concept as it moves towards a circular bioeconomy. These tools play a critical role in steering the development and implementation of biorefinery technologies and practices by considering the social facets of sustainability. In this context, we will examine four essential concepts: (1) Social aspects in decision-making and sustainability, (2) Social Life Cycle Assessment, (3) Social Life Cycle Assessment and Sustainable Development Goals (SDGs), and (4) The Social Foundations.

These social tools provide a comprehensive perspective on the biorefinery concept, enabling a more accurate understanding of its potential impacts and benefits in various social dimensions. Moreover, these tools facilitate informed decision-making, guiding stakeholders towards the most socially sustainable outcomes. This comprehensive approach ensures that the biorefinery projects address environmental and economic concerns and contribute to the overall well-being of communities and societies.

16.4.3.1 Social Aspects in Decision-Making and Sustainability

The biorefinery concept, aiming to achieve a circular bioeconomy, has evolved in response to a growing need for sustainable solutions to environmental and resource challenges [116]. While environmental and economic dimensions are often emphasized, social aspects are equally vital for biorefineries' success [117]. Social aspects in decision-making and sustainability ensure that biorefineries contribute to environmental, economic, and social well-being.

Social aspects encompass social equity, community participation, public acceptance, employment opportunities, and occupational health and safety. By incorporating these aspects into decision-making processes; stakeholders can ensure biorefinery projects respect human rights, promote social inclusion, and generate shared value for all parties involved.

16.4.3.2 Social Life Cycle Assessment (S-LCA)

S-LCA evaluates a product or process's social impacts throughout its life cycle, from raw material extraction to disposal or recycling [118]. Building upon the traditional Life Cycle Assessment (LCA) framework, S-LCA adds a social dimension to the analysis, helping stakeholders identify, quantify, and assess potential social risks and benefits associated with biorefinery projects.

S-LCA uses social impact categories and indicators to assess a product or process's social performance. Indicators may include working conditions, fair wages, community engagement, and cultural heritage preservation. Evaluating these social factors,

S-LCA offers valuable insights for decision-makers to optimize biorefinery projects to minimize negative social impacts and maximize positive ones.

16.4.3.3 Social Life Cycle Assessment and Sustainable Development Goals (SDGs)

The SDGs are 17 global goals established by the United Nations in 2015 to guide nations towards a more sustainable future [119]. These goals cover various social, economic, and environmental dimensions of sustainability, including poverty reduction, quality education, clean energy, and climate action.

S-LCA can effectively align biorefinery projects with the SDGs. Conducting an S-LCA allows decision-makers to identify social impacts that align with specific SDGs, prioritizing interventions and strategies to address these impacts [120]. For instance, an S-LCA might reveal that a biorefinery project could create jobs, reduce poverty (SDG 1), and promote gender equality (SDG 5) by employing men and women in the local community.

Integrating the SDGs into S-LCA enables stakeholders to optimize their projects for social sustainability and demonstrate their commitment to the global sustainable development agenda.

16.4.3.4 The Social Foundations

Introduced by Kate Raworth in her book "Doughnut Economics: Seven Ways to Think Like a 21st-Century Economist" [121], the Social Foundations comprise nine social dimensions that societies should aim to fulfill to ensure the well-being of their citizens. These dimensions encompass food, health, education, income and work, peace and justice, political voice, social equity, gender equality, and housing. Integrating the Social Foundations into biorefinery project evaluations provides stakeholders with a comprehensive framework for assessing social sustainability within the context of a circular bioeconomy.

By incorporating the Social Foundations into biorefinery evaluations, stakeholders can ensure that projects contribute to the well-being of local communities and society. This holistic approach identifies potential risks and opportunities while fostering shared value across diverse social dimensions.

As the biorefinery concept advances, the significance of incorporating social tools like the Social Foundations will only increase. By addressing these nine social dimensions, biorefinery stakeholders can formulate projects that adhere to the principles of a circular bioeconomy and promote a more sustainable and equitable future for everyone.

Considering the social dimensions of sustainability, alongside environmental and economic considerations, enables a more comprehensive evaluation of biorefineries. This integrated approach supports the development of projects that generate shared value for all stakeholders, ranging from local communities to global society.

16.5 Final Remarks

Biorefineries represent an emerging technology encompassing a wide range of biomass sources, converting them into energy, biofuels, and value-added products. The diversity of available biomass types regionally corresponds to the diversity of biorefining configurations and the range of products that can be derived. Challenges remain in increasing conversion yields and identifying the most profitable products for each type of biomass and socio-economic context. However, researchers are gaining experience in processing biomasses of different types productively. However, as the world evolves, new technologies must not only be economically competitive. They must also prioritize the planet, care for it, and deliver tangible and measurable social benefits. Therefore, the design and evaluation of novel biorefineries require a comprehensive and holistic approach. The use of environmental and social tools not only aids in identifying potential risks and opportunities but also fosters the creation of shared value among all stakeholders, ranging from local communities to global society. As the concept of biorefineries continues to evolve, the significance of these environmental and social tools will inevitably grow. By prioritizing the environmental-social dimensions of sustainability, stakeholders can steer the development of biorefineries that embody the principles of a circular bioeconomy, thereby fostering a better, more sustainable future for all.

Acknowledgements This work was supported by the Grupos Interdisciplinarios de Investigación (GII), Engeneering Institute-UNAM project "Cambio de Paradigma: Residuos Como Materia Prima Para Conciliar El Eje Agua-Energía-Ambiente-Seguridad Alimentaria", the Consejo Estatal de Ciencia y Tecnología (COECYT, Coahuila, Mexico) with the "Fondo Destinado a Promover el Desarrollo de la Ciencia y la Tecnología en el Estado de Coahuila (FONCYT)" Project "Sustainable cultivation of *Spirulina platensis* on a large scale for use in livestock feed"—COAH-2022-C19-C123.

References

1. Rexen F, Munck L (1984) Cereal crops for industrial use in Europe, EUR 9617 EN, Report prepared for The Commission of the European Communities, Brussels. https://archive.org/details/cerealcropsforin0000rexe/page/78/mode/1up?q=refinery. Accessed 28 May 2023
2. Koukios EG (1985). Biomass refining: a non-waste approach. In: Hall DO, Myers N, Margaris NS (eds) Economics of ecosystems management. Tasks for vegetation science, Springer, Dordrecht, pp 233–244
3. Munk L (1993) On the utilization of renewable plant resources. In: Hayward MD, Bosemark NO, Romagosa I, Cerezo M (eds) Plant breeding. Principles and prospects. Springer Dordrecht, pp 500–522
4. Rexen F (1990) A new project in Denmark—the biorefinery at Bornholm. In: Munck L, Rexen F (eds) Agricultural refineries—a bridge from farm to industry, EUR 11583 EN. The Commission of the European Communities, Brussels, pp 115–122
5. Gylling M (1993) Technical report. The whole crop biorefinery project. Midterm assessment. Concept- status- perspectives. U.S Department of Energy. Available from: https://www.osti.gov/etdeweb/biblio/10149032. Accessed 14 May 2023

6. Papatheofanous MG, Koullas DP, Koukios EG et al (1995) Biorefining of agricultural crops and residues: effect of pilot-plant fractionation on properties of fibrous fractions. Biomass Bioenerg 8:419–426
7. Audsley E, Sells JE (1997) Determining the profitability of a wholecrop biorefinery. In: Campbell GM, Webb C, McKee SL (eds) Cereal: novel uses and processes. Springer, New York, NY, pp 191–220
8. Bagger CL, Sorensen H, Sorensen JC (1998) High-quality oils, proteins and bioactive products for food and non-food purposes based on biorefining of cruciferous oilseed crops. In: Guéguen J, Popineau Y (eds) Plant Protein from European Crops. Springer, Berlin, Heidelberg, pp 272–278
9. Carlsson R (1998) Status quo of the utilisation of green biomass. In: Soyers K, Kamm B, Kamm M (eds) Die Grune Bioraffinerie. Verlag Gesellschaft fur okologische Technologie, Berlin, pp 39–45
10. Kiel P (1998) Technology-development of the Green Biorefinery. In: Soyers K, Kamm B, Kamm M (eds) Die Grune Bioraffinerie. Verlag Gesellschaft fur okologische Technologie, Berlin, pp 101–107
11. Andersen M, Kiel P (2000) Integrated utilization of green biomass in the green biorefinery. Ind Crops Prod 11:129–137
12. Starke I, Holzberger A, Kamm B, Kleinpeter E (2000) Qualitative and quantitative analysis of carbohydrates in green juices (wild mix grass and alfalfa) from a green biorefinery by gas chromatography/mass spectrometry. Fresenius J Anal Chem 367:65–72
13. Danner H, Madzingaidzo L, Holzer M et al (2000) Extraction and purification of lactic acid from silages. Bioresour Technol 75:181–187
14. Mielenz JR (2001) Ethanol production from biomass: technology and commercialization status. Curr Opin Microbiol 4:324–329
15. Kamm B, Kamm B (2004) Principles of biorefineries. App Microbiol Biotechnol 64:137–145
16. Fernando S, Adhikari S, Chandrapal C et al (2006) Biorefineries: current status, changes, and future direction. Energy Fuels 20:1727–1737
17. Kamm B, Gruber PR, Kamm M (eds) (2006) Biorefineries-industrial process and products. Wiley, Weinheim
18. IEA. IEA bioenergy Task 42 on biorefineries: https://www.ieabioenergy.com/wp-content/uploads/2022/09/IEA-Bioenergy-Task-42-Global-biorefinery-status-report-2022-220712.pdf. Accessed 26 May 2023
19. Renewable Fuels Association (RFA) (2023) Industry statistics. Available from: http://www.ethanolrfa.org/resources/industry/statistics. Accessed 12 May 2023
20. USDA. Barros S (2022) Biofuels annual. USDA. Available from: https://apps.fas.usda.gov/newgainapi/api/Report/DownloadReportByFileName?fileName=Biofuels%20Annual_Sao%20Paulo%20ATO_Brazil_BR2022-0047.pdf. Accessed 25 May 2023
21. Campuzano R, González-Martínez S (2016) Characteristics of the organic fraction of municipal solid waste and methane production: a review. Waste Manag 54:3–12
22. Hernández C, Escamilla-Alvarado C, Sánchez A et al (2019) Wheat straw, corn stover, sugarcane, and Agave biomasses: chemical properties, availability, and cellulosic-bioethanol production potential in Mexico. Biofuel Bioprod Biorefin 13(5):1143–1159
23. Lymperatou A, Engelsen TK, Skiadas IV, Gavala HN (2023) Prediction of methane yield and pretreatment efficiency of lignocellulosic biomass based on composition. Waste Manage 155:302–310
24. Bajpai P (2016) Structure of lignocellulosic biomass. In: Pretreatment of lignocellulosic biomass for biofuel production. SpringerBriefs in Molecular Science(). Springer, Singapore. https://doi.org/10.1007/978-981-10-0687-6_2
25. Mohanty A, Rout PR, Dubey B et al (2022) A critical review on biogas production from edible and non-edible oil cakes. Biomass Conv Bioref 12:949–966
26. Arias A, Feijoo G, Moreira MT (2023) Macroalgae biorefineries as a sustainable resource in the extraction of value-added compounds. Algal Res 69:102954

27. Makkar HPS, Tran G, Heuzé V et al (2016) Seaweeds for livestock diets: a review. Anim Feed Sci Technol 212:1–17
28. Lee H, Sohn YJ, Jeon S et al (2023) Sugarcane wastes as microbial feedstocks: A review of the biorefinery framework from resource recovery to production of value-added products. Bioresour Technol 376:128879
29. Ladakis D, Stylianou E, Ioannidou SM et al (2022) Biorefinery development, techno-economic evaluation and environmental impact analysis for the conversion of the organic fraction of municipal solid waste into succinic acid and value-added fractions. Bioresour Technol 354:127172
30. Mosquera-Toscano HG, González-Barceló O, Valdez-Vazquez I et al (2023) Ethanol and methane production from the organic fraction of municipal solid waste in a two-stage process. Bioenergy Res
31. Sari TP, Sirohi R, Krishania M et al (2022) Critical overview of biorefinery approaches for valorization of protein rich tree nut oil industry by-product. Bioresour Technol 362:127775
32. del Pilar Garcia-Mendoza M, Espinosa-Pardo FA, Savoire R et al (2021) Recovery and antioxidant activity of phenolic compounds extracted from walnut press-cake using various methods and conditions. Ind Crop Prod 167:113546
33. Rudke AR, da Silva M, de Andrade CJ et al (2022) Green extraction of phenolic compounds and carrageenan from the red alga *Kappaphycus alvarezii*. Algal Res 67:102866
34. Ruper R, Rodrigues KF, Thien VY et al (2022) Carrageenan from *Kappaphycus alvarezii* (Rhodophyta, Solieriaceae): metabolism, structure, production, and application. Front Plant Sci 13:859635
35. Paone E, Tabanelli T, Mauriello F (2020) The rise of lignin biorefinery. Curr Opin in Green Sustain Chem 24:1–6
36. González-Chavez J de J, Arenas-Grimaldo C, Amaya-Delgado L et al (2022) Sotol bagasse (*Dasylirion* sp.) as a novel feedstock to produce bioethanol 2G: Bioprocess design and biomass characterization. Ind Crop Prod 178:0–1
37. Choi S, Song CW, Shin JH et al (2015) Biorefineries for the production of top building block chemicals and their derivatives. Metab Eng 28:223–239
38. García-Villalba WG, Rodríguez-Herrera R, Ochoa-Martínez LA et al (2023) Comparative study of four extraction methods of fructans (agavins) from Agave durangensis: heat treatment, ultrasound, microwave and simultaneous ultrasound-microwave. Food Chem 415:135767
39. Escobedo-Flores Y, Chavez-Flores D, Salmeron I et al (2018) Optimization of supercritical fluid extraction of polyphenols from oats (*Avena sativa* L.) and their antioxidant activities. J Cereal Sci 2018:80
40. Waldbauer K, Mckinnon R, Kopp B (2017) Apple Pomace as potential source of natural active compounds. Planta Med 83:994–1010
41. Farías-Campomanes AM, Rostagno MA, Meireles MAA (2013) Production of polyphenol extracts from grape bagasse using supercritical fluids: yield, extract composition and economic evaluation. J Supercrit Fluid 77:70–78
42. Thomas M, Badr A, Desjardins Y et al (2018) Characterization of industrial broccoli discards (*Brassica oleracea* var. italica) for their glucosinolate, polyphenol and flavonoid contents using UPLC MS/MS and spectrophotometric methods. Food Chem 245:1204–1211
43. dos Reis LCR, de Oliveira VR, Hagen MEK et al (2015) Carotenoids, flavonoids, chlorophylls, phenolic compounds and antioxidant activity in fresh and cooked broccoli (*Brassica oleracea* var. Avenger) and cauliflower (*Brassica oleracea* var. Alphina F1). Lwt 4
44. Mittal R, Ranade VV (2023) Intensifying extraction of biomolecules from macroalgae using vortex based hydrodynamic cavitation device. Ultrason Sonochem 94:106347
45. Trujillo LE, Arrieta JG, Enriquez GA et al (2000) strategies for fructan production in transgenic sugarcane (*Saccharmu* spp L.) and Sweet Potato (/pomoea batata L.) plants expressing the Acetobacter diazotrophicus levansucrase. In: Arencibia AD (ed) Developments in plant genetics and breeding. Elsevier, pp 194–198
46. Mudgil D (2017) The interaction between insoluble and soluble fiber. Dietary fiber for the prevention of cardiovascular disease. In: Samaan RA (ed) Dietary fiber for the prevention of cardiovascular disease. Elsevier, pp 35–59

47. Kaur G, Grewal J, Jyoti K et al (2018) Oral controlled and sustained drug delivery systems: Concepts, advances, preclinical, and clinical status. In: Grumezescu AM (ed) Drug targeting and stimuli sensitive drug delivery systems (Chapter 15). Elsevier, pp 567–626
48. Chen C (2016) Sinapic acid and its derivatives as medicine in oxidative stress-induced diseases and aging. Oxid Med Cell Longev 2016:3571614
49. Naveed M, Hejazi V, Abbas M et al (2018) Biomedicine & Pharmacotherapy Chlorogenic acid (CGA): a pharmacological review and call for further research. Biomed Pharmacother 97:67–74
50. Martínez-Hernández GB, Gómez PA, Pradas I et al (2011) Postharvest Biology and Technology Moderate UV-C pretreatment as a quality enhancement tool in fresh-cut Bimi® broccoli. Postharvest Biol Technol 62:327–337
51. Ni N (2014) Sinapic acid and its derivatives: natural sources and bioactivity. Compr Rev Food Sci Food Saf 13:34–51
52. Gudiño I, Martín A, Casquete R et al (2022) Evaluation of broccoli (*Brassica oleracea* var. italica) crop by-products as sources of bioactive compounds. Sci Hortic 304:111284
53. Werpy T, Petersen G (2004) Top value added chemicals from biomass volume I. Us Nrel. Medium: ED; Size: p 76
54. Solis-Sánchez JL, Alcocer-García H, Sanchez-Ramirez E et al (2022) Innovative reactive distillation process for levulinic acid production and purification. Chem Eng Res Des 183:28–40
55. Sanford K, Chotani G, Danielson N et al (2016) Scaling up of renewable chemicals. Curr Opin Biotechnol 38:112–122
56. Pellis A, Gardossi L (2019) Integrating computational and experimental methods for efficient biocatalytic synthesis of polyesters, 1st edn. In: Enzymatic Polymerizations, vol 627. Elsevier Inc., p 23–55
57. Rodrigues AG (2016) Secondary metabolism and antimicrobial metabolites of *Aspergillus*. New and future developments in microbial biotechnology and bioengineering. Elsevier B.V., pp 81–94
58. Wang Z, Zhang R, Yang Q, et al (2021) Recent advances in the biosynthesis of isoprenoids in engineered Saccharomyces cerevisiae, 1st ed. vol. 114, Advances in Applied Microbiology. Elsevier Inc., pp 1–35
59. Noreen A, Sultana S, Sultana T et al (2020) Natural polymers as constituents of bionanocomposites. In: Zia KM, Jabeen F et al. (eds) Bionanocomposites, pp 55–85
60. Chandel AK, Garlapati VK, Singh AK et al (2018) The path forward for lignocellulose biorefineries: Bottlenecks, solutions, and perspective on commercialization. Bioresour Technol 264:370–381
61. Marques C, Tarek R, Sara M et al (2016) Sorbitol production from biomass and its global market. In: Brar SK, Sarma SJ, Pakshirajan K (eds) Platform chemical biorefinery. Elsevier Inc., pp 217–227
62. Newman AW, Vitez M, Mueller RL et al (1999) Sorbitol. In: Brittain H (ed) Analytical profiles of drug substances and excipients. Academic Press, pp 459–502
63. Summary N (2014) Carbohydrates, alcohols, and organic acids. In: Dierna A (ed) Nutrient Metabolism. Elsevier Ltd., pp 187–242
64. Umai D, Kayalvizhi R, Kumar V et al (2022) Xylitol: Bioproduction and applications—a review. Front Sustain 3:1–16
65. Shendurse AM, Dantiwada S (2016) Glucose: properties and analysis, 1st ed. Encyclopedia of Food and Health. Elsevier Ltd., pp 239–247
66. Katsampa P, Valsamedou E, Grigorakis S et al (2015) A green ultrasound-assisted extraction process for the recovery of antioxidant polyphenols and pigments from onion solid wastes using Box-Behnken experimental design and kinetics. Ind Crop Prod 77:535–543
67. Alchera F, Ginepro M, Giacalone G (2022) Microwave-assisted extraction of polyphenols from blackcurrant by-products and possible uses of the extracts in active packaging. Food 11(18):2727

68. Frausto-Torres LG, Vázquez-Núñez É, Molina-Guerrero CE (2022) Technoeconomic analysis of broccoli biorefineries for polyphenol extraction and biobutanol production. Renew Energ Biomass Sustain 4(1):23–37
69. Mahmud R, Moni SM, High K et al (2021) Integration of techno-economic analysis and life cycle assessment for sustainable process design—a review. J Clean Prod 317:128247
70. Gutiérrez Ortiz FJ (2020) The Journal of Supercritical Fluids Techno-economic assessment of supercritical processes for biofuel production. J Supercrit Fluid 160:104788
71. Contreras-Zarazúa G, Martin-Martin M, Sánchez-Ramirez E et al (2022) Furfural production from agricultural residues using different intensified separation and pretreatment alternatives. Economic and environmental assessment. Chem Eng Process Process Intensif 171:108569
72. Sanchez A, Ayala OR, Hernadez-Sanchez P et al (2020) An environment-economic analysis of hydrogen production using advanced biorefineries and its comparison with conventional technologies. Int J Hydrogen Energ 45(51):27994–28006
73. López-Ortega MG, Guadalajara Y, Junqueira TL et al (2021) Sustainability analysis of bioethanol production in Mexico by a retrofitted sugarcane industry based on the Brazilian expertise. Energ 232:121056
74. García-Velásquez C, van der Meer Y (2023) Mind the Pulp: environmental and economic assessment of a sugar beet pulp biorefinery for biobased chemical production. Waste Manag 155:199–210
75. Towler G, Sinnott R (2008) Chemical engineering design: principles, practice and economics of plant and process design. Elsevier, Burlington, MA, USA
76. Sanchez A, Magaña G, Gomez D et al (2014) Bidimensional sustainability analysis of ligno-cellulosic ethanol production processes. Method and case study. Biofuel Bioprod Biorefin 8:670–685
77. Teigeiro S, Solar-Pelleteir L, Bernard S et al (2018) Circular economy in Quebec: economic opportunities and impacts. Report prepared by Group on Globalization and Management of Technology (Polytechnique Montréal) and Institut de l'environnement, du développement durable et de l'économie circulaire at Université de Montréal (I-EDDEC) for the: Conseil du patronat du Québec, Quebec Business Council on the Environment and Éco Entreprises Québec
78. García Bustamante C, Masera Cerutti O (2016) Estado del arte de la bioenergía en México. Red temática de Bioenergía (RTB) del Conacyt. México
79. Rolón Rodríguez I (2019) Análisis de Ciclo de Vida de la energía eléctrica generada a partir de biomasa residual de origen urbano. Universidad Nacional Autónoma de México. Accesded 02 Mar 2023 https://132.248.9.195/ptd2019/octubre/0796997/Index.html
80. Yoshida H, ten Hoeve M, Christensen TH et al (2018) Life cycle assessment of sewage sludge management options including long-term impacts after land application. J Clean Prod 174:538–547
81. Güereca LP, Ochoa Sosa R, Gilbert HE et al (2015) Life cycle assessment in Mexico: overview of development and implementation. Int J Life Cycle Assess 20(3):311–317
82. ISO (2006a) ISO 14040 Environmental management—life cycle assessment—principles and framework. International Organization for Standardization.https://doi.org/10.1136/bmj.332.7550.1107
83. ISO (2006b) ISO 14044 Environmental management—life cycle assessment—requirements and guilelines. International Organization for Standardization.https://doi.org/10.1007/s11367-011-0297-3
84. Khoshnevisan B, Rafiee S, Tabatabaei M et al (2028) Life cycle assessment of castor-based biorefinery: a well to wheel LCA. Int J Life Cycle Assess 23:1788–1805
85. Aristizábal Marulanda V, García-Velásquez CA, Cardona Alzate CA (2021) Environmental assessment of energy-driven biorefineries: the case of the coffee cut-stems (CCS) in Colombia. Int J Life Cycle Assess 26:290–310
86. Wiloso EI, Heijungs R, de Snoo GR (2012) LCA of second generation bioethanol: a review and some issues to be resolved for good LCA practice. Renew Sustain Energ Rev 16:5295–5308

87. Daylan B, Ciliz N (2016) Life cycle assessment and environmental life cycle costing analysis of lignocellulosic bioethanol as an alternative transportation fuel. Renew Energ 89:578–587
88. Nemecek T, Schnetzer J, Reinhard J (2016) Updated and harmonised greenhouse gas emissions for crop inventories. Int J Life Cycle Assess 21:1361–1378
89. Sebastião D, Gonçalves MS, Marques S et al (2016) Life cycle assessment of advanced bioethanol production from pulp and paper sludge. Bioresour Technol 208:100–109
90. Shuai W, Chen N, Li B et al (2016) Life cycle assessment of common reed (Phragmites australis (Cav) Trin. ex Steud) cellulosic bioethanol in Jiangsu Province China. Biomass Bioenerg 92:40–47
91. Zucaro A, Forte A, Basosi R et al (2016) Life cycle assessment of second generation bioethanol produced from lowinput dedicated crops of Arundo donax L. Bioresour Technol 219:589–599
92. Ishikawa S, Hoshiba S, Hinata T et al (2006) Evaluation of a biogas plant from life cycle assessment (LCA), international congress series. Elsevier, New York, pp 230–233
93. Van Stappen F, Mathot M, Decruyenaere V et al (2016) Consequential environmental life cycle assessment of a farm-scale biogas plant. J Environ Manag 175:20–32
94. Wang Q-L, Li W, Gao X et al (2016) Life cycle assessment on biogas production from straw and its sensitivity analysis. Bioresour Technol 201:208–214
95. Collet P, Flottes E, Favre A et al (2017) Techno-economic and life cycle assessment of methane production via biogas upgrading and power to gas technology. Appl Energ 192:282–295
96. Tabatabaie SMH, Murthy GS (2017) Effect of geographical location and stochastic weather variation on life cycle assessment of biodiesel production from camelina in the northwestern USA. Int J Life Cycle Assess 22:867–882
97. Requena JS, Guimaraes AC, Alpera SQ et al (2011) Life cycle assessment (LCA) of the biofuel production process from sunflower oil, rapeseed oil and soybean oil. Fuel Process Technol 92:190–199
98. Caldeira C, Queirós J, Noshadravan A et al (2016) Incorporating uncertainty in the life cycle assessment of biodiesel from waste cooking oil addressing different collection systems. Resour Conserv Recycl 112:83–92
99. Fernández-Tirado F, Parra-López C, Romero-Gámez M (2016) Life cycle assessment of biodiesel in Spain: comparing the environmental sustainability of Spanish production versus Argentinean imports. Energy Sustain Dev 33:36–52
100. Harris TM, Hottle TA, Soratana K et al (2016) Life cycle assessment of sunflower cultivation on abandoned mine land for biodiesel production. J Clean Prod 112(Part 1):182–195
101. Karka P, Papadokonstantakis S, Kokossis A (2019) Environmental impact assessment of biomass process chains ay early design stages using decision trees. Int J Life Cycle Assess 24:1675–1700
102. Hunt R, Boguski T, Weitz K, Sharma A (1998) Case studies examining LCA streamlining techniques. Int J Life Cycle Assess 3:36–42
103. Schulz M, Short M, Peters G (2012) A streamlined sustainability assessment tool for improved decision making in the urban water industry. Integr Environ Assess Manag 8:183–193
104. Moreau V, Bage G, Marcotte D, Samson R (2012) Statistical estimation of missing data in life cycle inventory: an application to hydroelectric power plants. J Clean Prod 37:335–341
105. Eckelman M (2016) Life cycle inherent toxicity: a novel LCA-based algorithm for evaluating chemical synthesis pathways. Green Chem 18:3257–3264
106. Tula A, Babi D, Bottlaender J, Eden M, Gani R (2017) A computer-aided software tool for sustainable process synthesis-intensification. Comput Chem Eng 105:74–95
107. Young DM, Cabezas H (1999) Designing sustainable processes with simulation: the waste reduction (WAR) algorithm. Comput Chem Eng 23(10):1477–1491
108. Young D, Scharp R, Cabezas H (2000) The waste reduction (WAR) algorithm: environmental impacts, energy consumption, and engineering economics. Waste Manag 20(8):605–615
109. Saling P, Kicherer A, Dittrich-Krämer B, Wittlinger R, Zombik W, Schmidt I, Schmidt S (2002) Eco-efficiency analysis by BASF: the method. Int J Life Cycle Assess 7(4):203–218
110. García BA, Luna D, Cobos A, Lameiras D, Ortiz-Moreno H, Güereca LP (2018) A methodological framework of eco-efficiency base don fuzzy logic and life cycle assessment applied to a Mexican SME. Environ Impact Assess Rev 68:38–48

111. Azapagic A, Perdan S (2000) Indicators of sustainable developement for industry: a general framework. Proc Saf Environ Prot 78:243–261
112. Ruiz-Mercado GJ, Smith RL, Gonzalez MA (2012a) Sustainability indicators for chemical processes: I. Taxonomy. Ind Eng Chem Res 51(5):2309–2328
113. Ruiz-Mercado GJ, Smith RL, Gonzalez MA (2012b) Sustainability indicators for chemical processes: II. Data needs. Ind Eng Chem Res 51(5):2329–2353
114. Blok K, Huijbregts M, Roes L et al (2013) A novel methodology for the sustainability impact assessment of new technologies. Accessed on 14 Jun 2023. https://dspace.library.uu.nl/bitstream/handle/1874/303231/26.pdf?sequence=1
115. Agostini A, Giuntoli J, Marelli L et al (2020) Flaws in the interpretation phase of bioenergy LCA fuel the debate and mislead policy makers. Int J Life Cycle Assess 25:17–35
116. Conteratto C, Artuzo FD, Benedetti Santos OI et al (2021) Biorefinery: a comprehensive concept for the sociotechnical transition toward bioeconomy. Renew Sustain Energy Rev 151:111527
117. Padilla-Rivera A, Merveille N (2022) Social circular economy indicators applied to wastage biorefineries. In: Jacob-Lopes E, Queiroz Zepka L, Costa Deprá M (eds) Handbook of waste biorefinery. Springer, Cham
118. UNEP (United Nations Environment Programme) (2020) Guidelines for social life cycle assessment of products and organizations 2020. https://www.lifecycleinitiative.org/wp-content/uploads/2021/01/Guidelines-for-Social-Life-Cycle-Assessment-of-Products-and-Organizations-2020-22.1.21sml.pdf
119. United Nations (UN) (2020) The sustainable development goals report 2020. https://unstats.un.org/sdgs/report/2020/
120. Hannouf MB, Padilla-Rivera A, Assefa G et al (2022) Methodological framework to find links between life cycle sustainability assessment categories and the UN Sustainable Development Goals based on literature. J Ind Ecol 1–19
121. Raworth K (2012) A safe and just space for humanity. Oxfam Discussion Paper. https://www.oxfam.org/en/research/safe-and-just-space-humanity. Accessed Apr 2023

MIX
Papier aus verantwortungsvollen Quellen
Paper from responsible sources
FSC® C105338

If you have any concerns about our products,
you can contact us on
ProductSafety@springernature.com

In case Publisher is established outside the EU,
the EU authorized representative is:
Springer Nature Customer Service Center GmbH
Europaplatz 3, 69115 Heidelberg, Germany

Printed by Libri Plureos GmbH
in Hamburg, Germany